Proceedings in Adaptation, Learning and Optimization

Volume 2

Series editors

Yew Soon Ong, Nanyang Technological University, Singapore
e-mail: asysong@ntu.edu.sg

Meng-Hiot Lim, Nanyang Technological University, Singapore
e-mail: emhlim@ntu.edu.sg

The role of adaptation, learning and optimization are becoming increasingly essential and intertwined. The capability of a system to adapt either through modification of its physiological structure or via some revalidation process of internal mechanisms that directly dictate the response or behavior is crucial in many real world applications. Optimization lies at the heart of most machine learning approaches while learning and optimization are two primary means to effect adaptation in various forms. They usually involve computational processes incorporated within the system that trigger parametric updating and knowledge or model enhancement, giving rise to progressive improvement. This book series serves as a channel to consolidate work related to topics linked to adaptation, learning and optimization in systems and structures. Topics covered under this series include:

- complex adaptive systems including evolutionary computation, memetic computing, swarm intelligence, neural networks, fuzzy systems, tabu search, simulated annealing, etc.

- machine learning, data mining & mathematical programming

- hybridization of techniques that span across artificial intelligence and computational intelligence for synergistic alliance of strategies for problem-solving

- aspects of adaptation in robotics

- agent-based computing

- autonomic/pervasive computing

- dynamic optimization/learning in noisy and uncertain environment

- systemic alliance of stochastic and conventional search techniques

- all aspects of adaptations in man-machine systems.

This book series bridges the dichotomy of modern and conventional mathematical and heuristic/meta-heuristics approaches to bring about effective adaptation, learning and optimization. It propels the maxim that the old and the new can come together and be combined synergistically to scale new heights in problem-solving. To reach such a level, numerous research issues will emerge and researchers will find the book series a convenient medium to track the progresses made.

More information about this series at http://www.springer.com/series/13543

Hisashi Handa · Hisao Ishibuchi
Yew-Soon Ong · Kay Chen Tan
Editors

Proceedings of the 18th Asia Pacific Symposium on Intelligent and Evolutionary Systems – Volume 2

 Springer

Editors
Hisashi Handa
Department of Infomatics
Faculty of Science and Technology
Kindai University
Higashi-Osaka
Japan

Hisao Ishibuchi
Department of Computer Science
and Intelligent Systems
Osaka Prefecture University
Osaka
Japan

Yew-Soon Ong
School of Computer Engineering
Nanyang Technological University
Singapore

Kay Chen Tan
Department of Electrical and Computer
Engineering
National University of Singapore
Singapore

ISSN 2363-6084
ISBN 978-3-319-38620-1
DOI 10.1007/978-3-319-13356-0

ISSN 2363-6092 (electronic)
ISBN 978-3-319-13356-0 (eBook)

Springer Cham Heidelberg New York Dordrecht London

Preface

This book contains a collection of the papers accepted for presentation in the 18th Asia Pacific Symposium on Intelligent and Evolutionary Systems (IES 2014), which was held in Singapore from 10th to 12th November 2014. IES 2014 was sponsored by the Memetic Computing Society and co-sponsored by the SIMTECH-NTU Joint Lab, and the Center for Computational Intelligence at the School of Computer Engineering, Nanyang Technological University, and supported by the National University of Singapore and Nanyang Technological University. The book covers the topics in intelligent systems and evolutionary computation, and many papers have demonstrated notable systems with good analytical and/or empirical results.

Hisashi Handa
Hisao Ishibuchi
Yew-Soon Ong
Kay-Chen Tan

Contents

X   Contents

A New Grammatical Evolution
Based on Probabilistic Context-free Grammar

Hyun-Tae Kim and Chang Wook Ahn*

Department of Computer Engineering, Sungkyunkwan University (SKKU)
2066 Seobu-Ro, Suwon 440-746, Republic of Korea
{arkii,cwan}@skku.edu

Abstract. This paper presents a new grammatical evolution (GE) that
generates automatic program under favor of probabilistic context-free
grammar. A population of individuals is evolved under genotypic integer
strings and a mapping process is utilized to translate from genotype (i.e.,
integer string) to phenotype (i.e., complete program). To efficiently handle this process, unlike the standard GE that employs a simple modulo
function, the probability concept is introduced to context-free grammar,
thereby choosing production rules according to assigned probabilities.
Moreover, any crossover and mutation are not employed for generating
new individuals. Instead, along the lines of estimation of distribution
algorithms that perform search using a probabilistic model of superior
individuals, a new population is created/evolved from probabilistic relationship between production rules. A comparative study on the standard
GE and the proposed GE is conducted; the performance achieved by
the both methods is comparable. Also, the experimental results firmly
demonstrate the effectiveness of the proposed approach.

Keywords: Grammatical Evolution, Probabilistic Context-free Grammar, Automatic Programming, Estimation of Distribution Algorithm.

1 Introduction

Grammatical evolution (GE) [1, 2] is an evolutionary algorithm that can generate
programs in an arbitrary language. GE creates and evolves a population of individuals, which composed by a variable-length integer string (called codon). Furthermore, GE adopts the mapping process with context-free grammars (CFGs)
to translate integer string (genotype) to a program (genotype). In the mapping
process, GE utilizes in general the Backus-Naur form (BNF), which is a widely
used context-free grammar, and can be easily modified to output program.

The mapping process between genotype and phenotype plays important role
in GE since the process selects a sort of rules to generate its program. On this
point, how the mapping process is successful (i.e., generate suitable program
to target problem) has an effect on the quality of solution. In the basic GE
approach, they apply a simple modulo function in the mapping process. After

---

* Corresponding author.

© Springer International Publishing Switzerland 2015   1
H. Handa et al. (eds.), *Proc. of the 18th Asia Pacific Symp. on Intell. & Evol. Systems – Vol. 2,*
Proceedings in Adaptation, Learning and Optimization 2, DOI: 10.1007/978-3-319-13356-0_1

mapping process, GE selects fittest individuals to propagate to next genetic operations, which is crossover and mutation. Note that GE employs crossover and mutation operators to the integer string of selected individuals, which is same as simple GA [3, 4].

In this paper, we proposed a new scheme for GE that replaces mapping process and existing genetic operations (i.e., crossover and mutation) with program generation using probabilistic model for given context-free grammar. To design a new mechanism, we introduce Probabilistic Context-free grammar (PCFG) for mapping process, along with a concept of Estimation of Distribution Algorithms (EDAs) [5, 6].

The probabilistic context-free grammar (also called stochastic context-free grammar) is similar to the CFG, but each production rule has probability. In other words, GE selects production rules with assigned probability. This characteristic is different with the CFG, which only considers selecting a rule with same probability. On this point, if GE assign appropriate probability to each production rule, then the algorithm can produce program that are close to optimal solution. To archive this goal, we adopt original GE to process of EDAs, which uses a probabilistic model of promising solution to lead exploration of the search space. For instance, we assign same probability to each production rule at first, and update the assigned probability as generation goes by. When our proposed scheme re-assigns the probabilities for given grammar, we consider the production rules of fittest individuals. After that, we create new program using grammar with updated probability, and substitute older population with newly generated individuals.

The remaining part of paper is organized as follows. The next section is review of existing approach to GE. In the section 3, the proposed algorithms are explained. The experiments and its results are shown in section 4. Finally, we conclude this paper and discuss about further research topics.

2   Background

2.1   Grammatical Evolution and Mapping Process

GE is an evolutionary automatic programming system that can generate complete programs in an arbitrary language. To archive this aim, GE introduces principles from genetic and molecular biology, which is the genotype to phenotype mapping with context-free grammar. The grammar in GE, a CFG that is written in Backus-Naur form, can be easily modified to output programs.

An example of mapping process employed by standard GE [1, 2] is shown in Fig. 1. In this example, the given CFG has non-terminals (<e>, <o> and <v>), which can be expended to other symbols. The grammar also has terminals that are symbols excluding non-terminals, and declares production rule that how a non-terminal expand to others.

The mapping begins with the start symbol <e>, which is usually the first symbol declared in the grammar. The GE reads codons of integer of individual and determine which production rule is chosen for each time. In this case, the

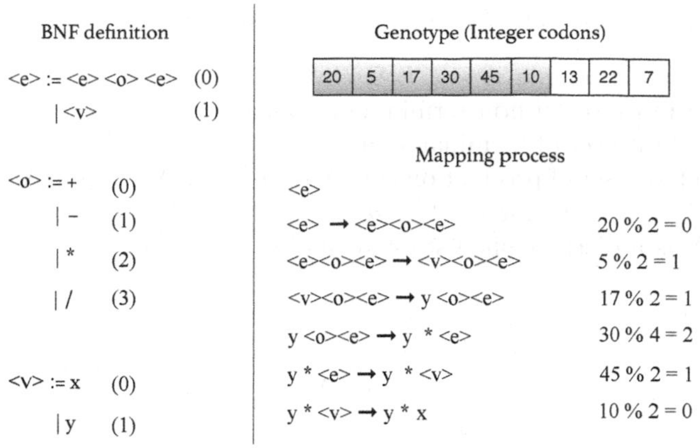

BNF definition

<e> := <e> <o> <e>   (0)
  | <v>   (1)

<o> := +   (0)
  | -   (1)
  | *   (2)
  | /   (3)

<v> := x   (0)
  | y   (1)

Genotype (Integer codons)

| 20 | 5 | 17 | 30 | 45 | 10 | 13 | 22 | 7 |

Mapping process

<e>

<e> → <e><o><e>   20 % 2 = 0

<e><o><e> → <v><o><e>   5 % 2 = 1

<v><o><e> → y <o><e>   17 % 2 = 1

y <o><e> → y * <e>   30 % 4 = 2

y * <e> → y * <v>   45 % 2 = 1

y * <v> → y * x   10 % 2 = 0

Fig. 1. An example of mapping process in standard GE

first codon is 20, and the non-terminal <e><o><e> has two alternative rules (<e><o><e> and <v>). The GE then applies modulo function for mapping n mod r, where n is the value of the codon read from individual, and r is the value of the number of possible rules. Therefore, we choose zero-th rule (i.e., 20 % 2 = 0), and non-terminal <e> is expanded to <e><o><e>. After that, the next codon is read to replace the leftmost non-terminal symbol <e><o><e> of the new expression. The mapping process performs from the leftmost (first) symbol of given expression, and it continues until the expression includes only the terminal. Note that if there are no codons left to read, then we read again given codons from leftmost (called wrapping). Moreover, as seen in Fig. 1, if the mapping process generate complete program before reading last integer codon (i.e., 7), we ignore the remaining part of codons.

As mentioned above, the original GE selects a production rule according to remainder, and the selection probability for all candidate rules is same. On this point, our proposed scheme attempts to treat mapping process with biased probability. It means that we assign higher probability to specific production rules, which are revealed many times in good individuals in former generation.

2.2 Probabilistic Context-free Grammar and EDAs Approach

In this study, we employ probabilistic context-free grammar (PCFG) to apply different probability to each production rule with regard to its usefulness, which means that how a production rule contributes to generate complete program is close to optimal solution. For instance, we give higher probability to a production rule, which was selected in best individual (i.e., complete program), than other production rules.

A probabilistic context-free grammar G consists of:

1. A context-free grammar $G = (N, \Sigma, R, S)$ where:
   (a) N is a finite set of non-terminal symbols.
   (b) Σ is a finite set of terminal symbols.
   (c) R is a finite set of production rule $X \rightarrow Y_1 Y_2 \ldots Y_n$, where $X \in N, n \geq 0$, and $Y_i \in (N \cup \Sigma)$ for $i = 1 \ldots n$.
   (d) $S \in N$ is a distinguished start symbol.

2. A parameter

$$p(\alpha \rightarrow \beta)$$

for each rule $\alpha \rightarrow \beta \in R$. The parameter $P(\alpha \rightarrow \beta)$ is a conditional probability of choosing rule $\alpha \rightarrow \beta$. For ant $X \in N$, a PCFG has the constraint

$$\sum_{\alpha \rightarrow \beta \in R : \alpha = X} p(\alpha \rightarrow \beta) = 1$$

Additionally, it has also $p(\alpha \rightarrow \beta) \geq 0$ for any $\alpha \rightarrow \beta \in R$.

A probabilistic context-free grammar is an extension of the normal context-free grammar, and it employs a parameter related to probability of choosing production rule.

```
<e> := <e> <o> <e>   (0.50)
     | <v>           (0.50)


<o> := +    (0.25)        <v> := x   (0.50)
     | -    (0.30)             | y   (0.50)
     | *    (0.20)
     | /    (0.25)
```

Fig. 2. A simple example of PCFG

Fig. 2 shows a simple example of probabilistic context-free grammar which has same rules in Fig. 1. For the grammar in Fig. 2, the program x has probability $0.5 \times 0.5 = 0.25$, which can be seen as $p(\text{<e>} \rightarrow \text{<v>}) \times p(\text{<v>} \rightarrow \text{x})$. In our designed scheme, we assign same probability belong with the number of production rules at first, such as the probability of each production rules of non-terminal <e> is identical to each other. After evaluation of individuals, the initial probability is updated from selected individuals, and the process goes on every generation.

Estimation of distribution algorithms (EDAs) is a new research field of evolutionary computation [5, 6]. In EDAs, the crossover and mutation operators are omitted, which are used in the GA. The existing recombination and evolution can sometimes lead inferior quality of solution since they disrupt of partial solution that propagates so far. In this regards, EDAs captures specific interactions among the variables of individual, and it consider joint probability distribution associated with the individuals. As a result, new population is generated by sampling the probability distribution, which is estimated from a selected individuals the previous generation. In this study, we employ EDAs approach to our new algorithm since we utilize PCFG to generate program code. Unlike standard GE, the probability of production rules is changed as generation goes by in our algorithm. Therefore, the approach of EDAs is more suitable than standard GE process, which applies GA's genetic operators. We will discuss about combination PCFG and EDAs in next section.

3   Proposed Algorithm

In this paper, a new scheme for GE is proposed using probabilistic context-free grammar instead of context-free grammar. The characteristic of PCFG is that each production rule has probability to be expanded, and the grammar chooses a rule with regard to assigned probability. Furthermore, we employ a concept of estimation of distribution algorithms that create and evolve population by means of probabilistic model, which considers specific interactions among the variables of individuals. In our algorithm, we update given probability through every generation. In other words, we assign same probability for each production rule at first, and the probability is replaced to new value after evaluation. Fig. 3 shows the overall process of our designed algorithm. Note that proposed algorithm selects half of population as fittest individuals for every generation.

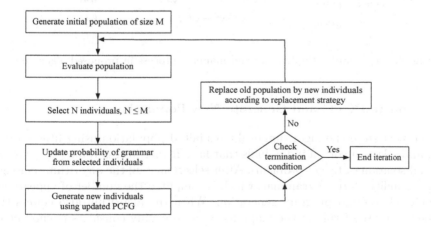

Fig. 3. The overall process of proposed algorithm

3.1   Program Generation Using PCFG

As mentioned above, we considers probability for choosing production rules in the process of generation of phenotype (complete program). For initial stage of our algorithm, the probability of each production rule is same. Moreover, the assigned value is computed as $1/N_r$, where the value N_r is the number of production rules. Unlikely genotype of standard GE [1, 2], our algorithm treats a genotype that consists of probability vector, which is generated randomly with range of 0 and 1. Fig. 4 shows an example of individual and mapping process with PCFG. As seen in Fig. 4, each production rule in non-terminal <o> has identical value (0.25) to others. After the initial probability is assigned, we perform the mapping process, which is similar to original GE. This mapping operation continues in every generation until termination condition is satisfied. Note that the mapping genotype to phenotype in initial generation is same as existing process that employs modulo function, because the probability for each rule is same.

BNF definition		Genotype (probability vector)

<e> := <e> <o> <e> (0.50)		0.13 0.75 0.68 0.80 0.91 0.12 0.23 0.77
| <v> (0.50)		

		Mapping process
<o> := + (0.25)	<e>	
| − (0.25)	<e> → <e><o><e>	0.13
| * (0.25)	<e><o><e> → <v><o><e>	0.75
| / (0.25)	<v><o><e> → y <o><e>	0.68
	y <o><e> → y * <e>	0.80
<v> := x (0.50)	y * <e> → y * <v>	0.91
| y (0.50)	y * <v> → y * x	0.12

Fig. 4. An example of individual and mapping process in proposed algorithm

3.2   Update PCFG and Generate New Population

For the next step, the algorithm evaluates initial population using fitness function, and selects a sort of individuals that have higher fitness value (i.e., close to optimal solution of target problem). After selection step, the algorithm re-assign the probability of given grammar, which is computed the number of appearance of each rule though program generation. When the algorithm also counts the frequency of use of rule in the mapping process, it only considers phenotype of selected individuals. The algorithm computes the parameter of PCFG to update probability of each production rules as the number of appearance of terminal is divided by the total number of occurrence of non-terminal.

For instance, the non-terminal <o> has four production rules, and the initial probability of each rules are same (0.25) as seen in Fig. 4. If the terminals in non-terminal <o> are appeared total 100 times, and occurrence of each terminal is counted as 10, 25, 40 and 15 through mapping process, then the probability to choose terminal '−' is 0.15 (15 / 100 = 0.15). In this regard, the algorithm updates the parameter of given PCFG.

After that the algorithm generates new population based on newly updated grammar. The offsprings (new population) can choose a sort of rules with higher probability that appeared many times in fittest individuals from previous generation. It means that our algorithm is able to reserve and propagate superior characteristics to next generation. Moreover, we omit the common genetic operators (crossover and mutation), and apply a concept of EDAs that finds a solution using probabilistic interrelations of given grammar.

4   Experiments

4.1   Experimental Setup

In this section, we illustrate the comparison of the performance between standard GE [1] to our algorithm, and analyzed the results to prove the proposed algorithm can archive better (or comparable) search capability with regard to original GE. We carried out the experiments to following problem domains; symbolic regression and Santa Fe trail problem.

Table 1. GE parameters adopted for each test problem

Parameter	Value
Generation	50 (standard GE)
	100 (proposed algorithm)
Population size	500
Initialization	Random
Size of codon(integer)	10 (symbolic regression)
	15 (Santa Fe Trail)
Maximum codon wraps	5
Selection method	Tournament selection (pair-wise)
Replacement strategy	Elitist replacement (0.5)
Probability of crossover (P_c)	0.9
Probability of mutation (P_m)	0.01

Table 1 presents the evolutionary parameters adopted for all test problems described below. We conducted each experiment 100 evolutionary runs and all the results are average value of runs. In the proposed algorithm, it evaluates half of population (i.e., selected individuals) for every generation. In this regard, we used double number of generation with regard to standard GE, therefore both algorithm performed same number of evaluation step even though they run different generations. The wrapping, as described in Section 2.1, was adopted up to 5 times for all experiments. Additionally, we conducted pair-wise tournament selection[4] and Elitist replacement[7]. Note that the parameters related to crossover and mutation are only available to standard GE.

4.2   Symbolic Regression

The object of Symbolic regression[8, 9] is to find some mathematical expression in symbolic form that represents a given set of input and output pairs. Additionally, the problem chooses 20 points in the range of -1 to 1 as the input values. We conducted this experiment using the particular function below

$$f(x) = x^4 + x^3 + x^2 + x$$

The context-free grammar G used in this problem is given below

$N = \{$ expr, op, preop, var $\}$
$\Sigma = \{$ Sin, Cos, Exp, Log, +, -, /, *. x, 1.0, (,) $\}$
$S = \{$ expr $\}$

and production rule R is shown in Table 2.

Table 2. BNF rules for Symbolic regression

(1)	<expr>::= <expr><op><expr>
	| (<expr><op><expr>)
	| <preop> (<expr>)
	| <var>
(2)	<op>::= +
	| −
	| *
	| /
(3)	<preop>::= Sin
	| Cos
	| Exp
	| Log
(4)	<var>::= x
	| 1.0

The fitness for this problem is root mean square error (RMSE) of sum, which is taken over 20 fitness cases. It indicates the error between the created program (i.e., phenotype) and target function. Note that the individual that has smaller fitness value than others is better individual.

4.3 Santa Fe Trail Problem

The Santa Fe Trail problem[8, 9] is to find a computer program to control artificial ants, which collect food in the given space, such as 32 x 32 grid region and 90 foods. This problem is a deceptive problem with many local and global optima. The BNF definition of context-free grammar G is given below

N = { code, line, expr, condition, op }
Σ = { food_ahead(), else(), left(), right(), move(), if, {, }, (,), }
S = { code }

and production rule R is shown in Table 3.

Table 3. BNF rules for Santa Fe Trail

(1)	\<code\>::= \<line\>
	| \<code\>\<line\>
(2)	\<line\>::= \<condition\>
	| \<op\>
(3)	\<condition\>::= if(food_ahead())
	{\<line\>}
	else
	{\<line\>}
(4)	\<op\>::= left()
	| right()
	| move()

As shown in Table 3, this problem has three type of movements (or steps), such as left(), right() and move(). The artificial ant starts in the upper left cell (0, 0), and it can only move in vertical and horizontal directions. Also, this problem has maximum energy for movement, which is consumed one energy for every movement. We specified maximum energy of ant as 400 for this problem. Additionally, this problem adopts an assumption that the ant in this problem has a sensor to find food. In this regard, the terminal food_ahead() is available, which the ant examines the existence of food in front of its position. The fitness for this problem is the number of remaining food after the simulation of phenotype (i.e., programming codes). Note that the smaller fitness value indicates the individual generates a program that has higher capability for searching foods.

4.4 Results

The convergence history of the fitness and its standard deviation of both average of population and best individual is shown in Fig. 5 and Fig. 6. From the experimental results, our proposed algorithm can generate solutions that are comparable with standard GE for all of test problems described here. For the symbolic regression problem, our approach can improve the quality of solution since the process of updating probability of production rule can guide the mapping process to adopt structure of selected individual's grammar.

As shown in Fig. 6, the original GE is better than our proposed algorithm, because the convergence speed of our approach is faster than comparative algorithm. It indicates that our approach, biased probability in the mapping process, can lead premature convergence for this problem, whereas it performs better in the symbolic regression problem. However, the gap of performance between both algorithm is not critical.

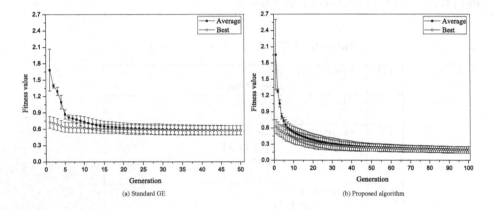

(a) Standard GE   (b) Proposed algorithm

Fig. 5. Experimental result for symbolic regression

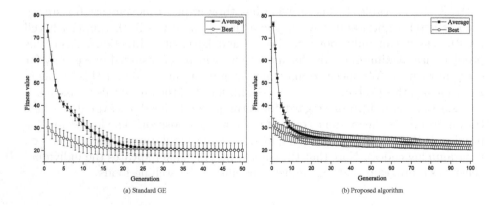

(a) Standard GE   (b) Proposed algorithm

Fig. 6. Experimental result for Santa Fe Trail problem

5   Conclusions

In this paper, we presented a new approach to grammatical evolution that adopts probabilistic context-free grammar. To archive this objective, we applied biased probability in the mapping process. It means that the algorithm chooses production rules based on different probability of each component, whereas the grammar in standard grammatical evolution has same probability for all of production rules. . In this regard, we need an appropriate evolutionary mechanism to utilize probabilistic context-free grammar. Therefore, we omitted the common genetic operators (crossover and mutation), and employed the concept of estimation of distribution algorithms that creates and evolves a population of individuals by means of probabilistic modeling.

The experiments were conducted with two test problems; symbolic regression and Santa Fe Trail. It was proven that the designed algorithm has the capability to generate complete program to solve the target problems. The experimental result for symbolic regression problem shows that our approach has higher convergence speed and quality of solution than standard GE. In the Santa Fe Trail problem, our algorithm performed comparable to conventional GE even though the problem has characteristics of real world problem, such as many local optima, large search spaces. Moreover, future work will investigate the extension of proposed mapping process and apply to other real world problems.

Acknowledgement. This work was supported by the National Research Foundation of Korea (NRF) grant funded by the Korea government (MSIP) (NRF-2012R1A2A2A01013735).

References

[1] O'Neill, M., Ryan, C.: Grammatical evolution. IEEE Transactions on Evolutionary Computation 5, 349–358 (2001)

[2] Ryan, C., Collins, J.J., Neill, M.O.: Grammatical evolution: Evolving programs for an arbitrary language. In: Banzhaf, W., Poli, R., Schoenauer, M., Fogarty, T.C. (eds.) EuroGP 1998. LNCS, vol. 1391, pp. 83–96. Springer, Heidelberg (1998)

[3] Fogel, D.B.: Evolutionary Computation: Toward a New Philosophy of Machine Intelligence, 3rd edn. Wiley-IEEE Press (2005)

[4] Goldberg, D.E., Holland, J.H.: Genetic algorithms and machine learning. Machine Learning 3, 95–99 (1988), doi:10.1023/A:1022602019183

[5] Larrañaga, P., Lozano, J.A.: Estimation of distribution algorithms: A new tool for evolutionary computation, vol. 2. Springer (2002)

[6] Paul, T.K., Iba, H.: Linear and combinatorial optimizations by estimation of distribution algorithms. In: 9th MPS Symposium on Evolutionary Computation, IPSJ, Japan (2002)

[7] Lima, C.F., Pelikan, M., Goldberg, D.E., Lobo, F.G., Sastry, K., Hauschild, M.: Influence of selection and replacement strategies on linkage learning in boa. In: IEEE Congress on Evolutionary Computation, CEC 2007, pp. 1083–1090. IEEE (2007)
[8] Koza, J.R.: Genetic programming II: automatic discovery of reusable programs. MIT Press (1994)
[9] Koza, J.R.: Genetic programming III: Darwinian invention and problem solving, vol. 3. Morgan Kaufmann (1999)

An Adaptive Cauchy Differential Evolution Algorithm with Population Size Reduction and Modified Multiple Mutation Strategies

Tae Jong Choi and Chang Wook Ahn*

Department of Computer Engineering, Sungkyunkwan University (SKKU)
2066 Seobu-Ro, Suwon 440-746, Republic of Korea
{gry17,cwan}@skku.edu

Abstract. Adapting control parameters is an important task in the literature of the differential evolution (DE) algorithm. A balance between Exploration and Exploitation plays a large role in the performance of DE. A dynamic population sizing method can help maintaining the balance. In this paper, we improved an adaptive differential evolution (ACDE) algorithm by attaching the modified population size reduction (4MPSR) method. 4MPSR method reduces the population size gradually and uses four mutation strategies with different ranges of the scaling factor. In short, 4MPSR method has better Exploration during the early stage and Exploitation during the late stage. ACDE algorithm performs well in solving various benchmark problems. However, ACDE algorithm adapts two control parameters, the scaling factor and the crossover rate but uses a fixed population size. By attaching 4MPSR method to ACDE algorithm, all of the control parameters can be adapted and, hence, the performance can be improved. We compared the proposed algorithm with some state-of-the-art DE algorithms in various benchmark problems. The performance evaluation results showed that the proposed algorithm is significantly improved for solving both the unimodal problems and the multimodal problems. And the proposed algorithm obtained the better final solutions than the state-of-the-art DE algorithms.

Keywords: Differential Evolution Algorithm, Adaptive Parameter Control, Population Size, Mutation Strategy, Global Numerical Optimization.

1 Introduction

The differential evolution (DE) algorithm is a population-based stochastic search algorithm [1],[2]. Although the structure of the DE algorithm is simple, the performance of global optimization is more effective and robust than other Evolutionary Algorithms (EAs) [3]. In order to improve the performance maintaining a balance between Exploration and Exploitation is important, e.g., Exploration should be strengthened during the early stage of the optimization and during the

---

* Corresponding author.

© Springer International Publishing Switzerland 2015   13
H. Handa et al. (eds.), *Proc. of the 18th Asia Pacific Symp. on Intell. & Evol. Systems – Vol. 2*,
Proceedings in Adaptation, Learning and Optimization 2, DOI: 10.1007/978-3-319-13356-0_2

late stage, Exploitation should be intensified [4]. A dynamic population sizing method is suitable to this purpose. If many individuals consist the population, then Exploration is strengthened. On the other hand, Exploitation is intensified if the population contains few individuals. In order to automatically adapt the population size, a dynamic population sizing method is required. However, significantly less work has been researched in terms of the subject [5].

In this paper, we improved an adaptive differential evolution (ACDE) algorithm [6] by attaching the modified population size reduction (4MPSR) method. Due to the Cauchy distribution, ACDE algorithm allocates not only the current good control parameter value, but also far from the current good value which may be suitable in the next generation to each individual. As a result, ACDE algorithm performs well in solving various benchmark problems. However, ACDE algorithm adapts two control parameters, the scaling factor and the crossover rate but uses a fixed population size. Maintaining a suitable population size is important to obtain the better final solutions [4],[5],[7]-[9].

The population size reduction (PSR) method [4] is a dynamic population sizing method. PSR method shows its effectiveness by maintaining a high population size during the early stage and a small population size during the late stage. 4MPSR method is an extension of PSR method by using four mutation strategies with different ranges of the scaling factor. More specifically, 4MPSR method uses the most Exploration mutation strategy during the early stage. After that, the method uses the stronger Exploitation mutation strategy at each step. Therefore, 4MPSR method has better Exploration during the early stage and Exploitation during the late stage. With a small increase of computational complexity, 4MPSR method can provide the DE algorithm a more suitable balance between Exploration and Exploitation.

We compared the proposed algorithm with some state-of-the-art DE algorithms in various benchmark problems. The performance evaluation results showed that the proposed algorithm was significantly improved for solving the unimodal problems, as well as the multimodal problems, and obtained the better final solutions than the state-of-the-art DE algorithms.

2   Related Work

2.1   DE Algorithm

The DE algorithm is a population-based stochastic search algorithm. NP individuals consist the population. D variables consist an individual. There are four main operators, Initialization, Mutation, Crossover and Selection operator. Initialization operator is for initializing the population. Mutation and Crossover operator is for generating the child individuals. And Selection operator compares the parent individuals with their corresponding child individuals. After that, the operator produces the population of the next generation. A parent and a child individual are denoted by $X_{i,G} = \{x_{i,1,G}, \ldots, x_{i,D,G}\}$ and $V_{i,G} = \{v_{i,1,G}, \ldots, v_{i,D,G}\}$, respectively.

Initialization. Initialization operator is for initializing the population. The minimum and the maximum search bound are denoted by $X_{min} = \{x_{min,1}, \ldots, x_{min,D}\}$ and $X_{max} = \{x_{max,1}, \ldots, x_{max,D}\}$, respectively. All of individuals are initialized as follows:

$$x_{i,j,0} = x_{min,j} + rand_{i,j}(0,1) \cdot (x_{max,j} - x_{min,j}) \tag{1}$$

where $rand_{i,j}(0,1)$ denotes a uniformly distributed random number within the range $[0,1]$.

Mutation. Mutation operator is the first operator for generating the child individuals. The operator generates the mutant individuals. A mutant individual is denoted by $V_{i,G} = \{v_{i,1,G}, \ldots, v_{i,D,G}\}$ and generated as follows:

$$V_{i,G} = X_{r_1,G} + F \cdot (X_{r_2,G} - X_{r_3,G}) \tag{2}$$

where $X_{r_1,G}$, $X_{r_2,G}$ and $X_{r_3,G}$ denote some donor individuals where $r_1, r_2, r_3 \in [1, 2, \cdots, NP]$ and $r_1 \neq r_2 \neq r_3 \neq i$. And F denotes the scaling factor which is one of the control parameters. This mutation operator is called DE/rand/1 strategy.

Crossover. Crossover operator is the second operator for generating the child individuals. At first, the operator chooses a random integer within the range $[1, D]$, denoted by j_{rand}. The random integer is to ensure that at least one dimension of the child individual is generated to that of its corresponding mutant individual's. A child individual is generated as follows:

$$u_{i,j,G} = \begin{cases} v_{i,j,G} & \text{if } rand_{i,j}(0,1) \leq CR \text{ or } j = j_{rand} \\ x_{i,j,G} & \text{otherwise} \end{cases} \tag{3}$$

where CR denotes the crossover rate which is one of the control parameters. This crossover operator is called Binomial crossover.

Selection. Selection operator compares the parent individuals with their corresponding child individuals. After that, the operator produces the population of the next generation. More specifically, if a child individual's fitness is lower than or equal to that of its corresponding parent individual's, then the child individual survives as an individual in the next generation and the parent individual is discarded. Otherwise, the parent individual remains, and the child individual is abandoned. Selection operator is formulated as follows:

$$X_{i,G+1} = \begin{cases} U_{i,G} & \text{if } f(U_{i,G}) \leq f(X_{i,G}) \\ X_{i,G} & \text{otherwise.} \end{cases} \tag{4}$$

where $f(X)$ denotes an objective function.

2.2 jDE

Brest et al. [10],[11] proposed an adaptive DE algorithm which adapts two control parameters, known as jDE algorithm. jDE algorithm encapsulates the scaling factor and the crossover rate in each individual as augmented dimensions, denoted by F_i and CR_i. The suitable control parameters lead to better individuals which, in turn, are more likely to survive and generate child individuals and, hence, propagate the control parameters. jDE algorithm contains four additional control parameters, τ_1, τ_2, F_l, and F_u. The first two control parameters determine whether F_i and CR_i should be updated or not. The last two parameters restrict the range of F_i. The control parameters F_i and CR_i are adapted in every generation as follows:

$$F_{i,G+1} = \begin{cases} F_l + rand_2(0,1) \cdot F_u & \text{if } rand_1(0,1) < \tau_1 \\ F_{i,G} & \text{otherwise} \end{cases} \tag{5}$$

$$CR_{i,G+1} = \begin{cases} rand_4(0,1) & \text{if } rand_3(0,1) < \tau_2 \\ CR_{i,G} & \text{otherwise} \end{cases} \tag{6}$$

where $rand_j(0,1)$, $j \in \{1,2,3,4\}$ denote uniformly distributed random numbers within the range $[0,1]$, $\tau_1 = 0.1$, $\tau_2 = 0.1$, $F_l = 0.1$, and $F_u = 0.9$.

2.3 dynNP-DE

Brest et al. [4] proposed the population size reduction (PSR) method which reduces the population size gradually and attached it with jDE algorithm, known as dynNP-DE algorithm. PSR method contains an additional control parameter $pmax$. The parameter denotes the number of different population sizes. A set of the generations G_R which executes the population size reduction operator is as follows:

$$G_R = \{gen_1, gen_1 + gen_2, \cdots, \sum_{p=1}^{pmax-1} gen_p\} \tag{7}$$

where

$$gen_p = \lfloor \frac{maxnfeval}{pmax \cdot NP_p} \rfloor \tag{8}$$

where $maxnfeval$ denotes the maximum evaluation counters. The population size reduction operator is as follows:

$$NP_{G+1} = \begin{cases} \frac{NP_G+1}{2} & \text{if } G = G_R \\ NP_G & \text{otherwise} \end{cases} \tag{9}$$

where NP_{G+1} and NP_G denote the reduced population size and the current population size, respectively. dynNP-DE algorithm uses an additional Selection operator to produces the population of the reduced population size. The Selection operator is as follows:

$$X_{i,G} = \begin{cases} X_{\frac{NP}{2}+i,G} & \text{if } f(X_{\frac{NP}{2}+i,G}) \leq f(X_{i,G}) \text{ and } G = G_R \\ X_{i,G} & \text{otherwise} \end{cases} \tag{10}$$

2.4 jDE$_{NP,MM}$

Zamuda and Brest [12] proposed the population size reduction with the multiple mutation strategies (PSRMM) method which reduces the population size gradually and uses two multiple mutation strategies, DE/rand/1 and DE/best/1. And the authors attached it with jDE algorithm, known as jDE$_{NP,MM}$ algorithm. At first, PSRMM method chooses a random number within the range $[0, 1]$, denoted by s. The first mutation strategy DE/rand/1 is applied when $NP \geq 100$ or $s \leq 0.75$. Otherwise, the second mutation strategy DE/best/1 is applied. PSRMM method is formulated as follows:

$$V_{i,G} = \begin{cases} X_{r_1,G} + F \cdot (X_{r_2,G} - X_{r_3,G}) & \text{if } NP \geq 100 \text{ or } s \leq 0.75 \\ X_{best,G} + F \cdot (X_{r_1,G} - X_{r_2,G}) \text{ otherwise} \end{cases} \qquad (11)$$

3   ACDE

In this paper, we improved ACDE algorithm by attaching 4MPSR method. In this section, we review ACDE algorithm. Choi et al. [6] proposed an adaptive DE algorithm which adapts two control parameters, known as ACDE algorithm. ACDE algorithm encapsulates the scaling factor and the crossover rate in each individual as augmented dimensions, denoted by F_i and CR_i. The control parameters are adapted by using the Cauchy distribution. More specifically, the mean value of the Cauchy distribution is the mean value of the successfully evolved individuals' control parameter in the previous generation and the scale value is 0.1. Threfore, Due to the Cauchy distribution, ACDE algorithm allocates not only the current good control parameter value, but also far from the current good value which may be suitable in the next generation to each individual. The control parameter F_i and CR_i are adapted in every generation as follows:

$$F_{i,G+1} = C(F_{avg,G}, 0.1) \qquad (12)$$

$$CR_{i,G+1} = C(CR_{avg,G}, 0.1) \qquad (13)$$

where $C(x_0, \gamma)$ denotes the Cauchy distribution. And $F_{avg,G}$ and $CR_{avg,G}$ denote the mean value of the successfully evolved individuals' scaling factor and crossover rate in G generation. After that, the scaling factor is truncated to the interval $[0.1, 1]$ and the crossover rate is truncated to the interval $[0, 1]$. In order to calculate the mean value, ACDE algorithm maintains the success memory of the scaling factor and the crossover rate.

4   Population Size Reduction and Modified Multiple Mutation Strategies

Although ACDE algorithm shows the good performance of global optimization by adapting two control parameters, the algorithm uses a fixed population size.

By attaching a dynamic population sizing method to ACDE algorithm all of the control parameters can be adapted and, hence, the performance of global optimization can be improved. At first, we review some related work briefly and explain the motivation of this paper.

As we mentioned in Section 1, in order to improve the performance maintaining a balance between Exploration and Exploitation is important. PSR method [4] was proposed to this purpose. In PSR method, maintaining a high population size during the early stage for strengthening Exploration and a small population size during the late stage for intensifying Exploitation. As a result, the authors showed that a dynamic population sizing method is suitable to maintain a balance between Exploration and Exploitation properties. PSRMM method [12] was proposed for improving PSR method. In PSRMM method, the multiple mutation strategies operator which uses DE/rand/1 and DE/best/1 is added to PSR method. Generally, DE/rand/1 strategy performs the slow convergence but finds a promising region easily. On the other hand, DE/best/1 strategy performs the fast convergence but gets stuck in a local optimum easily. Therefore, in PSRMM method, DE/rand/1 strategy is used during the first half of the maximum evaluation counters (due to $NP_{init} = 200$ and $pmax = 4$) or three out of four. Otherwise, DE/best/1 strategy is used. As a result, the authors showed that a dynamic population sizing method with multiple mutation strategies operator can improve the effectiveness and robustness of the performance.

In this paper, we modified PSRMM method by using four mutation strategies with different ranges of the scaling factor, called 4MPSR method. 4MPSR method is organized as two operators. The first is the population size reduction operator, and the second is the modified multiple mutation strategies operator. The population size reduction operator is same as PSR method [4]. The multiple mutation strategies operator uses four mutation strategies. The strategies are as follows: DE/rand/1, DE/rand2, DE/current-to-best/1 and DE/curren-to-best/2.

DE/rand/2 strategy

$$V_{i,G} = X_{r_1,G} + F \cdot (X_{r_2,G} - X_{r_3,G} + X_{r_4,G} - X_{r_5,G}) \qquad (14)$$

DE/current-to-best/1 strategy

$$V_{i,G} = X_{i,G} + F \cdot (X_{best,G} - X_{i,G} + X_{r_1,G} - X_{r_2,G}) \qquad (15)$$

DE/current-to-best/2 strategy

$$V_{i,G} = X_{i,G} + F \cdot (X_{best,G} - X_{i,G} + X_{r_1,G} - X_{r_2,G} + X_{r_3,G} - X_{r_4,G}) \quad (16)$$

DE/rand/1 strategy is presented in Equation (2). 4MPSR method uses a different mutation strategy at each step, i.e., DE/rand/2, DE/rand/1, DE/current-to-best/2 and DE/current-to-best/1 strategies are used at the first, the second, the third and the fourth step. Each step is an equal number of evaluation counters $\frac{mfeval}{4}$ where $mfeval$ denotes the maximum evaluation counters. DE/rand/2 strategy has strong Exploration property. Therefore, the mutation strategy is suitable to be used in the early stage. DE/rand/1 strategy has good Exploration property. We assumed that the population diversity of DE/rand/1 strategy is less than that of DE/rand/2 strategy because its perturbation vector is lower than DE/rand/2. Based on the reason, 4MPSR method uses DE/rand/1 strategy after DE/rand/2 strategy. DE/current-to-best/2 has good Exploitation property than DE/rand/1 but less than DE/current-to-best/1. Therefore, the mutation strategy is suitable to be used before DE/current-to-best/1 strategy. And DE/current-to-best/1 strategy has strong Exploitation property. Therefore, the mutation strategy is suitable to be used in the late stage. In short, 4MPSR method uses higher Exploration property mutation strategies during the early stage and higher Exploitation property mutation strategies during the late stage. 4MPSR method is formulated as follows:

$$
V_{i,G} = \begin{cases} \text{DE/rand/2} \cdots Equation(14) & \text{if } cfeval \leq mfeval \cdot 0.25 \\ \text{DE/rand/1} \cdots Equation(2) & \text{if } mfeval \cdot 0.25 < cfeval \leq mfeval \cdot 0.5 \\ \text{DE/current-to-best/2} \cdots Equation(16) & \text{if } mfeval \cdot 0.5 < cfeval \leq mfeval \cdot 0.75 \\ \text{DE/current-to-best/1} \cdots Equation(15) & \text{if } cfeval \leq mfeval \end{cases}
$$

$$(17)$$

where $cfeval$ denotes the current evaluation counter. We applied different ranges of scaling factor to each mutation strategy i.e., DE/rand/2 strategy for $F_i \in [0.1, 0.5]$, DE/rand/1 strategy for $F_i \in [0.1, 1]$, DE/current-to-best/2 strategy for $F_i \in [0.1, 1]$ and DE/current-to-best/1 strategy for $F_i \in [0.6, 1]$. In other words, we restricted $F_i \in [0.1, 0.5]$ and $F_i \in [0.6, 1]$ for slightly alleviating Exploration property of DE/rand/2 and Exploitation property of DE/current-to-best/1.

To sum up, 4MPSR method uses the most Exploration mutation strategy during the early stage. After that, the method uses the stronger Exploitation mutation strategy at each step. Therefore, 4MPSR method has better Exploration during the early stage and Exploitation during the late stage. With a small increase of computational complexity, 4MPSR method can provide the DE algorithm a more suitable balance between Exploration and Exploitation. An example of 4MPSR method is presented in Figure (1).

5   Performance Evaluation

5.1   Benchmark Problems

In order to carry out the performance of global optimization of each algorithm, we use various benchmark problems. The first thirteen benchmark problems are from [13],[14] and the rest benchmark problems are Extended f_{14} (F_{14}), Bohachevsky (F_{15}), and Schaffer (F_{16}). The characteristic of the benchmark problems is as follows: $F_1 - F_4$ are continuous unimodal problems, F_6 is a discontinuous step problem, F_7 is a noise quadratic problem, and F_5, $F_8 - F_{16}$ are continuous multimodal problems which the number of local optimums grows exponentially when the dimension grows. The detailed information of each benchmark problem can be found in [13],[14]. The benchmark problems are presented

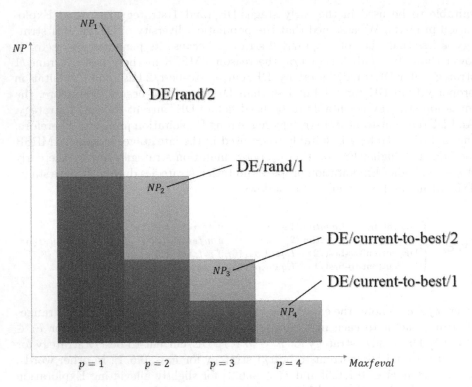

Fig. 1. An example of 4MPSR method, when $pmax = 4$

in Table 1. In the table, D, S and F_{min} denote the number of dimensions, the search bound and the global optimum, respectively.

The maximum evaluation counters is assigned by 150K for F_1, F_6, F_{12} and F_{13}, 200K for F_2 and F_{10}, 300K for F_7, F_{11}, F_{14} and F_{16}, 500K for F_3, F_4 and F_9, 900K for F_8, 2000K for F_5 and 100K for F_{15}.

5.2 Comparison between the Proposed Algorithm with Some State-of-the-Art DE Algorithms

We compared the proposed algorithm with some state-of-the-art DE algorithms. The six algorithms in comparison are listed as follows:

1) ACDE [6];
2) JADE without archive [15];
3) jDE [10];
4) SaDE [16];
5) DE/rand/1/bin with $F = 0.5$ and $CR = 0.9$ [1],[2];
6) ACDE$_{4MPSR}$ with $NP_{init} = 200$ and $pmax = 4$ (The proposed algorithm);

Table 1. Benchmark problems

Benchmark problems	D	S	F_{min}
$F_1(x) = \sum_{i=1}^{D} x_i^2$	30	$[-100, 100]^D$	0
$F_2(x) = \sum_{i=1}^{D} \lvert x_i \rvert + \prod_{i=1}^{D} \lvert x_i \rvert$	30	$[-10, 10]^D$	0
$F_3(x) = \sum_{i=1}^{D} (\sum_{j=1}^{i} x_j)^2$	30	$[-100, 100]^D$	0
$F_4(x) = max_i(\lvert x_i \rvert, 1 \leq i \leq D)$	30	$[-100, 100]^D$	0
$F_5(x) = \sum_{i=1}^{D-1} [100(x_{i+1} - x_i^2)^2 + (x_i - 1)^2]$	30	$[-30, 30]^D$	0
$F_6(x) = \sum_{i=1}^{D} (\lfloor x_i + 0.5 \rfloor)^2$	30	$[-100, 100]^D$	0
$F_7(x) = \sum_{i=1}^{D} i x_i^4 + random[0, 1)$	30	$[-1.28, 1.28]^D$	0
$F_8(x) = \sum_{i=1}^{D} -x_i sin(\sqrt{\lvert x_i \rvert})$	30	$[-500, 500]^D$	-12569.5
$F_9(x) = \sum_{i=1}^{D} [x_i^2 - 10cos(2\pi x_i) + 10]$	30	$[-5.12, 5.12]^D$	0
$F_{10}(x) = -20exp(-0.2\sqrt{\frac{1}{D}\sum_{i=1}^{D} x_i^2}) - exp(\frac{1}{D}\sum_{i=1}^{D} cos2\pi x_i)$ $+20 + exp(1)$	30	$[-32, 32]^D$	0
$F_{11}(x) = \frac{1}{4000}\sum_{i=1}^{D} x_i^2 - \prod_{i=1}^{D} cos(\frac{x_i}{\sqrt{i}}) + 1$	30	$[-600, 600]^D$	0
$F_{12}(x) = \frac{\pi}{D}\{10sin^2(\pi y_1) + \sum_{i=1}^{D-1}(y_i - 1)^2[1 + 10sin^2(\pi y_{i+1})]$ $+(y_D - 1)^2\} + \sum_{i=1}^{D} u(x_i, 10, 100, 4)$ $y_i = 1 + \frac{1}{4}(x_i + 1)$ $u(x_i, a, k, m) = \begin{cases} k(x_i - a)^m & , x_i > a \\ 0 & , -a \leq x_i \leq a \\ k(-x_i - a)^m & , x_i < -a \end{cases}$	30	$[-50, 50]^D$	0
$F_{13}(x) = 0.1\{sin^2(3\pi x_1) + \sum_{i=1}^{D-1}(x_i - 1)^2[1 + sin^2(3\pi x_{i+1})]$ $+(x_D - 1)^2[1 + sin^2(2\pi x_D)]\} + \sum_{i=1}^{D} u(x_i, 5, 100, 4)$	30	$[-50, 50]^D$	0
$F_{14}(x) = f_{14}(x_n, x_1) + \sum_{i=1}^{D-1} f_{14}(x_i, x_{i+1})$ $f_{14}(x, y) = (x^2 + y^2)^{0.25}[sin^2(50(x^2 + y^2)^{0.1}) + 1]$	30	$[-100, 100]^D$	0
$F_{15}(x) = \sum_{i=1}^{D-1}(x_i^2 + 2x_{i+1}^2 - 0.3cos(3\pi x_i) - 0.4cos(4\pi x_{i+1}) + 0.7)$	30	$[-15, 15]^D$	0
$F_{16}(x) = \sum_{i=1}^{D-1}(x_i^2 + x_{i+1}^2)^{0.25}[sin^2(50(x_i^2 + x_{i+1}^2)^{0.1}) + 1]$	30	$[-100, 100]^D$	0

All of the used parameter values for each algorithm were recommended parameter values in their authors. The population size NP was fixed by 100 for each algorithm except the proposed algorithm which was initialized by 200.

The performance evaluation results of the comparison between the proposed algorithm with the state-of-the-art DE algorithms are presented in Table 2. In the table, $MFeval$ denotes the maximum evaluation counters. And $Mean$ and Std denote the mean of final solutions and its standard deviation, respectively. All of the performance evaluation results were run 50 times, independently. For clarity, the best algorithm was marked in boldface type. The proposed algorithm performed the best performance for solving F_2, F_4, F_5, F_9, $F_{11} - F_{16}$, which were 10 out of 16 benchmark problems. ACDE algorithm was 8 out of 16, which were $F_9 - F_{16}$. JADE and jDE algorithm were 6 out of 16. In JADE algorithm, F_1, F_3, F_7, F_9 F_{12} and F_{13}. In jDE algorithm, F_8, F_9, F_{11}, $F_{14} - F_{16}$. SaDE algorithm was 2 out of 16, which were F_9 and F_{11}. Standard DE algorithm was 1 out of 16, which was F_{11}. As a result, the proposed algorithm was significantly improved for solving the unimodal problems, as well as the multimodal problems, and obtained the better final solutions than the state-of-the-art DE algorithms.

Table 2. Comparison between the proposed algorithm with some state-of-the-art DE algorithms

MFeval	ACDE		JADE		jDE		SaDE		DE		ACDE$_{AMPSR}$	
	Mean	Std	Mean	Std	Mean	Std	Mean	Std	Mean	Std	Mean	Std
F_1 150K	3.3E-36	2.7E-36	**1.1E-60**	**6.6E-60**	1.1E-28	1.3E-28	4.5E-20	1.0E-19	7.6E-14	5.6E-14	6.7E-56	1.7E-55
F_2 200K	2.0E-30	1.4E-30	4.2E-26	2.8E-25	7.1E-24	4.5E-24	2.1E-14	2.1E-14	2.2E-10	1.2E-10	**2.9E-40**	**1.7E-39**
F_3 500K	9.1E+00	2.7E+01	**4.6E-62**	**1.1E-61**	3.1E-10	5.2E-10	4.0E-36	1.8E-35	3.9E-11	4.3E-11	1.6E-23	9.5E-23
F_4 500K	2.6E-07	1.4E-06	3.9E-07	1.7E-07	1.0E-06	7.1E-06	9.6E-06	3.1E-06	1.3E-01	3.2E-01	**3.6E-10**	**1.6E-09**
F_5 2000K	1.2E+01	8.6E+00	8.0E-02	5.6E-01	2.1E+00	1.5E+00	1.1E+01	1.1E+01	8.0E-02	5.6E-01	**8.2E-28**	**1.0E-27**
F_6 150K	0.0E+00	0.0E+00	0.0E+00	0.0E+00	0.0E+00	0.0E+00	0.0E+00	0.0E+00	0.0E+00	0.0E+00	0.0E+00	0.0E+00
F_7 300K	3.0E-03	7.4E-04	**6.6E-04**	**2.9E-04**	3.6E-03	9.4E-04	4.6E-03	1.5E-03	4.6E-03	1.1E-03	1.3E-03	3.9E-04
F_8 900K	4.4E-12	7.2E-15	2.4E+00	1.7E+01	**3.8E-12**	**4.9E-14**	3.9E-12	2.2E-14	1.5E+03	5.5E+02	4.4E-12	2.3E-14
F_9 500K	**0.0E+00**	**0.0E+00**	**0.0E+00**	**0.0E+00**	**0.0E+00**	**0.0E+00**	**0.0E+00**	**0.0E+00**	6.8E+01	2.5E+01	**0.0E+00**	**0.0E+00**
F_{10} 200K	**3.1E-15**	**7.9E-31**	4.3E-15	1.7E-15	3.2E-15	5.0E-16	1.2E-13	1.6E-13	9.9E-11	5.1E-11	6.6E-15	5.0E-16
F_{11} 300K	**0.0E+00**	**0.0E+00**	1.1E-21	7.6E-21	**0.0E+00**	**0.0E+00**	**0.0E+00**	**0.0E+00**	**0.0E+00**	**0.0E+00**	**0.0E+00**	**0.0E+00**
F_{12} 150K	**1.6E-32**	**1.6E-47**	**1.6E-32**	**1.6E-47**	4.4E-30	5.5E-30	4.6E-19	8.9E-19	9.1E-15	8.4E-15	**1.6E-32**	**1.6E-47**
F_{13} 150K	**1.4E-32**	**8.2E-48**	**1.4E-32**	**8.2E-48**	5.9E-29	8.2E-29	2.4E-19	7.2E-19	4.0E-14	3.1E-14	**1.4E-32**	**8.2E-48**
F_{14} 300K	**0.0E+00**	**0.0E+00**	8.0E-07	7.1E-07	**0.0E+00**	**0.0E+00**	1.8E-05	5.4E-06	3.3E-03	2.8E-03	**0.0E+00**	**0.0E+00**
F_{15} 100K	**1.6E-15**	**9.9E-31**	1.7E-02	8.1E-02	**1.6E-15**	1.4E-18	3.8E-12	4.2E-12	3.9E-07	2.6E-07	**1.6E-15**	**9.9E-31**
F_{16} 300K	**0.0E+00**	**0.0E+00**	7.4E-07	7.6E-07	**0.0E+00**	**0.0E+00**	2.3E-05	7.9E-06	3.5E-03	2.6E-03	**0.0E+00**	**0.0E+00**

Table 3. Comparison between 4MPSR method with PSRMM method

	MFeval	$ACDE_{NP,MM}$		$ACDE_{4MPSR}$		$jDE_{NP,MM}$		jDE_{4MPSR}	
		Mean	Std	Mean	Std	Mean	Std	Mean	Std
F_1	150K	7.1E+02	2.9E+02	**6.7E-56**	**1.7E-55**	6.3E-50	3.7E-49	2.9E-40	1.8E-39
F_2	200K	8.1E+00	1.5E+00	2.9E-40	1.7E-39	**4.4E-42**	**2.9E-41**	9.6E-32	1.8E-31
F_3	500K	2.0E+04	5.3E+03	**1.6E-23**	**9.5E-23**	1.1E-17	5.0E-17	4.6E-20	1.6E-19
F_4	500K	2.9E+01	6.7E+00	**3.6E-10**	**1.6E-09**	4.0E-06	2.3E-06	3.6E-07	2.6E-07
F_5	2000K	1.5E+05	1.2E+05	8.2E-28	1.0E-27	1.7E-25	7.6E-25	**0.0E+00**	**0.0E+00**
F_6	150K	6.6E+02	2.2E+02	**0.0E+00**	**0.0E+00**	0.0E+00	0.0E+00	0.0E+00	0.0E+00
F_7	300K	7.7E-02	5.9E-02	**1.3E-03**	**3.9E-04**	1.4E-03	5.5E-04	1.4E-03	5.3E-04
F_8	900K	6.5E+02	2.0E+02	4.4E-12	2.3E-14	4.0E-12	4.4E-14	**3.9E-12**	**4.2E-14**
F_9	500K	3.5E+01	7.8E+00	**0.0E+00**	**0.0E+00**	0.0E+00	0.0E+00	0.0E+00	0.0E+00
F_{10}	200K	6.8E+00	9.0E-01	6.6E-15	5.0E-16	**3.2E-15**	**5.0E-16**	6.9E-15	1.4E-15
F_{11}	300K	7.4E+00	2.6E+00	**0.0E+00**	**0.0E+00**	0.0E+00	0.0E+00	5.3E-20	7.6E-21
F_{12}	150K	6.1E+03	3.9E+04	**1.6E-32**	**1.6E-47**	1.6E-32	1.6E-47	1.6E-32	1.6E-47
F_{13}	150K	9.4E+04	1.5E+05	**1.4E-32**	**8.2E-48**	1.4E-32	8.2E-48	1.4E-32	8.2E-48
F_{14}	300K	9.1E+01	8.4E+00	**0.0E+00**	**0.0E+00**	1.0E-09	2.1E-09	0.0E+00	0.0E+00
F_{15}	100K	6.7E+01	1.6E+01	**1.6E-15**	**9.9E-31**	1.6E-15	9.9E-31	1.6E-15	9.9E-31
F_{16}	300K	8.5E+01	9.1E+00	**0.0E+00**	**0.0E+00**	1.6E-08	1.0E-07	0.0E+00	0.0E+00

5.3   Comparison between 4MPSR Method with PSRMM Method

We compared 4MPSR method with PSRMM method. The four algorithms in comparison are listed as follows:

1) $jDE_{NP,MM}$ with $NP_{init} = 200$ and $pmax = 4$ [12];
2) jDE_{4MPSR} with $NP_{init} = 200$ and $pmax = 4$;
3) $ACDE_{NP,MM}$ with $NP_{init} = 200$ and $pmax = 4$;
4) $ACDE_{4MPSR}$ with $NP_{init} = 200$ and $pmax = 4$ (The proposed algorithm);

All of the used parameter values for each algorithm were recommended parameter values in their authors. The population size NP was initialized by 200 for each algorithm.

The performance evaluation results of the comparison between 4MPSR method with PSRMM method are presented in Table 3. The proposed algorithm ($ACDE_{4MPSR}$) performed the best performance for solving F_1, F_3, F_4, F_6, F_7, F_9, $F_{11} - F_{16}$, which were 12 out of 16 benchmark problems. jDE algorithm with 4MPSR method (jDE_{4MPSR}) was 9 out of 16, which were F_5, F_6, F_8, F_9, $F_{12} - F_{16}$. jDE algorithm with PSRMM method ($jDE_{NP,MM}$) was 8 out of 16, which were F_2, F_6, $F_9 - F_{13}$, F_{15}. ACDE algorithm with PSRMM method ($ACDE_{NP,MM}$) performed the worst performance. As a result, the proposed population size reduction and modified multiple mutation strategies method can improve the performance of global optimization not only ACDE algorithm but also jDE algorithm.

6   Conclusion

The control parameters play a large role in the differential evolution (DE) algorithm. Many researchers have attempted to design a good adaptive parameter control method. Adaptive Cauchy differential evolution (ACDE) algorithm is one of the results. ACDE algorithm allocates not only the current good control parameter value, but also far from the current good value which may be suitable in the next generation to each individual by using the Cauchy distribution. However, ACDE algorithm adapts two control parameters, the scaling factor and the crossover rate but uses a fixed population size. Recently, some researchers showed that adapting the population size is as important as adapting other two control parameters this is because maintaining a suitable population size can help to maintain a balance between Exploration and Exploitation, which is necessary to obtain a good performance.

Based on their results, in this paper, we improved ACDE algorithm by attaching the modified population size reduction (4MPSR) method. 4MPSR method is an extension of the population size reduction (PSR) method. 4MPSR method reduces the population size gradually and uses four mutation strategies with different ranges of the scaling factor. More specifically, 4MPSR method uses the most Exploration mutation strategy during the early stage. After that, the method uses the stronger Exploitation mutation strategy at each step. Therefore, 4MPSR method has better Exploration during the early stage and Exploitation during the late stage.

We compared the proposed algorithm with some state-of-the-art DE algorithms in various benchmark problems. The performance evaluation results showed that the proposed algorithm was significantly improved for solving the unimodal problems, as well as the multimodal problems, and obtained the better final solutions than the state-of-the-art DE algorithms. In addition, we compared 4MPSR method with PSRMM method in various benchmark problems. The performance evaluation results showed that 4MPSR method can improve the performance of global optimization of not only ACDE algorithm but also jDE algorithm. As a result, with a small increase of computational complexity, 4MPSR method can provide the DE algorithm a more suitable balance between Exploration and Exploitation.

Acknowledgement. This work was supported under the framework of international cooperation program managed by National Research Foundation of Korea (NRF-2013K2A1B9066056).

References

1. Storn, R., Price, K.: Differential evolution-a simple and efficient adaptive scheme for global optimization over continuous spaces. ICSI, Berkeley (1995)
2. Storn, R., Price, K.: Differential evolution–a simple and efficient heuristic for global optimization over continuous spaces. Journal of Global Optimization 11(4), 341–359 (1997)
3. Das, S., Suganthan, P.N.: Differential evolution: A survey of the state-of-the-art. IEEE Transactions on Evolutionary Computation 15(1), 4–31 (2011)
4. Brest, J., Maučec, M.S.: Population size reduction for the differential evolution algorithm. Applied Intelligence 29(3), 228–247 (2008)
5. Teo, J.: Exploring dynamic self-adaptive populations in differential evolution. Soft Computing 10(8), 673–686 (2006)
6. Choi, T.J., Ahn, C.W., An, J.: An adaptive Cauchy differential evolution algorithm for global numerical optimization. The Scientific World Journal 2013 (2013)
7. Elsayed, S.M., Sarker, R.A.: Differential Evolution with automatic population injection scheme for constrained problems. In: 2013 IEEE Symposium on Differential Evolution (SDE). IEEE (2013)
8. Goldberg, D.E., Deb, K., Clark, J.H.: Accounting for Noise in the Sizing of Populations. In: FOGA (1992)
9. Goldberg, D.E., Deb, K., Clark, J.H.: Genetic algorithms, noise, and the sizing of populations. Complex Systems 6, 333–362 (1991)
10. Brest, J., et al.: Self-adapting control parameters in differential evolution: A comparative study on numerical benchmark problems. IEEE Transactions on Evolutionary Computation 10(6), 646–657 (2006)
11. Brest, J., et al.: Performance comparison of self-adaptive and adaptive differential evolution algorithms. Soft Computing 11(7), 617–629 (2007)
12. Zamuda, A., Brest, J.: Population reduction differential evolution with multiple mutation strategies in real world industry challenges. In: Rutkowski, L., Korytkowski, M., Scherer, R., Tadeusiewicz, R., Zadeh, L.A., Zurada, J.M. (eds.) EC 2012 and SIDE 2012. LNCS, vol. 7269, pp. 154–161. Springer, Heidelberg (2012)

13. Yao, X., Liu, Y., Lin, G.: Evolutionary programming made faster. IEEE Transactions on Evolutionary Computation 3(2), 82–102 (1999)
14. Yao, X., et al.: Fast evolutionary algorithms. In: Advances in Evolutionary Computing, pp. 45–94. Springer, Heidelberg (2003)
15. Zhang, J., Sanderson, A.C.: JADE: adaptive differential evolution with optional external archive. IEEE Transactions on Evolutionary Computation 13(5), 945–958 (2009)
16. Qin, A.K., Suganthan, P.N.: Self-adaptive differential evolution algorithm for numerical optimization. In: The 2005 IEEE Congress on Evolutionary Computation, vol. 2. IEEE (2005)

A Bi-level Evolutionary Algorithm for Multi-objective Vehicle Routing Problems with Time Window Constraints

Abhishek Gupta[1], Yew-Soon Ong[1], Allan N. Zhang[2], and Puay Siew Tan[2]

[1] School of Computer Engineering, Nanyang Technological University, Singapore
{abhishekg,ASYSOng}@ntu.edu.sg
[2] Singapore Institute of Manufacturing Technology, A*STAR, Singapore
{nzhang,pstan}@simtech.a-star.edu.sg

Abstract. The presence of multiple, often conflicting, objectives is a naturally occurring scenario in real-world decision making. Such problems are characterized by the existence of a set of efficient solutions instead of a single optimum. Evolutionary algorithms (EAs), which employ a population based search mechanism over a bounded decision/solution space, have emerged as popular tools for concurrently obtaining a good approximation to the entire efficient set, instead of finding a single solution at a time. While EAs hardly impose any restrictions on the form of the objective functions to be optimized, their design is generally based on the requirement that the feasible solution space of the problem be fixed. However, a situation contrary to the aforementioned is seen to occur during a bi-level formulation of the vehicle routing problem. Moreover, the problem is general enough to conceivably arise in many real-world situations. In light of this fact, the present paper alleviates the stated requirement by incorporating a bi-level perspective into a multi-objective EA. Thereafter, a preliminary study is carried out on the multi-objective variant of the NP-hard vehicle routing problem with time window constraints (VRPTW), in order to demonstrate the efficacy of the proposed approach.

Keywords: Bi-level Programming, Multi-Objective Optimization, Evolutionary Algorithm, Vehicle Routing Problem with Time Windows.

1   Introduction

Evolutionary algorithms (EAs) represent a class of stochastic optimization methods which are based on the Darwinian principles of natural evolution. These methods employ a population of candidate solutions that search the decision space in tandem while imposing very few, if any, of the classical restrictions on the objective functions; i.e. those of continuity, differentiability or convexity. Such features, along with their ease of implementation, have made EAs indispensable to solving several real-world optimization problems which inevitably possess noisy, disjoint functional forms.

© Springer International Publishing Switzerland 2015   27
H. Handa et al. (eds.), *Proc. of the 18th Asia Pacific Symp. on Intell. & Evol. Systems − Vol. 2*,
Proceedings in Adaptation, Learning and Optimization 2, DOI: 10.1007/978-3-319-13356-0_3

An EA starts by randomly generating a population of solutions in a prescribed, bounded solution space. Each candidate solution is called an *individual* and each iteration loop of the algorithm is called a *generation*. Over each generation the so called "fit" solutions are *selected* as parents who undergo simple operations of *recombination* (*crossover*) or *mutation* to create a generation of hopefully "fitter" offspring solutions.

Despite the simplicity of the underlying mechanism, EAs have proven themselves as general and powerful search tools [1]. In addition to solving single-objective optimization problems (SOPs), their population based search procedure immediately lends itself to the design of schemes that concurrently obtain a complete representative set of efficient points that characterize a multi-objective optimization problem (MOP) [2].

MOPs are ubiquitous in most real-world decision making scenarios. Consequently, the natural ability of EAs to seamlessly tackle multiple objectives has earned them significant research interest over recent years. Several variants of multi-objective evolutionary algorithms (MOEAs) have been proposed; among them the Strength Pareto Evolutionary Algorithm (SPEA) [3] and the Non-dominated Sorting Genetic Algorithm (NSGA) [4] are arguably most popular in the present-day. These two algorithms vary primarily in terms of the fitness assignment procedure adopted for their individuals. Nevertheless, one feature that is common to all the suggested algorithms is that they assume the feasible solution space to remain fixed. This enables the individuals of a population, acting in concert, to thoroughly search the same, unchanging and bounded space. Although the assumption is a valid one for many problems of interest, one may conceive situations in which it no longer holds. As a first example, the immediate motivation for the present study is drawn from a multi-objective extension of a bi-level formulation of the vehicle routing problem (VRP); this problem shall be discussed in some detail through the course of this paper. Another, more intuitive example, is the hierarchical decision making process within a single organization with a prescribed set of objective functions. It may so happen that the set of choices/strategies available to the follower is itself dependent on the decision taken previously in time by the leader. In other words, the feasible solution space available to the follower is a function of the leader's decisions. It is important to note here that the described situation is distinct from a Stackelberg leader-follower game, in which the players have different, and often competing, objectives. In the present case, both players are on the same team and attempt to optimize the same objective functions. Yet, as the feasible solution space of the follower cannot be prescribed, MOEAs, at least in their present forms, are not suitably equipped to handle such problems.

In this paper, we propose enhancements to the popular NSGA-2 algorithm [4] so as to specifically tackle the type of problem described above. The improvements are particularly attuned to a multi-objective extension of the NP-hard vehicle routing problem with time window constraints (VRPTW), which has previously been presented in [9, 13, 14]. Moreover, the VRP has been shown to have a bi-level formulation in [5, 10]. However, to the best of the authors' knowledge, bi-level VRP formulations do not consider time window constraints and are also restricted to a single objective function. The latter is likely due to the lack of an established solution procedure for the multi-objective case, as mentioned above. This paper presents an improved MOEA which successfully relaxes both the stated restrictions, thereby making the bi-level approach more useful in practical settings.

The rest of the paper is organized as follows. In Section 2 we briefly describe a general MOP. In Section 3 we provide the particular class of MOPs which are of interest to the present study with a concrete mathematical description. Subsequently, the improved MOEA developed to handle the problem is described in Section 4. Section 5 contains a description of the multi-objective VRPTW and how it can be modelled as the aforementioned MOP. Moreover, the section presents some results from computational experiments carried out on Solomon's instances which incorporate time window constraints [6]. Finally, Section 6 provides some concluding remarks which summarize the contents of this work.

2   General Multi-objective Optimization

A general multi-objective minimization problem is one which attempts to find all solutions $x \in X$ such that the vector-valued mapping of the solution into the objective space, given by $f(x) = (f_1(x), f_2(x),...., f_m(x))$, is minimized. Here $X \subset R^n$ is the bounded feasible solution space of n dimensions, and m is the number of individual objectives to be minimized.

Consider two solutions x_1 and x_2, with $x_1, x_2 \in X$. Then x_1 is said to *dominate* x_2 if and only if,

$$\forall i \in \{1,2,...,m\}: f_i(x_1) \leq f_i(x_2) \text{ and } \exists j \in \{1,2,....,m\}: f_j(x_1) < f_j(x_2). \quad (1)$$

A solution x_0 is said to be *efficient* or *Pareto efficient* or *Pareto optimal* if $x_0 \in X$ and there exists no other feasible point x such that $f(x)$ dominates $f(x_0)$. Moreover, the images of the Pareto efficient solutions in the objective space constitute the *Pareto front* or *Pareto surface*.

From the definition of Pareto efficiency, it is clear that for an m (> 1) dimensional objective space, the corresponding solution space will generally have a set of Pareto efficient solutions instead of a single optimum.

3   A Class of Multi-objective Optimization Problems

In this section, we present a mathematical description of the class of MOPs that are of interest to the present study. It should be noted that in the formulation below the feasible solution space $X \subset R^n$ is instead split into two parts, $X \subset R^{n1}$ and $Y \subset R^{n2}$, where $n1 + n2 = n$. Moreover, for brevity the equality and inequality constraints are not specified.

$$\text{Minimize } f(x,y) = (f_1(x,y), f_2(x,y),....., f_m(x,y)),$$

$$\text{such that } x \in X; \quad X \subset R^{n1} \text{ and } y \in Y(x); \ Y(x) \subset R^{n2}. \quad (2)$$

The most significant feature of Eq. (2) which distinguishes it from standard MOPs is the fact that the domain of y is not fixed; rather it is a function of the chosen x. As a result, boundaries for the search space of y cannot be predefined, thereby positioning the problem beyond the scope of standard MOEAs.

In this study it is proposed that a possible method for tackling Eq. (2) is to first view it as a bi-level program, as follows;

$$Minimize\ f(x,y)\ |\ x \in X,$$

$$such\ that\ y \in argmin_{given\ parameter\ x} \{f(x,y)\ |\ y \in Y(x)\}. \tag{3}$$

In Eq. (3) X, which remains fixed, is considered to form a higher level solution space. On the other hand, Y forms the lower level solution space. Corresponding to a chosen x in X, a lower level (or nested) optimization problem must be solved to obtain an efficient set of y. As for a given x, Y is temporarily fixed, a standard MOEA is a viable option for each independent run of the lower level problem. Moreover, through the formulation of Eq. (3) y can be viewed as a function of x. As a result, the higher level problem reduces to finding the optimum combination $(x,\ y(x))$ that minimizes f.

It must be observed that an SOP ($m =1$) corresponding to Eq. (2) is comparatively easier to solve as for every x there generally exists a single optimum y at the lower level. As a result $y(x)$ is well defined. In contrast however, for the case of MOPs such a definition becomes vague as there exists a set of efficient y corresponding to any x. This feature makes the problem significantly more challenging. One must determine which, "if any", among all y in the efficient set are also solutions of the higher level optimization problem. It may so happen that only a select few, or indeed none, of the y corresponding to a particular x are efficient at the higher level.

In recent years, researchers have proposed MOEAs capable of handling bi-level problems [7]. However, these algorithms are generally based on the assumption that the objective functions at the two levels are distinct to each other. In other words, it is assumed that the two levels form part of competing decision making units, with each unit selfishly optimizing their own payoffs. The fact that in the present case both levels attempt to optimize the same, shared objectives, allows one to design specific search strategies that exploit this relationship. The subsequent section of the paper shall be devoted to the description of such an algorithm.

4   A Bi-level Multi-objective Evolutionary Algorithm

The MOEA is built on the popular NSGA-2 algorithm and directly borrows the concepts and procedures for non-domination sorting and crowding distance [4] without any alteration. Terms such as chromosome, gene, mutation and crossover, commonly referred to in the EA literature, are also used in their standard form. For the sake of brevity, these concepts will therefore not be discussed in this paper. Moreover, for simplicity of exposition, *it is assumed that the optimization problem is in fact bi-objective*. Nevertheless, the ideas shall be generalizable to higher dimensions.

4.1   The Higher Level

We begin the description assuming that the problem has already been reduced to the form expressed in Eq. (3). The first departure from a standard MOEA then occurs at the first step itself. In addition to randomly initializing a population of solutions belonging to the nl dimensional higher level solution space X, the chromosome is endowed with an additional gene 'w' $(0 < w < 1)$ which prescribes a scalarization parameter for the lower level problem. In other words, the chromosomes evolve in terms of both x and w. The value of w should be viewed as a weighting associated with f_1. Equivalently, the weighting associated with f_2 is $(1 - w)$.

At the higher level, the MOEA thus proceeds according to NSGA-2 while searching through a fixed and bounded $(nl + 1)$ dimensional space. During crossover operations special attention is paid to w. An offspring solution is considered to borrow w from the nearest of its two parents, and then adds a small random perturbation to it for the sake of exploration. For continuous optimization problems, it is possible to compute the Euclidean distance to each parent in the solution space. For the case of combinatorial problems, the Jaccard index can be used instead as a measure of the distance.

In addition to the above, a second mutation operation M_2 (i.e., in addition to the chosen primary mutation operator M_1) is defined in order to fine-tune w for an existing individual (without altering the remaining chromosome) in a manner that hopefully improves its performance at the higher level. This fine-tuning follows from two simple observations:

1. If a solution $(x^*, y^*) \in X \times Y$ corresponds to the global minimum of f_1, then $y^* =$ argmin $_{given\ x^*} \{f_1(y) \mid y^* \in Y\ (x^*)\}$. In other words, if an individual in its current state maps close to the global minimum of f_1, then the scalarization parameter w prescribed to the lower level should be close to 1, i.e., $(1 - w) \approx 0$.
2. If a solution $(x^*, y^*) \in X \times Y$ corresponds to the global minimum of f_2, then $y^* =$ argmin $_{given\ x^*} \{f_2(y) \mid y^* \in Y\ (x^*)\}$. In other words, if an individual in its current state maps close to the global minimum of f_2, then the scalarization parameter w prescribed to the lower level should be close to 0, i.e., $(1 - w) \approx 1$.

Based on the above arguments, we jump to the following heuristic conclusion: the position of a higher level individual in objective space, in relation to the currently known individual objective minima, can be used to estimate the portion of its corresponding lower level Pareto front that is most likely to contribute to the Pareto front at the higher level. It is proposed that this conclusion be quantified in the following simple manner for an individual I undergoing M_2 in the bi-objective case;

$$w_I = \left(1 - \frac{f_1^I - f_1^{min}}{\sum_{i=1}^{2}\left(f_i^I - f_i^{min}\right)}\right) + \varepsilon \cdot (\pm\,rand). \qquad (4)$$

In Eq. (4), f_i^{min} corresponds to the minimum value of objective f_i found up to the current generation, while f_i^I is the ith objective value of I prior to fine-tuning. If it so happens that $w_I = 0$ or 1 then we replace it by either ε or $(1 - \varepsilon)$, respectively. Here ε represents a very small positive number.

Ultimately, for every individual I at the higher level the following information is passed on to the lower level problem: *[some representation of x_I, w_I]* and the currently known $\{f_i^{min}, f_i^{max}; for\ i = 1, 2\}$.

4.2   The Lower Level

Once the lower level receives a scalarizing parameter from the higher level, it uses the information to convert the MOP into an SOP, thereby obtaining only a single optimum y_I^* corresponding to x_I. The scalarization procedure must however be appropriately chosen. The three most popular choices available are the *weighted sum method*, the *ε-constraint method*, and the *lexicographic weighted Chebyshev method*.

The weighted sum approach, although the most traditional, has a major drawback. Roughly stated, it can generate all Pareto efficient solutions only for convex Pareto fronts. To overcome this restriction, other alternatives must also be considered. On the other hand, the ε–constraint method and the Chebyshev method are both capable of handling non-convex fronts. Although both methods are equally compatible with the proposed algorithm, we consider the Chebyshev method to be the scalarization procedure in this study.

Under the above assumption, the mathematical program at the lower level takes the following form:

$$Minimize\ max\ \{w_I \times (f_1(x_I, y) - z_1), (1 - w_I) \times (f_2(x_I, y) - z_2)\},$$

$$such\ that\ y \in Y(x_I). \tag{5}$$

In Eq. (5) (z_1, z_2) denotes a reference point in objective space. Eq. (5) can be geometrically interpreted as finding the point on the Pareto front which lies nearest to a ray extending from the reference point with slope $w_I/(1 - w_I)$.

It must be noted that any preferred algorithm, EA or otherwise, can be used to solve Eq. (5). The resultant y_I^* is then returned to the higher level along with $(f_1(x_I, y_I^*), f_2(x_I, y_I^*))$, which constitutes the mapping of I into objective space.

The procedure described in Sections 4.1 and 4.2 are repeated on a finite sized population for a large number of generations until some stopping criterion is met.

It is emphasized here again that although the discussions so far have been presented for a bi-objective case, the ideas can easily be extended to higher dimensions. More importantly, through the novel idea of incorporating the objective function weights as genes undergoing evolution, an efficient MOEA for the desired class of problems emerges, one that stays honest to the spirit of traditional MOEAs by concurrently approximating the entire efficient set.

5   Application to the VRPTW

In this section we start with briefly introducing the vehicle routing problem with time windows and the objectives considered for optimization. Subsequently, a bi-level approach for tackling the problem is discussed, one that can be cast in the form presented in Eq. (2). Finally, the results achieved by running our prescribed algorithm on a representative set of Solomon's instances are presented and discussed.

5.1   The Vehicle Routing Problem with Time Windows

The VRP is an increasingly important problem to solve in the fields of transportation and logistics, primarily due to soaring costs associated with fuel consumption. The problem is simply to serve a set of customers, each having some demand, with a fleet of fixed capacity vehicles, in a manner that minimizes the cost incurred by the service provider. Each vehicle begins its journey from a particular depot, serves a subset of all the customers, and returns to the same depot before the end of the planning horizon. Despite the simplicity of the problem statement, it is NP-hard. EAs have emerged as one of the most popular methods for solving this problem as they demonstrate notable performance even for large scale problems [8].

The VRP is a combinatorial optimization problem set on a graph $G(V,E)$. Where $V = \{v_0, v_1, \dots v_i, v_{i+1}, \dots, v_k\}$ is the set of nodes (or vertices) and $E = \{(v_i, v_j)|v_i, v_j \in V; i \neq j\}$ is the set of arcs (or edges). We assume there to be only a single depot, represented by v_0. Each arc (v_i, v_j) is associated with a cost c_{ij}, with $c_{ij} = c_{ji}$, and a customer at node v_i has a demand d_i. For the case of VRPTW, a customer at node v_i is also associated with a time window $[a_i, b_i]$ specifying the period during which that particular customer most prefers to be served.

The purpose of solving a traditional VRPTW is to find the vehicle trip schedules which minimize the total cost $(C) = \left(\sum_{\forall j} F \cdot x_{0j} + \sum_{\forall i} \sum_{\forall j} x_{ij} \cdot c_{ij} \right)$, where $x_{ij} = 1$ if arc (v_i, v_j) is used by the trip schedule or 0 otherwise, and F is the fixed cost of a vehicle. The schedule must guarantee that the time window constraints set by the customers are met. Moreover, it is to be ensured that each customer is serviced by only a single vehicle and that at no point does the load carried by a vehicle exceed its capacity.

In the present study it is argued that in practice the customer time windows need not be perceived as hard constraints. Indeed, due to inherent traffic unpredictability the constraints might get violated regardless. Most importantly, it has been shown that relaxing the hard time windows can lead to significantly lower cost trip schedules [9]. As long as the constraint violations are not too large, soft time window constraints can be used as a legitimate means of service cost reduction. With this argument in mind, the present study expands the traditional VRPTW formulation by considering two objectives, as follows;

$$Minimize \ (C, \sum_{\forall i} max\{0, arr_i - b_i\}) . \tag{6}$$

In the above, arr_i indicates the arrival time of a vehicle at v_i, thus $arr_i - b_i$ is the waiting time of the customer. Note that if a vehicle arrives at v_i before a_i, then it must wait till a_i. We term the second objective considered in Eq. (6) *customer dissatisfaction*. Clearly, the two objectives are strongly competing as it is easy to conceive situations in which cost cuts lead to poor service quality.

5.2   A Bi-level Approach to the VRPTW

Bi-level approaches to the vehicle routing have previously been suggested in the lite-rature [5,10], but only for single objective versions and without any time window constraints. In the present paper we extend the work done in [10] to multiple objec-tives while also accounting for time window constraints. We begin this sub-section by very briefly describing the method adopted in [10]. It is then shown that the extension of the method to multiple objectives indeed takes the form of Eq. (2), thereby making the algorithm proposed in this study immediately applicable.

In [10], the higher level problem dealt with finding the optimum sequence of all the customers without any trip delimiters. In other words, if there are k customers then a possible solution chromosome at the higher level is $x \overset{\text{def}}{=} random\ permutation(1 : k)$.

The customer sequence in x is first used to rearrange the rows and columns of a symmetric cost matrix (D), where initially $D_{ij} = c_{ij}$. Subsequently, an *auxiliary graph* is created on the rearranged $D (= D')$ with arc costs represented by an upper triangu-lar matrix D^{aux} such that $D^{aux}_{i'j'} = F + \left(D'_{0(i'+1)} + D'_{(i'+1)(i'+2)} + \cdots + D'_{(j'-1)(j')} + D'_{(j')0} \right)$; here i' and j' are used to emphasize the rows and columns have been rearranged. In other words, an arc cost $D^{aux}_{i'j'}$ on the auxiliary graph is simply the total cost of an entire vehicle route, given by $\{v_0, v_{x_{i'}}, v_{x_{i'+1}}, \ldots. v_{x_{j'}}, v_0\}$.

In case a route violates the vehicle capacity constraint, then the corresponding arc of the auxiliary graph is artificially assigned a very large cost. Finally, the constructed D^{aux} is passed to the lower level which applies the Dijkstra's shortest path algorithm [11] on the auxiliary graph in order to obtain the optimum *splitting* of the chromo-some into a set of vehicle trips, one that minimizes the total service cost.

It is crucial to note here that the set of choices available at the lower level, i.e., the possible set of trip schedules, is itself a function of the chromosome at the higher level. For different permutations of the customers, the possible trip schedules are indeed entirely different. This only goes to show that for the case of multiple objec-tives, the approach discussed above exactly resembles Eq. (2). All the possible permu-tations of the customers is analogous to X while $\{splitting_1(x), splitting_2(x), \ldots.\}$, which is the set of all possible ways of splitting a particular chromosome x, is analog-ous to $Y(x)$.

5.2.1   Extension to the Bi-objective Case

In order to tackle time window constraints in this study (which forms the basis of our second objective), we construct an additional matrix T^{aux} for the previously described auxiliary graph. The second set of arc costs $T^{aux}_{i'j'}$ equals the total waiting time of customers on an entire vehicle route (the structure of T^{aux} is otherwise the same as D^{aux}). Both T^{aux} and D^{aux} are then passed to the lower level problem along with w_I(see Sections 4.1 and 4.2).

At the lower level, each element of D^{aux} and T^{aux} is first normalized using $\{f_i^{min}$ and $f_i^{max}, for\ i = 1, 2\}$ and then scaled as per the Chebyshev method in Eq. (5). In order to handle two objectives through Dijkstra's algorithm, a pair of cost la-bels is associated with each node of the graph, instead of a single label as in the case of single-objective shortest path problems. Based on the elements of D^{aux} and T^{aux}, the node labels are then updated repeatedly in a *lexicographic manner* to produce a bi-objective shortest path on the auxiliary graph; this also corresponds to a bi-objective optimum splitting of the higher level chromosome x. The optimally split x along with the corresponding values for f_1 and f_2 are then returned to the higher level.

Apart from the general framework of the MOEA, it is noted that in order to make the results obtained from our algorithm competitive with those available in the VRP literature, additional local search heuristics must be incorporated. The VRP is otherwise too hard to solve using the MOEA alone. In the present work, simple local search moves such as node swaps, customer reinsertion and two-opt swaps are carried out on offspring solutions at the higher level, with some probability. These moves are performed only after the offspring undergoes trip delimiting at the lower level.

5.3   Computational Experiments

The proposed EA has been implemented in MATLAB while the subroutines specific to VRPTW, including Dijktra's algorithm and the local search heuristics, have been coded in C++. The two environments are interfaced by compiling the C++ programs into MATLAB executable files.

For all test cases, an initial population of 24 individuals is generated completely at random. For subsequent evolution, the probability of crossover (order crossover [10]) is set at 60%. Randomization of a portion of the individual's chromosome is chosen as a simple mutation operator (M_1), which is performed with 25% probability. The probability of weight fine-tuning (M_2) is also set at 15%. Finally, the probability of performing local search on an offspring is set at 30%.

Table 1. Performance comparison results for a representative set of Solomon's instances (with 50 customers)

	Best-known [12]		Present algorithm	
	Vehicles	Distance	Vehicles	Distance
R101_50	12	1044.0	12	1046.7
R201_50	6	791.9	4	**861.94**
C101_50	5	362.4	5	363.25
C201_50	3	360.2	2	**444.96**
RC101_50	8	944	8	959
RC201_50	5	684.8	3	**898.49**

In order to confirm the validity of the present approach, a comparison study is carried out against medium sized versions of Solomon's instances (50 customers). Since the best-known solutions published in the literature are all for the single-objective case, a comparison can only be performed by extracting a single solution from the entire Pareto front, the one which fully satisfies the time window constraints (i.e. the chromosome with $w_l \approx 0$). Moreover, in the literature the total distance cost is generally considered to be a secondary objective, while minimizing the number of vehicles is the primary one. This too can be incorporated into the present approach by simply setting the vehicle fixed cost (F) to a very large number. The results thus obtained on a representative set of instances are presented in Table 1. Note that the MOEA is run for 30 seconds for the 50 customer case. It is clear from the tabulated results that the MOEA performs exceedingly well on such problems. Not only are the results of our present approach competitive to the best-known solutions of the respective problem instances, new best-known solutions are also found for 3 problem instances (R201_50, C201_50 and RC201_50).

With increasingly large instances, however, the probability of the obtained results falling marginally short of the best-known also increases. This is simply for the reason that the efforts of the MOEA are distributed over a set of solutions, so as to find a good approximation to the entire Pareto front. It does not focus on a single point in objective space, as is the case for all high performing single–objective algorithms recorded in the literature. Nevertheless, the MOEA is expected to consistently provide high quality solutions to real-world routing problems.

The above claim is established through Fig. 1, which depicts the final population of solutions obtained for 100 customer versions of the selected Solomon's instances. While solving these problems the value of F is again set to a very large number and the algorithm is allowed to run for approximately 4 minutes prior to termination. The solutions corresponding to C101_100 and C201_100 are not provided as the algorithm produces only a single optimum for these cases. On comparing the point intersecting the '*Total Distance*' axis (in each of Figs. 1a-d) to the single-objective best-known solution for the same problem (available in [12]), we find that our results are indeed competitive. This comparison is also tabulated in Table 2. Moreover, the reasonable run-time required to produce such solutions further encourages the application of the new approach in practical settings.

Table 2. Performance comparison for large Solomon's instances (100 customers)

	Best-known [12]		Present algorithm	
	Vehicles	Distance	Vehicles	Distance
R101_100	19	1645.79	19	1695.2
R201_100	4	1252.37	5	1317.1
C101_100	10	828.94	10	828.94
C201_100	3	591.56	3	591.56
RC101_100	14	1696.94	15	1731.2
RC201_100	4	1406.91	5	1486.5

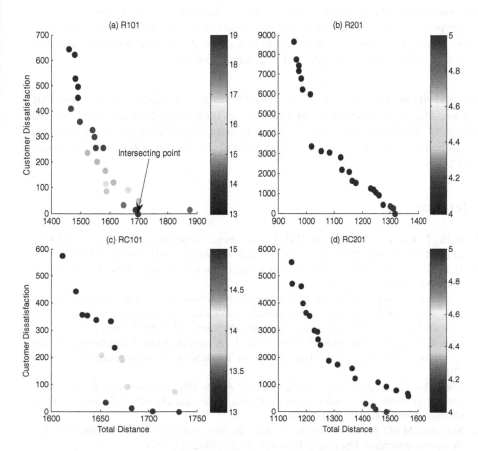

Fig. 1. Final population of solutions obtained after 1 run of the MOEA on 100 customer variants of Solomon's instances. The **color-bar** indicates number of vehicles used.

The most important observation to draw from Fig. 1, particularly from cases R101 and RC101 where customer dissatisfaction levels are low, is that small relaxations in the time window constraints can be used to appreciably reduce the total distance travelled as well as the number of vehicles required. In other words, the two most critical factors affecting service costs can be significantly reduced. This further goes on to reinforce the claim made previously that soft time window constraints can be used as a legitimate means of service cost reduction

6   Conclusions

This paper presents a preliminary study on the vehicle routing problem with time window constraints as a bi-level multi-objective problem. The inclusion of time window constraints in the bi-level approach and its subsequent extension to multiple objectives has been performed for the first time. An improved MOEA is designed for this purpose and is shown to provide high quality solutions at competitive run-times. Future research shall be aimed at including additional objective functions which enhance the utility of the developed tool in practical settings.

Acknowledgements. This work is partially supported under the A*Star-TSRP funding, Singapore Institute of Manufacturing Technology-Nanyang Technological University (SIMTech-NTU) Joint Laboratory and Collaborative research Programme on Complex Systems, and the Computational Intelligence Research Laboratory at NTU.

References

1. Back, T., Hammel, U., Shwefel, H.P.: Evolutionary Computation: Comments on the History and Current State. IEEE Transactions on Evolutionary Computation 1(1), 3–17 (1997)
2. Deb, K.: Multi-Objective Optimization Using Evolutionary Algorithms. John Wiley and Sons (July 2001)
3. Zitzler, E., Laumanns, M., Thiele, L.: SPEA2: Improving the strength Pareto evolutionary algorithm for multiobjective optimization. In: Evolutionary Methods for Design, Optimization and Control with Application to Industrial Problems, EUROGEN 2001 (2002)
4. Deb, K., Pratap, A., Agarwal, S., Meyarivan, T.: A Fast and Elitist Multiobjective Genetic Algorithm: NSGA-II. IEEE Transaction on Evolutionary Computation 6(2), 181–197 (2002)
5. Marinakis, Y., Migdalas, A., Pardalos, P.M.: A new bilevel formulation for the vehicle routing problem and a solution method using a genetic algorithm. Journal of Global Optimization 38(4), 555–580 (2007)
6. Solomon, M.M.: Algorithms for the Vehicle Routing and Scheduling Problems with Time Window Constraints. Operations Research 35(2), 254–265 (1987)
7. Deb, K., Sinha, A.: An Efficient and Accurate Solution Methodology for Bilevel Multi-Objective Programming Problems Using a Hybrid Evolutionary-Local-Search Algorithm. Evolutionary Computation 18(3), 403–449 (2010)
8. Baker, B.M., Ayechew, M.A.: A genetic algorithm for the vehicle routing problem. Computers and Operations Research 30(5), 787–800 (2003)
9. Zhou, Y., Wang, J.: A Local Search-Based Multiobjective Optimization Algorithm for Multiobjective Vehicle Routing Problem with Time Windows. IEEE Systems Journal, Early Access Articles (2014)
10. Prins, C.: Two memetic algorithms for heterogeneous fleet vehicle routing. Engineering Applications of Artificial Intelligence 22, 916–928 (2009)
11. Dijkstra, E.W.: A note on two problems in connexion with graphs. Numerische Mathematik 1(1), 269–271 (1959)
12. http://w.cba.neu.edu/~msolomon/problems.htm (retrieved)
13. Najera, A.G., Bullinaria, J.A.: An improved multiobjective evolutionary algorithm for the vehicle routing problem with time windows. Computers & Operations Research 38(1), 287–300 (2011)
14. Tan, K.C., Chew, Y.H., Lee, L.H.: A hybrid multiobjective evolutionary algorithm for solving vehicle routing problem with time windows. Computational Optimization and Applications 34(1), 115–151 (2006)

Analyzing Organization Structures and Performance through Agent-Based Socio-technical Modeling

Partha Dutta[1,*], Carla Pepe[2], and Hui Xi[1]

[1] Rolls-Royce Singapore Pte Ltd, Advanced Technology Centre, Singapore, Singapore
{partha.dutta,hui.xi}@rolls-royce.com
[2] Rolls-Royce Plc, Design Systems Engineering, Derby, United Kingdom
carla.pepe@rolls-royce.com

Abstract. Industry-scale projects often require large multi-functional teams to deliver complex inter-related tasks. However, to put together an organization structure to meet a project's targets most effectively is challenging. In this paper, we introduce a novel agent-based simulation system that models team member behaviour by combining socio-technical attributes to estimate time and cost outputs. It defines team-level performance as a function of individual attributes including new skill learning by exchanging help requests. It provides a framework for simulating arbitrary team configurations against given project task workflow and hence, evaluating the performances of different organization structures. Different modeling granularity levels (tasks, individuals, and teams), together with the flexibility with which the system allows rapid modeling and evaluation of organizational scenarios represent a step change from existing organizational and project analysis tools. It provides the foundation for resource managers to design optimized project teams in future.

Keywords: Organization modeling, Experimentation, Agent-based simulation, Decision-support.

1 Introduction

Having the right combination of team members is crucial for the success of large industry projects. It is common to find in such domains that several tens or even hundreds of individuals are involved in several interdependent tasks to achieve a certain delivery quality within a certain time.

It is apparent that a team member's expertise and the complexity of tasks are factors which influence the individual's performance. The less apparent factors are the social elements of human behaviour such as motivation, allegiance, and availability [1]. The interplay of these factors in projects with several team members performing several tasks presents a highly complex system. To ensure high

---

* The authors hereby thankfully acknowledge the contributions of Johnsen Kho in developing the organizational modeling system.

© Springer International Publishing Switzerland 2015
H. Handa et al. (eds.), *Proc. of the 18th Asia Pacific Symp. on Intell. & Evol. Systems – Vol. 2*,
Proceedings in Adaptation, Learning and Optimization 2, DOI: 10.1007/978-3-319-13356-0_4

quality and timely completion of projects, project managers need to do careful reasoning of how the social and technical aspects of team engagements influence project outputs.

Nevertheless, standard project management tools offer limited support in this respect. In addition, managers often use their past experience of team member performances to determine the 'best' team combination. While this intuitive approach can be sufficient for teams with few members performing few tasks, in large projects it fails to adequately capture the complex interactions between individuals and between individuals and tasks [2]. Hence, managers are left significantly under-equipped and with limited reliable foresight about the eventual performance of their teams. The situation increases the likelihood for projects suffering delays and resulting budget over-runs.

Against this background, we argue that a robust simulation of large project teams, that incorporates models of team members, tasks, and inter-relationships between them, could provide effective decision support to the resource management problem. To implement such a simulation platform, we have selected the agent-based modeling paradigm [3], which has widely been used to perform 'bottom-up' modeling of complex distributed systems such as humans with individual behaviour interacting with one another to deliver assigned objectives [7], [3], [4]. Agent-based modeling offers system designers a robust, scalable, and transparent mechanism for representing and analysing the emergent behaviour of such systems (which are difficult to analyse using methods from finite element analysis or analytical modeling).

In this paper, we have implemented a sophisticated platform that allows endusers to define individual and team models, as well as task workflows in a highly flexible manner, run simulations of large project teams (thousands of team members and tasks) within a rich interface providing a host of performance evaluation, analysis, and *what if* study tools to test project resourcing strategies. To define our team level aggregation of individual models, we have extended the individual behaviour models developed in HIPARSYS [7] and VDT (or commercially well known as SimVision [5], [8], [10], both of which assumed only one team member per task (and thus their use for resource management simulation is significantly impaired). Therefore, this work represents a first step towards building a resource management tool that represents realistic project teams by combining both technical and social aspects of organizational behaviour, to mitigate the challenge faced by managers in resourcing high performing teams.

In the rest of the paper, Section 2 describes the organizational model formalization. Section 3 summarizes the different types of agents used to model the organization. Section 4 describes the evaluation metrics for organizational performance. The simulator is briefly discussed in Section 5. Experimental results are presented in Section 6. Section 7 concludes and presents future research directions.

2   The Organizational Model

This section describes our model of an engineering organization consisting of individuals jointly performing concurrent tasks. The model is based on information captured during interviews conducted with the organization practitioners. Overall, our model is made of individual agents who work in teams (there can be several teams working on different tasks at a time) on tasks which may have interdependencies – some tasks can be performed only after others are completed. Although our organization model components (e.g. individual members, teams, and tasks) are generic, we have selected a specific engineering design project to instantiate our model against and perform experimental analysis. The task workflow of this example project will be described in more details in Section 6.1.

2.1   Task

A project is made of a set I of tasks. Each task $i \in I$ is associated with complexity level com_i describing the difficulty of the task. The complexity parameter could take values from the set of discrete values $\{VL, L, M, H, VH\}$[1]. This range is typically sufficient to explore the difficulty nature of any engineering and design projects.

2.2   Organization Member

We represent the set of m organization members with $\mathfrak{J} = \{1, \ldots, m\}$. Each member $j \in \mathfrak{J}$ has attributes of capability (cap_i^j), availability (ava_i^j), allegiance (all_i^j), and motivation (mot_i^j) while performing task $i \in I$. cap_i^j represents the skill and experience of the individual with respect to task i, ava_i^j describes the proportion of time the individual has available for the task, all_i^j determines the individual's level of commitment / priority towards the task, and mot_i^j represents how motivated / excited the individual is in completing the task[2]. Note, an individual may work on multiple tasks and perform differently on these tasks. Each of these attributes takes values from the same range as that of task complexity.

2.3   Project Workflow

A project workflow is a collection of tasks with ordering constraints and is represented as a directed acyclic graph[3]. We express the workflow as $W = (I, K)$, where I is the set of tasks (as described above) that make up the workflow, and $K \subseteq I \times I$ is a strict partial order over the tasks, where member $k \mapsto i \in K$, in which $k, i \in I$, specifies that task k must successfully complete before i can

---

[1] VL = Very Low, L = Low, M = Medium, H = High, and VH = Very High.
[2] The attributes have been selected based on information gathered by interviewing organization leaders.
[3] No task iterations (i.e. rework) are assumed in this phase of the research.

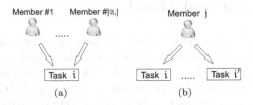

Fig. 1. (a) A team of $|\mathfrak{A}_i|$ members delivering task i and (b) individual j working on different tasks (i and i′ which could take place simultaneously)

start. The tasks with no predecessor task in the project are considered as *initial* tasks that would be executed at the start of the workflow.

Task i has its start and end times represented respectively as $\mathfrak{p}_i$ and $\mathfrak{q}_i$ such that its execution period is assumed to cover a certain bounded time interval $[\mathfrak{p}_i, \mathfrak{q}_i]$, where $\mathfrak{p}_i, \mathfrak{q}_i \in \mathfrak{R}^+$, and $\mathfrak{p}_i < \mathfrak{q}_i$ as $\mathfrak{p}_i = \max\{\mathfrak{q}_k\}$, where $k \mapsto i \in K, \forall k \in I$, and $\mathfrak{q}_i = \mathfrak{p}_i + \beta_i$ where $\beta_i > 0$ and β_i is the time required to complete the task (explained in Section 4.1). In another word, a task starts as soon as all its predecessor tasks are completed. The start time of each of the initial tasks is t_0 which is the workflow start time.

2.4   Team-Level Aggregation

A key contribution of this paper is the transformation of the simplistic organization model, introduced in [7], where every task was performed by a single individual and each member performs only one task at any time.

In all of [7], [8], [9], [10], each task is handled by a single individual. In contrast, in our organization model, more than one individuals can be assigned to a task (see Figure 1). Because of this addition, rules are derived in order to determine the *composite* capability, availability, and motivation attributes of a team of individuals $\mathfrak{A}_i \subseteq \mathfrak{J}$ in performing task $i \in I$. We note here two special entities in each team. The *task owner* ($\mathfrak{l\_a}_i \in \mathfrak{A}_i$) with overall responsibility and the *technical leader* ($\mathfrak{l\_b}_i \in \mathfrak{A}_i$) who provides specific technical advise to deliver the task. The two are represented by the 2-tuple $\mathfrak{L}_i = (\mathfrak{l\_a}_i, \mathfrak{l\_b}_i)$. The composite capability and motivation of the team are defined as the identified task owner's capability and motivation correspondingly: $\mathfrak{cap}^{(\mathfrak{A}_i)} = \mathfrak{cap}^{\mathfrak{l\_a}_i}$ and $\mathfrak{mot}^{(\mathfrak{A}_i)} = \mathfrak{mot}^{\mathfrak{l\_a}_i}$. The team availability is also defined as the task owner's availability: $\mathfrak{ava}^{(\mathfrak{A}_i)} = \mathfrak{cap}^{\mathfrak{l\_a}_i}$ [4]

Furthermore, it is assumed that each individual j is a finite resource with a workload limit $\mathfrak{B}^j$, such that if the total complexity of tasks in which j is involved in exceeds this threshold, j's actual performance (for each task) would degrade significantly. We do not evaluate this feature in the experiments reported in this paper. However, our ongoing research aims to utilize this as a means of deriving optimized allocation of teams to projects.

---

[4] These rules are proposed after discussing organization modeling with social scientists from *Leeds University Business School*.

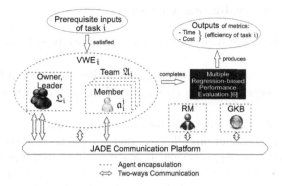

Fig. 2. Several organization members (team $\mathfrak{A}_i$) are performing task i. The agents communicate via JADE Communication Platform.

3 Agent Types

In this section, we present the different types of agents used to model the organization described in the previous section.

- **Virtual Working Environment (VWE):** The VWE agent for task i monitors the required individuals and inputs of the task, and would only execute the task if each of them is available (see Figure 2).
- **Individual:** Each individual agent $a_i^j \in \mathfrak{A}_i$, assigned to task j enters the working environment or leaves depending on whether the task has started or not.
- **Task Owner:** The main role of a task owner agent $\mathfrak{l}\_a_i$ is to initiate a 'learning protocol' which helps team members with low capability levels to improve and perform tasks better. This protocol will be described in Section 4.2.
- **Technical Leader:** The technical leader agent $\mathfrak{l}\_b_i$ provides assistance to individual team members when a learning protocol has been issued (details are provided in Section 4.2).
- **Global Knowledge Base:** The GKB agent is the de-facto fall-back to ask assistance from (when a learning protocol has been issued) if the technical leader agent is unavailable. The GKB agent is assumed to be always available but with low capability.
- **Resource Manager:** The RM agent is responsible for allocating tasks to individuals. End-users would assume the role of this agent.

4 Evaluation of Organizational Performance

The metrics chosen for evaluating organization performance and the learning protocol is described here.

4.1 Performance Evaluation Criteria

The following metrics are used to measure organization performance:

- **Time:** $\beta(i, \mathfrak{A}_i)$ is the time elapsed between the start and end of task i.
- **Cost:** $\gamma(i, \mathfrak{A}_i)$ determines the cost incurred during completing task i.

The β, and γ valuation functions determine the goodness of a certain *organization structure*. These functions allow comparisons to be made between different organization structures (e.g. whether assigned team of individuals $\mathfrak{A}_i$ for task i completes the task faster compared to an alternative team $\mathfrak{A}_i' \neq \mathfrak{A}_i$, or whether $\mathfrak{A}_i$ finish it in a shorter amount of time compared to $\mathfrak{A}_k$ performing task k).

We now describe the methods of combining the attributes used to describe tasks, individual members, and teams (described in Section 2) to define the valuation functions. These relationships have been derived from data on team behaviour gathered by interviewing engineering team leaders and members (in total 35 teams and 216 individuals were involved from two multi-national companies), and subjecting the captured information to multiple regression analyses. More details of this empirical modeling activity can be found in [7].

The various performance criteria are now defined.

Time: The *time* taken to accomplish task i is calculated as:

$$\beta(i, \mathfrak{A}_i) = (a1 \cdot \mathfrak{com}_i + b1 \cdot \mathfrak{cap}^{(\mathfrak{A}_i)} + c1 \cdot \mathfrak{mot}^{(\mathfrak{A}_i)} + d1) + \delta(i, \mathfrak{A}_i), \quad (1)$$

where system parameters $a1$, $b1$, and $c1$ are correspondingly weighting factors for the task's complexity, the team's composite capability, and composite motivation, while $d1$ is a constant[5]. $a1$, $b1$, and $c1$ are such that the time increases with task complexity but it decreases with capability and motivation. $\delta(i, \mathfrak{A}_i)$ is the length of the learning process (if required) as will be explained shortly.

Cost: The *cost* of performing task i is calculated as:

$$\gamma(i, \mathfrak{A}_i) = \frac{1}{\beta(i, \mathfrak{A}_i)} \quad (2)$$

This is because, in typical engineering work environments, cost output is typically inversely correlated with the time taken to complete the task.

Note that the performance metrics currently represent relative values. The purpose of the current version of the simulator is therefore to allow comparison studies of the differences in the computed metrics (particularly the time output) when team composition and attributes are varied.

---

[5] In [7], these parameters are calculated to be 0.0668, −0.4719, −0.3262, and 3.0355 respectively.

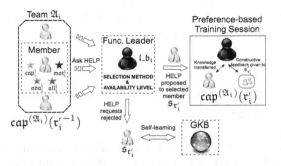

Fig. 3. The flow of the rth instance of the iterative learning process undergone by team $\mathfrak{A}_i$ before they are able to accomplish task i

4.2 Learning Protocol

For tasks with complexity higher than the capability of the team members assigned to it, the latter undergoes an information gathering or *learning* phase, in order to progressively increase the team capability in accomplishing the tasks. This is depicted in Figure 3 where team members request for help from the team leader. A help request may or may not be answered by the leader, depending on his availability. If it rejects a request, the *selected* help seeking agent ($\mathfrak{s}_{\mathfrak{r}_i^r} \in \mathfrak{A}_i$) forwards its query to the GKB agent.

We represent the learning phase by the vector: $\mathfrak{R}_i = (\mathfrak{r}_i^1, \ldots, \mathfrak{r}_i^u)^6$. Note, if the leader accepts a request, it first prioritises its responses among the set of requesting agents. The following priority schemes have been implemented:.

- **Uniform priority:** one team member is randomly chosen from the set of help requesters with probability $\frac{1}{|\mathfrak{A}_i|}$ during any instance of the iterative process;
- **Capability-based preference:** the leader selects the agent with a specified capability level (which is a parameter in our experiments to simulate different preference types of the leader) with the highest probability. A normal distribution is used to model this prioritisation.

After receiving help (i.e. after one learning iteration), the individual capability level of every learner j $\in \mathfrak{A}_i$ increases by:

$$\Delta cap^j = \frac{e + f \cdot (cap^{l\_b_i} - cap^j)}{g} \tag{3}$$

where constant $e > 0$ is to ensure that there is no negative increment. Parameters f and g govern the weighting factors for the individual j's capability

---

6 Through such learning process, the team members cooperate and coordinate [11].

improvement[7]. Due to improvements in the individual capability levels, the team's composite capability improves by:

$$\Delta cap^{(\mathfrak{A}_i)}(\mathfrak{r}_i^r, \mathfrak{s}_{\mathfrak{r}_i^r}) = \begin{cases} cap^{\mathfrak{s}_{\mathfrak{r}_i^r}} \cdot \Delta cap^{\mathfrak{s}_{\mathfrak{r}_i^r}} + \dfrac{\sum_{\mathfrak{s}'_{\mathfrak{r}_i^r} \in \mathfrak{A}_i \setminus \mathfrak{s}_{\mathfrak{r}_i^r}} cap^{\mathfrak{s}'_{\mathfrak{r}_i^r}} \cdot \Delta cap^{\mathfrak{s}'_{\mathfrak{r}_i^r}}}{h} & \text{if leader helps } @\mathfrak{r}_i^r \\ cap^{\mathfrak{s}_{\mathfrak{r}_i^r}} \cdot \Delta cap^{\mathfrak{s}_{\mathfrak{r}_i^r}} & \text{otherwise} \end{cases} \tag{4}$$

where h determines that only a fraction of the knowledge increment from each of the non-selected team members $\mathfrak{s}'_{\mathfrak{r}_i^r} \in \mathfrak{A}_i \setminus \mathfrak{s}_{\mathfrak{r}_i^r}$ would increase the team's composite capability in solving the task.

If the leader rejects a help request (e.g. when it has low availability), the requesting agent would consult the GKB agent. In this case, the individual and team capability increments would be lower (as we assume that GKB has very low capability) and therefore more learning iterations would be needed.

At the end of the rth learning session, the team's capability can be calculated using:

$$cap^{(\mathfrak{A}_i)}(\mathfrak{r}_i^r) = cap^{(\mathfrak{A}_i)} + \sum_{l=1}^{r} \Delta cap^{(\mathfrak{A}_i)}(\mathfrak{r}_i^l, \mathfrak{s}_{\mathfrak{r}_i^l}) \tag{5}$$

An additional consequence of the learning process is that the technical leader's availability decreases as he spends time in training. His availability, at the end of the process, is given by:

$$ava^{l\_b_i}(\mathfrak{r}_i^r) = ava^{l\_b_i} - \sum_{l=1}^{r} \begin{cases} man^{l\_b_i} & \text{if leader helps } @\mathfrak{r}_i^l \\ 0 & \text{otherwise} \end{cases} \tag{6}$$

where $man_{l\_b_i}$ determines how well the leader manages its time to provide help to its team members – the smaller the value of $man_{l\_b_i}$, the better is the leader's time management skill, or, the higher the number of help requests it can handle.

Finally, the total learning time is given by:

$$\delta(i, \mathfrak{A}_i) = \begin{cases} 0 & \text{if } com_i <= cap^{(\mathfrak{A}_i)} \\ \sum_{r=1}^{u} \dfrac{1}{rate(ava^{\mathfrak{s}_{\mathfrak{r}_i^r}})} & \text{otherwise} \end{cases} \tag{7}$$

5   The Organizational Simulator

Simulation-based approach has been widely used to gain insights into the behaviour of engineering and physical systems. Less often done is, however, using simulations to gain insights into the behaviour of social systems involving teams of individuals and their interactions jointly undertaking concurrent tasks. Within this context, agent-based modeling is exploited in order to investigate the use

---

[7] Using [7] as guidance, these parameters are set to be 15, 1.75, and 200 respectively.

of agent technology for simulating the actions and interactions of different autonomous entities within our organizational model defined in Section 2, with a
view for assessing the emerging social phenomena on the system as a whole.

The simulation tool is built using Java^{TM} and open source libraries such
as *jfreechart* for plotting charts and *junit* for unit- and system-levels testings.
It allows simulation experiments to be conducted through a highly interactive,
visually-rich interface by way of: 1)setting parameter values (e.g. task, individual,
team, workflow related variables, among others), 2) evaluating multiple organization structures simultaneously to efficiently evaluate and compare multiple
scenarios, 3) performing run-time analysis and spot checks by pausing, resum-
ing, and tracing back in time, and 4) allowing *what-if* analysis and recording
variety of individual-level and team-level performance factors through intuitive
charts and plots.

6 Empirical Evaluation

The aims of the experimental evaluations are to: 1) test the stability of the
generated simulation results, and 2) conduct relative performance analyses of
real-world teams involved in an engineering design project against a number of
organizational scenarios. These alternative organizations are selected to demonstrate the ability of our simulation system to model realistic team configurations
and therefore be able to answer typical resource management queries.

It is expected that the organizational tool will produce small variations in
terms of time and cost due to the introduced randomness of communication
activities in the underlying model. Here, the experimental setup is first described.
Following that the actual evaluations are discussed.

6.1 Parameters Setup

The chosen organization test case, a real-world engineering design project, consists of 41 tasks performed by a total of 27 members. Figure 4 describes the input
configurations. It shows the number of team members (on the left Y-axis) and
their capability distribution per task as well as the complexity of each task. For
example, task #3 is acted upon by 4 team members with one member having
capability level 'M', one 'H', and two 'VH'. The complexity of this task (the
right Y-axis) is 'M'.

The workflow of the project is *complete* (i.e. all tasks will ultimately execute
as the work yield of a task will potentially trigger a few other tasks to begin).
Three tasks have no predecessors and hence will initiate the workflow. There are
seven tasks with complexity higher than the capabilities of their owners.

Experiments with the same input configuration are repeated 1000 times for
each of the organizational scenarios described in Section 6.3 to ensure the statistical significance of the output.

6.2   Organizational Scenarios

In this experiment, the real-world scenario is benchmarked against an ideal case which could have been achieved by delegating every task to the most capable teams (in which all members have the highest attribute levels). In the ideal case, the teams are very effective, hence, no learning is required and the tasks are completed quickly. This scenario represents the highest level of organization performance achieved by an ideal team which cannot be expected to exist in practice.

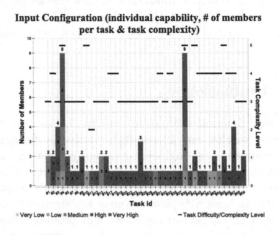

Fig. 4. Number of organization members per task, their individual capability, and the task complexity

In addition to the ideal, the actual test case is compared against several other organizational scenarios, enumerated below:

- **Case 0 - Actual case:** which is the real-world teams (as shown in Figure 4).
- **Case 1 - Optimal case:** which is the ideal scenario.
- **Case 2 - Increase in high capability members:** where each team is given two additional members with the highest attributes.
- **Case 3 - Large increase in high capability members:** where each team is given four additional members with the highest attributes.
- **Case 4 - Increase in low capability members:** each team provided with two additional members with the lowest attributes.
- **Case 5 - Large increase in low capability members:** each team provided with four additional members with the lowest attributes.

Overall, we hypothesize that teams with higher mean attributes will complete tasks in less time. However, the advantage gained by adding more high performing individuals may not increase monotonically. On the other hand, increasing too many low performing individuals may cause excessive help information exchange leading to poor team performance in term of time.

Table 1. Simulation results showing the completion times of representative tasks for different organizational scenarios (the less time required, the better). Each of the tasks written in bold has complexity higher than the capability of its owner.

Task id	Different Organizational Scenarios (Case #)											
	0		1		2		3		4		5	
	μ	σ	μ	σ	μ	σ	μ	σ	μ	σ	μ	σ
#1	11.53	0.00	9.43	0.00	11.53	0.00	11.53	0.00	11.53	0.00	11.53	0.00
#4	**34.44**	**1.31**	**9.73**	**0.00**	**33.10**	**1.31**	**32.60**	**1.29**	**44.01**	**4.06**	**50.78**	**5.62**
#9	16.27	0.00	9.43	0.00	16.27	0.00	16.27	0.00	16.27	0.00	16.27	0.00
#10	**37.44**	**0.56**	**9.73**	**0.00**	**27.53**	**1.08**	**27.54**	**1.14**	**69.14**	**4.20**	**75.70**	**3.29**
#35	13.99	0.00	9.43	0.00	13.99	0.00	13.99	0.00	13.99	0.00	13.99	0.00
#39	**34.45**	**1.38**	**9.73**	**0.00**	**32.23**	**1.30**	**32.62**	**1.25**	**39.04**	**3.11**	**42.92**	**4.22**
#40	16.33	0.00	9.43	0.00	16.33	0.00	16.33	0.0	16.33	0.00	16.33	0.00
#42	**54.28**	**1.71**	**9.73**	**0.00**	**47.23**	**2.53**	**42.11**	**2.14**	**87.35**	**5.81**	**106.94**	**6.54**
#47	13.59	0.00	9.58	0.00	13.59	0.00	13.59	0.00	13.59	0.00	13.59	0.00
#51	**39.90**	**0.79**	**9.58**	**0.00**	**29.84**	**1.32**	**30.81**	**1.43**	**51.67**	**2.06**	**57.58**	**2.68**
#52	16.46	0.00	9.58	0.00	16.46	0.00	16.46	0.00	16.46	0.00	16.46	0.00
#54	**20.58**	**0.00**	**9.58**	**0.00**	**19.47**	**0.31**	**19.23**	**0.28**	**22.33**	**0.69**	**23.33**	**0.94**
#55	**21.49**	**0.26**	**9.73**	**0.00**	**21.08**	**0.15**	**20.94**	**0.15**	**24.23**	**1.09**	**25.18**	**1.20**
#56	15.53	0.00	9.58	0.00	15.53	0.00	15.53	0.00	15.53	0.00	15.53	0.00
#59	12.61	0.00	9.43	0.00	12.61	0.00	12.61	0.00	12.61	0.00	12.61	0.00

6.3   Experimental Results

The result of the simulation process is summarized in Table 1 which shows the time performance – mean and standard error – achieved by the organizational cases enumerated in the previous section on various tasks.

Performance of Actual and Ideal Organizations (Cases 0 and 1). Table 1 shows that on average the teams in *case 0* requires 50% more time than that required by the ideal teams *case 1*. For more complex (difficult) tasks, the teams in *case 0* take at least twice as long as in *case 1* (e.g. in task #55, the teams in *cases 1* and *0* take 9.73 and 21.49±0.26 time respectively, while in task #4, the team in *case 1* takes 9.73 time and the team in *case 0* takes 34.44±1.31 which is more than three times longer). This is because the teams in the ideal scenario (*case 1*) are the most capable, available, and motivated. They require no help (i.e. no learning process is initiated) during the project, hence, saving significant amount of time. This is also the reason why the standard deviation errors of the time performance of these teams are always zero, indicating that they complete tasks in exactly the same time every time.

Performance of High Capability Teams (Cases 2 and 3). Adding two and four members (for *cases 2* and *3*, respectively) with the highest attributes to each task could expectedly reduce the total project completion time relative to *case 0*. However, this improvement in time performance comes at a cost of *work overloading*. Table 2 shows the daily activity of various organizational members for the different organizational cases[8]. The last row of the table highlights the average project completion times (in days).

In particular, the table shows that the members in *case 0* are on average involved in only one task or none in this project, on a daily basis. A short busy period is experienced by member #003 for a number of days, when he is involved in performing 3-4 tasks simultaneously. This is because, in the project workflow we modeled, this member is involved in a total of sixteen tasks of which several are overlapping in time. Comparatively, the members in *cases 2* and *3* experience much more busy timetables. This is also one of the reasons why these organizations finish the project slightly earlier.

Diminishing Returns from Resource Allocation. Despite the previous observation on expediting project by allocating more tasks to individuals, *resource over-allocation* could have a diminishing return in term of time improvement, hence each additional member yields smaller and smaller improvement in time performance. This is verified from the results explained in the last row of Table 2 which shows the average number of days taken to complete the project by organizations of *cases 0, 2,* and *3*. We observe that the project under *case 3* finishes just six days earlier compared to *case 2* which finishes twenty four days earlier compared to *case 0*.

Communication Overload and The Learning Process (Cases 4 and 5). The scenarios in which low capability members (capability level 'VL') are additionally assigned into every task (*cases 4* and *5*) perform poorly as the leaders spend significant amount of time to train these individuals via the learning process. Table 3 shows, for organizations of *cases 4* and *5*, the average numbers of help communications exchanged between the teams and the GKB agent overwhelm those between the teams and their respective team leaders. For each row in the table (i.e. task i), $\mathcal{H}_i^{l\text{-}b}$ represents the number of learning instances (i.e. helps) provided by the respective leader, while $\mathcal{H}_i^{GKB}$ represents that of self-learning instances (when the leader does not answer help requests during the task, recall Section 4.2). Also shown is the number of times inexperienced members (that is $|\mathfrak{S}_i^-|$ where $\mathfrak{S}_i^- = \{\mathfrak{s}_{\mathfrak{r}_i^r}|cap^{\mathfrak{s}_{\mathfrak{r}_i^r}} = \text{VL or } cap^{\mathfrak{s}_{\mathfrak{r}_i^r}} = \text{L}, \forall \mathfrak{r}_i^r \in \mathfrak{R}_i\}$) are selected from team $\mathfrak{A}_i$ and offered help by the leader, while performing task i. $|\mathfrak{S}_i^+|$ represents the number of times relatively more experienced members are selected, where $\mathfrak{S}_i^+ = \{\mathfrak{s}_{\mathfrak{r}_i^r}|cap^{\mathfrak{s}_{\mathfrak{r}_i^r}} = \text{M or } cap^{\mathfrak{s}_{\mathfrak{r}_i^r}} = \text{H or } cap^{\mathfrak{s}_{\mathfrak{r}_i^r}} = \text{VH}, \forall \mathfrak{r}_i^r \in \mathfrak{R}_i\}$.

More specifically for task #42, the help communications between the assigned team and its leader are only 31% and 26% of the total communications, for *cases 4* and *5* respectively. This is due to the fact that the leader is quickly overloaded with help-requests from the extra low capability individuals who occupy close to 75% ($\frac{4.64}{6.25} * 100\%$) and 81% ($\frac{5.17}{6.39} * 100\%$) of the leader's time. Compared to *case 0*, the low capability individuals and team capability increments (recall

---

[8] In our experiments, we perform detailed monitoring of the activities of each individual over each day of a project. Only summarised data are presented due to space restrictions.

Table 2. Simulation results showing organizational members' daily activities over the project duration for different organizational scenarios. For each row, *Max* represents the maximum number of tasks on a day the member is involved in. μ and σ describe the mean number of tasks and its standard deviation respectively, $\#d^0$, and $\#d^{1|2}$, and $\#d^{3|4}$ describe the number of days the member is doing 0 task, 1-2 tasks, and 3-4 tasks correspondingly

Mem.		Case #					
Id		0	1	2	3	4	5
#001	Max	1	1	3	3	1	2
	μ	0.13	0.17	0.34	0.70	0.14	0.52
	σ	0.33	0.37	0.63	0.91	0.35	0.64
	$\#d^0$	301	142	235	167	379	274
	$\#d^{1|2}$	43	28	82	123	62	214
	$\#d^{3|4}$	0	0	3	24	0	0
#003	Max	4	3	4	4	4	5
	μ	1.13	1.33	1.54	1.62	1.08	1.08
	σ	1.01	1.03	1.12	1.10	1.11	1.12
	$\#d^0$	101	42	39	59	177	174
	$\#d^{1|2}$	202	100	217	175	200	252
	$\#d^{3|4}$	41	28	64	80	64	62
#017	Max	2	2	3	2	2	2
	μ	0.84	0.68	1.06	1.19	0.97	1.09
	σ	0.73	0.75	0.74	0.76	0.70	0.71
	$\#d^0$	125	83	74	66	115	103
	$\#d^{1|2}$	219	87	242	248	326	385
	$\#d^{3|4}$	0	0	4	0	0	0
#024	Max	3	3	3	3	3	3
	μ	0.31	0.41	0.39	0.68	0.37	0.49
	σ	0.76	0.86	0.86	0.94	0.78	0.79
	$\#d^0$	281	128	254	184	339	321
	$\#d^{1|2}$	42	28	45	109	81	146
	$\#d^{3|4}$	21	14	21	21	21	21
Duration (d)		344	170	320	314	441	488

Equations 3 and 4 respectively) are lower in these cases and therefore, more learning iterations are needed. For *case 0*, the team, on average, undergoes 15.33 learning iterations. While for *cases 4* and *5*, the team undergoes 20.38 and 24.17 iterations (i.e. there are 33% and 58% more learning iterations respectively).

As a result, the average project completion times under these scenarios are much longer than the others (recall the last row of Table 2) due to the additional learning time ($\delta(i, \mathfrak{A}_i)$ as in Equation 7) which is proportional to ($\mathcal{H}_i^{\text{l-b}} + \mathcal{H}_i^{\text{GKB}}$).

Table 3. Simulation results showing the different types of help communications (i.e. learning iterations) which occur in representative tasks, for different organizational scenarios

Task Id		Case # 0		1		2		3		4		5	
		μ	σ	μ	σ	μ	σ	μ	σ	μ	σ	μ	σ
#10	$\mathcal{H}_i^{1-b}$	6.25≈32%	0.57	-	-	5.75≈43%	0.64	5.50≈49%	0.66	6.60≈21%	0.50	6.63≈20%	0.49
	$|\mathfrak{G}_i^-|$	6.25≈32%	0.57	-	-	2.09≈16%	1.03	0.37≈3%	0.60	6.60≈21%	0.50	6.63≈20%	0.49
	$|\mathfrak{G}_i^+|$	0.00≈0%	0.00	-	-	3.66≈27%	1.03	5.13≈46%	0.80	0.00≈0%	0.00	0.00≈0%	0.00
	$\mathcal{H}_i^{GKB}$	13.47≈68%	0.98	-	-	7.55≈57%	1.42	5.60≈51%	1.28	25.38≈79%	2.09	25.92≈80%	1.85
	$|\mathfrak{T}_i^-|$	13.47≈68%	0.98	-	-	2.94≈22%	1.20	0.38≈4%	0.61	25.38≈79%	2.09	25.92≈80%	1.85
	$|\mathfrak{T}_i^+|$	0.00≈0%	0.00	-	-	4.61≈35%	1.36	5.22≈47%	1.24	0.00≈0%	0.00	0.00≈0%	0.00
#42	$\mathcal{H}_i^{1-b}$	5.97≈39%	0.60	-	-	5.74≈45%	0.62	5.59≈48%	0.63	6.25≈31%	0.56	6.39≈26%	0.56
	$|\mathfrak{G}_i^-|$	3.12≈20%	0.90	-	-	1.62≈13%	0.96	0.72≈6%	0.77	4.64≈23%	1.09	5.17≈21%	1.04
	$|\mathfrak{G}_i^+|$	2.85≈19%	0.89	-	-	4.12≈32%	0.99	4.87≈42%	0.87	1.61≈8%	0.98	1.22≈5%	0.90
	$\mathcal{H}_i^{GKB}$	9.36≈61%	1.06	-	-	7.03≈55%	1.34	6.02≈52%	1.30	14.13≈69%	2.09	17.78≈74%	2.34
	$|\mathfrak{T}_i^-|$	4.63≈30%	0.87	-	-	1.75≈14%	1.17	0.81≈7%	0.84	10.17≈50%	2.67	13.95≈58%	2.97
	$|\mathfrak{T}_i^+|$	4.73≈31%	1.21	-	-	5.28≈41%	1.27	5.21≈45%	1.24	3.96≈19%	1.36	3.83≈16%	1.37

7   Conclusions and Future Work

In this paper, we have presented a novel approach to modeling real-world organizations which significantly extends previous related work [7], [8], [9], [10]. In particular, we present ways of modeling both individual member and team level characteristics in terms of socio-technical attributes that are representative of real-world organizations. In addition, using these attributes we have defined both individual and team level performance metrics to enable end-users (e.g. resource managers in organizations) to evaluate the effectiveness of different organization structures. Finally, we have introduced a learning protocol (also similar to real organization behaviour) to let inexperienced team members learn the skills to perform complex tasks by seeking help. The impact of such knowledge sharing on the organization's performance is modeled.

The above features have been modeled as a multi-agent system simulator that allows end-users to create, modify, run, analyze, and visualize complex organizational structures including the tasks undertaken by the organization members and their inter-dependencies. Several experiments are conducted to evaluate a number of realistic organization scenarios. These studies show that increasing very high capability members in a team improve the time performance but at a cost of extra man-hours. On the other hand, we observe that additional members with very low attribute values can be beneficial under certain situations but high communication overhead becomes an issue as they learn the skills to perform the tasks by asking help from others. We are validating the observed results and conclusions drawn from our simulation studies against data obtained from completed projects in our organization. This activity is crucial in order to ensure total quality in the underlying organizational models [12], [13].

As part of the future work, we envisage the following extensions:

- **Optimal organization structure:** we will minimize the sum of the execution period of every task. That is, we wish to solve $\mathfrak{A}^* = \arg\min_{\{\mathfrak{J}\}} \sum_{i=0}^{n} \beta(i, \mathfrak{A}_i)$ while satisfying the workload constraint on each individual $j \in \mathfrak{J}$.
- **Improving the organization models:** as highlighted in Section 2.3, the incorporation of work iterations into our organizational model would be beneficial.

Acknowledgments. This work forms part of Strategic Investment in Low carbon Engine Technology (SILOET) Project - P3: Integrated Decision Support Systems - which is a collaborative Research & Technology project supported by the Technology Strategy Board and Rolls-Royce plc.

References

1. Cherns, A.: Principles of Sociotechnical Design Revisited. Human Relations 40(3), 153–161 (1987)
2. Brocco, M., Groh, G.: Team Recommendation in Open Innovation Networks. In: 3rd ACM Conference on Recommender Systems, New York City, pp. 365–368 (2009)
3. Jennings, N.R.: An Agent-Based Approach for Building Complex Software Systems. Communications of the ACM 44(4), 35–41 (2001)
4. Bonabeau, E.: Agent-based Modeling: Methods and Techniques for Simulating Human Systems. National Academy of Sciences of the United States of America 99(3), 7280–7287 (2002)
5. Chen, X., Zhan, F.: Agent-based Modeling and Simulation of Urban Evacuation: Relative Effectiveness of Simultaneous and Staged Evacuation Strategies. Operational Research Society 59(1), 25–33 (2008)
6. Quillinan, T.B., Brazier, F., Aldewereld, H., Dignum, F., Dignum, V., Penserini, L., Wijngaards, N.: Developing Agent-based Organizational Models for Crisis Management. In: 8th International Conference on Autonomous Agents and Multiagent Systems, Budapest, pp. 45–52 (2009)
7. Crowder, R., Hughes, Sim, Y.W., Robinson, M.: An Agent Based Approach to Modelling Design Teams. In: 17th International Conference on Engineering Design, Stanford, pp. 91–102 (2009)
8. Jin, Y., Levitt, R.E.: The Virtual Design Team: A Computational Model of Project Organizations. Computational and Mathematical Organization Theory 2(3), 171–195 (1996)
9. Levitt, R.E., Thomsen, J., Christiansen, T.R., Kunz, J.C., Jin, Y., Nass, C.: Simulating Project Work Processes and Organizations: Toward a Micro-Contingency Theory of Organizational Design. Management Science 45(11), 1479–1495 (1999)
10. Louie, M.A., Carley, K.M., Haghshenass, L., Kunz, J.C., Levitt, R.E.: Model Comparisons: Docking OrgAhead and SimVision. In: North American Association for Computational Social and Organization Sciences, Pittsburgh (2003)
11. Emery, F.E., Trist, E.L.: Socio-technical Systems. Management Science, Models and Techniques 2, 83–97 (1960)
12. Balci, O.: A Methodology for Certification of Modeling and Simulation Applications. ACM Transactions on Modeling and Computer Simulation 11(4), 352–377 (2001)
13. Galan, J.M., Izquierdo, L.R., Izquierdo, S.S., Santos, J.I., Olmo, R.D., Paredes, A.L., Edmonds, B.: Errors and Artefacts in Agent-Based Modelling. Artificial Societies and Social Simulation 12(1), 1 (2009)

A New Sign Distance-Based Ranking Method for Fuzzy Numbers

Kok Chin Chai[1], Kai Meng Tay[1], and Chee Peng Lim[2]

[1] Universiti Malaysia Sarawak, Kota Samarahan, Sarawak, Malaysia
[2] Centre for Intelligent Systems Research, Deakin University, Australia
kcchai@live.com, kmtay@feng.unimas.my

Abstract. In this paper, a new sign distance-based ranking method for fuzzy numbers is proposed. It is a synthesis of geometric centroid and sign distance. The use of centroid and sign distance in fuzzy ranking is not new. Most existing methods (e.g., distance-based method [9]) adopt the Euclidean distance from the origin to the centroid of a fuzzy number. In this paper, a fuzzy number is treated as a polygon, in which a new geometric centroid for the fuzzy number is proposed. Since a fuzzy number can be represented in different shapes with different spreads, a new dispersion coefficient pertaining to a fuzzy number is formulated. The dispersion coefficient is used to fine-tune the geometric centroid, and subsequently sign distance from the origin to the tuned geometric centroid is considered. As discussed in [5-9], an ideal fuzzy ranking method needs to satisfy seven reasonable fuzzy ordering properties. As a result, the capability of the proposed method in fulfilling these properties is analyzed and discussed. Positive experimental results are obtained.

Keywords: Geometric centroid, sign distance, fuzzy numbers, reasonable ordering properties, dispersion coefficient.

1   Introduction

Fuzzy ranking attempts to order a set of fuzzy numbers. The importance of fuzzy ranking in many application domains has been highlighted in the literature, e.g., decision-making [1], data analysis [2], risk assessment [3] and artificial intelligence [4]. Indeed, many fuzzy ranking methods have also been developed [1][5][6]. These methods can be categorized into three categories [5-6]; i.e., (1) transforming a set of fuzzy numbers into crisp numbers and subsequently ranking the crisp numbers; (2) mapping a set of fuzzy numbers to crisp numbers based on a pre-defined reference set(s) for comparison; (3) ranking through pairwise comparison of fuzzy numbers.

In this paper, our focus is on the first category, which is straightforward as compared with the other two categories. Nevertheless, the challenge is to develop a method that satisfies a set of reasonable ordering properties [5-9], as detailed in Section 2.2. Recently, a number of methods, e.g., deviation degree-based [8][10], distance-based [9][11], centroid-based [12], and area-based [13], have been proposed. However, many of these methods [8][10-13] do not have analysis pertaining to fulfillment

© Springer International Publishing Switzerland 2015   55
H. Handa et al. (eds.), *Proc. of the 18th Asia Pacific Symp. on Intell. & Evol. Systems – Vol. 2*,
Proceedings in Adaptation, Learning and Optimization 2, DOI: 10.1007/978-3-319-13356-0_5

of the reasonable ordering properties [5-9]. Besides that, fuzzy ranking is complicated when fuzzy numbers are presented as different shapes with different spreads,
e.g., as illustrated in Fig. 1 [8].

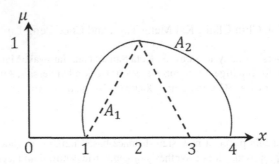

Fig. 1. Triangular (i.e., A_1) and generalized (i.e., A_2) fuzzy numbers [8]

In this paper, a new fuzzy ranking method that deals with fuzzy numbers (Definition 2), which is a synthesis of geometric centroid and sign distance, is proposed.
The proposed method is capable of handling more complex fuzzy numbers i.e., generalized fuzzy numbers, as illustrated in Fig. 1. The method can be summarized into a
few steps, as follows: (1) Fuzzy numbers are firstly converted into normalized fuzzy
numbers (i.e., a set of fuzzy numbers available in space $R^* \in [-1,1]$). (2) A discretization technique [14], i.e., slicing the support of each normalized fuzzy number into
N slices vertically at even intervals, is employed. The intersections between the
normalized fuzzy number and the N vertical slices are called intersection points or
discretized points. (3) Using the discretized points, geometric centroid of the normalized fuzzy number is then computed using Bourke's method [15]. To deal with
more complex fuzzy numbers, as depicted in Fig. 1, a new dispersion coefficient of
the normalized fuzzy numbers is proposed. The dispersion coefficient attempts to
fine-tune the geometric centroid. (4) Finally, sign distance between the origin and
the tuned geometric centroid is measured. In this paper, the results obtained from the
sign distances are used to order a set of fuzzy numbers. The fulfilment of the seven
reasonable ordering properties as stated in [5-9] is further examined. To evaluate the
proposed method, benchmark examples from [8] are used.

The rest of the paper is organized as follows. In Section 2, some background studies are presented. In Section 3, the proposed ranking method is explained. In Section 4, an experimental study and the associated results are presented. Finally,
concluding remarks are given in Section 5.

2   Background

2.1   Definitions

The following mathematical notations are widely used. R denotes a set of real numbers, μ represents a fuzzy number, and $\mu(x)$ for its membership function, $\forall x \in R$.
Definition 1, as follows, is considered.

Definition 1 [16]: A fuzzy number, A is a fuzzy set such that $\mu : R \rightarrow [0,1]$ which satisfies the following properties:

- A is represented as $(a_1,\ a_2,\ a_3,\ a_4)$
- μ is upper semi-continuous
- $a_1,\ a_2,\ a_3$ and a_4 are real numbers, in such $a_1 \leq a_2 \leq a_3 \leq a_4$. μ is strictly increasing for interval $[a_1, a_2]$ and strictly decreasing for interval $[a_3, a_4]$.
- $\mu(x) = 1$, for $a_2 \leq x \leq a_3$,
- $\mu(x) = 0$ for x not in the interval of $[a_1, a_4]$,
- Support of A, i.e., $sup(A) = \{x \in R | \mu(x) > 0\}$.
- μ is expressed as follows:

$$\mu(x) = \begin{cases} \mu^L(x), & a_1 \leq x \leq a_2, \\ 1, & a_2 \leq x \leq a_3, \\ \mu^R(x), & a_3 \leq x \leq a_4, \\ 0, & otherwise. \end{cases} \tag{1}$$

where $\mu^L : [a_1, a_2] \rightarrow [0,1]$ and $\mu^R : [a_3, a_4] \rightarrow [0,1]$ are left and right membership functions of fuzzy number u. If $\mu^L(x)$ and $\mu^R(x)$ are linear functions and $a_2 < a_3$, then A is a trapezoidal fuzzy number. If $\mu^L(x)$ and $\mu^R(x)$ are linear functions, $a_2 = a_3$, then A is a special case of a trapezoidal fuzzy number i.e., a triangular fuzzy number.

Definition 2 [17]: Normalized fuzzy numbers (i.e., $A_i^*, i = 1,2,3,\ldots,m$) are a set of fuzzy numbers in space $R^* \in [-1,1]$. Consider a set of fuzzy numbers A_i where $i = 1,2,3,\ldots,m$ in space $R \in (-\infty, \infty)$. A_i^* is obtained with Eq. (2).

$$A_i^* = (\frac{a_{i1}}{k}, \frac{a_{i2}}{k}, \frac{a_{i3}}{k}, \frac{a_{i4}}{k}) = (a_{i1}^*, a_{i2}^*, a_{i3}^*, a_{i4}^*) \tag{2}$$

where $k = \max(|a_{ij}|, 1)$, $|a_{ij}|$ is the absolute value of a_{ij}, $1 \leq i \leq m$ and $1 \leq j \leq 4$. As an example, $A_1^* = (a_{11}^*, a_{12}^*, a_{13}^*, a_{14}^*)$ is illustrated in Fig.2, and centroids of A_1 is written as $cX_{A_1^*}, cY_{A_1^*}$.

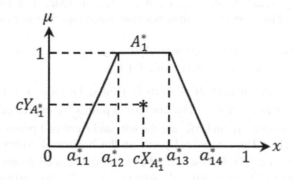

Fig. 2. The membership function of a normalized trapezoidal fuzzy number

2.2   Seven Reasonable Ordering Properties

Consider three fuzzy numbers, i.e., A_1, A_2, and A_3 in space R. An ideal fuzzy ranking method should satisfy seven reasonable ordering properties, as follows. The first six reasonable ordering properties (i.e., **P1-P6**) are introduced in [5-6], while **P7** is introduced in recent literatures [7-9].

P1: If $A_1 \succcurlyeq A_2$ and $A_2 \succcurlyeq A_1$, then $A_1 \sim A_2$.
P2: If $A_1 \succcurlyeq A_2$ and $A_2 \succcurlyeq A_3$, then $A_1 \succcurlyeq A_3$.
P3: If $A_1 \cap A_2 = \emptyset$ and A_1 is on the right of A_2, then $A_1 \succcurlyeq A_2$.
P4: The order of A_1 and A_2 is not affected by other fuzzy numbers under comparison.
P5: If $A_1 \succcurlyeq A_2$, then $A_1 + A_3 \succcurlyeq A_2 + A_3$.
P6: If $A_1 \succcurlyeq A_2$, then $A_1 A_3 \succcurlyeq A_2 A_3$.
P7: If $A_1 \succcurlyeq A_2 \succcurlyeq A_3$ then $-A_1 \preccurlyeq -A_2 \preccurlyeq -A_3$.

This paper focuses on fulfillment of these seven reasonable ordering properties for trapezoidal fuzzy numbers. Note that the addition and multiplication operators in **P5** and **P6** are based on fuzzy arithmetic operators as follows:

Fuzzy Addition operator [11]:

$$A_1 \oplus A_2 = (a_{11}, a_{12}, a_{13}, a_{14}) \oplus (a_{21}, a_{22}, a_{23}, a_{24})$$
$$= (a_{11} + a_{21}, a_{12} + a_{22}, a_{13} + a_{23}, a_{14} + a_{24}) \qquad (3)$$

Fuzzy Multiplication [11]:

$$A_1 \otimes A_2 = (a_{11}, a_{12}, a_{13}, a_{14}) \otimes (a_{21}, a_{22}, a_{23}, a_{24})$$
$$= (a_{11} \times a_{21}, a_{12} \times a_{22}, a_{13} \times a_{23}, a_{14} \times a_{24}) \qquad (4)$$

Fuzzy Image [7]-[9]:
Fuzzy image of $A_1 = (a_{11}, a_{12}, a_{13}, a_{14})$ is $-A_1 = (-a_{14}, -a_{13}, -a_{12}, -a_{11})$.

3   The Proposed Methodology

Consider m fuzzy numbers, A_i, i.e., $A_1, A_2, ..., A_m$, in space $R \in [-\infty, \infty]$, which have to be ranked. The proposed method is summarized into five steps as follows:

Step 1: Transform each fuzzy number $A_i = (a_{i1}, a_{i2}, a_{i3}, a_{i4})$ into a normalized fuzzy number $A_i^* = (a_{i1}^*, a_{2i}^*, a_{i3}^*, a_{i4}^*)$ using Eq. (2).

Step 2: Discretize the support of A_i^* into N points, i.e., x_{ij}^*, $j = 1,2,3 ..., N$, and obtain $\mu_{A_i^*}(x_{ij}^*)$, where $x_{ij}^* \in R^*$. The discretized points are expressed in a sequence of $x_{i1}^*, x_{i2}^*, ..., x_{iN}^*$, where x_{i1}^* and x_{iN}^* are the left- and right-end points of A_i^*, i.e., a_{i1}^* and a_{i4}^*, respectively. The discretized points in the horizontal component, i.e., x_{ij}^*, are computed using Eq. (5). On the other hand, the discretized points in the vertical component, i.e., $\mu_{A_i^*}(x_{ij}^*)$, are computed using Eq. (1). The discretized points for A_i^* are expressed in Eqs. (6) and (7) as follows.

$$x^*_{ij} = x^*_{i1} + (j - 1)\left(\frac{x^*_{iN} - x^*_{i1}}{N-1}\right), \tag{5}$$

$$x^* = \begin{bmatrix} x^*_{11} & \cdots & x^*_{1N} \\ \vdots & \ddots & \vdots \\ x^*_{m1} & \cdots & x^*_{mN} \end{bmatrix}, \tag{6}$$

$$\mu_{A^*_i}(x^*) = \begin{bmatrix} \mu_{A^*_i}(x^*_{11}) & \cdots & \mu_{A^*_i}(x^*_{1N}) \\ \vdots & \ddots & \vdots \\ \mu_{A^*_i}(x^*_{m1}) & \cdots & \mu_{A^*_i}(x^*_{mN}) \end{bmatrix}, \tag{7}$$

where $i = 1, 2, \ldots, m$, $j = 1, 2, 3, \ldots, N$.

Step 3: Compute the centroid of A^*_i (i.e., $cX_{A^*_i}, cY_{A^*_i}$). In this paper, the geometric centroid [15] is adopted, whereby $cX_{A^*_i}$ and $cY_{A^*_i}$ are obtained with Eqs. (8) and (9), as follows.

$$cX_{A^*_i} = \frac{1}{6L_{A^*_i}}\sum_{j=1}^{N}\left(x^*_{ij} + x^*_{i(j+1)}\right)\left(x^*_{ij}\mu_{A^*_i}(x^*_{i(j+1)}) + x^*_{i(j+1)}\mu_{A^*_i}(x^*_{ij})\right), \tag{8}$$

$$cY_{A^*_i} = \begin{cases} \frac{1}{6L_{A^*_i}}\sum_{j=1}^{N}[\left(\mu_{A^*_i}(x^*_{ij}) + \mu_{A^*_i}(x^*_{i(j+1)})\right)\left(x^*_{ij}\mu_{A^*_i}(x^*_{i(j+1)}) - x^*_{i(j+1)}\mu_{A^*_i}(x^*_{ij})\right)], & if\, x^*_{i1} \neq x^*_{iN} \\ \frac{\mu_{A^*_i}(x^*_i)}{2}, & if\, x^*_{i1} = x^*_{iN} \end{cases}, \tag{9}$$

where $L_{A^*_i} = \frac{1}{2}\sum_{j=1}^{N}\left(x^*_{ij}\mu_{A^*_i}(x^*_{i(j+1)}) - x^*_{i(j+1)}\mu_{A^*_i}(x^*_{ij})\right)$, $i = 1,2,3,\ldots,m, j = 1,2,3,\ldots,N$.

Step 4: Refine the centroid (i.e., $cX_{A^*_i}, cY_{A^*_i}$) using Eqs. (10)-(13) to obtain the new centroid, i.e., $cX^*_{A^*_i}, cY^*_{A^*_i}$.

$$\overline{X}_{A^*_i} = \frac{1}{N}\sum_{j=1}^{N}x^*_{ij}, \tag{10}$$

$$\delta_{A^*_i} = \sqrt{\frac{1}{N-1}\sum_{j=1}^{N}\left(x^*_{ij} - \overline{X}_{A^*_i}\right)^2}, \tag{11}$$

$$cX^*_{A^*_i} = cX_{A^*_i}, \tag{12}$$

$$cY^*_{A^*_i} = \frac{H_i}{2} - \delta_{A^*_i}cY_{A^*_i}, \tag{13}$$

where $\delta_{A^*_i}$ denotes a dispersion coefficient of A^*_i, and $H_i = \max_{j=1,2,\ldots,N}\{\mu_{A^*_i}(x^*_{ij})\}$, $i = 1, 2, 3, \ldots, m$.

Step 5: Compute the ordering index, $I_{A^*_i}$, using Eqs. (14)-(15) as follows.

$$\varphi_{A^*_i} = \begin{cases} 1, & sign(cX^*_{A^*_i}) \geq 0 \\ -1, & sign(cX^*_{A^*_i}) < 0 \end{cases}, \tag{14}$$

$$I(A^*_i) = \varphi_{A_i}\sqrt{cX^{*2}_{A^*_i} + cY^{*2}_{A^*_i}}, \tag{15}$$

where $\varphi_{A_i^*}$ indicates the position of the fuzzy numbers. $I(A_i^*)$ is the distance between point $(cX_{A_i^*}^*, cY_{A_i^*}^*)$ and the origin $(0,0)$. A larger ordering index (i.e., $I(A_i^*)$)
indicate a higher the ranking order.

4   Experimental Study

In this section, benchmark examples from [8] are used to evaluate the proposed method for ranking fuzzy numbers. An analysis of P1-P7 [5-9] is further reported.

4.1   Ranking Generalized Fuzzy Numbers

An example from [8], as depicted in Fig.1, is considered. Fuzzy membership functions for A_1 and A_2, are as follows.

$$\mu_{A_1}(x) = \begin{cases} x - 1, & 1 \leq x \leq 2, \\ 3 - x, & 2 \leq x \leq 3, \\ 0, & otherwise, \end{cases} \tag{16}$$

$$\mu_{A_2}(x) = \begin{cases} [1 - (x-2)^2]^{1/2}, & 1 \leq x \leq 2, \\ [1 - \frac{1}{4}(x-2)^2]^{1/2}, & 2 \leq x \leq 4, \\ 0, & otherwise. \end{cases} \tag{17}$$

The fuzzy numbers, A_1 and A_2 in Fig. 1, are transformed into normalized fuzzy
numbers (i.e., A_1^*, A_2^* respectively), as illustrated in Fig.3.

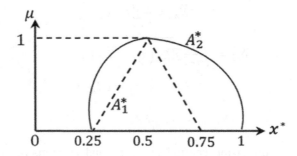

Fig. 3. Normalized triangular (i.e., A_1^*) and generalized (i.e., A_2^*) fuzzy numbers [8]

With the proposed method, $I(A_1^*) = 0.669$ and $I(A_2^*) = 0.723$ i.e., $I(A_1^*) <
I(A_2^*)$; therefore $A_1^* \preccurlyeq A_2^*$ and $A_1 \preccurlyeq A_2$. The ranking results are in agreement with
those in [8].

4.2   Analysis of the Seven Reasonable Ordering Properties

Three fuzzy numbers A_1, A_2, and A_3, are considered, where A_1^*, A_2^*, and A_3^* are normalized fuzzy numbers. An ideal fuzzy ranking method should comply with the seven reasonable ordering properties [5-9] as follows:

P1: $I(A_1^*) \geq I(A_2^*)$ occurs if $A_1 \succcurlyeq A_2$ and $I(A_2^*) \geq I(A_1^*)$ occurs if $A_2 \succcurlyeq A_1$. Therefore, $I(A_1^*) = I(A_2^*)$ is always true, if $A_1 \sim A_2$.

P2: $I(A_1^*) \geq I(A_2^*)$ occurs if $A_1 \succcurlyeq A_2$ and $I(A_2^*) \geq I(A_3^*)$ occurs if $A_2 \succcurlyeq A_3$. Therefore, $I(A_1^*) \geq I(A_3^*)$ is always true, for $A_1 \succcurlyeq A_2 \succcurlyeq A_3$.

P3:   If $A_1 \cap A_2 = \emptyset$, and if A_1 is on the right side of A_2, then $I(A_1^*) > I(A_2^*)$ is always true.

P4: $I(A_1^*)$ and $I(A_2^*)$ are computed separately to order A_1 and A_2. Therefore, the ranking outcome is not affected by other fuzzy numbers under comparison.

P5:   If $A_1 \succcurlyeq A_2$, then $I(A_1^*) \geq I(A_2^*)$. An addition of A_3 to A_1 and A_2 changes the centroids of A_1 and A_2 correspondingly. fore,$I(A_1 + A_3^*) \geq I(A_2 + A_3^*)$, if $A_1 + A_3 \succcurlyeq A_2 + A_3$.

P6:   If $A_1 \succcurlyeq A_2$, $I(A_1^*) \geq I(A_2^*)$. Similar to P5, a multiplication of A_3 to A_1 and A_2 changes the centroids of A_1 and A_2 correspondingly. Therefore, $I(A_1A_3^*) \geq I(A_2A_3^*)$, if $A_1A_3 \succcurlyeq A_2A_3$.

P7:   If $A_1 \succcurlyeq A_2 \succcurlyeq A_3$, then $I(A_1^*) \geq I(A_2^*) \geq I(A_3^*)$. Consider $\varphi_{A_1^*} = \varphi_{A_2^*} = \varphi_{A_3^*} = 1$ for A_1, A_2 and A_3. For $-A_1, -A_2$ and $-A_3$, $\varphi_{-A_1^*} = \varphi_{-A_2^*} = \varphi_{-A_3^*} = -1$. Therefore, $-I(-A_1^*) \leq -I(-A_2^*) \leq -I(-A_3^*)$, for $-A_1 \preccurlyeq -A_2 \preccurlyeq -A_3$.

The proof for **P1-P4** is straightforward. Analysis of **P5**, **P6** and **P7** are more complicated, and are further illustrated with an example in [8]. The details are presented in the following section.

4.3   An Empirical Study

Consider an example in [8], $A_1 = (2,4,4,6); A_2 = (1,5,5,6)$, and $A_3 = (3,5,5,6)$ as shown in Fig. 4(a). $A_1 + A_3, A_2 + A_3, A_1A_3$, and A_2A_3 (as depicted in Fig. 4(b) and Fig. 4(c)) are computed using fuzzy addition and fuzzy multiplication as summarized in Eqs. (3) and (4). Fuzzy images of A_1, A_2, and A_3 are presented in Fig. 4(d). Fuzzy numbers in Fig. 4 are normalized and presented in Fig. 5.

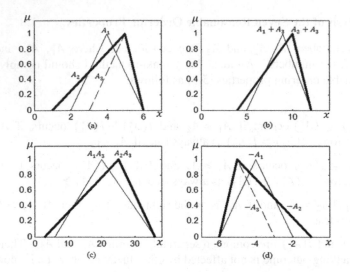

Fig. 4. (a) Three fuzzy numbers, A_1, A_2, and A_3, (b) fuzzy addition, $A_1 + A_3$ and $A_2 + A_3$, (c) fuzzy multiplication, $A_1 A_3$ and $A_2 A_3$, and (d) fuzzy images $-A_1, -A_2$ and $-A_3$

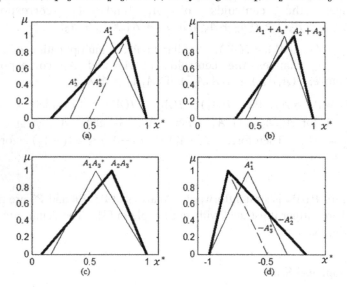

Fig. 5. Normalized fuzzy numbers of fuzzy numbers in Fig.4

The objective of this section is to analyze **P5, P6,** and **P7** empirically. In this paper, $N = 11$ is considered. The experimental results are summarized in Table 1. Columns "$I(A_1^*)$", "$I(A_2^*)$", and "$I(A_3^*)$" show the ordering indexes for A_1, A_2, and A_3, respectively. Columns "$I(A_1 + A_3^*)$" and "$I(A_2 + A_3^*)$" show the ordering indexes for $A_1 + A_3$ and $A_2 + A_3$, respectively. Columns "$I(A_1 A_3^*)$" and "$I(A_2 A_3^*)$" show the ordering indexes for $A_1 A_3$ and $A_2 A_3$, respectively. Columns "$I(-A_1^*)$", "$I(-A_2^*)$" and "$I(-A_3)$" show the ordering indexes for $-A_1$, $-A_2$, and $-A_3$, respectively.

Table 1. Ranking results of the proposed method

$I(A_1^*)$	$I(A_2^*)$	$I(A_3^*)$	$I(A_1 + A_3^*)$	$I(A_2 + A_3^*)$	$I(A_1 A_3^*)$	$I(A_2 A_3^*)$	$I(-A_1^*)$	$I(-A_2^*)$	$I(-A_3^*)$
0.791	0.782	0.872	0.827	0.822	0.687	0.684	-0.791	-0.782	-0.872
$A_2 \lessgtr A_1 \lessgtr A_3$			$A_2 + A_3 \lessgtr A_1 + A_3$		$A_2 A_3 \lessgtr A_1 A_3$		$-A_3 \lessgtr -A_1 \lessgtr -A_2$		

P5, **P6** and **P7** can be observed from Table 1. With the proposed method, $I(A_1^*) = 0.791$, $I(A_2^*) = 0.782$, $I(A_3^*) = 0.872$; as such $A_2 \lessgtr A_1 \lessgtr A_3$; $I(A_1 + A_3^*) = 0.827$, and $I(A_2 + A_3^*) = 0.822$; therefore $I(A_1 + A_3^*) > I(A_2 + A_3^*)$ is satisfied. On the other hand, $I(A_1 A_3^*) = 0.687$ and $I(A_2 A_3^*) = 0.684$; therefore $I(A_1 A_3^*) > I(A_2 A_3^*)$ is satisfied. Lastly, $I(-A_1^*) = -0.791$, $I(-A_2^*) = -0.782$ and $I(-A_3^*) = -0.872$; therefore $-A_3 \lessgtr -A_1 \lessgtr -A_2$. In short, **P5**, **P6** and **P7** are satisfied.

5   Concluding Remarks

In this paper, a new fuzzy ranking method is proposed. It constitutes a solution for ranking fuzzy numbers. The proposed method has been empirically analyzed using benchmark examples. The seven reasonable ordering properties i.e., **P1-P7** [5-9], have been satisfied empirically. The rationale and implications of the proposed method have also been analyzed and discussed.

For future work, the proposed method can be extended to measure similarity between fuzzy numbers. Application of the proposed method to decision making [17] will also be investigated.

References

1. Zhang, F., Ignatius, J., Lim, C.P., Zhao, Y.: A new method for ranking fuzzy num-bers and its application to group decision making. Appl. Math. Model. 38, 1563–1582 (2014)
2. Flaig, A., Barner, K.E., Arce, G.R.: Fuzzy ranking: theory and applications. Signal Proces. 80, 1017–1036 (2000)
3. Wang, Y.M., Chin, K.S., Poon, G.K.K., Yang, J.B.: Risk evaluation in failure mode and effects analysis using fuzzy weighted geometric mean. Expert Syst. Appl. 36, 1195–1207 (2009)
4. Lin, F.T.: Fuzzy job-shop scheduling based on ranking level ( ) interval-valued fuzzy numbers. IEEE Trans. Fuzzy Syst. 10, 510–522 (2002)
5. Wang, Z.X., Liu, Y.J., Fan, Z.P., Feng, B.: Ranking LR fuzzy number based on deviation degree. Inform. Sciences 179, 2070–2077 (2009)
6. Wang, Y.M., Luo, Y.: Area ranking of fuzzy numbers based on positive and negative ideal points. Comput. Math. Appl. 58, 1769–1779 (2009)
7. Asady, B.: The revised method of ranking LR fuzzy number based on deviation degree. Expert Syst. Appl. 37, 5056–5060 (2010)
8. Yu, V.F., Chi, H.T.X., Shen, C.: Ranking fuzzy numbers based on epsilon-deviation degree. Appl. Soft Comput. 13, 3621–3627 (2013)

9. Abbasbandy, S., Asady, B.: Ranking of fuzzy numbers by sign distance. Inform. Sciences 176, 2405–2416 (2006)
10. Wang, Z.X., Liu, Y.J., Fan, Z.P., Feng, B.: Ranking LR fuzzy number based on deviation degree. Inform. Sciences 179, 2070–2077 (2009)
11. Chen, S.J., Chen, S.M.: Fuzzy risk analysis based on the ranking of generalized trapezoidal fuzzy numbers. Appl. Intell. 26, 1–11 (2007)
12. Dat, L.Q., Yu, V.F., Chou, S.Y.: An improved ranking method for fuzzy numbers based on the centroid-index. J. Intell. Fuzzy Syst. 14, 413–419 (2012)
13. Deng, Y., Zhenfu, Z., Qi, L.: Ranking fuzzy numbers with an area method using radius of gyration. Comput. Math. Appl. 51, 1127–1136 (2006)
14. Chai, K.C., Tay, K.M., Lim, C.P.: A new fuzzy ranking method using fuzzy prefer-ence relations. In: IEEE International Conference on Fuzzy Systems (FUZZ), pp. 1–5 (2014)
15. Bourke, P.: Calculating the area and centroid of a polygon (1988), http://www.seas.upenn.edu/~sys502/extra_materials/Polygon%20Area%20and%20Centroid.pdf
16. Abbasbandy, S., Hajjari, T.: A new approach for ranking of trapezoidal fuzzy numbers. Comput. Math. Appl. 57, 413–419 (2009)
17. Chen, S.M., Sanguansat, K.: Analyzing fuzzy risk based on a new fuzzy ranking method between generalized fuzzy numbers. Expert Syst. Appl. 38, 2163–2171 (2011)
18. Chai, K.C., Tay, K.M.: A perceptual computing-based approach for peer assessment. In: IEEE International Conference on System of Systems Engineering (SoSE), pp. 1–6 (2014)

Application of Precise-Spike-Driven Rule in Spiking Neural Networks for Optical Character Recognition

Qiang Yu, Sim Kuan Goh, Huajin Tang, and Kay Chen Tan

Department of Electrical and Computing Engineering,
National University of Singapore, 4 Engineering Drive 3, Singapore 117576

Abstract. Recently, the Precise-Spike-Driven (PSD) rule has been proposed to train neurons to associate spatiotemporal spike patterns. The PSD rule is not limited to association, but can also be used for recognition of spike patterns. This paper presents a new approach with the PSD rule for optical character recognition (OCR). A new encoding method is also proposed to convert the external images into spike patterns. This is an essential step for spiking neurons to process the image patterns. The simulation results show that our approach can perform the task well, and the applied PSD rule is also benchmarked with other learning rules on the given task.

Keywords: Spiking neural networks (SNNs), optical character recognition, encoding, learning.

1 Introduction

Spiking neurons have received significant attentions during these several decades as they are more biologically plausible and computationally powerful than traditional neurons such as perceptrons. Like biological neurons, spiking neurons are capable of processing spikes (or called as action potentials), which are believed to be the principal feature of information transmission in neural systems. With these building units of spiking neurons, the brain presents remarkably cognitive abilities such as recognition. However, it still remains unclear about how these spiking neurons are involved in the recognition of external stimuli like images. A system with functional parts of both encoding and learning would be feasible to tackle the problem of recognizing images [1].

Neural coding refers to the scheme by which the information of external stimuli is represented by spikes. Recently, increasing evidence shows that neural systems convey information through the precise timing of spikes, which supports the hypothesis of the temporal code [2]. Examples of experimental observations include the retina [3], the lateral geniculate nucleus [4] and the visual cortex [5]. According to the temporal code, the complex external stimuli can thus drive different sensory neurons to spike at different time, with this fact forming a spatiotemporal representation of the external stimuli. However, due to the complexity of

© Springer International Publishing Switzerland 2015

H. Handa et al. (eds.), *Proc. of the 18th Asia Pacific Symp. on Intell. & Evol. Systems – Vol. 2,*
Proceedings in Adaptation, Learning and Optimization 2, DOI: 10.1007/978-3-319-13356-0_6

processing the spatiotemporal spike patterns, the development of efficient learning algorithms is demanded.

Many learning algorithms [6–11] have been proposed for spiking neurons to process spatiotemporal spike patterns. The tempotron rule [6] has shown how a neuron can give a binary response to precise-spike spatiotemporal patterns. However, the tempotron is unable to produce precise-timing spikes that can be further served as the information-carrying input for subsequent neurons. By contrast, the SpikeProp rule [7] can train the neuron to perform a spatiotemporal classification by emitting single spikes at the desired firing time. Nevertheless, in its original form, the SpikeProp rule cannot train the neuron to reproduce multiple-spike trains. Several learning algorithms, such as ReSuMe [8], Chronotron [9], SPAN [10] and PSD [11], have been proposed to address this issue. Among these four rules, without complex error calculation, the PSD rule is simple and efficient from the computational point of view, and yet biologically plausible [11]. In the classification of spatiotemporal patterns, the PSD rule can even outperform the efficient tempotron rule [11]. Moreover, the PSD rule is not limited to the classification, but can also train the neuron to associate the spatiotemporal spike patterns with the desired spike trains.

In order to apply the temporal learning rules, a proper encoding method is required for converting the external stimuli into spike patterns. In this paper, we present a new encoding method for encoding the patterns. The PSD rule is then used to train neurons to learn the encoded spike patterns. The system with both the encoding and learning are consistently integrated for an optical character recognition (OCR) task.

The rest of this paper is organized as follows. In section 2, detailed descriptions are presented about the methods used in our integrated system, including the sensory encoding method and the PSD learning rule. Simulation results and discussions are given in section 3, and followed by a conclusion in section 4.

2   Methods

In this section, the whole system for recognition is described, including encoding and learning. The encoding neurons are used to generate spatiotemporal spike patterns that represent the external stimuli. The learning neurons focus on recognizing each input stimulus.

2.1   Neural Encoding Method

Neural encoding considers how to generate a set of specific activity patterns that represent the information of external stimuli. The specific activity patterns considered in this paper are in a spatiotemporal form where precise timing of spikes is used for carrying information.

An increasing body of evidence shows that action potentials are related to the phases of the intrinsic subthreshold membrane potential oscillations ($SMOs$) [12–14]. These observations support the hypothesis of a phase code [15–17]. Such a coding method can encode and retain information with high spatial and temporal selectivity [15]. Following the coding methods presented in [15, 17], we propose a simple phase encoding method. Our encoding mechanism is presented in Figure 1.

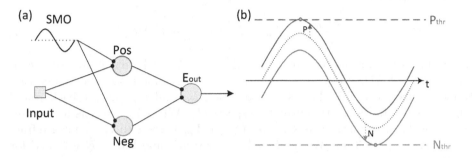

Fig. 1. Illustration of the phase encoding method. (a) shows the structure of an encoding unit. Each encoding unit contains a positive neuron (Pos), a negative neuron (Neg) and an output neuron (E_{out}). The encoding unit receives signals from an input and a subthreshold membrane potential oscillation (SMO). (b) shows the dynamics of the encoding. A positive (negative) input will drive the membrane potential upwards (downwards) from the SMO. Whenever the membrane potential crosses the threshold (P_{thr} or N_{thr}), the neuron (Pos or Neg) will fire. The firing of either the Pos neuron or the Neg neuron will immediately trigger the firing of the E_{out} neuron.

Each encoding unit contains a positive neuron (Pos), a negative neuron (Neg) and an output neuron (E_{out}). Each encoding unit is connected to an input signal and a SMO. A positive (negative) input will cause an upward (downward) shift from the SMO. The firing of either the Pos neuron or the Neg neuron will immediately cause the firing of the E_{out} neuron. The SMO for the i-th encoding unit is described as:

$$SMO_i = M\cos(\omega t + \phi_i) \tag{1}$$

where M is the magnitude of the SMO, ω is the phase angular velocity and ϕ_i is the initial phase. ϕ_i is defined as:

$$\phi_i = \phi_0 + (i-1) \cdot \Delta\phi \tag{2}$$

where ϕ_0 is the reference phase and $\Delta\phi$ is the phase difference between nearby encoding units. We set $\Delta\phi = 2\pi/N_{en}$ where N_{en} is the number of encoding units.

2.2   The PSD Rule

The PSD rule [11] is recently proposed for processing spatiotemporal spike patterns. This rule is not only able to train the neurons to associate spatiotemporal spike patterns with desired spike trains, but also able to train the neurons to perform the classification of spatiotemporal patterns. As the PSD rule is simple and efficient, we use it to train the learning neurons for recognition.

We choose the leaky integrate-and-fire (LIF) model to be the neuron model of our learning neurons. The dynamics of each neuron evolves according to the following equation:

$$\tau_m \frac{dV_m}{dt} = -(V_m - E) + (I_{syn} + I_{ns})R_m \qquad (3)$$

where V_m is the membrane potential, $\tau_m = R_m C_m$ is the membrane time constant, $R_m = 1\ M\Omega$ and $C_m = 10\ nF$ are the membrane resistance and capacitance, respectively, E is the resting potential, I_{ns} and I_{syn} are the background current noise and synaptic current, respectively. When V_m exceeds a constant threshold V_{thr}, the neuron is said to fire, and V_m is reset to V_{reset} for a refractory period $t_{ref} = 3\ ms$. We set $E = V_{reset} = 0\ mV$ and $V_{thr} = E + 18\ mV$ for clarity.

The synaptic current is in the form of:

$$I_{syn}(t) = \sum_i w_i \cdot I_{PSC}^i(t) \qquad (4)$$

where w_i is the synaptic weight of the i-th afferent neuron, and I_{PSC}^i is the un-weighted postsynaptic current (PSC) from the corresponding afferent.

$$I_{PSC}^i(t) = \sum_{t^j} K(t - t^j)H(t - t^j) \qquad (5)$$

where t^j is the time of the j-th spike from the i-th afferent neuron. $H(t)$ is the Heaviside function, and K denotes a normalized kernel as:

$$K(t - t^j) = V_0 \cdot \left(\exp(\frac{-(t - t^j)}{\tau_s}) - \exp(\frac{-(t - t^j)}{\tau_f}) \right) \qquad (6)$$

where V_0 is a normalization factor such that the maximum value of the kernel is 1, τ_s and τ_f are the slow and fast decay constants respectively, and their ratio is fixed at $\tau_s/\tau_f = 4$. We choose $\tau_s = 10\ ms$ in this paper.

The PSD rule is in the form of:

$$\frac{dw_i(t)}{dt} = \eta[s_d(t) - s_o(t)]I_{PSC}^i(t) \qquad (7)$$

where η is the learning rate. $s_d(t)$ and $s_o(t)$ are the desired and the actual output spike trains respectively. The spike train is described as:

$$s(t) = \Sigma_f \delta(t - t^f) \qquad (8)$$

where t^f is the firing time of the f-th spike, and $\delta(x)$ is the Dirac function: $\delta(x) = 1$ (if $x = 0$) or 0 (otherwise).

By integrating Eq. (7), we get:

$$\Delta w_i = \eta \left[\sum_g \sum_f K(t_d^g - t_i^f) H(t_d^g - t_i^f) - \sum_h \sum_f K(t_o^h - t_i^f) H(t_o^h - t_i^f) \right] \quad (9)$$

This equation describes a trial learning where the weight modification is performed at the end of the pattern presentation. We use Eq. (9) to train the learning neurons.

In order to measure the distance between two spike trains, we used the van Rossum metric [18] but with a different filter function as described in Eq. (6). The spike distance is given as:

$$Dist = \frac{1}{\tau} \int_0^{\infty} [f(t) - g(t)]^2 dt \quad (10)$$

where τ is a free parameter (we set $\tau = 10 \ ms$), $f(t)$ and $g(t)$ are filtered signals of the two spike trains that are considered for distance measurement.

Combined with the previous encoding method, the PSD rule is applied to solve the OCR task.

3   Simulation Results and Discussions

This section presents the performance of our system on a given OCR task. A set of optical characters with images of digits 0-9 is used. Each image has a size of 20×20 black/white (B/W) pixels, and each would be destroyed by a reversal noise where each pixel is randomly reversed with a probability denoted as the noise level. Some clean and noisy samples are demonstrated in Figure 2.

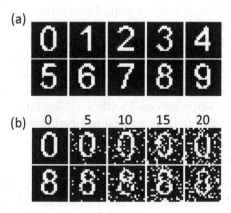

Fig. 2. Illustration of the OCR samples. (a) shows the template images. (b) shows image samples with different levels of reversal noise.

The phase encoding method illustrated in Figure 1 is used to convert the digit images into spatiotemporal spike patterns. Each pixel acts as an input to each encoding unit, with a W/B pixel causing a positive/negative shift from the SMO. Through a fine tuning of the values of M, P/N and P_{thr}/N_{thr}, we set the encoded spikes to occur at peaks of the $SMOs$. The number of encoding units is equal to the number of pixels which is 400 here. We set the oscillation period of the $SMOs$ to be 200 ms which corresponds to a frequency of 5 Hz. An encoded sample is demonstrated in Figure 3.

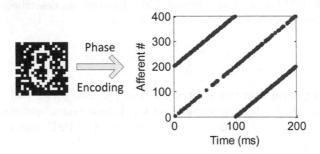

Fig. 3. Demonstration of the phase encoding method with a given sample. Each dot denotes a spike.

We select 10 learning neurons trained by the PSD rule, with each learning neuron corresponding to one category. The learning parameters in the PSD rule are set to be $\eta = 0.06$ and $\tau_s = 10$ ms. All the learning neurons are trained to fire a target spike train with the corresponding category. The target spike train is set to be evenly distributed over the time window T_{max} (200 ms here) with a specified number of spikes n. The firing time of the i-th target spike: $t_i = i/(n+1) \cdot T_{max}$, $i = 1, 2...n$. We choose $n = 4$ by default, otherwise will be stated. For the recognition, a relative confidence criterion is used [11] with the PSD rule, where the incoming pattern is represented by the neuron that fires the most closest spike train to its target spike train.

In our experiments, three noisy scenarios are considered: (1) spike jitter noise where a Gaussian jitter with a standard deviation (denoted as the jitter strength) is added into each encoded spike; (2) reversal noise (as illustrated in Figure 2(b)) where each pixel is randomly reversed with a probability denoted as the noise level; (3) combined noise of both jitter and reversal noises.

3.1   Spike Jitter Noise

In this scenario, the templates of the digit images are firstly encoded into spatiotemporal spike patterns. After that, jitter noises are added to generate noisy patterns. The learning neurons are trained for 100 epochs with a jitter strength of 2 ms. In each learning epoch, a training set of 100 patterns, with 10 for each category, is generated. After training, a jitter range of 0-8 ms is used to investigate the generalization ability. The number of the testing patterns for each jitter strength is set to 200. The PSD rule is applied with different numbers of target spikes (n =1, 2, 4, 6, 8, 10). All the results are averaged over 100 runs.

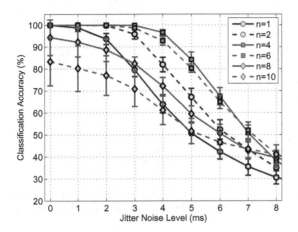

Fig. 4. The performance of the PSD rule with different numbers of target spikes under the case of jitter noise

Figure 4 shows the effects of the number of the target spikes on the learning performance of the PSD rule. As can be seen from Figure 4, when n is low (e.g. 1, 2), the recognition performance is also relatively low. An increasing number of the target spikes can improve the recognition performance significantly (see $n = 1, 2 \rightarrow n = 4, 6$). However, a further increase in the number of target spikes ($n = 6 \rightarrow n = 8, 10$) would reduce the recognition performance. The reasons for this phenomenon are due to the local temporal features associated with each target spike. For small number of target spikes, the neurons make decision based on a relatively less number of temporal features. This small number of features only covers a part range of the whole time window, which inevitably leads to a lower performance compared to a more number of spikes. However, when the number of spikes continues increasing, an interference of local learning processes [8] occurs and increases the difficulty of the learning. Thus, a higher number of spikes normally cannot lead to a better performance due to the interference.

Figure 5 shows the performance of different learning rules for the same classification task. We use a similar approach for the perceptron rule as in [19, 20], where the spatiotemporal spike patterns are transformed into continuous states by a low-pass filter. The target spike trains are separated into bins of size t_{smp}, with $t_{smp} = 2\ ms$ being the sampling time. The target vectors for the perceptron contain values of 0 and 1, with 1 (or 0) corresponding to those bins that contain (or not contain) a target spike in the bin. The input vectors for the perceptron are sampled from the continuous states with t_{smp}. The input pattern will be classified by the winning perceptron that has the closest output vector to the target vector.

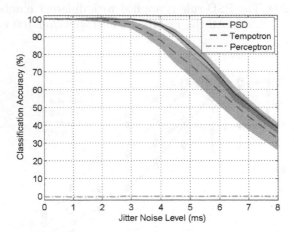

Fig. 5. The performance of different rules under the case of jitter noise. The PSD rule uses $n = 4$ target spikes. The PSD rule outperforms the other two rules in the considered task.

As can be seen from Figure 5, the PSD rule outperforms both the tempotron rule and the perceptron rule. The inferior performance of the perceptron rule can be explained. The complexity of the classification for the perceptron rule depends on the dimensionality of the feature space and the number of input vectors for decisions. A value of $t_{smp} = 2\ ms$ will generate 100 input vectors for each input pattern. These 100 points in 400-dimensional space are to be classified into 1 or 0. This can increase the difficulty for the perceptron rule, let alone considering a large number of input patterns from different categories. Without separating the time window into bins, the spiking neurons by their nature are more powerful than the traditional neurons such as the perceptron. Both the PSD rule and the tempotron rule are better than the perceptron rule. The PSD rule is better than the tempotron rule since the PSD rule makes a decision based on a combination of several local temporal features over the entire time window, but the tempotron rule only makes a decision by firing one spike or not based on one local temporal feature.

3.2   Reversal Noise

In this scenario, the reversal noise is used for generating noisy patterns as illustrated in Figure 2(b). The learning neurons are trained for 100 epochs with a reversal noise level randomly drawn from the range of 0-10% in each learning epoch. Meanwhile, a training set of 100 noisy patterns, with 10 for each category, is generated for each learning epoch. After training, another number of 100 noisy patterns are generated and used to test the generalization ability.

As can be seen from Figure 6, the performances of all the three rules decrease with the increasing noise level. The performance of the PSD rule again outper-

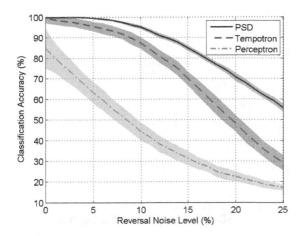

Fig. 6. The performance of different rules under the case of reversal noise. The PSD rule uses $n = 4$ target spikes. The PSD rule outperforms the other two rules even when the noise level is high.

forms the other two rules as in the previous scenario. Spiking neurons trained by the PSD rule can obtain a high classification accuracy (around 85%) even when the reversal noise reaches a high level (15%). The performance of the perceptron rule in this scenario is much better than that in the previous scenario. This is because of the type of the noise. The performance of the perceptron rule is quite susceptible to the changes in state vectors. Every spike of the input spatiotemporal spike patterns in the case of spike jitter noise suffers a change, while in the case of reversal noise, a change only occurs with a probability of the reversal noise level. This is to say, the elements in a filtered state vector have a less chance to change under the reversal noise than that under the jitter noise. Thus, the performance of the perceptron rule under the reversal noise is better than that under the jitter noise.

3.3 Combined Noise

In this scenario, the jitter noise and the reversal noise are combined together to evaluate performance of our system. Again, the learning neurons are trained for 100 epochs. In each epoch, a random reversal noise level chosen from 0-10% is used, as well as a jitter noise level of 2 ms. After training, a reversal noise level of 10% and a jitter noise level of 4 ms is used to investigate the generalization ability.

Figure 7 shows that the combined noise has a stronger impact on the performance than each single noise alone on the performance. This is expected since the effects of the two noises are combined. The perceptron rule still has a poor performance due to the jitter noise. The PSD rule still performs the best with a high average accuracy and a low deviation.

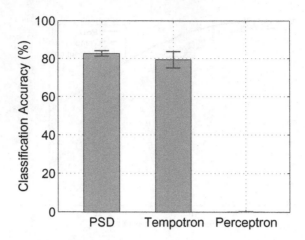

Fig. 7. Robustness of different rules against the combination of the jitter and reversal noises. A 10% reversal noise and a 4 *ms* jitter noise are used for testing.

The results in this section demonstrate the recognition with the applied PSD rule is robust to different noisy sensory inputs. The performance of the PSD rule outperforms both the Tempotron rule and the Perceptron rule for the given OCR task.

4   Conclusion

In this paper, a biologically plausible network of spiking neurons is proposed for optical character recognition. The system is consistently integrated with functional parts of sensory encoding and learning. The system operates in a temporal framework, where the precise timing of spikes is considered for information processing and cognitive computing. The recognition performance of the system is robust to different noisy sensory inputs. Simulation results also show that the applied PSD rule outperforms both the tempotron rule and the perceptron rule for the given OCR task.

Acknowledgments. This work was supported by the Singapore Ministry of Education Academic Research Fund Tier 1 under the project R-263-000-A12-112.

References

1. Yu, Q., Tang, H., Tan, K.C., Li, H.: Rapid feedforward computation by temporal encoding and learning with spiking neurons. IEEE Transactions on Neural Networks and Learning Systems 24(10), 1539–1552 (2013)

2. Panzeri, S., Brunel, N., Logothetis, N.K., Kayser, C.: Sensory neural codes using multiplexed temporal scales. Trends in Neurosciences 33(3), 111–120 (2010)
3. Uzzell, V.J., Chichilnisky, E.J.: Precision of spike trains in primate retinal ganglion cells. Journal of Neurophysiology 92(2), 780–789 (2004)
4. Reinagel, P., Reid, R.C.: Temporal coding of visual information in the thalamus. The Journal of Neuroscience 20(14), 5392–5400 (2000)
5. Mainen, Z.F., Sejnowski, T.J.: Reliability of spike timing in neocortical neurons. Science 268(5216), 1503–1506 (1995)
6. Gütig, R., Sompolinsky, H.: The tempotron: a neuron that learns spike timing-based decisions. Nature Neuroscience 9(3), 420–428 (2006)
7. Bohte, S.M., Kok, J.N., La Poutre, H.: Error-backpropagation in temporally encoded networks of spiking neurons. Neurocomputing 48(1), 17–37 (2002)
8. Ponulak, F., Kasinski, A.: Supervised learning in spiking neural networks with ReSuMe: sequence learning, classification, and spike shifting. Neural Computation 22(2), 467–510 (2010)
9. Florian, R.V.: The chronotron: a neuron that learns to fire temporally precise spike patterns. PloS One 7(8), e40233 (2012)
10. Mohemmed, A., Schliebs, S., Matsuda, S., Kasabov, N.: Span: Spike pattern association neuron for learning spatio-temporal spike patterns. International Journal of Neural Systems 22(04) (2012)
11. Yu, Q., Tang, H., Tan, K.C., Li, H.: Precise-Spike-Driven Synaptic Plasticity: Learning Hetero-Association of Spatiotemporal Spike Patterns. PloS One 8(11), e78318 (2013)
12. Llinas, R.R., Grace, A.A., Yarom, Y.: In vitro neurons in mammalian cortical layer 4 exhibit intrinsic oscillatory activity in the 10-to 50-Hz frequency range. Proceedings of the National Academy of Sciences 88(3), 897–901 (1991)
13. Jacobs, J., Kahana, M.J., Ekstrom, A.D., Fried, I.: Brain oscillations control timing of single-neuron activity in humans. The Journal of neuroscience 27(14), 3839–3844 (2007)
14. Koepsell, K., Wang, X., Vaingankar, V., Wei, Y., Wang, Q., Rathbun, D.L., Sommer, F.T.: Retinal oscillations carry visual information to cortex. Frontiers in Systems Neuroscience 3 (2009)
15. Nadasdy, Z.: Information encoding and reconstruction from the phase of action potentials. Frontiers in Systems Neuroscience 3 (2009)
16. Kayser, C., Montemurro, M.A., Logothetis, N.K., Panzeri, S.: Spike-phase coding boosts and stabilizes information carried by spatial and temporal spike patterns. Neuron 61(4), 597–608 (2009)
17. Hu, J., Tang, H., Tan, K.C., Li, H., Shi, L.: A spike-timing-based integrated model for pattern recognition. Neural Computation 25(2), 450–472 (2013)
18. Rossum, M.: A novel spike distance. Neural Computation 13(4), 751–763 (2001)
19. Maass, W., Natschläger, T., Markram, H.: Real-time computing without stable states: A new framework for neural computation based on perturbations. Neural Computation 14(11), 2531–2560 (2002)
20. Xu, Y., Zeng, X., Zhong, S.: A new supervised learning algorithm for spiking neurons. Neural Computation 25(6), 1472–1511 (2013)
21. Bair, W., Koch, C.: Temporal precision of spike trains in extrastriate cortex of the behaving macaque monkey. Neural Computation 8(6), 1185–1202 (1996)
22. VanRullen, R., Guyonneau, R., Thorpe, S.J.: Spike times make sense. Trends in Neurosciences 28(1), 1-4 (2005)

A Hybrid Differential Evolution Algorithm – Game Theory for the Berth Allocation Problem

Nasser R. Sabar, Siang Yew Chong, and Graham Kendall

The University of Nottingham Malaysia Campus,
Jalan Broga, 43500 Semenyih, Selangor, Malaysia
{Nasser.Sabar,siang-yew.chong,Graham.Kendall}@nottingham.edu.my

Abstract. The berth allocation problem (BAP) is an important and challenging problem in the maritime transportation industry. BAP can be defined as the problem of assigning a berth position and service time to a given set of vessels while ensuring that all BAP constraints are respected. The goal is to minimize the total waiting time of all vessels. In this paper, we propose a differential evolution (DE) algorithm for the BAP. DE is a nature-inspired meta-heuristic that has been shown to be an effective method to addresses continuous optimization problems. It involves a population of solutions that undergo the process of selection and variation. In DE, the mutation operator is considered the main variation operator responsible for generating new solutions. Several mutation operators have been proposed and they have shown that different operators are more suitable for different problem instances and even different stages in the search process. In this paper, we propose an enhanced DE that utilizes several mutation operators and employs game theory to control the selection of mutation operators during the search process. The BAP benchmark instances that have been used by other researchers are used to assess the performance of the proposed algorithm. Our experimental results reveal that the proposed DE can obtain competitive results with less computational time compared to existing algorithms for all tested problem instances.

Keywords: Differential evolution, berth allocation problem, meta-heuristics, optimization.

1 Introduction

Maritime transportation has experienced a tremendous growth of container usage over the last two decades [1], [2]. Port managers face great challenges in providing effective and efficient services. The berth allocation problem (BAP) is one of the main challenges confronting port managers. Providing an efficient solution to the BAP plays an important role in improving port effectiveness [2]. BAP seeks to assign, for each vessel, a berth position and service time on the selected berth. The goal is to minimize the total waiting time of all vessels as far as possible [1].

BAP is an NP-hard problem [1]. Small instances can be solved optimality using exact methods. However, they become impractical as the size of the instances in-

© Springer International Publishing Switzerland 2015

H. Handa et al. (eds.), *Proc. of the 18th Asia Pacific Symp. on Intell. & Evol. Systems – Vol. 2*,
Proceedings in Adaptation, Learning and Optimization 2, DOI: 10.1007/978-3-319-13356-0_7

creases [2]. As such, researchers have utilized meta-heuristic algorithms to deal with large-scale instances as they can often provide good quality solutions within realistic computational times. Examples of meta-heuristic algorithms being utilized for BAP include: tabu search [1], clustering search [3] particle swarm optimization [4] and hybrid column generation approach [5].

In this paper, we propose a differential evolution (DE) algorithm for the BAP. DE is a nature-inspired population-based meta-heuristic that has been demonstrated to be efficient and effective for many hard continuous optimization problems. DE operates on a population of solutions and iteratively improves them. In each iterative step or generation, a new solution is generated using two variation operators: mutation and crossover. The mutation operator in DE is considered the primary variation operator and several mutation operators have been introduced. However, it is not known in advance which operator should be used as different operators work well for different problem instances with different characteristics and in different stages of the search process [6], [7], [8], [9]. In this paper, we propose an enhanced DE that utilizes a variation operator with multiple mutation operators for the BAP. We utilize game theory to provide a mechanism to control the selection of mutation operators throughout the search process of DE. The performance of the proposed algorithm has been assessed using the existing BAP benchmark instances [1]. The experimental results reveal that the proposed algorithm can obtain competitive results with less computational time when compared with existing algorithms.

2   Problem Description

Bierwirth and Meisel [2] has classified BAP into two types according to the berth type and the vessel's arrival time. The berth type is categorized as discrete if the quay has been partitioned into a set of berth sections and continuous if the quay is not partitioned. The vessel's arrival time is categorized as dynamic if the vessels can arrive at any time during the planning horizon, and static if all vessels have arrived at the port before the berth planning begins. We focus on the discrete dynamic BAP [2], [10]. For this BAP, there are a set of berth sections with predefined lengths and a set of vessels. Each vessel has an arrival time, priority, vessel length and handling time [2]. Some of them can be allocated to any berths based on vessel lengths while others can only be allocated to a subset of berths. The vessel handling time is different from one berth to another. The overall goal is to allocate for each vessel a berth section and service time (berthing time) on the allocated berth while respecting the following constraints [1]:

- Each vessel is allocated to exactly one berth.
- There is no more than one vessel allocated to the same berth at the same time (same service time).
- Each berth can handle at most one vessel at any given time.

The main role of the optimization algorithm is to minimize the total waiting time of all vessels which is calculated as follows (objective function) [1]:

$$\min \sum_{i \in n} \ \sum_{k \in m} v_i \left| T_i^k - a_i + + t_i^k \sum_{j \in n} x_{ij}^k \right| \qquad (1)$$

where

- n : number of vessels
- m : number of berths
- v_i : the priority of vessel i
- T_i^k : the berting time of a vessel i at berth k.
- a_i : the arrival time of vessel i.
- t_i^k : the handling time of vessel i at berth k.
- x_{ij}^k : decision variable, $x_{ij}^k = 1$ if vessel j is serviced by berth k after the vessel i and $x_{ij}^k = 0$ otherwise.

3   Proposed Methodology

In this section, we first present the basic DE algorithm that is followed by the proposed approach to enhance DE for the BAP problem.

3.1   Basic Differential Evolution Algorithm

The DE algorithm was proposed in [11] to deal with continuous optimization problems (real-valued fitness functions). It belongs to a class of nature-inspired, population-based meta-heuristic algorithms [12]. A general DE algorithm starts with a population of solutions and then applies evolutionary operators of variation (mutation and crossover) and selection to improve the population of solutions iteratively over a certain number of generations. For every solution in the population, DE generates a new solution using the mutation operator that randomly selects three different solutions from the current population and combines them according to a prescribed operation. The new generated solution is then combined with the parent solution using the crossover operator to generate an offspring. The selection step first calculates the fitness of the offspring and then replaces it with the parent solution if it has a better fitness. This process is repeated for a predefined number of generations.

Over the years, many DE variants that use different mutation or crossover operators have been proposed. A general DE scheme use the notation *DE/x/y/z*, where *x* represents the base solution to be mutated, y defines the number of different solutions to be used to perturb x, and z denotes the crossover type, *binomial* or *exponential* [11], [12]. A well-known DE variant is the "*DE/rand/1/bin*", where "*DE*" is Differential Evolution, "*rand*" means the solutions will be randomly selected, "1" indicates the number of pairs of solutions and "*bin*" indicates that binomial crossover will be used. The basic steps of the *DE/rand/1/bin* are as follows [11]:

Step 1: Randomly generate a population of solutions, *NP*.

Step 2: Calculate the fitness, f, of the population.

Step 3: For each parent solution (x_i^G) in the current population *NP* (i is the solution index and G is the current generation) generate a new solution (m_i^G) using (2):

$$m_{i,j}^G = x_{1,j}^G + F * (x_{2,j}^G - x_{3,j}^G), \forall j \in \{1,...,n\} \tag{2}$$

where j represents the index of current decision variable, n is the maximum number of decision variables in a given problem instance, F is the scaling factor ($F \in [0, 1]$) and x_1^G, x_2^G and x_3^G are three randomly chosen solutions from the current population where $x_1^G \neq x_2^G \neq x_3^G$.

Step 4: Apply the crossover operator to combine the solution (m_i^G) generated by the mutation operator with the parent solution (x_i^G) based on the crossover rate *CR* ($CR \in [0, 1]$) in order to generate a new offspring (m_i^{G+1}) as follows (3):

$$m_{i,j}^{G+1} = \begin{cases} m_{i,j}^G & if \ Rand(j) \leq CR \ or \ j = Rnd(i) \\ x_{i,j}^G & if \ Rand(j) > CR \ and \ j \neq Rnd(i) \end{cases} \tag{3}$$

$$\forall j \in \{1,...,n\}, \forall i \in \{1,...,| \ NP \ |\}$$

where *Rand(j)* is a random number (*Rand(j)* $\in [0, 1]$) selected for the jth decision variable, *Rnd(i)* is a random decision variable index (*Rnd(i)* $\in \{1,..., n\}$). *Rnd* ensure that m_i^{G+1} gets at least one decision variable from m_i^G.

Step 5: Calculate the fitness of m_i^{G+1} and compare it with x_i^G. Replace x_i^G with m_i^{G+1} if m_i^{G+1} fitness is better than x_i^G as follows (4):

$$x_i^{G+1} = \begin{cases} m_i^{G+1} & if \ f(m_i^{G+1}) \leq f(x_i^G) \\ x_i^G & if \ f(m_i^{G+1}) > f(x_i^G) \end{cases} \tag{4}$$

$$\forall i \in \{1,...,| \ NP \ |\}$$

Step 6: If the termination criterion is satisfied (the number of generations), stop and return the best solution. Otherwise, go to *Step 3*.

4   The Proposed Algorithm

The variety of problem instances having different characteristics, or those with complex structures (e.g., different sections having different structures), makes it challenging to know in advance the best variation operator to use (i.e., blackbox optimization) [12]. Each one has its own strength and weakness and may work well for certain instances or at a certain stages in the search process. Consequently, several DE frameworks that utilize a set of mutation operators have been proposed [6], [7], [8], [9]. These frameworks seek to combine the strength of several mutation operators in one framework that can effectively solve the given problem instances [13], [14], [15]. In this paper, we propose a DE algorithm that utilizes several mutation operators to solve the BAP. The proposed DE makes use of game theory in the design of the selection mechanism that chooses the operator to be used at the current search stage. In the following, we first describe the solution representation and the population generation method, and later the utilized mutation operators and proposed selection mechanism.

4.1   Solution Representation and the Population Generation Method

DE was originally proposed to solve continuous optimization problems [12]. To deal with combinatorial optimization problems such as BAP, a suitable solution representation or a decoding scheme is needed to convert the real numbers into integers [12]. In BAP, each vessel has to be assigned to a berth section and being given a service time on the selected berth. This implies that we need to deal with the assignment problem that is responsible for assigning for each vessel a berth section and the scheduling problem that assigns a service time for each vessel. In this paper, we avoid modifying the DE mutation operator by having a decimal representation of BAP solutions $a_0.a_1a_2...$ where the integer part a_0 represents assigned berth sections while the fraction parts $a_1a_2...$ represent the order of this vessel on this berth [16]. Figure 1 shows an example of BAP solution representation. Here, an instance of BAP has 6 vessels ($n=6$) that are needed to be assigned to 3 berths. If we assign numbers for the vessels from 1 to 6, the decision variables will be (1, 2, 3, 4, 5, 6), as shown in the first row of Table 1. Next, we generate for each decision variable a random number r ($r \in [1, 3]$)), as shown in second row of Table 1.

Table 1 can be decoded into a BAP solution as follows: vessel 1 is assigned to the second berth and is second in order on this berth, vessel 2 is assigned to the first berth and is first in order on this berth, vessel 3 is assigned to the first berth and is second in order on this berth, and so on. Next, on each berth we sort the assigned vessels in an ascending order based on their arrival time. In this paper, the initial population of solutions of DE is randomly generated by assigning each decision variable a random value between 1 and the maximum number of vessels in a given problem instance. The generated solutions are assigned fitness values using equation (1).

Table 1. BAP solution representation

Vessel index	1	2	3	4	5	6
Decision variables	2.2	1.1	1.2	3.2	3.1	2.1

4.2 Mutation Operators and the Selection Mechanism

In this paper, the proposed DE makes use of the following mutation operators [11], [12]:

- M_1: DE/rand/1/bin, $m_i = x_1 + F * (x_2 - x_3)$
- M_2: DE /best /1/bin, $m_i = x_{best} + F * (x_1 - x_2)$
- M_3: DE /parent-to- best/1, $m_i = x_i + F * (x_{best} - x_i) + F * (x_1 - x_2)$
- M_4: DE /best /2/ bin, $m_i = x_{best} + F * (x_1 - x_2) + F * (x_3 - x_4)$

where *"parent"* (x_i) is the parent solution to be perturbed, *"best"* indicates the best solution in the population and x_1, x_2, x_3 and x_4 are randomly selected solutions from the current population where $x_1 \neq x_2 \neq x_3 \neq x_4 \neq x_i \neq x_{best}$

In this paper, each parent solution is associated with a set of mutation operators and, at each generation, one of them is selected to generate a new solution. We employ a game theoretic concept to design the selection mechanism of the mutation operators for each parent. Each strategy is associated with a profile that keeps the history of the strategy performance. As in [17], we model the selection mechanism in the context of a two-player game. The DE population of solutions is divided into two to play two-player game. Each solution represents a player with a set of mutation operators $(M_1, M_2, M_3$ and $M_4)$ representing strategies that can be played. Each strategy will be assigned a payoff representing the improvement obtained by the selected strategy. According to the strategy probability distribution, each player selects one strategy to play against another strategy. Based on the obtained result, the player will update the strategy profile and the payoff. In this paper, the strategy profile keeps the accumulated payoff of each one and it is updated at every generation. The payoff of each strategy is calculated as follows: let S [] be the array of the probability of selecting the strategy, f_p and f_n represents the fitness values of the parent and generated solutions, NS represents the number of strategies. Then, if the application of the i-th strategy improves the fitness value of the parent solution, the payoff of the i-th strategy is updated as follows: $S[i]=S[i]+\Delta$ where $\Delta=(f_p - f_n)/ (f_p + f_n)$, $\forall j \in \{1,...,NS\}$ and $j \neq i$, $S[j]=S[j]-(\Delta/(NS-1))$. Otherwise (if the solution cannot be improved), $S[i]=S[i]-|(\Delta*\alpha)|$ where $\alpha=Current\_Genetraion/Total\_Generations$, $\forall j \in \{1,...,NS\}$ and $j \neq i$, $S[j]=S[j]+(|\Delta|*\alpha/(NS-1))$. Initially, the selection probability of each strategy is set to $1/NS$.

5   Experimental Setup

The BAP benchmark instances that have been introduced in [1] are used to validate the performance of the proposed algorithm. The benchmark involves 30 different instances (denoted as $i1$ to $i30$); each instance has 30 vessels and 13 berths [1]. In all instances, the vessel characteristics such as length and the arrival time as well as berth lengths are different. Table 2 shows the parameter settings of proposed algorithm. These settings were determined based on preliminary experiments. In this paper, we executed the proposed algorithm 31 times for each instance using different random seeds.

Table 2. The Parameter Settings

#	Parameter	Value
1	No. Of generations	500
2	Population size, NP	20
3	Scaling Factor, F	0.1
4	Crossover Rate, CR	0.4

6   The Computational Results

We have carried out two types of experiments. The goal of the first one is to evaluate the impact of the proposed multi-mutation operators on the performance of DE in solving BAP through a comparison with a standard, baseline DE (*DE/rand/1/bin*). The goal of second experiment is to compare the results of the proposed algorithm against the state of the art algorithms.

6.1   The Computational Comparisons of DE with and without the Game Theory Concept

In this section, we compare the computational results of DE with and without the game theory concept (denoted as DEGT and DE, respectively) using the same parameter settings, stopping condition and computer resources. Both algorithms (DEGT and DE) are executed for 31 independent runs and the results are compared using the Wilcoxon statistical test with a significance level of 0.05. The p-value of DEGT against DE for all instances is presented in Table 3. In this table, "+" indicates DEGT is statistically better than DE (p-value < 0.05), "-" indicates DE is statistically better than DEGT (p-value > 0.05), and "=" indicates both DEGT and DE have the same performance (p-value = 0.05). As Table 3 reflects, DEGT is statistically better than DE on 21 instances and preforms the same as DE on 2 out of 30 tested instances. This table also reveals that on 7 out of 30 tested instances, DEGT is not statistically better than DE. Although the results show that DEGT is not statistically better than DE on all tested instances, the overall finding justifies the benefit of integrating the game theory concept with DE algorithm. Indeed, the use of the game theory concept can effectively enhance the performance of DE to obtain very good results for all tested instances.

Table 3. The p-value of DEGT compared to DE

DEGT vs.	DE
Instance	*p*-value
i01	-
i02	-
i03	-
i04	-
i05	=
i06	=
i07	-
i08	-
i09	+
i10	+
i11	+
i12	-
i13	+
i14	+
i15	+
i16	+
i17	+
i18	+
i19	+
i20	+
i21	+
i22	+
i23	+
i24	+
i25	+
i26	+
i27	+
i28	+
i29	+
i30	+

6.2   The Computational Comparisons of DEGT with State of the Art Algorithms

In this section, we compare the computational results of DEGT with the current state of the art algorithms. The algorithms that we compare against are:

- Generalized set partition programming (GSPP) [18].
- Tabu search (TS) algorithm [1].
- Column generation (CG) algorithm [5].
- Clustering search (CS) [3].
- Particle swarm optimization (PSO) [4].

Table 4 gives the results of DEGT over 31 runs as well as the compared algorithms. For each instance, we present the best obtained results (best objective value) and the computational time (seconds) obtained by DEGT and the compared algorithms. In Table 4, the third column (Opt.) indicates the optimal value for each instance [18], the last row represents the average overall instances (Avg.) and boldfont indicates the best obtained results.

Table 4. The results of DEGT compared to the state of the art methods

Inst.	DEGT		GSPP		TS	CG		CS		PSO	
	Best	Time	Opt.	Time	Best	Best	Time	Best	Time	Best	Time
i01	**1409**	10.2	**1409**	17.92	1415	**1409**	74.61	**1409**	12.47	**1409**	11.11
i02	**1261**	6.4	**1261**	15.77	1263	**1261**	60.75	**1261**	12.59	**1261**	7.89
i03	**1129**	7.1	**1129**	13.54	1139	**1129**	135.45	**1129**	12.64	**1129**	7.48
i04	**1302**	6.01	**1302**	14.48	1303	**1302**	110.17	**1302**	12.59	**1302**	6.03
i05	**1207**	4.2	**1207**	17.21	1208	**1207**	124.7	**1207**	12.68	**1207**	5.84
i06	**1261**	7.4	**1261**	13.85	1262	**1261**	78.34	**1261**	12.56	**1261**	7.67
i07	**1279**	6.5	**1279**	14.6	**1279**	**1279**	114.2	**1279**	12.63	**1279**	7.5
i08	**1299**	8.9	**1299**	14.21	**1299**	**1299**	57.06	**1299**	12.57	**1299**	9.94
i09	**1444**	3.8	**1444**	16.51	**1444**	**1444**	96.47	**1444**	12.58	**1444**	4.25
i10	**1213**	4.4	**1213**	14.16	**1213**	**1213**	99.41	**1213**	12.61	**1213**	5.2
i11	**1368**	7.2	**1368**	14.13	1378	1369	99.34	**1368**	12.58	**1368**	10.52
i12	**1325**	10.3	**1325**	15.6	**1325**	**1325**	80.69	**1325**	12.56	**1325**	12.92
i13	**1360**	10.7	**1360**	13.87	**1360**	**1360**	89.94	**1360**	12.61	**1360**	11.97
i14	**1233**	6.1	**1233**	15.6	**1233**	**1233**	73.95	**1233**	12.67	**1233**	7.11
i15	**1295**	5.7	**1295**	13.52	**1295**	**1295**	74.19	**1295**	13.8	**1295**	8.3
i16	**1364**	6.8	**1364**	13.68	1375	1365	170.36	**1364**	14.46	**1364**	8.48
i17	**1283**	4.6	**1283**	13.37	**1283**	**1283**	46.58	**1283**	13.73	**1283**	5.66
i18	**1345**	6.2	**1345**	13.51	1346	**1345**	84.02	**1345**	12.72	**1345**	8.02
i19	**1367**	9.6	**1367**	14.59	1370	**1367**	123.19	**1367**	13.39	**1367**	11.42
i20	**1328**	10.4	**1328**	16.64	**1328**	**1328**	82.3	**1328**	12.82	**1328**	12.28
i21	**1341**	6.5	**1341**	13.37	1346	**1341**	108.08	**1341**	12.68	**1341**	7.11
i22	**1326**	5.7	**1326**	15.24	1332	**1326**	105.38	**1326**	12.62	**1326**	7.94
i23	**1266**	6.7	**1266**	13.65	**1266**	**1266**	43.72	**1266**	12.62	**1266**	7.25
i24	**1260**	4.3	**1260**	15.58	1261	**1260**	78.91	**1260**	12.64	**1260**	5.67
i25	**1376**	6.2	**1376**	15.8	1379	**1376**	96.58	**1376**	12.62	**1376**	7.13
i26	**1318**	5.8	**1318**	15.38	1330	**1318**	101.11	**1318**	12.62	**1318**	7.44
i27	**1261**	4.0	**1261**	15.52	**1261**	**1261**	82.86	**1261**	12.64	**1261**	6.16
i28	**1359**	9.8	**1359**	16.22	1365	1360	52.91	**1359**	12.71	**1359**	11.52
i29	**1280**	7.1	**1280**	15.3	1282	**1280**	203.36	**1280**	12.62	**1280**	8.11
i30	**1344**	5.4	**1344**	16.52	1351	**1344**	71.02	**1344**	12.58	**1344**	7.13
Avg	**1306.8**	6.80	**1306.8**	14.98	1309.7	1306.9	93.99	**1306.8**	12.79	**1306.8**	8.17

As can be seen from Table 4, DEGT obtained the optimal values for all tested instances. In particular, DEGT best results are the same as those produced by the GSPP. With respect to individual comparisons, DEGT best results are the same as CS, PSO, TS and CG on 30, 30 18 and 24 out of 30 tested instances, respectively. DEGT obtained better results than TS on 18 and CG on 4 instances. In addition, the average result of all instances (last row in Table 4) of DEGT is better (or the same) than the compared algorithms.

As for the computational time comparisons, Table 4 shows that, on all tested instances, the computational time of DEGT is lower than GSPP, CS, PSO, TS and CG. The overall results demonstrate that DEGT is an effective and efficient algorithm for the BAP as it obtained very good results for all tested instances within a small computational time when compared to previously reported algorithms.

7   Conclusion

In this paper, we have presented a DE algorithm to solve the BAP. DE is a population based algorithm that seeks to improve the population of solutions through the use of mutation operator(s), crossover operator and selection rule. To further enhance the performance of the DE, we coupled it with a several mutation operators in order to combine strength of different operators in one framework. Game theory is used to control the selection of which mutation operator should be used at any decision point. The computational results are carried out using the existing BAP benchmark instances. The obtained results reveal that the proposed algorithm obtained very good results when compared to DE without game theory as well as the state of the art algorithms. In addition, the computational time of proposed algorithm is lower than the compared algorithms, indicating that proposed algorithm is an effective algorithm for the berth allocation benchmark instances.

References

1. Cordeau, J.-F., Laporte, G., Legato, P., Moccia, L.: Models and tabu search heuristics for the berth-allocation problem. Transportation Science 39(4), 526–538 (2005)
2. Bierwirth, C., Meisel, F.: A survey of berth allocation and quay crane scheduling problems in container terminals. European Journal of Operational Research 202(3), 615–627 (2010)
3. de Oliveira, R.M., Mauri, G.R., Nogueira Lorena, L.A.: Clustering Search for the Berth Allocation Problem. Expert Systems with Applications 39(5), 5499–5505 (2012)
4. Ting, C.-J., Wu, K.-C., Chou, H.: Particle swarm optimization algorithm for the berth allocation problem. Expert Systems with Applications 41(4), 1543–1550 (2014)
5. Mauri, G.R., Oliveira, A.C.M., Lorena, L.A.N.: A hybrid column generation approach for the berth allocation problem. In: van Hemert, J., Cotta, C. (eds.) EvoCOP 2008. LNCS, vol. 4972, pp. 110–122. Springer, Heidelberg (2008)
6. Mallipeddi, R., Suganthan, P.N., Pan, Q.-K., Tasgetiren, M.F.: Differential evolution algorithm with ensemble of parameters and mutation strategies. Applied Soft Computing 11(2), 1679–1696 (2011)
7. Qin, A.K., Suganthan, P.N.: Self-adaptive differential evolution algorithm for numerical optimization. In: The 2005 IEEE Congress on Evolutionary Computation, pp. 1785–1791. IEEE (2005)
8. Brest, J., Bošković, B., Greiner, S., Žumer, V., Maučec, M.S.: Performance comparison of self-adaptive and adaptive differential evolution algorithms. Soft Computing 11(7), 617–629 (2007)
9. Zhang, J., Sanderson, A.C.: JADE: adaptive differential evolution with optional external archive. IEEE Transactions on Evolutionary Computation 13(5), 945–958 (2009)
10. Sabar, N.R., Kendall, G., Ayob, M.: An Exponential Monte-Carlo Local Search Algorithm for the Berth Allocation Problem. In: 10th International Conference on the Practice and Theory of Automated Timetabling (PATAT 2010), York, UK, August 26-29, 2014, pp. 544–548 (2014)
11. Storn, R., Price, K.: Differential evolution–a simple and efficient heuristic for global optimization over continuous spaces. Journal of Global Optimization 11(4), 341–359 (1997)

12. Das, S., Suganthan, P.N.: Differential evolution: A survey of the state-of-the-art. IEEE Transactions on Evolutionary Computation 15(1), 4–31 (2011)
13. Sabar, N.R., Ayob, M., Kendall, G., Rong, Q.: Grammatical Evolution Hyper-Heuristic for Combinatorial Optimization Problems. IEEE Transactions on Evolutionary Computation 17(6), 840–861 (2013), doi:10.1109/TEVC.2013.2281527
14. Sabar, N.R., Ayob, M., Kendall, G., Qu, R.: A Dynamic Multiarmed Bandit-Gene Expression Programming Hyper-Heuristic for Combinatorial Optimization Problems. IEEE Transactions on Cybernetics PP(99), 1 (2014), doi:10.1109/TCYB.2014.2323936
15. Sabar, N.R., Ayob, M., Kendall, G., Qu, R.: The Automatic Design of Hyper-heuristic Framework with Gene Expression Programming for Combinatorial Optimization problems. IEEE Transactions on Evolutionary Computation PP(99), 1 (2014), doi:10.1109/TEVC.2014.2319051
16. Sabar, N.R., Kendall, G.: Aircraft Landing Problem using Hybrid Differential Evolution and Simple Descent Algorithm. Paper presented at the 2014 IEEE Congress on Evolutionary Computation (CEC 2014), pp. 520–527 (2014)

Hybrid Differential Evolution and Gravitational Search Algorithm for Nonconvex Economic Dispatch

Luong D. Le[1], Loc D. Ho[1], Dieu N. Vo[2], and Pandian Vasant[3]

[1] Ho Chi Minh City University of Technology (Hutech), Vietnam
ledinhluong@gmail.com, hdloc@hcmhutech.edu.vn
[2] Department of Power Systems, HCMC University of Technology, Vietnam
vndieu@gmail.com
[3] Department of Fundamental and Applied Sciences, Universiti Teknologi PETRONAS
pvasant@gmail.com

Abstract. The hybrid differential evolution and gravitational search algorithm (DEGSA) to solve economic dispatch (ED) problems with non-convex cost functions is presented in this paper with various generator constraints in power systems. The proposed DEGSA method is an improved differential evolution method based on the gravitational search algorithm scheme. The DEGSA method has the flexible adjustment of the parameters to get a better optimal solution. Moreover, an effective constraint handling framework in the method is employed for properly handling equality and inequality constraints of the problems. The proposed DEGSA has been tested on three systems with 13, 15, 40 units and the obtained results from the DEGSA algorithm have been compared to those from other methods in the literature. The result comparison has indicated that the proposed DEGSA method is more effective than many other methods for obtaining better optimal solution for the test systems. Therefore, the proposed DEGSA is a very favorable method for solving the non-convex ED problems.

Keywords: Differential Evolution, Gravitational Search Algorithm, Hybrid Meta-Heuristic, Nonconvex Economic Dispatch.

1 Introduction

In general, economic dispatch (ED) problem is one of the most important in the operation of power systems. Previous efforts on solving ED problems have employed various mathematical programming methods and optimization techniques such as linear programming, quadratic programming, gradient based method, Lagrange relaxation, etc. Many algorithms have bean used to solve the ED problems by approximating the cost function of each generator by a single quadratic function. In practical power system operation conditions, many thermal generating units, especially those units have valve-point effects, prohibited operating zones, have formed a nonlinear ED problem with many local optima and multiple constraints in nature, which prevents the classical methods from obtaining the global optima. This is a

© Springer International Publishing Switzerland 2015

H. Handa et al. (eds.), *Proc. of the 18th Asia Pacific Symp. on Intell. & Evol. Systems – Vol. 2,*
Proceedings in Adaptation, Learning and Optimization 2, DOI: 10.1007/978-3-319-13356-0_8

complicated, real-world and non-convex optimization problem since it contains the discontinuous values at each boundary forming multiple local optimal. Therefore, the classical solution methods have difficulty in dealing with this problem.

In recent years, many methods have been applied to solve the non-convex ED problem such as evolutionary programming (EP) for solving the ED problem with multiple fuel cost function has been discussed in [1, 2], genetic algorithm (GA) for soling the ED problem with many types of fuel cost function, GA for solving the ED problem with valve-point effects was proposed in [3], tabu search algorithm (TSA) for solving the ED problem with multiple minima [4], TSA for solving the ED problem consider valve-point effects was proposed in [5], simulated annealing (SA) and the hybrid GA/SA for dealing with classical ED problems [6,7], particle swarm optimization (PSO) with improvements and a new PSO hybrid with local search [8], etc. However, these methods have large number of iterations to achieve solution and easily affected by the relevant control parameters. One powerful algorithm from evolutionary computation due to its excellent convergence characteristics and few control parameters is differential evolution. There have been many applications of DE for solving ED problems such as an improved DE (IDE) based on cultural algorithm and diversity measure has been discussed in [9] to solve two problems with valve point effects, a DE algorithm with a specially designed repair operation was proposed in [10] to solve the ED problem with different constraints, a new algorithm by combining a chaotic differential evolution (CDE) and quadratic programming has been discussed in [11] to solve the ELD problem with valve point effects. Recently, a new heuristic search algorithm, namely gravitational search algorithm (GSA), motivated by the gravitational law and laws of motion has been proposed by Rashedi et al. [12], [13]. They have been applied successfully in solving various nonlinear functions. The obtained results confirm the high performance and efficiently of proposed method in these problems. GSA has a flexible and well-balanced mechanism to enhance exploration and exploitation abilities [13].

This paper proposes a hybrid differential evolution and gravitational search algorithm (DEGSA) algorithm by combining the mechanisms of both GSA and DE. We had some experiments on the classical GSA meta-heuristics with slow convergence and the growth of search space dimensionality. In the present work, we have replaced the pitch adjustment operation in original GSA with a mutation strategy borrowed from the realm of the DE algorithms. After that, the memory consideration and the enhanced pitch adjustment operation are both employed to strengthen the exploration ability. Compared with the classical GSA, the use of differential mutation and crossover can enhance the exploitation in the DEGSA. The DEGSA algorithm may inherit elements from as many individuals as its number of dimensions when generating a new individual to enhance the exploration ability. The proposed DEGSA has been tested on three systems including 13, 40 unit system with valve effects, 15 unit system with prohibited operating zones, ramp rate constraints, and power loss. The numerical results from the proposed method are compared to those from many other methods in the literature.

2   Nonconvex Economic Dispatch Problem

2.1   Classical Economic Dispatch Problem

The main purpose of the ED problem is to minimize the total fuel cost subject to the constraints of a power system. This problem is formulated as follows:

$$C = Min \sum_{i=1}^{N} F_i(P_i) \tag{1}$$

$$F(x) = \sum_{i-1}^{N_G} (a_i + b_i P_{Gi} + c_i P_{Gi}^2) \tag{2}$$

where $F_i(P_i)$ is the problem objective and the fuel cost function of the ith unit, a_i, b_i, c_i are the fuel cost coefficients of the ith unit, N_G is the number of generators, and P_i is the power generated by the ith unit.

subject to

- Power balance constraints

$$\sum_{i=1}^{N} P_i = P_D + P_{Loss} \tag{3}$$

where P_D is the system load demand and P_{Loss} is the transmission loss. The power loss can be approximately calculated by using Kron's loss formula:

$$P_L = \sum_{i=1}^{N_G} \sum_{j=1}^{N_G} P_{Gi} B_{ij} P_{Gj} + \sum_{i=1}^{N_G} P_{Gi} B_{i0} + B_{00} \tag{4}$$

where B_{ij}, B_{i0}, B_{00} are power loss coefficients or B-coefficients.

- Generating capacity limits

$$P_{i,min} \leq P_i \leq P_{i,max} \tag{5}$$

where $P_{i,min}$ and $P_{i,max}$ are the minimum and maximum power outputs of the ith unit.

- Ramp rate constraints
  The actual operating range of all the online units is restricted by their corresponding ramp rate limits. The ramp-up and ramp-down constraints can be written as follows:

$$P_i - P_i^0 \leq UR_i \text{ as power increases} \tag{6}$$

$$P_i^0 - P_i \leq DR_i \text{ as power decreases} \tag{7}$$

where P_i^0 is the previous power output of the ith generating unit; UR_i and DR_i are the up-ramp and down-ramp limits of generator i, respectively.

To combine the ramp rate limits into power output limits constraints, the generator limits (5) can be rewritten as:

$$\max\{P_{i,min}, P_i^0 - DR_i\} \leq P_i \leq \min\{P_{i,max}, P_i^0 + UR_i\} \tag{8}$$

2.2   Economic Dispatch Problem with Valve-Point Effects

The ED with valve point loading effects (VPE) is a non-smooth and non-convex problem with the cost function include the ripple curve. The generating units with valve point effects steam turbines exhibit a greater variation in the fuel-cost functions. Since the valve point results in the ripples, the cost function becomes more nonlinear. Therefore, the equation (2) should be replaced by (9) for considering the valve-point effects. The sinusoidal functions are thus added to the quadratic cost functions as following.

$$F_i(P_i) = a_i + b_i P_i + c_i P_i^2 + \left| e_i \times \sin(f_i \times (P_{i,\min} - P_i)) \right| \tag{9}$$

where e_i and f_i are the fuel cost coefficients of the ith unit with valve point effects.

2.3   ED Problem Considering Prohibited Operating Zones

In some cases, the operating range of a generating unit is not always available due to limited physical operation. Units may have prohibited operating zones due to generator have some faults in operation or associated auxiliaries. Such faults may lead to instability in certain ranges of generator power output [6]. Thus, for generating units with prohibited operating zones, there are additional constraints on the unit operating range as follows:

$$P_i \in \begin{cases} P_{i,\min} \le P_i \le P_{i,1}^l \\ P_{i,k-1}^u \le P_i \le P_{i,k}^l \\ P_{i,pz_i}^u \le P_i \le P_{i,\max} \end{cases} \tag{10}$$

where $k = 2, 3, \ldots, pz_i$; $i = 1, 2, \ldots, np_z$; $P_{i,k}^l$ and $P_{i,k}^u$ are the lower and upper bounds of prohibited operating zone of unit i respectively. Here, pz_i is the number of prohibited zones of unit i and np_z is the number of units which have prohibited operating zones.

3   Hybrid Differential Evolution and Gravitational Search Algorithm

3.1   Differential Evolution

In 1995, Price and Storn proposed a new evolutionary algorithm for global optimization and named it as differential evolution (DE) [14]. This method has few parameters for tuning make the algorithm quite popular in the literature. DE obtains solutions to optimization problems using three basic operations including mutation, crossover, and selection. The steps of DE including the operations are described as follows:

Step 1: mutation

Choose the target vector $\mathbf{x}_{i,g}$ ($= \mathbf{x}_{0,g}$) and basic vector $\mathbf{x}_{r0,g}$ ($= \mathbf{x}_{2,g}$).

Randomly select two vector components $\mathbf{x}_{r1,g}$ ($= \mathbf{x}_{3,g}$) and $\mathbf{x}_{r2,g}$ ($= \mathbf{x}_{Np-2,g}$.).

Calculate the value for a mutant vector:

$$v_{i,g} = x_{r0,g} + F.(x_{r1,g} - x_{r2,g}) \tag{11}$$

Step 2: crossover

Select new vector from the target vector and mutant vector according to the following rules:

$$u_{i,g} = u_{j,i,g} = \begin{cases} v_{j,i,g} & if\,(rand_j(0,1) \le Cr\ or\ j = j_{rand}) \\ x_{j,i,g} & otherwise \end{cases} \tag{12}$$

Step 3: selection

Calculate the value of the objective function with the new vector is created, compare this value to the objective function value of the first selected target vector to decide whether to select a new vector or not.

$$x_{i,g+1} = \begin{cases} u_{i,g} & if\ f\left(u_{i,g}\right) \le f\left(x_{i,g}\right) \\ x_{i,g} & otherwise \end{cases} \tag{13}$$

3.2   Gravitational Search Algorithm

Gravitation search algorithm is a stochastic, population-based search method introduced by Rashedi and Hossein (2009). The mechanism of GSA got inspired by the law of Newtonian gravity: "In the universe, every particle attracts every other particle with a force, and the force is directly proportional to the product of their masses and inversely proportional to the square of the distance between them". A GSA algorithm maintains a population of individuals, where each individual represents a possible solution. Each of N agents is initialized as thus:

$$X_i = (x_i^1, x_i^2 ..., x_i^d, ..., x_i^n) \quad \text{with } i = 1,2,...,N \tag{14}$$

where n is dimension of the problem, and also the position of the ith agent in the dth dimension. At start point of the solution, agents are situated randomly. At specific time, a gravitational force is defined as thus:

$$F_{ij}^d(t) = G(t).\frac{M_{pi}(t)xM_{aj}(t)}{R_{ij}(t)+\varepsilon}(x_j^d(t) - x_i^d(t)) \tag{15}$$

where M_{aj} is the active gravitational mass, M_{pi} is the passive gravitational mass, $G(t)$ is gravitational constant at time t, ε is a small constant, and $R_{ij}(t)$ is the Euclidian distance between two particles:

$$R_{ij}(t) = \| X_i(t), X_j(t) \|_2 \tag{16}$$

The total force that acts on a given individual i in a dimension d is a randomly sum of dth components of the forces exerted from other agents:

$$F_i^d(t) = \sum_{j=1, j\neq i}^{N} rand_j F_{ij}^d(t) \tag{17}$$

where $rand_j \in [0,1]$ is a random number.

The acceleration rate of the individual i at time t, and in the direction d^{th}, denoted by $a_i^d(t)$ can be calculated by:

$$a_i^d(t) = \frac{F_i^d(t)}{M_{ii}(t)} \tag{18}$$

where M_{ii} is the inertial mass of individual i.

Furthermore, the position and velocity of the individual can be updated as follows:

$$v_i^d(t+1) = rand_i . v_i^d(t) + a_i^d(t) \tag{19}$$

$$x_i^d(t+1) = x_i^d(t) + v_i^d(t+1) \tag{20}$$

where $rand_i \in [0,1]$ is used to give a stochastic characteristic to the algorithm.

Assuming the equality of the gravitational and inertia mass, the value of mass are calculated using the map of fitness. We update by the following equation:

$$M_{ai} = M_{pi} = M_{ii} = M_i, \qquad i = 1, 2, ... N \tag{21}$$

$$m_i(t) = \frac{fit_i(t) - worst(t)}{best(t) - worst(t)} \tag{22}$$

$$M_i(t) = \frac{m_i(t)}{\sum_{j=1}^{N} m_j(t)} \tag{23}$$

where $fit_i(t)$ denotes the fitness value of the agent i at time t and best(t), worst(t) are defined as follow:

$$bets(t) = minfit_j(t), \qquad j = 1, 2, ... N \tag{23}$$

$$worst(t) = maxfit_j(t), \quad j = 1, 2, ... N \tag{24}$$

3.3   Hybrid Differential Evolution and Gravitational Search Algorithm

The overall procedure of the proposed algorithm can be summarized as in Fig. 1.

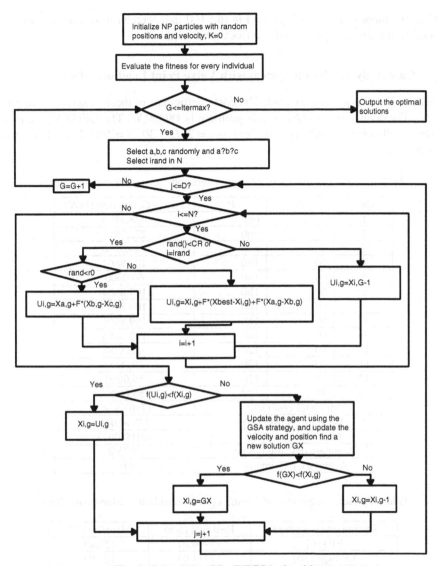

Fig. 1. Flow-chart of the DEGSA algorithm

4   Numerical Results

The proposed DEGSA algorithm has been applied to ED problems in three different power systems including 13-unit system, 40-unit system with valve-point effects, and 15-unit system with prohibited operating zones, ramp rate limits, and transmission network losses. The proposed DEGSA method is coded in Matlab version 7.13 and run on an Intel CPU Core2duo T6600 2.2 GHz processor with 2.0 GB of RAM. The stopping criterion of the algorithm is the maximum number of iterations. For each system, the proposed DEGSA method is run 100 independent trials and the obtained optimal results are compared to those from other methods.

4.1 Case study 1: 13-Unit System with Valve Point Loading Effects

Consider a thirteen generators case. The cost coefficients of these generators are given in [15]. The demanded load P_D of this problem is 1800MW. The DEGSA parameters are set as follows: $G_0 = 600$, $\alpha = 10$, $final\_per = 2$, $D = 30$, $N = 500$, $F = 0.1$, $Cr = 0.5$ and $r_0 = 0.1$.

Table 1. Optimal dispatch of 13-unit system considering valve-point effects

Unit	$P_{i,min}$	$P_{i,max}$	Power Generation (MW)
1	0	680	628.3185
2	0	360	222.7917
3	0	360	149.5576
4	60	180	109.8664
5	60	180	60.0000
6	60	180	109.8664
7	60	180	109.8665
8	60	180	109.8663
9	60	180	109.8665
10	40	120	40.0000
11	40	120	40.0000
12	55	120	55.0000
13	55	120	55.0000
Min total cost ($/h)			17963.8342
Mean total cost ($/h)			17994.046
Max total cost ($/h)			18100.0622
Standard deviation ($/h)			27.7558
Average CPU time (s)			9.61

Table 2. Result comparison of 13-unit system considering valve-point effects

Optimization method	Best cost ($/h)	Meantime (s)
EP [16]	17994.07	157.43
EP-SQP [16]	17911.03	121.93
PSO [16]	18030.72	77.37
PSO-SQP [16]	17969.93	33.97
UHGA [17]	17964.81	15.33
GSA	17968.97	16.0982
DEGSA	17963.8342	9.61

Table 1 gives the power output values of individual generators of 13 unit system, minimum, mean, maximum cost, standard deviation and average CPU time. In Fig.2 show that convergence characteristic curve of the best case with valve point effect for GSA and DEGSA method. From the compared results in Table 2, It shows that the DEGSA has succeeded in finding a global optimal solution. As visualized from the

Fig.2 and Table 2, it gives that the proposed DEGSA method of optimization is more efficient when compared with other optimization methods. The optimum active power is in their secure values and is far from the min and max limits. It is also clear from the optimum solution that the DEGSA easily prevent the violation of all the active constraints. Also total cost in 100 independent runs of DEGSA method for this case study is shown in Fig.3.

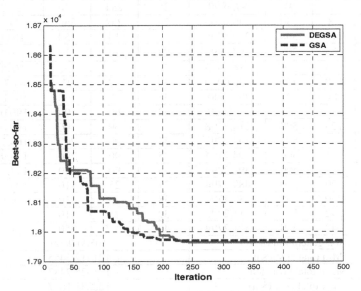

Fig. 2. Convergence nature of GSA and DEGSA in tested case of 13-unit system

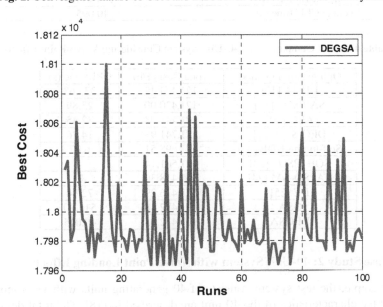

Fig. 3. Total cost in 100 independent runs of DEGSA method in tested case of 13-unit system

Table 3. Optimal dispatch of 40-unit system

Unit	$P_{i,min}$	$P_{i,max}$	Output (MW)	Unit	$P_{i,min}$	$P_{i,max}$	Output (MW)
1	36	114	110.7999	21	254	550	523.2794
2	36	114	110.7999	22	254	550	523.2794
3	60	120	97.3999	23	254	550	523.2794
4	80	190	179.7331	24	254	550	523.2794
5	47	97	87.7999	25	254	550	523.2794
6	68	140	140.0000	26	254	550	523.2794
7	110	300	259.5996	27	10	150	10.0000
8	135	300	284.5997	28	10	150	10.0000
9	135	300	284.5996	29	10	150	10.0000
10	130	300	130.0000	30	47	97	87.7999
11	94	375	94.0000	31	60	190	190.0000
12	94	375	94.0000	32	60	190	190.0000
13	125	500	214.7598	33	60	190	190.0000
14	125	500	394.2793	34	90	200	164.7998
15	125	500	394.2794	35	90	200	199.9996
16	125	500	394.2793	36	90	200	194.3983
17	220	500	489.2794	37	25	110	109.9999
18	220	500	489.2794	38	25	110	110.0000
19	242	550	511.2793	39	25	110	109.9998
20	242	550	511.2794	40	242	550	511.2793
Total generation (MW)				10500.0000			
Min total cost ($/h)				121412.5455			
Mean total cost ($/h)				121625.7483			
Max total cost ($/h)				122231.1617			
Standard deviation ($/h)				155.9395			
Average CPU time (s)				40.095			

Table 4. Result Comparison of 40-Unit System Considering Valve-Point Effects

Optimization method	Total Cost ($/h)	CPU time (s)
UHGA [17]	121,424.48	333.68
SA-PSO [19]	121,430.00	23.89
ABC [20]	121,441.03	32.45
DEC-SQP [21]	121,741.98	14.26
Self-tuning HDE [22]	121,698.51	6.07
SOH_PSO [23]	121,501.14	-
FAPSO-NM [24]	121,418.30	40
DE [25]	121,416.29	72.94
GSA	121,472.48	511.20
DEGSA	121,412.55	40.095

4.2   Case Study 2: 40-Unit System with Valve Point Loading Effects

In this example, the test system consists of 40 generating units with valve-point effects and the characteristics of the 40 unit are described in [18]. The total demand of the system is 10500 MW neglecting power loss and ramp rate constraints. The DEGSA parameters are set as follows: $G_0 = 6000$, $\alpha = 10$, *final_per* = 2, $D = 100$, $N = 1000$; $\alpha = 5$, $F = 0.1$, $Cr = 0.5$ and $r_0 = 0.1$.

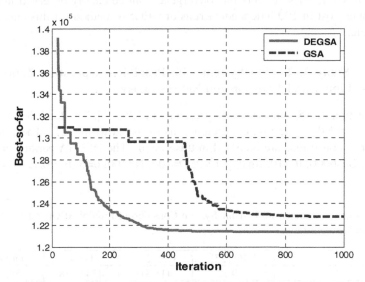

Fig. 4. Convergence nature of GSA and DEGSA in tested case of 40-unit system

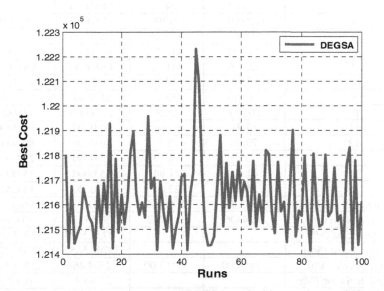

Fig. 5. Total cost in 100 independent runs of DEGSA method in tested case of 40-unit system

Table 3 shows the best optimal solution with total power generation, the minimum, mean, maximum cost achieved by the proposed method for the system. The best total cost from the DEGSA method is compared to that from other methods as given in

Table 4. The result comparison has shown that the proposed DEGSA is better than many other methods for this system. Fig. 4 shows the comparison of the two graphs – GSA method and DEGSA method. The DEGSA method has the convergence faster than GSA method. These effects on convergence can be clearly observed in this figure. Also total cost in 100 independent runs of DEGSA method for this case study is shown in Fig.5.

4.3   Case Study 3: 15-Unit System with Prohibited Operating Zones, Ramp Rate Limits, and Transmission Network Losses

The information of 15 unit system is presented in [26, 27]. The load demand of the system is 2630 MW. The loss coefficients matrix is shown in [27]. The power loss and ramp rate constraints are included in this system. The DEGSA parameters are set as follows: G_0=6000, α =10, final_per =2, D=100, N=100; α=5, F = 0.1, Cr = 0.5 and r_0 = 0.1.

Table 5. Result Comparison of 15-Unit System Considering Prohibited Operating Zones and Transmission Network Losses

Unit	PSO [27]	GA [27]	GSA	DEGSA
1	439.12	415.31	454.9978	454.9980
2	407.97	359.72	380.0000	380.0000
3	119.63	104.42	130.0000	130.0000
4	129.99	74.98	130.0000	130.0000
5	151.07	380.28	170.0000	170.0000
6	459.99	426.79	460.0000	460.0000
7	425.56	341.32	430.0000	430.0000
8	98.56	124.79	69.5911	72.2117
9	113.49	133.14	61.0644	58.4538
10	101.11	89.26	160.0000	160.0000
11	33.91	60.06	80.0000	80.0000
12	79.96	50	80.0000	80.0000
13	25	38.77	25.0000	25.0000
14	41.41	41.94	15.0000	15.0000
15	35.61	22.64	15.0000	15.0000
Total power (MW)	2662.41	2668.44	2660.6533	2660.6628
Total loss (MW)	32.4306	38.2782	30.6533	30.6635
Max total cost ($/h)			32727.5527	32707.7178
Mean total cost ($/h)			32706.5652	32705.2765
Min total cost ($/h)	32858.00	33113.00	32704.466	32704.4536
Standard deviation ($/h)			22.9006	0.9335
Average CPU time (s)	2.74	4.95	2.401	6.859

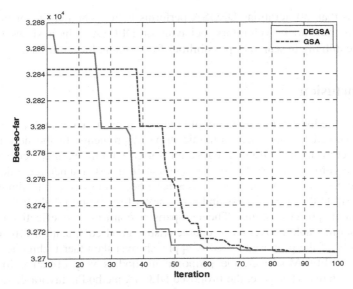

Fig. 6. Convergence nature of GSA and DEGSA in tested case of 15-unit system

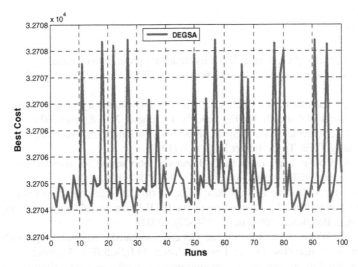

Fig. 7. Total cost in 100 independent runs of DEGSA method in tested case of 15-unit system

The best solution, total power loss, total cost, standard deviation and average CPU time obtained by the DEGSA method for the 15-unit system with prohibited operating zones, ramp rate limits, and transmission network losses are compared to those from genetic algorithm (GA) [27], particle swarm optimization (PSO) [27] and GSA as shown in Table 5. Obviously, the minimum cost obtained by the proposed DEGSA method is better than that from both PSO, GA and GSA.

The convergence behavior of GSA and DEGSA methods are shown in Fig.6. From the figure it can be observed that initially the GSA method appears to converge faster but after less than 30 iterations DEGSA performs better. The reason for this can be attributed to the initial exploratory behavior of DEGSA, which slows down the convergence initially in search of possibilities for avoiding local minima.

5   Conclusion

In this paper, the hybrid differential evolution and gravitational search algorithm had been implemented to solve the economic dispatch problems. In the new improved method, the conventional DEGSA algorithm is used with the variance coefficients to speed up the convergence to the global solution in a fast manner regardless of the shape of the cost function. Three test cases have been considered, the simulation results demonstrate the effectiveness and robustness of the proposed algorithm to solve ED problem in power systems. The comparison confirms the effectiveness high-quality solution, stable convergence characteristic, good computation efficiency and the superiority of the proposed DEGSA approach over the other techniques in terms of solution quality. Moreover, the proposed method is very effective for solving large-scale systems. Therefore, the proposed DEGSA method is favorable for solving economic dispatch problems with non-convex cost functions.

References

1. Yang, H.T., Yang, P.C., Huang, C.L.: Evolutionary programming based economic dispatch for units with non-smooth fuel cost functions. IEEE Trans. Power Syst. 11(1), 112–118 (1996)
2. Sinha, N., Chakrabarti, R., Chattopadhyay, P.K.: Evolutionary programming techniques for economic load dispatch. IEEE Trans. Evol. Comput. 7(1), 83–94 (2003)
3. Walters, D.C., Sheble, G.B.: Genetic algorithm solution of economic dispatch with valve point loading. IEEE Trans. Power Syst. 8(3) (1993)
4. Lin, W.M., Cheng, F.S., Tsay, M.T.: Improved tabu search for economic dispatch with multiple minima. IEEE Trans. Power Syst. 17(1), 108–112 (2002)
5. Khamsawang, S., Boonseng, C., Pothiya, S.: Solving the economic dispatch problem with tabu search algorithm. In: Proceeding of the IEEE International Conference on Industrial Technology, Bangkok, Thialand, pp. 274–278 (2002)
6. Wong, K.P., Wong, Y.W.: Genetic and Genetic/Simulated – Annealing approaches to economic dispatch. IEE Proc. Gener. Transm. Distrib. 141(5), 507–513 (1994)
7. Wong, K.P.: Solving power system optimization problems using simulated annealing. Engng. Applic. Artif. Intell. 8(6), 665–670 (1996)
8. Selvakumar, A.I., Thanushkodi, K.: A new particle swarm optimization solution to non-convex economic dispatch problem. IEEE Trans. Power Syst. 22(1), 42–51 (2007)
9. Coelho, L.D.S., Souza, R.C.T., Mariani, V.C.: Improved differential evolution approach based on cultural algorithm and diversity measure applied to solve economic load dispatch problems. Math. Comput. Simul. 79, 3136–3147 (2009)
10. Noman, N., Iba, H.: Differential evolution for economic load dispatch problems. Electr. Power Syst. Res. 78, 1322–1331 (2008)

11. Coelho, L.D.S., Mariani, V.C.: Combining of chaotic differential evolution and quadratic programming for economic dispatch optimization with valve-point effect. IEEE Trans. Power Syst. 21, 989–996 (2006)
12. Rashedi, E., Nezamabadi-pour, H., Saryazdi, S.: Filter modeling using gravitational search algorithm. In: Engineering Applications of Artificial Intelligence (to be published, 2010) (accepted for publication)
13. Victorie, T.A.A., Jeyakumar, A.E.: Hybrid PSO-SQP for economic dispatch with valve-point effect. Electric Power Systems Research 71, 51–59 (2004)
14. Luong, L.D., Dieu, V.N., Hop, N.T., Dung, L.A.: A Hybrid Differential Evolution and Harmony Search for Nonconvex Economic Dispatch Problems. In: 2013 IEEE 7th International Power Engineering and Optimization Conference (PEOCO), June 3-4, pp. 238–243 (2013)
15. Chen, C.H., Yeh, S.N.: Particle Swarm Optimization for Economic Power Dispatch with Valve-Point Effects. In: 2006 IEEE PES Transmission and Distribution Conference and Exposition Latin America, Venezuela, August 15-18 (2006)
16. Victoire, T.A.A.: Hybrid PSO-SQP for economic dispatch with valve point effect. Elec. Power Syst. Res. 71(1), 51–59 (2004)
17. He, D.-K., Wang, F.-L., Mao, Z.-Z.: Hybrid genetic algorithm for economic dispatch with valvepoint effect. Elec. Power Syst. Res. 78, 626–633 (2008)
18. Chaturvedi, K.T., Pandit, M., Srivastava, L.: Self-organizing hierar-chical particle swarm optimization for nonconvex economic dispatch. IEEE Trans. Power Systems 23(3), 1079–1087 (2008)
19. Chen, Y.-P., Peng, W.-C., Jian, M.-C.: Particle swarm optimization with recombination and dynamic linkage discovery. IEEE Trans. Systems, Man, and Cybernetics - Part B: Cybernetics 37(6), 1460–1470 (2007)
20. Hemamalini, S., Simon, S.P.: Artificial bee colony algorithm for economic load dispatch problem with non-smooth costfunctions. Electric Power Components and Systems 38(7), 786–803 (2010)
21. dos Santos Coelho, L., Mariani, V.C.: Combining of chaotic differ-ential evolution and quadratic programming for economic dispatch optimization with valve-point effect. IEEE Trans. Power Systems 21(2), 989–996 (2006)
22. Wang, S.-K., Chiou, J.-P., Liu, C.-W.: Non-smooth/non-convex economic dispatch by a novel hybrid differential evolution algorithm. IET Gener. Transm. Distrib. 1(5), 793–803 (2007)
23. Chaturvedi, K.T., Pandit, M., Srivastava, L.: Self-organizing hierar-chical particle swarm optimization for nonconvex economic dispatch. IEEE Trans. Power Systems 23(3), 1079–1087 (2008)
24. Niknam, T.: A new fuzzy adaptive hybrid particle swarm optimization algorithm for non-linear, non-smooth and non-convexeconomic dispatch problem. Applied Energy 87(1), 327–339 (2010)
25. Noman, N., Iba, H.: Differential evolution for economic load dis-patch problems. Electric Power Systems Research 78(8), 1322–1331 (2008)
26. Safari, A., Shayeghi, H.: Iteration particle swarm optimization procedure for economic load dispatch with generator constraints. Expert Systems with Applications 38(5), 6043–6048 (2011)
27. Gaing, Z.L.: Particle swarm optimization to solving the economic dispatch considering the generator constraints. IEEE Trans. Power Syst. 18(3), 1187–1195 (2003)

Automatic Evolutionary Music Composition Based on Multi-objective Genetic Algorithm

Jae Hun Jeong and Chang Wook Ahn*

Department of Computer Engineering, Sungkyunkwan University (SKKU)
2066 Seobu-Ro, Suwon 440-746, Republic of Korea
{a12gjang,cwan}@skku.edu

Abstract. This paper presents a multi-objective approach for making melody compositions in evolutionary music. There exist several methods to generate music based on computer algorithms, but they cannot deal with multi-dimensional aspects such as a trade-off relation effectively. Our approach generates a set of melodies that maximize two fitness functions, which represent a trade-off between stability and tension. It includes a pre-defined chord progression and rhythm to initialize and evaluate population. Multi-objective genetic algorithm is applied to the music composition, it is able to successfully compose melody lines based on the chord progression.

Keywords: Evolutionary music composition, Multi-objective genetic algorithm, Generative music.

1   Introduction

Evolutionary computation (EC) is a field that encompasses various techniques or methods inspired by the evolutionary process of nature. Algorithms which belong to EC are Evolutionary Algorithm (EA), Genetic Algorithm (GA), Genetic Programming (GP) etc., that search massive space of data structure that represents a potential solution of the problem at hand. For example, elaborate design of an engine of an airplane, generating beautiful image, scheduling time table of exam or composition of musical pieces can be applications of the EC algorithms [1]. These tasks are powered by an evolutionary process such as natural selection, recombination and mutation. The basic idea is that better solutions have high possibility to survive to the next generation and they are modified by the genetic operators to bring the diversity in the population.

In ancient days, ECs were mainly applied to industry optimization. EC system has become more and more popular to many researchers. Then, the range of the application has widened into creative and artistic areas such as music composition [1]. This field is called Evolutionary Music. Since 1950s, various artificial intelligence techniques including grammatical representation, probabilistic methods, neural networks, symbolic rule-based systems, constraint programming and

---

* Corresponding author.

H. Handa et al. (eds.), *Proc. of the 18th Asia Pacific Symp. on Intell. & Evol. Systems – Vol. 2*,
Proceedings in Adaptation, Learning and Optimization 2, DOI: 10.1007/978-3-319-13356-0_9

evolutionary algorithms have been used for music composition [2]. Especially, genetic algorithm, which is a branch of the evolutionary algorithm, is recognized as an appropriate tool in this field. The process of continual improvement and innovation that are carried out by a combination of genetic operators (mutation and recombination together with selection) is very useful at a creative search of art and music [3].

Some studies [4, 5] have generated pitch sequences without rhythm. While, some researches [6–8] have evolved just rhythm sequences excluding pitch. Few researches [9, 11, 12] have evolved pitch and duration sequences simultaneously.

Evolutionary music composition can be divided into two groups that are automatic and interactive. Automatic evolutionary systems for composing music generate musical pieces without human intervention, while interactive system involves human mentor in the fitness evaluation phase [10].

Most studies, in the field of evolutionary music composition, have only focused on optimization of the single objective function to generate music. Even, if musical aesthetic judgments are typically multi-dimensional, it has been carried out by optimizing the sum of total fitness values of every objective function regardless of relations between them.

The main purpose of this study is to build a music composition system that enables to deal with the trade-off relation of music. We designed two fitness functions that are related to stability and tension in the music. Non-dominated sorting genetic algorithm (NSGA-II) that have become standard approach is applied to compose melody by optimizing multi objective functions simultaneously.

The overall structure of the study takes the form of five sections, including this introductory section. Section two begins by explaining the background of this research briefly. Section 3 covers the overall process of our approach. Section 4 presents the experimental results. Finally, section 5 concludes with a summary.

2 Background

2.1 Music Terminology

In this section, the basic terms of music that are used in this paper are described. This paper does not explain common or less important terms.

Pitch is a basic concept of music that represents high and low of sound. From an objective point, pitch can be considered as a frequency of an appropriate sound wave. For example, a note A4 is determined by sound waves of 440Hz. If the frequency of certain musical note is two times higher than another note, this interval is called octave. For example, the frequency of the A5 is 880Hz which is double of the frequency of A4. There are twelve notes between octave (C, C#, D, D#, E, F, F#, G, G#, A, A#, B).

Melody consists of pitches arranged in a horizontal sequence, but also duration of each pitch. Series of pitch durations can be considered as a melodys rhythm. Rhythm refers to timing, both in terms of how long sound events last and when they scheduled to occur.

Tonality is a musical system in which pitch or chords are arranged so as to induce a hierarchy of perceived stabilities and attractions. The pitch or chord with the greatest stability is called the tonic. In the context of the tonal organization, a chord or note is said to be consonant when it implies stability, and dissonant when it implies instability. A dissonant chord is in tension against the tonic, and implies that the music is distant from that tonic chord. Resolution is the process by which the harmonic progression moves from dissonant to consonant chords or notes.

Table 1. Diatonic chords of the C Major scale

Chord	Chord tone	Non-chord tone	
		Tension note	Avoid note
CM7	C, E, G, B	D, A	F
Dm7	D, F, A, C	E, G	B
Em7	E, G, B, D	A	F, C
FM7	F, A, C, E	G, B, D	-
G7	G, B, D, F	A, E	C
Am7	A, C, E, G	B, D	F
Bm7b5	B, D, F#, A	E, G	C

In music, a chord means a harmonic set of three or more notes sounding simultaneously. The most common chords are triads that consist of three distinct notes. Seventh chord, extended chords, or added tone chord are the chords which have additional notes except triad. Major and minor triad are the most common chord and the next most things are augmented and diminished triads.

Member notes of the chord are called chord tone and any others are called non-chord tone. Non-chord tone can be divided to available tension note and avoid note by harmonics. For example, as we can see in the Table 1, if a piece of music implies C Major seventh chord, then notes C, E, G, B are chord tones which mean members of that chord, while notes D, A are available tension note and note F is an avoid note.

Chord tone is very stable when it sounds with chord sound. On the other hand, most non-chord tones are unstable, which are required to resolve to a chord tone in conventional ways. Especially, avoid notes are dissonant relative to the harmony implied by the chord.

2.2   Multi-objective Genetic Algorithm

Genetic algorithm is a branch of evolutionary algorithms. It simulates the evolutionary process of the organism to solve difficult problems focusing on gene. It has been applied to a wide range of applications. Due to its suitability for composing music better than any other algorithms, it has a special position in the field of evaluating music [3].

Multi-objective Genetic Algorithms (MOGA) can optimize more than one objective function that has trade-off relation simultaneously, while simple genetic algorithm can only optimize single objective function. In multi-objective optimization, typically there does not exist a solution that minimizes all objective functions simultaneously. To solve multi-objective optimization problem, we have to pay attention to Pareto optimal solution. Because, a solution dealing with the trade-off relation cannot be improved in any of the objective without degrading at least one of the other objectives.

There are various multi-objective approach algorithms. We apply non-dominated sorting genetic algorithm II (NSGA-II) which is widely known as a standard approach to compose musical pieces.

$$f_i(x^1) \leq f_i(x^2) \, for \, all \, indices \, i \in \{1, 2, \cdots, k\} \tag{1}$$

$$f_j(x^1) < f_j(x^2) \, for \, at \, lest \, one \, index \, j \in \{1, 2, \cdots, k\} \tag{2}$$

In mathematical terms, if the Equation (1) and (2) are satisfied, solution x^1 dominates solution x^2. A set of solution which is not dominated by any other solutions is called a Pareto front. The first Pareto front is assigned rank 1 and then removed from the population. Then, rank 2 is assigned to the Pareto front set of the population except rank 1 Pareto optimal set. This process is repeated until the population becomes empty [13].

Chromosomes that have higher rank value are considered as better solutions in the selection phase of the evolutionary process. If, two chromosomes have the same rank, then the chromosome with the lower density is considered as a better solution [13].

3   Evolutionary Music Composition Based on Multi-objective Genetic Algorithms

3.1   Representation

In the EC system, to solve the problem at hand, appropriate representation of real world problem has a decisive effect to ability to find a solution. When we deal with musical tasks such as making melody line based on genetic algorithm, array data structure can be an effective genetic representation.

Table 2. Mapping table of pitch and rest

Rest	Hold	$\cdots$	B3	C4	C#4	D4	D#4	E4	F4	F#4	G4	G#4	$\cdots$
-1	0	$\cdots$	59	60	61	62	63	64	65	66	67	68	$\cdots$

As shown in the Table 2, notes are mapped to the integer number. For example, note C4 can be mapped to integer value 60. Pitch duration is expressed by a number of successive zeroes on the right side of the pitch value in the chromosome. For example, in Figure 1, four bar melody of four-four time is represented based on the array whose length is 32. One element of the array means eight-note duration as a unit length. For example, if there is no zero on the right side of the pitch value, duration of the pitch becomes an eighth note. If there are three consecutive zeroes on the right side of the pitch value, duration of the pitch becomes a half note that is four times of an eighth note.

| -1 | 60 | 64 | 71 | 69 | 0 | 0 | 76 | 74 | 71 | 72 | 60 | 62 | 0 | 0 | 0 | -1 | 64 | 76 | 74 | 72 | 0 | 0 | 71 | 69 | 67 | 79 | 77 | 76 | 0 | 0 | 0 |

Fig. 1. Example melody and its genetic representation based on the array

3.2 Genetic Operators

Genetic operators, which are crossover and mutation, contribute to maintain the diversity of the population in execution of the genetic algorithm. The crossover operator varies the solution by exchanging existing parts between two parent chromosomes. The mutation operator prevents from becoming too similar to each other by altering one or more gene value in a chromosome to avoid local minima. The points of the crossover and mutation are selected randomly [14].

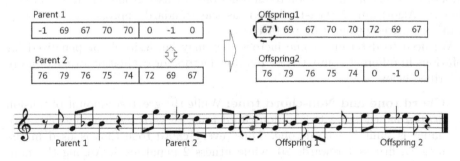

Fig. 2. Exapmle of crossover and mutation

In Figure 2, we can see that the right side of the selected point of the parent chromosome is exchanged by crossover. The first element of the offspring 1 is mutated to the 67 (G4) from the rest (-1). These new chromosomes compose population of the next generation.

3.3   Fitness Function

Two fitness evaluation functions are designed in order to deal with the trade-off between stability and tension of the melody given chord progression. First fitness function is defined as a stability function. Chord tone is considered to a desirable trait from stability point. For example, melody of most childrens songs or meditation music are composed by chord tone to make the listener comfortable or relax. Also, when the melody line shows the gradual motion to a lower pitch, it can make sound stable. The maximization of fitness function which is designed as stability function will represent the evaluation of this sort of harmony and guideline to compose melody.

The second fitness function to be maximized, named tension function, gives higher fitness values to a melody which contain dissonance on their chord. On the contrary to the first function, gradual motion to a higher pitch of the melody is evaluated as a good part because it may produce tension on the melody.

It would be possible to find best solution which satisfies both objectives if we already knew sufficient information about the preference of user to the music from tension and stability points. We will be able to assign proper weight values to each objective based on preference information. However, it is difficult task to get this information in advance and to determine the set of weight values.

With a multi-objective approach, the user obtains a set of melodies that are evaluated as a good solution at the same level. Then the user picks a melody which satisfies preference of himself after listening to the set of melodies. Through this process, we can involve users preference to the final output.

Because two fitness values have to be assigned to every chromosome, to decide rank of the solutions in relation to all functions, we used Non-dominated Sorting Genetic Algorithm (NSGA-II) known as the standard approach to deal with multi-objective optimization problem.

We need to determine some factors that may be desirable or penalized for solutions in different context. The relevant factors for evaluation of melody are described below.

- **Chord tone and Non-chord tone:** While the greatest amount of tension or dissonance is created by non-chord tone, the greatest amount of consonance is built by chord tone. Thus, every chord tone that is used in the melody, fitness 1 rewards 20, while fitness 2 penalizes 5. Among the non-chord tones, available tension note creates acceptable tension. Avoid notes are more dissonant especially, and notes that are not part of scale weaken tonality of the music or may create critical tension. Therefore, fitness 1 penalizes 10, 20 and 30, while fitness 2 rewards 20, 5 and penalizes 5 to the available tension, avoid note and non-scale notes.
- **Resolution of Tension:** In music, tension is the perceived need for relaxation or release created by a listener's expectation. Most tension can resolve up or down to the neighboring chord tone. If the non chord tone moves to the closest chord tone, fitness 1 rewards 10 and fitness 2 rewards 30.
- **Motion:** Tension may also be produced through gradual motion to a higher pitch. The other way, gradual motion to a lower pitch gives a sense of stabil-

ity to a listener. If contour of the melody shows ascending motion, fitness 1 rewards 10 and fitness 2 rewards 15. If contour of the melody shows descending motion, fitness 1 rewards 10 and fitness 2 rewards 15. When contour of the melody shows stepwise motion after leap (more than major or minor third interval), both of fitness 1 and fitness 2 reward 20.

- **Interval:** Traditionally perfect interval is considered perfectly consonant. Within a diatonic scale, all unison and octaves are perfect. Most fourths and fifths are also perfect, with five and seven semitones respectively. Consonances from perfect intervals are rewarded 5 by fitness 1 and penalized 3 by fitness 2. Both of fitness 1 and fitness 2 penalize 10 to the interval which is greater than octave, because it is extremely undesirable in the melody.

Table 3. Parameters of the fitness evaluation

		Fitness 1	Fitness 2
1. Chord tone		+30	-5
2. Available tension note		-10	+20
3. Avoid note		-20	+5
4. Non-scale note		-30	-10
5. Resolution of non-chord tone		+10	+30
6. Motion	6.1 Ascending	+15	+20
	6.2 Descending	+20	+15
	6.3 Stepwise motion after leap	+20	+20
7. Interval	7.1 Perfect	+10	-5
	7.2 Greater than octave	-20	-20

Table 3 shows a brief scoring system that is described above. If we desire to obtain a certain kind of result such as specific genre or style of the famous musician, target melodic features can be used to evaluate melody [15]. Empirical measures such as the Zipfs law and fractal analysis can also be used to improve aesthetic aspects of melody [16, 17].

4   Experiments and Results

This section represents a some melodies which are obtained by the proposed method. Before starting algorithm, the user determines key, chord progression and rhythm pattern of the melody. The key and chord progression of this experiment were the C Major and "Am7, FM7, CM7, G7" that is common in the pop music. And the rhythm pattern of music score, in Figure 1, was used to initialize population.

These experiments were carried out to show the ability of the proposed method to compose pleasant melody given chord progression and rhythm pattern. The parameters of the algorithm are configured by 100 population size, 100 iterations, 90% crossover and 1% mutation.

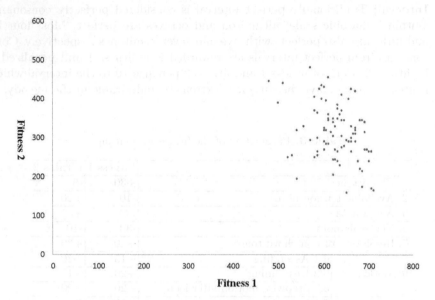

Fig. 3. Evolution of the Pareto fronts

Table 4. Fitness values of the compositions shown in Figure 4

	Melody	1	2	3	4	5	6	7	Sum
1)	Fitness1	510	-30	0	0	30	120	80	710
	Fitness2	-85	60	0	0	90	140	-40	165
2)	Fitness1	450	-50	0	0	50	185	50	685
	Fitness2	-75	100	0	0	150	195	-25	345
3)	Fitness1	360	-80	0	0	80	200	30	570
	Fitness2	-60	160	0	0	240	215	-150	480

Figure 3 shows example of the evolution of the Pareto fronts of the experiment as repeat generations. Among a set of melodies of the final generation, two extreme cases and one middle case between them are selected for analysis. Table 4. shows the fitness values of the three cases in detail. The first melody whose fitness value is (710, 165) is an extreme case for the fitness 1 that means the stability function. The second melody whose fitness value is (685, 345) is a middle case between two extreme cases. And the third melody whose fitness value is (570, 480) is the an extreme case for the fitness 2 that means the tension function.

Fig. 4. Compositions which are obtained by the proposed method

As shown in Figure 4, three melodies look similar to each other, because of convergence through the evolutionary process. However, they have different parts from stability and tension point. The first melody shows highest chord tone usage, while the third melody shows highest tension note usage. At the first bar implied Am7 chord, the first melody consists of only chord tone, while the second and third melody have 25% and 37.5% of the tension part each. At the fourth bar implied G7 chord, the first melody consists of all chord tone, while the second and third melody have 12.5% of the tension part.

Fig. 5. Eight bar melody obtained by the proposed method

Our approach can be extended to compose longer melody by giving an expanded chord progression. Figure 5 shows the composition which was obtained by our approach based on the same rhythm pattern with Figure 4 and different chord progression that is "CM7, CM7, F7, G7, F7, CM7, G7, CM7".

5   Conclusion and Future Work

A music composition method based on multi-objective genetic algorithm is presented in this paper. The proposed method produced the set of pleasant melodies of given tonality, chord progression and rhythm pattern, according to the two fitness values which are related to stability and tension. The multi-objective approach enables to generate a set of feasible solutions. Among the set, the user can choose one as a final result by the preference of himself.

In the previous works for generating music automatically, there exists the issue of originality of the composition. Because most previous works have used similarity to existing target music to compose music. Also, It has not effectively explored the multidimensional space of compositions, because it is based on simple EA. Our approach has produced interesting and pleasant compositions as shown in the Figure 4, 5 without involving target music. Therefore, it enables user expects to obtain unpredictable creative musical results without controversy over originality. In addition, it could generate various compositions that encompass trade-off relation between stability and tension. It can be a meaningful tool which aids the process of music composition from a practical point of view. It makes non-musicians who do not have a knowledge or experiment of making music can compose music by themselves. Also, to the musician, it can be an assistant to cooperate musical tasks.

The method that we proposed is a novel way of generating melody dealing with the trade-off relation of music. In the future work, we will research to improve quality of the results by designing better fitness functions from aesthetic of music point. Also, we will enable chord progression to evolve with melody, while proposed method in this paper evolves melody given fixed chord progression. It would be a more ideal process of the melody creation.

Acknowledgement. This work was supported by the National Research Foundation of Korea (NRF) grant funded by the Korea government (MSIP) (NRF-2012R1A2A2A01013735).

References

1. Husbands, P., Copley, P., Eldridge, A., Mandelis, J.: An introduction to evolutionary computing for musicians. In: Evolutionary Computer Music, pp. 1–27. Springer, London (2007)
2. Fernández, J.D., Vico, F.: AI methods in algorithmic composition: a comprehensive survey. arXiv preprint arXiv:1402.0585 (2014)
3. Gartland-Jones, A., Copley, P.: The suitability of genetic algorithms for musical composition. Contemporary Music Review 22(3), 43–55 (2003)
4. Ralley, D.: Genetic algorithms as a tool for melodic development. Urbana 101, 61801 (1995)
5. Johanson, B., Poli, R.: GP-music: An interactive genetic programming system for music generation with automated fitness raters. University of Birmingham, Cognitive Science Research Centre (1998)
6. Horowitz, D.: Generating rhythms with genetic algorithms. In: AAAI, vol. 94 (1994)
7. Burton, A.R.: A hybrid neuro-genetic pattern evolution system applied to musical composition. Diss. University of Surrey (1998)

8. Tokui, N., Iba, H.: Music composition with interactive evolutionary computation. In: Proceedings of the 3rd International Conference on Generative Art, vol. 17(2) (2000)
9. Biles, J.: GenJam: A genetic algorithm for generating jazz solos. In: Proceedings of the International Computer Music Conference. International Computer Music Accociation (1994)
10. Miranda, E.R., Biles, J.A. (eds.): Evolutionary computer music. Springer (2007)
11. Jacob, B.: Composing with genetic algorithms (1995)
12. Marques, M., et al.: Music composition using genetic evolutionary algorithms, pp. 714–719 (2000)
13. Deb, K., et al.: A fast and elitist multiobjective genetic algorithm: NSGA-II. IEEE Transactions on Evolutionary Computation 6(2), 182–197 (2002)
14. Whitley, D.: A genetic algorithm tutorial. Statistics and Computing 4(2), 65–85 (1994)
15. Jensen, J.H.: Evolutionary music composition: A quantitative approach (2011)
16. Manaris, B., Vaughan, D., Wagner, C., Romero, J., Davis, R.B.: Evolutionary music and the zipf-mandelbrot law: Developing fitness functions for pleasant music. In: Raidl, G.R., et al. (eds.) EvoWorkshops 2003. LNCS, vol. 2611, pp. 522–534. Springer, Heidelberg (2003)
17. Dodge, C.: Profile: A Musical Fractal. Computer Music Journal, 10–14 (1988)

Integration of Spatial and Spectral Information by Means of Sparse Representation-Based Classification for Hyperspectral Imagery

Sen Jia, Yao Xie, and Zexuan Zhu*

College of Computer Science and Software Engineering,
Shenzhen University, Shenzhen, China
Shenzhen Key Laboratory of Spatial Information Smarting Sensing and Services,
Shenzhen University, China
{senjia,zhuzx}@szu.edu.cn, 924300533@qq.com

Abstract. Recently, sparse representation-based classification (SRC), which assigns a test sample to the class with minimum representation error via a sparse linear combination of all the training samples, has successfully been applied to hyperspectral imagery. Meanwhile, spatial information, that means the adjacent pixels belong to the same class with a high probability, is a valuable complement to the spectral information. In this paper, we have presented a new spatial-neighborhood-integrated SRC method, abbreviated as SN-SRC, to jointly consider the spectral and spatial neighborhood information of each pixel to explore the spectral and spatial coherence by the SRC method. Experimental results have shown that the proposed SN-SRC approach could achieve better performance than the other state-of-the-art methods, especially with limited training samples.

Keywords: Hyperspectral imagery, sparse representation-based classification, spatial neighborhood.

1 Introduction

Through imaging the same area on the surface of the Earth at hundreds of different wavelength channels simultaneously, a data cube, called hyperspectral imagery (HSI), can be acquired with each spatial pixel corresponding to an essentially continuous radiance spectrum, which makes possible the remote identification of ground materials [14]. As a major application of hyperspectral data analysis, pixel-oriented classification has been widely addressed [3,13,16]. However, due to the extremely high spectral dimensionality of the data, the small

---

* This work was jointly supported by grants from National Natural Science Foundation of China (61271022 and 61471246), Guangdong College Excellent Young Teacher Training Program (Yq2013143 and Yq2013141), Shenzhen Scientific Research and Development Funding Program (JCYJ20120613113106357, JCYJ20130329115450637, JCYJ20140418095735628, and KQC201108300045A). All correspondence should be addressed to Zexuan Zhu.

sample size scenario (it is very difficult and time consuming to collect sufficient training samples in practice) is one crucial problem that limits the performance of many existing classification methods [11, 17].

Recently, Wright et al. have presented a sparse representation-based classification (SRC) scheme, which has demonstrated striking performance in face recognition [18]. Concretely, after a test sample has been sparsely coded over the training set as a whole, the classification is performed by checking which class yields the least reconstruction error. As a powerful classification framework, SRC has also been developed for hyperspectral imagery classification. Chen et al. [4,5] directly combined SRC with the contextual information for HSI classification, but a large number of training samples are needed for the method to work well, and several parameters should be pre-determined before conducting the classification. Haq et al. [7,8] proposed a fast homotopy-based SRC approach to handel the small sample size problem, which is based on the assumption that the samples within the same class are similar. But the assumption may not be completely met in practice due to the high variations of the spectral signature caused by various factors, such as atmospheric conditions, sensor noise, etc. Even worse, the measured spectra between different classes could be highly similar [10]. Hence, it is difficult to effectively address the small sample size problem by SRC using only the spectral information.

In this paper, a spatial-neighborhood-integrated SRC method, abbreviated as SN-SRC, is proposed for hyperspectral imagery classification. Due to the spatial homogeneity of surface materials, SN-SRC jointly considers the spectral and spatial neighborhood information of each pixel to explore the spatial coherence by the SRC method. More precisely, training samples are used to encode each test sample to get the spectral-based sparse representation, while the neighborhood pixels of the test sample are used to encode it to get the spatial-based sparse representation. Then both the spectral and spatial information are considered together to identify the sample. After all the test samples have been computed by the above steps, the classification map can be obtained. Compared with the map obtained by SRC method, the spatial information has been incorporated into the classification map. Obviously, in order to acquire a stable and consistency map, the spectral-spatial procedure can be repeated several times and the classification accuracy can be gradually improved. Extensive experiments on real hyperspectral data set have demonstrated that the proposed SN-SRC method could achieve a better performance than the other state-of-the-art methods.

The rest of this paper is organized as follows. Section 2 reviews the problem formulation and solution for the standard sparse representation. Section 3 describe the proposed algorithm, SN-SRC, in detail. Experiment was run on one real hyperspectral data set and the results are shown in Section 4. Section 5 concludes the paper with a summary of the proposed work.

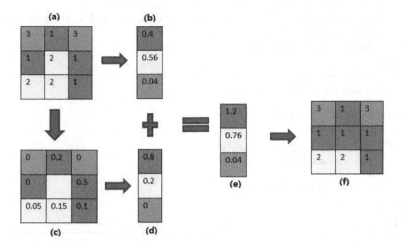

Fig. 1. One-step example of the proposed SN-SRC. (a) the classification map $\mathbf{L}_0$ of one center pixel and the eight neighborhood pixels obtained by SRC on the whole training set; (b) the spectral-based representation vector $\mathbf{P}_0$; (c) the representation coefficient of the center pixel by SRC on the eight neighborhood pixels; (d) the spatial-based representation vector $\mathbf{P}_1$; (e) the combined representation vector $(\mathbf{P}_0 + \mathbf{P}_1)$; (f) the new classification map $\mathbf{L}_1$ (the label of the central pixel (2) has been corrected to 1 due to the incorporation of the spatial information).

2 Sparse Representation-Based Classification (SRC)

Let $\mathbf{A}$ be the training samples with all p classes, $\mathbf{A} = [\mathbf{A}_1, \mathbf{A}_2, ..., \mathbf{A}_p] \in \mathbb{R}^{l \times n}$, l denote the feature number of each sample, n denote the number of training samples, and $\mathbf{A}_i \in \mathbb{R}^{l \times n_i}$ stacks the training samples of the i-th class. For an unknown signal vector $\mathbf{x}$, the linear model can be represented as:

$$\mathbf{x} = \mathbf{A}\mathbf{s} \qquad (1)$$

What we hope is that, the calculated $\mathbf{s}$ is sparsest, and its nonzero items correspond to a single class of $\mathbf{A}$, so that we can easily classify the test samples. The sparse representation of $\mathbf{x}$ can be obtained by solving the following optimization problem:

$$(l_0): \quad \widehat{\mathbf{s}} = \arg\min \|\mathbf{s}\|_0 \quad \text{subject to} \quad \mathbf{A}\mathbf{s} = \mathbf{x} \qquad (2)$$

where $\| \cdot \|_0$ counts the number of nonzero elements, which is the best sparsity measure available. Solving the optimization equation 1turns out to be a non-convex NP-hard problem, and is computationally intractable [15]. On the other hand, the theory of compressed sensing has showed that if the solution of $\mathbf{s}$ is sufficiently sparse, the l_0-norm minimization problem can be relaxed and solved by optimizing the following objective function [2, 6]:

$$(l_1): \quad \widehat{\mathbf{s}} = \underset{s \in \mathbb{R}^n}{\arg\min} \ \|\mathbf{x} - \mathbf{A}\mathbf{s}\|_2^2 + \lambda\|\mathbf{s}\|_1 \qquad (3)$$

Algorithm 1. One-step-update of SN-SRC

Input: the classification map $\mathbf{L}_0$, the representation vector $\mathbf{P}_0$, the test set $\mathbf{X} \in$
$\quad\quad\quad\mathbb{R}^{l \times N}$, the class number p, λ;
Output: the classification map $\mathbf{L}_1$;
1: **BEGIN**
2: **for** $i = 1$ to N **do**
3: $\quad \mathbf{s}_1 = \underset{\mathbf{s}_1 \in \mathbb{R}^8}{\arg \min} \|\mathbf{X}_i - \mathbf{H}(i)\mathbf{s}_1\|_2^2 + \lambda\|\mathbf{s}_1\|_1$;
4: $\quad$ neighbor_label $= \mathbf{L}_0(\mathbf{H}(i))$;
5: $\quad$ **for** $j = 1$ to p **do**
6: $\quad\quad \mathbf{P}_1(j,i) = \displaystyle\sum_{\text{neighbor\_label}(k)==j} \mathbf{s}_1(k)$;
7: $\quad$ **end for**
8: $\quad \mathbf{L}_1(i) = \underset{j=1,\ldots,p}{\max} (\mathbf{P}_0(i) + \mathbf{P}_1(i))$;
9: **end for**
10: **END**

---

where $\| \cdot \|_1$ denotes the l_1-norm, which is the sum of the absolute value of the vector, and λ is a scalar weight that quantifies the relative importance of minimizing $\mathbf{s}$. Unlike the l_0-norm, the l_1-norm is convex and can be cast as a linear program. In the paper [1], l_1-homotopy method is given to solve equation (2) effectively. The l_1-norm minimization is more efficient than l_0-norm.

3   Spatial-Neighborhood-Integrated SRC (SN-SRC)

Concerning SRC, the test samples are operated one by one when calculating the sparse representation coefficients. In fact, this kind of operation has omitted the spatial information among samples, which is valuable because it is very likely that two neighborhood pixels belong to the same class and they should have similar spectral structure. Hence, the spatial information has been combined with the spectral information by SRC in this paper.

3.1   One-Step-Update Procedure of SN-SRC

SRC has been firstly applied on the hyperspectral data to get a classification map (denoted as $\mathbf{L}_0$), and the "initial" spectral-based representation vector $\mathbf{P}_0(i)$ of the i-th test pixel can be obtained, where $\mathbf{P}_0 \in \mathbb{R}^{p \times N}$, $i \in 1, 2, 3, ..., N$ (N is the number of test samples). Secondly, the neighborhood pixels $\mathbf{H}(i)$ ($\mathbf{H} \in \mathbb{R}^{N \times 8}$) of the i-th pixel have been used to represent the central one (here the eight-neighborhood is adopted for simplicity), and the coefficients corresponding to the same class have been summed up to form the spatial-based representation vector $\mathbf{P}_1(i)$ ($\mathbf{P}_1 \in \mathbb{R}^{p \times N}$) of the i-th test pixel. At last, the two vectors ($\mathbf{P}_0(i)$ and $\mathbf{P}_1(i)$) are added together to yield a new weight vector, and the new classification map ($\mathbf{L}_1$) can be obtained. It is worth to point out that the maximal

---

Algorithm 2. Iterative-updating process of SN-SRC

Input: the training set $\mathbf{A} \in \mathbb{R}^{l \times n}$, the test set $\mathbf{X} \in \mathbb{R}^{l \times N}$, the class number p, λ, ϵ;
Output: the classification map $\mathbf{L}_{final}$;
1: **BEGIN**
2: **for** $i = 1$ to N **do**
3:    $\mathbf{s}_0 = \arg\min_{\mathbf{s}_0 \in \mathbb{R}^n} \|\mathbf{X}_i - \mathbf{A}\mathbf{s}_0\|_2^2 + \lambda \|\mathbf{s}_0\|_1$;
4:    **for** $j = 1$ to p **do**
5:       $\mathbf{P}_0(j, i) = \sum_{class(k)==j} \mathbf{s}_0(k)$;
6:    **end for**
7:    $\mathbf{L}_0(i) = \max_{j=1,\ldots,p} \mathbf{P}_0(i))$;
8: **end for**
9: $t = 0$;
10: **repeat**
11:    $t = t + 1$;
12:    $\mathbf{L}_t = \text{One-step-update}(\mathbf{L}_{t-1}, \mathbf{P}_{t-1}, \mathbf{X}, p, \lambda)$;
13: **until** $|\mathbf{L}_t - \mathbf{L}_{t-1}| < \epsilon$;
14: $\mathbf{L}_{final} = \mathbf{L}_t$;
15: **END**

---

sparse representation coefficient is used to classify the pixel instead of the minimal representation error adopted in the standard SRC method [18]. Detailed description of the one-step-update procedure of the proposed algorithm SN-SRC has been given in Algorithm 1, and Figure 1 shows an example of the one-step-update procedure of SN-SRC. Suppose there are only three classes, and the initial spectral-based classification map is given in Figure 1(a). After applying the one-step-update procedure, the label of the central pixel (i.e., the 2-th class) has been corrected to the 1-th class due to the incorporation of the spatial information (as displayed in the final classification map Figure 1(f)), indicating the effectiveness of the proposed approach.

3.2   Iterative-Updating Process of SN-SRC

Obviously, it can be easily found from Figure 1 that the obtained classification map $\mathbf{L}_1$ is much different from that of $\mathbf{L}_0$, that is, the neighborhood labels of each pixel has been changed. Therefore, in order to further increase the spatial coherence of each class and eventually improve the classification accuracy, the one-step-update procedure can be repeatedly applied on the obtained classification map $\mathbf{L}_t$ in each iteration until the difference between $\mathbf{L}_t$ and L_{t-1} is less than a predefined small value. Algorithm 2 gives the iterative-updating process of SN-SRC.

4   Experimental Results

After presenting the SN-SRC approach in the last section, in this section, one real hyperspectral imagery has been used to evaluate the performance of the proposed approach.

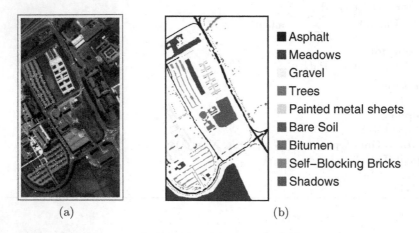

(a)   (b)

Fig. 2. The ROSIS Pavia data set. (a) False color composition of the scene. (b) Ground-truth map containing 9 mutually exclusive land-cover classes.

The hyperspectral data set was acquired by the ROSIS sensor during a flight campaign over Pavia, northern Italy. The number of spectral bands is 103(with spectral range from 0.43 to 0.86 μm). The image size in pixels is 610×340, with very high spatial resolution of 1.3 m/pixel. Figure 2(a) shows a false color composition of the image, while Figure 2(b) shows nine ground-truth classes of interest, which comprise urban features, as well as soil and vegetation features.

In order to illustrate the performance of the proposed approaches with small sample size, 5, 10, 15 labeled training sets are used respectively, and the remaining samples are used as the test set. Meanwhile, the classification results are compared with three state-of-the-art methods, i.e., the standard SRC, support vector machine with linear kernel (SVM), and MLRsub(Subspace Projection-Based Multinomial Logistic Regression Classifier) [9, 12]. For SVM, the parameters are selected by ten-fold cross validation. MLRsub is a subspace-based classification method which has shown good accuracy for hyperspectral classification. Each experiment has been repeated for 10 times with randomly selected training sets and the mean results have been reported. When solving the l_1-norm problem, l1-homotopy method is applied for its high performance. Moreover, to further increase the efficiency of the proposed methods, the loop structure in the program flow has been employed by using the MATLAB parallel Computing toolbox (here 8 workers have been set up in the pool).

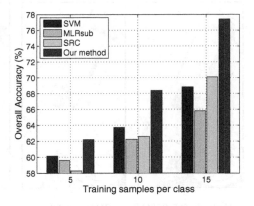

Fig. 3. Overall classification accuracies obtained for the ROSIS Pavia University Data Set using SN-SRC (our approach with 10 iterations), SRC, SVM and MLRsub, with 5, 10, 15 samples per class respectively.

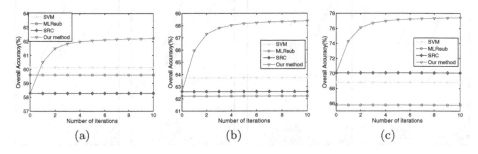

Fig. 4. Overall classification accuracies obtained for the ROSIS Pavia University Data Set using SN-SRC, SRC, SVM, MLRsub with different number of iteration. (a): 5 samples per class; (b): 10 samples per class; (c): 15 samples per class.

Figure 3 displays the overall classification accuracy (OA) of the four compared methods, i.e., SVM, MLRsub, SRC and the proposed approach SN-SRC. OA is the sum of the correctly classified samples divided by the total number of test samples. Obviously, the greater the OA, the better the results. It can be seen from the figure that the classification accuracies improve as the number of the training samples increase. The results obtained by our SN-SRC method are more accurate than those of the other three ones. It can be observed that SN-SRC is a better alternative than the SVM, MLRsub, and the SRC methods.

Moreover, we run the iterative-updating scheme of SN-SRC to improve the classification map. The iterative-updating process has been repeated for several times until the distribution is stable. In Figure 4, the initial values of SN-SRC correspond to the values of SRC due to the connection between the two kinds of methods. Clearly, the performance of the other three methods (SRC, MLRsub and SVM) do not change during the iterative-updating process because the spatial information has not been incorporated. It can be seen from the figure that

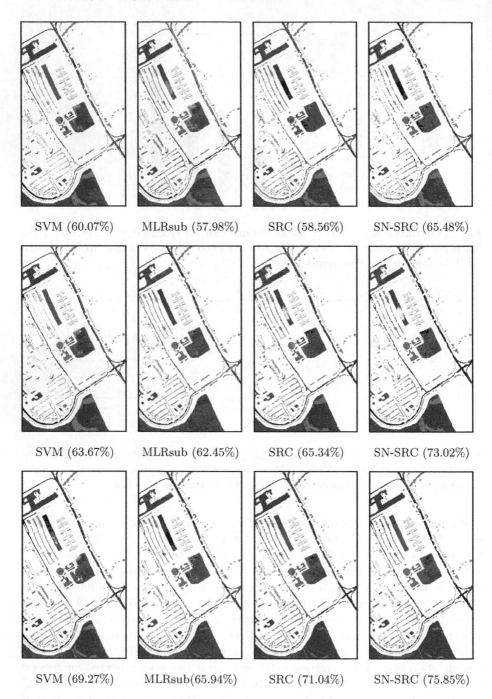

SVM (60.07%) MLRsub (57.98%) SRC (58.56%) SN-SRC (65.48%)

SVM (63.67%) MLRsub (62.45%) SRC (65.34%) SN-SRC (73.02%)

SVM (69.27%) MLRsub(65.94%) SRC (71.04%) SN-SRC (75.85%)

Fig. 5. ROSIS Pavia University data set : classification maps using SVM, MLRsub, SRC, and SN-SRC with different samples(5, 10, 15 respectively, from top to bottom row) per class.

the curve of SN-SRC gradually increases with the progression of the iterative-updating scheme, and eventually reach a stable state, indicating the effectiveness of the proposed spatial-neighborhood-integrated SRC method for hyperspectral classification.

Figure 5 shows the classification maps on the test set obtained from the various techniques with different training sample size (the unlabeled pixels in the scenario are masked and represented in white). It can be visually seen that the maps of SN-SRC are in better accordance with the real map than the others. Concretely, there are many scattered points that are incorrectly estimated in the maps of SRC, in which some are isolated points. On the contrary, through taking into account the spatial information, most isolated points have been corrected to the right class in the maps of SN-SRC.

5   Conclusion

In this paper, we have developed a new approach, called SN-SRC, for spatial-neighborhood-integrated hyperspectral classification. By incorporating the spatial neighborhood information into the central spectral pixel under SRC framework, the spectral and spatial information has been integrated together. Experiments on a real hyperspectral data set have demonstrated the superiority of the proposed algorithm.

References

1. Asif, M.: Primal dual pursuit: a homotopy based algorithm for the Dantzig selector. Master's thesis, Georgia Institute of Technology (2008)
2. Candes, E.J., Tao, T.: Decoding by linear programming. IEEE Transactions on Information Theory 51(12), 4203–4215 (2005)
3. Chang, C.I.: Hyperspectral imaging: techniques for spectral detection and classification, vol. 1. Springer (2003)
4. Chen, Y., Nasrabadi, N.M., Tran, T.D.: Classification for hyperspectral imagery based on sparse representation. In: 2010 2nd Workshop on Hyperspectral Image and Signal Processing: Evolution in Remote Sensing (WHISPERS), pp. 1–4. IEEE (2010)
5. Chen, Y., Nasrabadi, N.M., Tran, T.D.: Hyperspectral image classification using dictionary-based sparse representation. IEEE Transactions on Geoscience and Remote Sensing 49(10), 3973–3985 (2011)
6. Donoho, D.L., Tsaig, Y.: Fast solution of-norm minimization problems when the solution may be sparse. IEEE Transactions on Information Theory 54(11), 4789–4812 (2008)
7. Haq, Q.S.u., Tao, L., Sun, F., Yang, S.: A fast and robust sparse approach for hyperspectral data classification using a few labeled samples. IEEE Transactions on Geoscience and Remote Sensing 50(6), 2287–2302 (2012)
8. ul Haq, Q.S., Shi, L., Tao, L., Yang, S.: Hyperspectral data classification via sparse representation in homotopy. In: 2010 2nd International Conference on Information Science and Engineering (ICISE), pp. 3748–3752. IEEE (2010)

9. Krishnapuram, B., Carin, L., Figueiredo, M.A., Hartemink, A.J.: Sparse multinomial logistic regression: Fast algorithms and generalization bounds. IEEE Transactions on Pattern Analysis and Machine Intelligence 27(6), 957–968 (2005)
10. Landgrebe, D.: Hyperspectral image data analysis. IEEE Signal Processing Magazine 19(1), 17–28 (2002)
11. Lee, M.A., Prasad, S., Bruce, L.M., West, T.R., Reynolds, D., Irby, T., Kalluri, H.: Sensitivity of hyperspectral classification algorithms to training sample size. In: First Workshop on Hyperspectral Image and Signal Processing: Evolution in Remote Sensing, WHISPERS 2009, pp. 1–4. IEEE (2009)
12. Li, J., Bioucas-Dias, J.M., Plaza, A.: Spectral-spatial hyperspectral image segmentation using subspace multinomial logistic regression and markov random fields. IEEE Transactions on Geoscience and Remote Sensing 50(3), 809–823 (2012)
13. Lu, D., Weng, Q.: A survey of image classification methods and techniques for improving classification performance. International Journal of Remote Sensing 28(5), 823–870 (2007)
14. Manolakis, D., Marden, D., Shaw, G.A.: Hyperspectral image processing for automatic target detection applications. Lincoln Laboratory Journal 14(1), 79–116 (2003)
15. Natarajan, B.K.: Sparse approximate solutions to linear systems. SIAM Journal on Computing 24(2), 227–234 (1995)
16. Plaza, A., Benediktsson, J.A., Boardman, J.W., Brazile, J., Bruzzone, L., Camps-Valls, G., Chanussot, J., Fauvel, M., Gamba, P., Gualtieri, A., et al.: Recent advances in techniques for hyperspectral image processing. Remote Sensing of Environment 113, S110–S122 (2009)
17. Prasad, S., Bruce, L.M.: Overcoming the small sample size problem in hyperspectral classification and detection tasks. In: IEEE International Geoscience and Remote Sensing Symposium, IGARSS 2008, vol. 5, p. V-381. IEEE (2008)
18. Wright, J., Yang, A.Y., Ganesh, A., Sastry, S.S., Ma, Y.: Robust face recognition via sparse representation. IEEE Transactions on Pattern Analysis and Machine Intelligence 31(2), 210–227 (2009)

FQZip: Lossless Reference-Based Compression of Next Generation Sequencing Data in FASTQ Format

Yongpeng Zhang, Linsen Li, Jun Xiao, Yanli Yang, and Zexuan Zhu*

College of Computer Science and Software Engineering,
Shenzhen University, Shenzhen, China 518060
zhuzx@szu.edu.cn

Abstract. High-throughput DNA sequence data generated by next generation sequencing (NGS) technologies have brought tremendous stress in data storage and transmission. Data compression serves as a candidate solution to mitigate this pressure. In this paper, a lossless referenced-based compression framework namely FQZip is proposed for NGS data in FASTQ format. Particularly, the three components namely metadata, sequence reads, and quality scores in FASTQ files are compressed independently with specific coding schemes. The sequence reads are aligned to a reference genome and then arithmetic coding, Huffman coding, and LZMA are adopted to store the indispensable alignment results. The metadata and quality scores are stored with other simple yet efficient compression mechanisms. Experimental results on real-world NGS data indicate that FQZip obtains superior compression ratio to other state-of-the-art NGS data compression methods.

Keywords: Next generation sequencing, DNA sequence compression, Reference-based compression, FASTQ, Big data.

1 Introduction

The advance of next generation sequencing (NGS) technologies has remarkably decreased the cost of DNA sequencing to the level of $1,000 for human genomes [1, 2]. The increasing of NGS data volume outpaces the improvement of disk storage capacity [3]. The storage and transmission of NGS data have posed great challenges to downstream analyses. Data compression techniques have been widely adopted to solve this 'big data' issue [3–5].

The majority of NGS platforms produce raw data in FASTQ format [6], which could be further aligned against some reference genomes and converted to SAM/BAM format [7] or VCF format [8]. This study focuses on the compression of raw NGS data in FASTQ format, as it contains the most complete information of the sequencing and is the most compatible format to the downstream applications.

---

* Corresponding author.

© Springer International Publishing Switzerland 2015   127
H. Handa et al. (eds.), *Proc. of the 18th Asia Pacific Symp. on Intell. & Evol. Systems – Vol. 2*,
Proceedings in Adaptation, Learning and Optimization 2, DOI: 10.1007/978-3-319-13356-0_11

An FASTQ file is a text file containing millions of NGS records, each of which consists of metadata (including read name, platform, project ID, etc.), a sequence short read, and a quality score sequence. The three components usually are compressed independently. The metadata and quality scores are compressed as normal text, while the sequence reads, mainly composed of {'A','C','G','T'}, need more specific methods. General-purpose text compression tools like gzip [9] and bzip2 [10] can be adopted to compress the reads but their compression performance is unsatisfactory for not being able to take into account the biological nature of the sequences. Specifically-designed read compression methods are widely categorized into reference-free and reference-based methods [5].

Reference-free methods directly compress the sequence reads with specific coding scheme by taking into account the biological nature of the data. For example, Deorowicz et al. [11] proposed a reference-free compression method DSRC for FASTQ data where input reads are divided into blocks and compressed via LZ77 coding. Jones et al. [12] implemented a reference-free compression method for reads in a tool Quip, using arithmetic coding and high order Markov chains. Bonfield et al. [13] presented Fqzcomp using order-N context model and arithmetic coder to compress NGS data in FASTQ format.

Reference-based methods do not store the original read data but instead the alignment results, like aligned positions and mismatches, of the reads to some reference genomes. For instance, CRAM [14] was implemented based on the alignment and encoding procedure. In addition, a secondary compression frame using de Bruijn graph based assembly approach was also proposed in CRAM to assemble contigs which can be used as references for the compression for unmapped reads. Quip [12] also implements a standard reference-based compression (denoted as Quip -r) and also has a lightweight de novo assembly component to generate references from the target data in case there is no references externally provided. HUGO [15] aligns short reads against a reference genome and stored the exactly mapped reads like other reference-based methods. For the inexactly mapped or unmapped reads, it iteratively shortens the reads so as to realign them against different reference genomes using an adaptive scheme until most of the reads are mapped exactly.

Reference-based methods tend to have superior compression ratio to reference-free methods on sequence reads, especially when suitable reference genomes are readily available. In this paper, we propose a lossless reference-based method namely FQZip for the compression of NGS data in FASTQ format. Particularly, the short reads are aligned to a reference genome using Burrows Wheeler Aligner (BWA) [16]. Indispensable information is extracted and stored from the alignment results so as to recover the nucleotide sequences. The metadata and quality scores are separated from the reads and compressed with Lempel-Ziv-Markov chain-Algorithm (LZMA) [17], run-length coding [18], and/or arithmetic coding [19]. Experimental results on seven real-world NGS data sets indicate that FQZip obtains better compression ratios than other state-of-the-art NGS data compression methods.

The remainder of this paper is organized as follows. Section 2 describes the details of FQZip. Section 3 presents the experimental results of FQZip and other compared methods on seven real-world NGS data sets in FASTQ format. Finally the conclusion is given in Section 4.

2   Method

An FASTQ file usually consists of millions of NGS records. Each record, as shown in Fig. 1, contains four lines. The first, second, and fourth lines of a record correspond to the metadata, sequence short read, and quality scores, respectively. The third line usually is a simple repeat of the metadata which can be ignored in compression. Given an input FASTQ file, FQZip first separates the metadata, short reads, and quality scores into three data streams and then compress them independently according to their own characteristics. Particularly, the repeats in metadata are identified and compressed with LZMA, while the redundancy in quality scores is handled by run-length coding and arithmetic coding. The compression of sequence reads is more complicated. They are aligned against a homologous reference genome using BWA [16] and a few fields are extracted from the resultant SAM [7] file to represent the original reads. The extracted fields are then compressed using arithmetic coding, Huffman coding, and LZMA. Note that different coding schemes are empirically selected for different data components to optimize the overall compression ratio. The outputs from the three data streams are finally packed into a single file. The work flow of FQZip is summarized in Fig. 2.

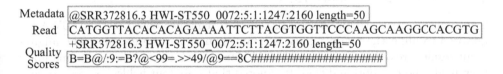

Fig. 1. An NGS record in an FASTQ file

2.1   Metadata

The metadata in an FASTQ file is used to uniquely identify a sequence read. As shown in Fig. 1, the metadata of a sequence read is a line located before the read and started with character '@'. A metadata line typically contains three parts delimited by blank spaces. The first part is the read ID, e.g.,'SRR372816.3'. The second part normally is a character string to record the information such as instrument name, flow cell identifier and tile coordinates. The last part including a keyword 'length=' indicates the length of the read. The first and third parts of the metadata are highly similar or identical for every read, so we just need to store one copy of the them. The most differences come from the second part, which contains fewer repeats across different reads. The second part is extracted for every record and compressed using LZMA.

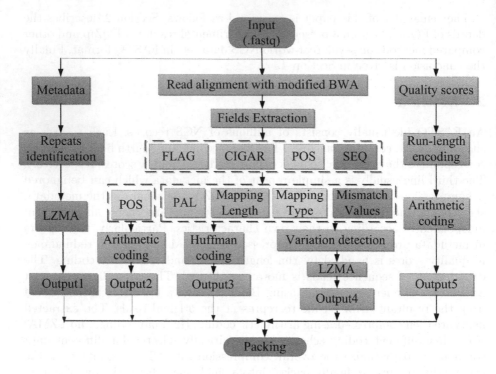

Fig. 2. The framework of FQZip

2.2 Nucleotide Sequence Read

The sequence read stream is compressed with reference-based method. BWA is adopted to align the reads against a homologous reference genome. The alignment results are output and stored in SAM format, which contains eleven mandatory fields and one optional field. It is noted that only four fields, i.e., read bases (SEQ), mapping positions (POS), mapping flag (FLAG) and the mapping result (CIGAR) are sufficient to recover the original nucleotide sequences, assuming the same reference genome is available. For the sake of efficiency, BWA is modified to output only these four fields and specific coding schemes are used to compress the alignment results. A read may be aligned to multiple places on the reference genome, but for the purpose of compression, we need only the first one.

According to the information contained in fields SEQ, POS, FLAG, and CIGAR, a read can be encoded as '[POS] <PAL> <MLength> <MType> <MisValues>...' The descriptions of the coding elements are provided in Table 1. The alignment position POS is mandatory and the other elements are optional. PAL is set to 0 if a palindrome or reverse complement is detected otherwise it is omissible. MLength denotes the number of matches or mismatches in the alignment. Whether it is match or mismatch is specified by the following element MType which could be M (Match), I (Insertion), D (Deletion), or S (Substitution). If mismatches are identified, the mismatch values, i.e., one or multiple bases in {'A','C','G','T'}, should be recorded in MisValues.

Table 1. Read Mapping Results

Encoding elements	Description
POS	The start position on the reference where the read is aligned.
PAL	PAL=0: palindrome
MLength	The number of matches/mismatches in the alignment
MType	M: Match; I: Insertion; D: Deletion; S: Substitution
MisValues	One or multiple bases in {'A','C','G','T'}.

Examples of read encoding are shown in Fig. 3. The first read Read1 is exactly mapped to the reference, so only the mapped position POS=2 is needed to record the read. If there are mismatches in the alignment, more information should be included. For example, Read2 is mapped to position 7, i.e., POS=7, from which eight matches '$8M$' are detected followed by two insertions '$2I$' of 'AG', four matches '$4M$', one deletion '$1D$', and finally the other three matches '$3M$'. To record a palindrome like Read3, PAL=0 is imposed before encoding other mapping results. Note that BWA allows a small of number of approximate matches in each alignment, so a match identified by BWA is not necessary an exact match. For example, a ten-match '$10M$' in Read4 is identified by BWA from position 9 to 18. However, there is actually a substitution of 'A' for 'T' in position 17, which is defined as a variation in this study. All variations should be identified and stored to ensure a lossless compression. Accordingly, a read is reconstructed from the codes and compared to the original sequence in the SEQ field to identify the variations.

As shown in Fig. 2, after encoding the alignment results, the mapped positions and variations of all reads are stored in two separate files and compressed with arithmetic coding and LZMA, respectively. The other information including PAL, MLength, MType, and MisValues is combined together and compressed using Huffman coding.

2.3 Quality Scores

Quality scores are ASCII characters accompanying the sequence reads. A quality score indicates the probability of a corresponding base being correctly determined. Due to the larger alphabet and the quasi-random distributions, quality scores are harder to compress [20, 21]. Yet, it is observed that quality scores could also contain long repeats of the same character, e.g., there are 22 continued '#' shown in Fig. 1. If long repeats (i.e., more than four characters) are detected, run-length coding is first used to encode the scores. Afterward, arithmetic coding is further applied to compress the data.

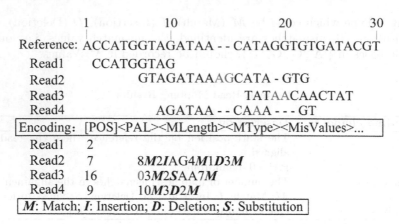

Fig. 3. Examples of read encoding

3   Experimental Results

In this section, the performance of the proposed method FQZip is tested on seven real-world NGS data sets ranging from 1.41 GB to 10.99 GB. All data sets were downloaded from National Center for Biotechnology Information (NCBI) Sequence Read Archive [22] in FASTQ format. The detailed information of the data sets is provided in Table 2.

Table 2. Seven NGS data sets used for performance evaluation

Data	Species	Read Length	# Unmapped Reads	Size (GB)	Reference
ERR008613	*E. coli*	2×100	5,840 (0.04%)	6.34	NC_000913
ERR231645	*E. coli*	51	24,473 (0.11%)	10.99	NC_000913
ERR022075	*E. coli*	2×101	133,518 (2.10%)	1.41	NC_000913
SRR801793	*L. pneumophila*	2×100	424,130 (7.85%)	2.75	NC_018140
SRR022866	*S. aureus*	2×76	3,757,989 (29.41%)	5.49	NC_003923
SRR352384	*S. cerevisiae*	2×76	22,182,981 (85.22%)	9.88	NC_001136.10
SRR327342	*S. cerevisiae*	138	13,568,326 (90.23%)	5.74	NC_001147

Other state-of-the-art NGS data compression methods, including reference-free methods DSRC [11], Quip [12] and Fqzcomp [13] and reference-based methods CRAM [14] and Quip -r [12], are considered for comparison. All reference-free methods support FASTQ input files, whereas CRAM and Quip -r accept only SAM/BAM format, so the FASTQ files are converted to BAM format using SAMtools [7] for both CRAM and Quip -r. All methods are compared in terms of compression ratio, i.e., the compressed file size divided by the original file size. Brief descriptions of the compression methods are provided in Table 3.

Table 3. Overview of NGS data compression methods

Method	Input data format	Brief Description
Quip	FASTQ, SAM, BAM	Reference-free method with Markov chain model and arithmetic coding
DSRC	FASTQ	Reference-free method with LempelCZiv and Huffman coding
Fqzcomp	FASTQ	Reference-free method with context model, arithmetic coding, and deltas coding
CRAM	BAM	Reference-based method with Huffman coding
Quip -r	SAM, BAM	Reference-based compression with arithmetic coding
FQZip	FASTQ	Reference-based method with LZMA, arithmetic coding, Huffman coding, and run-length coding

Table 4. Compression ratios of the methods on the seven NGS data sets

	Quip	DSRC	Fqzcomp	CRAM	Quip -r	FQZip
ERR008613	0.211	0.283	0.234	0.230	0.224	**0.195**
ERR231645	0.139	0.164	0.136	0.125	**0.111**	0.121
ERR022075	0.174	0.227	0.193	0.261	0.168	**0.154**
SRR801793	0.185	0.234	0.202	0.267	0.178	**0.172**
SRR022866	0.235	0.326	0.238	0.279	0.272	**0.232**
SRR352384	0.115	0.145	0.126	0.132	0.134	**0.111**
SRR327342	0.189	0.242	0.202	0.262	**0.186**	0.198
Average	0.167	0.215	0.176	0.194	0.170	**0.159**

All methods are tested on a cluster running 64-bit Red Hat operating system with 32-core 3.1GHz Intel(R) Xeon(R) CPU. The compression ratios of the methods on the seven data sets are tabulated in Table 4. As the input of CRAM and Quip -r is in BAM format, the compression ratios of these two methods are calculated by dividing the compressed file size by the uncompressed SAM file size. Reference-based methods CRAM and Quip -r tend to obtain better compression ratios than reference-free methods in sequence reads. However, considering the whole FASTQ file, i.e., including metadata and quality scores, they do not show obvious superiority to reference-free methods. DSRC does not work as well as other methods, because it sacrifices compression ratio for random access capabilities. It is observed that FQZip obtains lower compression ratios on all test data sets except ERR231645 and SRR327342. The average compression ratio over all data sets also demonstrates the superiority of FQZip to other methods.

The compression ratios of FQZip on the three components of FASTQ files are reported in Table 5. It is shown that FQZip obtains best compression ratios on metadata. The compression ratios of quality scores are higher than the other two components, due to the larger alphabet and the quasi-random distributions. The compression of quality scores remains the major challenge and opportunity to achieve substantial reduction on the storage space of NGS data. Lossy compression could be applied to solve the problem by clustering quality scores with similar values [5]. Yet the use of lossy compression must minimize the loss of accuracy in the corresponding downstream NGS data analyses.

Table 5. The compression ratios of FQZip on the three components of FASTQ files

	Metadata	Sequence Reads	Quality Scores
ERR008613	0.002	0.053	0.385
ERR231645	0.047	0.042	0.418
ERR022075	0.052	0.039	0.316
SRR801793	0.046	0.056	0.369
SRR022866	0.048	0.089	0.540
SRR352384	0.050	0.126	0.130
SRR327342	0.029	0.113	0.420
Average	0.039	0.074	0.368

4 Conclusion

The coming of $1,000 human genome milestone brings us 'big data' challenges on genomics. Efficient data compression methods can help to mitigate this challenge. It is for this consideration, we developed a lossless reference-based compression method FQZip for raw NGS data in FASTQ format. FQZip compresses the metadata, sequence reads, and quality scores in FASTA files independently with different coding schemes. The experimental results on real-world NGS data sets demonstrate the superiority of FQZip in terms of compression ratio to the other state-of-the-art NGS data compression methods including Quip, Fqzcomp, DSRC, CRAM, and Quip -r. Because FQZip is mainly designed to optimize the compression ratio, it spends more time in identifying redundancy than the other methods. Nevertheless, since the compression of NGS data is normally off-line and the data could be compressed once and reused forever, it is acceptable to sacrifice some compression time for a better compression ratio.

Acknowledgment. This work was partially supported by the National Natural Science Foundation of China [61171125 and 61205092], the Guangdong Foundation of Outstanding Young Teachers in Higher Education Institutions [Yq2013141], the Shenzhen Scientific Research and Development Funding Program [JCYJ20130329115450637, KQC201108300045A, and ZYC201105170243A], and the Guang-dong Natural Science Foundation [S2012010009545].

References

1. Pennisi, E.: Will Computers Crash Genomics? Science 331, 666–668 (2011)
2. Kozanitis, C., Heiberg, A., Varghese, G., Bafna, V.: Using Genome Query Language to Uncover Genetic Variation. Bioinformatics 30, 1–8 (2014)
3. Kahn, S.D.: On the Future of Genomic Data. Science. 331, 728-729 (2011)
4. Giancarlo, R., Rombo, S.E., Utro, F.: Compressive Biological Sequence Analysis and Archival in the Era of High-Throughput Sequencing Technologies. Briefings in Bioinformatics 15, 390–406 (2014)

5. Zhu, Z., Zhang, Y., Ji, Z., He, S., Yang, X.: High-throughput DNA Sequence Data Compression. Briefings in Bioinformatics (2013), doi:10.1093/bib/bbt087
6. Cock, P.J.A., Fields, C.J., Goto, N., Heuer, M.L., Rice, P.M.: The Sanger FASTQ File Format for Sequences with Quality Scores, and the Solexa/Illumina FASTQ Variants. Nucleic Acids Research 38, 1767–1771 (2010)
7. Li, H., Handsaker, B., Wysoker, A., Fennell, T., Ruan, J., Homer, N., Marth, G., Abecasis, G., Durbin, R.: The Sequence Alignment/Map Format and SAMtools. Bioinformatics 25, 2078–2079 (2009)
8. Danecek, P., Auton, A., Abecasis, G., Albers, C.A., Banks, E., DePristo, M.A., Handsaker, R.E., Lunter, G., Marth, G.T., Sherry, S.T., McVean, G., Durbin, R.: The Variant Call Format and VCFtools. Bioinformatics 27, 2156–2158 (2011)
9. http://www.gzip.org/
10. http://www.bzip.org/
11. Deorowicz, S., Grabowski, S.: Compression of DNA Sequence Reads in FASTQ Format. Bioinformatics 27, 860–862 (2011)
12. Jones, D.C., Ruzzo, W.L., Peng, X., Katze, M.G.: Compression of Next-Generation Sequencing Reads Aided by Highly Efficient De Novo Assembly. Nucleic Acids Research 40, 171 (2012)
13. Bonfield, J.K., Mahoney, M.V.: Compression of FASTQ and SAM Format Sequencing Data. Plos One 8, e59190 (2013)
14. Fritz, M.H.Y., Leinonen, R., Cochrane, G., Birney, E.: Efficient Storage of High Throughput DNA Sequencing Data Using Reference-Based Compression. Genome Research 21, 734–740 (2011)
15. Li, P., Jiang, X., Wang, S., Kim, J., Xiong, H., Ohno-Machado, L.: HUGO: Hierarchical mUlti-reference Genome cOmpression for Aligned Reads. Journal of the American Medical Informatics Association 21, 363–373 (2014)
16. Li, H., Durbin, R.: Fast and Accurate Short Read Alignment with Burrows-Wheeler transform. Bioinformatics 25, 1754–1760 (2009)
17. http://7-zip.org/sdk.html
18. Storer, J.A.: Data Compression: Methods and Theory. Computer Science Press, Inc., New York (1988)
19. Rissanen, J., Langdon, G.G.: Arithmetic coding. IBMJ Res. Dev. 23, 149–162 (1979)
20. Popitsch, N., von Haeseler, A.: NGC: lossless and lossy compression of aligned high-throughput sequencing data. Nucleic Acids Research 41 (2013)
21. Wan, R., Anh, V.N., Asai, K.: Transformations for the Compression of FASTQ Quality Scores of Next-Generation Sequencing Data. Bioinformatics 28, 628–635 (2012)
22. http://www.ncbi.nlm.nih.gov/

Emergent Effects from Simple Mechanisms in Supply Chain Models

Gabriel Yee[1], Yew Soon Ong[2], and Puay Siew Tan[1]

[1] Singapore Institute of Manufacturing Technology, Singapore
(qmyee,pstan)@simtech.a-star.edu.sg
[2] Nanyang Technological University, Singapore
asysong@ntu.edu.sg

Abstract. It has been widely recognised that today's supply chains are a lot more complex than those in the past. This recognition has prompted calls for supply chains to be recognized and be studied as complex adaptive systems to make them more relevant to the needs of the modern-day, real-world supply chain. In the context of supply chain modelling, these calls are already answered by a number of prominent works that study the emergence of supply chain topology from undefined supply chain structures. In this paper, we present another form of emergence and demonstrate that it can be induced by simple mechanisms when a supply chain is modelled as a complex adaptive system.

Keywords: supply chain, complex adaptive systems, agent-based simulation, emergence.

1   Introduction

Just over a decade ago, Choi et al. [1] wrote for a need to recognize supply chains as complex adaptive systems and to highlight that traditional, deterministic means of supply chain management is reaching its limits in the face of modern-day challenges. The same sentiment was also echoed in works of Surana et al. [2] and Pathak et al. [3]. Reasons, as it was given, for this need stems from the increasing difficulty in the efforts needed to optimize operations, increasingly global operations, and increasingly demanding customers. It was also recognized that firms' efforts to manage supply chains, in the face of such complexity, have often led to frustration and helplessness, thus prompting a call to develop a new mental model that accurately reflects the true underlying complexity and dynamism of the modern day supply chain [1].

In recognizing supply chains as complex adaptive systems, one has to confront with a cornerstone notion in complexity theory that is controversial as it is remarkable – emergence. The debate on the meaning and significance of emergence is not new ([4-8] as cited in [9]) and is seeing a revival in recent years, particularly across the social sciences ([10-17] as cited in [9]). It is theorized that this is due to the increasing attention given to the study of self-organizing and adaptive properties of complex adaptive systems and a renewed interest in and debate over the emergentist hypotheses in philosophy [9]. Nevertheless, in recognizing supply chains as complex adaptive systems, we concern ourselves with emphasizing that, given appropriate conditions, emergence should also occur in supply chains.

As of writing, the predominant direction for modelling supply chains as complex adaptive systems are pioneered by the works of Pathak [18-20] and Li [21, 22]. Emergence, as demonstrated by their works, is the coming into existence of a supply chain topology that is dependent on the three foci for the dynamics of a supply chain that Choi et al. [1] presented. However, we postulate that another form of emergence exists. This form of emergence, the contribution of this work, is the coming into existence of a new power structure within a single tier or echelon of a supply chain. We demonstrate this form of emergence by studying a simple model of monopolistic competition[1] where boundedly rational agents are imbued with simple adaptive mechanisms and products are differentiated by selling prices. Our simulation results show that, under suitable conditions, a structure of higher power concentration can emerge purely out of competitive actions.

2   The Model

Similar to the works of Pathak and Li, we employ the use of an agent-based simulation model. We create the model, using NetLogo [23], as a generic model of a three tier supply chain in which firms, or nodes as they are named in the model, trade and jostle for advantage against each other. The traded goods are modelled to be homogeneous and non-perishable, and all nodes act as price takers. Within the model also exists order agents and a pair of economic agents whose supporting roles and behaviors are described in the following sub-sections.

2.1   Economic Agent Behaviors

The economic agents are modelled to represent the economic demand and supply that is outside of the supply chain. Due to their naming, it is important to explicitly mention that these agents are not *homo economicus*, but instead function as a source and sink for goods in the supply chain. Situated at the most upstream end of the supply chain and as a source for goods, the role of Source economic agent is to introduce goods into the supply chain and it does so by selling the full amount of goods requested by the most upstream nodes at a pre-set price. In line with the works of Pathak & Dilts [18] and Pathak et al. [19], this would be synonymous with having a munificent environment.

At the most downstream end of the supply chain, the role of the Sink economic agent is to remove goods from the supply chain and it does so by buying goods from

---

[1] For the sake of brevity and focus, we avoid discussions about market structure. The interested reader will find more knowledge on the topic in any standard economics textbook or, with our recommendation, see Hal Varian *Intermediate microeconomics: a modern approach.* 2009, WW Norton & Company New York.

the most downstream nodes at the price offered by the node. The quantity demanded q_{sink} by the Sink from the node offering a price p is determined by applying an exponential demand function with increasing returns to scale ([24-28] as cited in [29]) as represented by

$$q_{sink} = Q^0 e^{-\beta_0 p} \tag{1}$$

For the purposes of our experimental runs, we set $Q^0 = 50$, and $\beta_0 = 0.01$ such that quantity demanded by the Sink will be 50 when the price is at its minimum $p = 0$. Eq 1, after simplification, can be expressed as

$$q_{sink} = 50 \cdot e^{-0.01p} \tag{2}$$

2.2 Order Agents

The order agents function as carriers of the goods that are traded between supply chain nodes and store information such as the time for the different lifecycle stages of the order, quantity of goods that is requested or shipped, the price at which the goods were bought at, and the time delay before the goods are to be delivered. Fundamentally, the order agents do not play a significant role in the outcome of any simulation run but are modelled to facilitate the collection of run-time statistics.

2.3 Node Behaviors

Each node in the model is modelled to be only eligible to trade with its immediate upstream and downstream tiers at discrete time periods and is subjected to limited visibility of the population of eligible nodes. All nodes begin with only one upstream supplier but will gradually 'discover' new upstream supplier nodes with a pre-set probability ρ. Upon a successful 'discovery', both upstream supplier and downstream consumer will gain visibility of each other and be eligible to commence trading. Bounded rationality is imposed on the nodes by restricting its decision-making information to only consist of locally-known downstream demand for its goods, its history of profit earned, and prices shared to it from its visible upstream supplier nodes.

Each node is modelled to have simple trading behavior. At each time period, nodes will determine a replenishment quantity using a standard (s S) inventory policy and place orders with its upstream supplier nodes, fulfil orders that it has received from its downstream nodes, then receive orders that had completed delivery at the end of the period. At regular intervals κ, the nodes will perform two extra actions at the beginning of the time period: modify its attributes to increase its ability to amass profits and evaluate its preferred upstream supplier.

Every node is profit-chasing and would adapt to amass more profits. To ensure that nodes are boundedly rational, they are not modelled with the ability to calculate the profit-maximizing combination of price and handling capacity as they do not know their downstream customer demand function. Instead they would adapt only using information from the customers' purchasing behavior, by means of a simple mechanism that makes use of its profit history. Over its immediate past two intervals t

and $t - 1$, the node will determine if its currently adopted strategy is increasing its profits π, i.e. $\pi(t) > \pi(t - 1)$. If $\pi(t) > \pi(t - 1)$, then the node would know that its current adopted strategy is increasing its profits and would continue to revise its selling price and handling capacity in the same direction as its strategy in the previous interval. If, however, $\pi(t) \leq \pi(t - 1)$, the node would know that its strategy is not leading to an increase of profits and would revise its selling price p and handling capacity c in the opposite direction. By modifying these internal attributes, nodes execute their strategy by revising their increments to selling price Δp and increments to handling capacity Δc. Adapted from Li et al. [21] and Li et al. [22], the revisions to selling price and handling capacity are expressed as

$$p(t + 1) = p(t) + \Delta p \qquad (3)$$

$$c(t + 1) = c(t) - \Delta c \qquad (4)$$

where $sgn(\Delta p) * sgn(\Delta c) = 1$, $c(t + 1) \geq 1$, and $p(t + 1) \geq$ the weighted average of its cost of goods sold.

Whenever a replenishment quantity is demanded by a node, it will exhibit preferential treatment by placing an order with its elected preferred supplier node. However, since all nodes are modelled to be capacitated, there exist possibilities where the entirety of the replenishment quantity would not be fulfilled by a node that has reached its capacity limit. In such a situation, the customer nodes would then attempt to place orders with its other visible upstream supplier nodes beginning from the ones having the lowest selling prices until its entire replenishment quantity is fulfilled.

Since each node, when placing orders, is a price taker, its profitability is affected by the price that is offered by its choices of upstream supplier nodes. Hence, it experiences a motivation to adapt and elect preferred suppliers that offer lower prices. To mimic real-life loyalty arising from preference, a maximal level that can be experienced is thus required when considering whether to replace preferred upstream supplier nodes. This maximal level, introduced as a rigidity threshold α, is part of the nodes' re-evaluation criteria in selecting preferred upstream supplier nodes. In this manner, nodes will seek out the lowest selling price p_L amongst their non-preferred suppliers and compare it with the selling price offered by their current preferred supplier p_P. If the ratio of price difference between the two suppliers exceeds the rigidity threshold, i.e.$(p_P - p_L) / p_P > \alpha$, then the supplier node offering p_L will be elected to be the new preferred supplier.

As the final activity before the end of the time period, the node will calculate its profits and store the value as part of its profit history. Each node calculates its profits by subtracting its expenses from its proceeds. Proceeds for this matter would be the product of the quantity of goods that were sold and the price at which they were sold at. The nodes expenses consist of two components, expenses from purchases – a product of the quantity of goods requested and the price at which they were sold at – and a variable cost function that is related to the handling capacity of the node. The variable cost function, an adaption of the Total Cost [30] and Cobb-Douglas production [31] function with constant returns to scale[2], is expressed as[3]

$$f(c) = P_K K + P_L \frac{c^2}{A^2} \tag{5}$$

where $P_K K$ represents the fixed cost, P_K the price per unit of capital employed, K the number of units of capital employed, $P_L \frac{c^2}{A^2}$ the total variable cost, P_L the price per unit of labor used, c the handling capacity of the node also used in Eq 4, and A represents the total factor productivity. As a replacement for output, the entire handling capacity of the node is used instead of just the utilized capacity as nodes are modelled to allocate the entire amount of resources in anticipation for its entire capacity to be utilized. Unutilized capacity is intentionally modelled to 'hurt' the profitability of the nodes so that nodes are motivated to adapt towards being 'fitter' competitors. For the purposes of our experimental runs, we set $P_K = 1$, $K = 1$, $P_L = 10$, $A = 100$. Eq 5, after simplification, can be expressed as

$$f(c) = 1 + 0.001 \cdot c^2 \tag{6}$$

At this point, the astute reader might note that traditional supply chain costs such as backorder costs, lost sale costs, inventory or holding costs etc. are not taken into account in our model. This decision was purposefully made as the addition of such considerations hold little significance to the purpose and adds little value to our model.

2.4   Power Concentration

Typically, performance measurements for supply chain simulation models are adapted from metrics suggested by the SCOR model [32]. However, given the nature of the changes that our work focuses on, the usefulness of such metrics are very much limited as there is little change to the operating processes in the simulated supply chain.

To track the structural change in the supply chain, we find the most appropriate measure to be the Herfindahl-Hirschman Index (HHI) [33-35]. As an economic concept, the HHI takes into account the size distribution of firms within a market, taking on a value close to zero when the market is occupied by large number of firms with relatively equal power and rising to a maximum of 10,000 points when a single firm controls the market. In our model, the size distribution needed for the HHI can be related directly to the handling capacity of the nodes. Nodes with higher handling capacity would be able to satisfy greater demanded quantities of goods, and thus are 'bigger' and would naturally have the ability to command a larger market share. The Herfindahl–Hirschman Index H is expressed as

$$H = \sum_i h_i^2 \tag{7}$$

---

[2] In standard microeconomics, returns to scale are purely imposed by technology. For simplicity, we assume all nodes have the employ similar technology with constant returns to scale.

[3] We direct the interested reader to Appendix A for the mathematical derivation for the adaptation of the total cost and Cobb-Douglas production functions.

where h_i represents the market share of the ith node. We dictate the market share h of a node as the ratio of the quantity of goods demanded of it over the total quantity of similar goods demanded by all nodes in the model within a time period.

3   Results

3.1   Premises

When performing the simulation runs, the following values for common parameters are enforced: random seed for NetLogo = 100, number of Tier 1 nodes = 10, number of Tier 2 nodes = 50, number of Tier 3 nodes = 250, p_{source} = 10, κ = 3. Each simulation run goes through 500 time steps or periods, a value that was found to be sufficiently long enough to show the desired behaviors of each varied parameter.

It is also important to note that data from Tier 3 nodes were purposefully left of out consideration. This is because of the nature of how the model implements demand in the most downstream tier, in this case, purchases from the Sink economic agent. Unlike Tier 1 and Tier 2 nodes that receive their demand for goods subject to preferential treatment, the demand for Tier 3 nodes is dictated by Eq. (1), hence causing a demand volume that is unaffected by the decisions of other Tier 3 nodes. Due to this implementation, the market share, and corresponding HHI value, for the Tier 3 nodes would not change and thus do not provide any observation or trends worthy of interest.

Another issue to highlight would be that the results from Tier 1 and Tier 2 nodes would be markedly different from each other. This because the number of Tier 1 and Tier 2 nodes within the supply chain differ and thus should have different run-time dynamics and behaviors even though the similar trends can be observed.

To conduct the simulation runs, the BehaviorSpace feature in NetLogo was used and set to varying values of the probability to discover new upstream suppliers ρ, capacity growth increment Δc, price growth increment Δp, and rigidity threshold α. At the end of each run, a special snippet of code is executed to export the required data from the outcome of each simulation, and then later combined to present the results of the simulation run.

3.2   Analysis

Observation 1: The probability to discover new upstream suppliers ρ influences the length of time required for high concentration to emerge.

Fig. 1 presents a trend showing that when nodes have a greater chance of discovering a new upstream supplier ρ, the length of time required for higher concentration to occur is shorter. Thus, it appears that having a higher ρ allows nodes to be able to discover the more attractive suppliers in a shorter length of time, thus increasing their visibility of the supply chain, and consequentially leading to the more attractive suppliers being discovered within a shorter length of time.

Upon a more thorough examination of the results for a wider range of values for ρ, it was also found that results of $\rho > 1.0$ were uncannily similar even over different random seeds. This finding suggests that there exists a special event experienced by nodes during the simulation run. We postulate this special event to be the event in which either the node discovers the most attractive suppliers and/or one or more of its visible suppliers evolve to become the most attractive suppliers. When a significant number of nodes experiences this event, a 'tipping point' would occur in the supply chain whereby the attractive supplier nodes will begin to experience an amplification of their attractiveness as they begin on their evolutionary track towards lower prices and higher handling capacities. Higher values of ρ should contribute to a faster discovery, but the results suggests that there will be a continued increase to the effect after $\rho > 1.0$ but will eventually hit a peak and then decrease at a very slow pace.

Observation 2: The magnitude at which capacity changes Δc influences the length of time required for high concentration to emerge.

When the simulation run is initialized, nodes at different tiers have different starting capacity values - $c_{T1Node} = 1000$, $c_{T2Node} = 100$, $c_{T3Node} = 10$. With the exception of the Tier 3 nodes that sell directly to the Sink economic agent, every node is evenly allocated five downstream customer nodes and has the capacity to support up to ten. These values are sufficient for nodes to survive and function given the initial conditions but are insufficient for long-term survivability if the nodes do not adapt to the changing conditions.

In Fig. 2, it can be observed that with greater incremental magnitudes to ty Δc, the faster the supply chain will evolve to a structure of higher power concentration. We postulate that this behavior is expected as a node's ability to gain market share is tied to the amount of goods that it can handle and sell. Thus, logically, for a node to command a large market share, its capacity must first grow to a level in which it is able to handle the larger volume needed to maintain the larger market share. Consequentially, this would also imply that the emergence of a higher power concentration structure is dependent on the emergence of a firm that is able to handle the large sales volume that is requested of it.

If the growth is constrained to a small value, then naturally, it would take a longer length of time for a firm with a large handling capacity to emerge. Additionally, with sufficiently low values of Δc, larger number of nodes within the same tier would co-evolve to be able to support the demand required of the tier, thus making it possible for the supply chain to never achieve a structure of higher power concentration.

Observation 3: The magnitude at which prices changes Δp influences the length of time required for high concentration to emerge.

In Fig. 3, it can be observed that with greater incremental magnitudes to price Δp, the shorter the length of time before the supply chain evolves to a structure of higher power concentration. We postulate that the reason for this behavior is that with greater incremental magnitude, there would be a greater magnitude of price differences exhibited by the nodes and a greater chance that a node will switch to a supplier that offers a more attractive selling price, thus resulting in a shorter length of time for the disparity between supplier nodes to emerge.

A secondary behavior that requires acknowledgement is the observation that values of Δp increases the disparity between nodes at a significantly faster rate as compared to any of the other simulation runs. The effect of this behavior can be seen in Fig. 4 where the gap between attractive nodes with lower prices and unattractive nodes with higher prices can be seen. The explanation behind this secondary behavior is that the rational, price-taking nodes are modelled to increase their prices because the weighted average of their cost of goods sold increases when suppliers evolve to increase prices. This effect, when compounded from Tier 1 nodes to Tier 3 nodes can, potentially, be huge and drastically increase prices across the supply chain. A significant side effect that was also observed was that the rising prices cause lower quantities of goods to be demanded by the Sink economic agent as presented in Fig. 5. This side effect also influences the length of time needed for the structure of high concentration of power to emerge as the attractive supplier nodes would not have to grow their handling capacity too greatly to satisfy their downstream customer nodes' demand.

Observation 4: The rigidity threshold, or habituation, towards suppliers α influences the length of time required for high concentration to emerge.

In Fig. 6 and Fig. 7, an obvious trend is that the larger the value of α, the longer the length of time that the supply chain will evolve to a structure of higher power concentration. The trend is more visibly obvious in Fig. 7 where the HHI values are consistently lower and the intervals of consistent HHI values become more increasingly spaced out. These results suggest that the relationship between the rigidity threshold and the length of time needed for emergence is not likely to be linear.

We postulate that the explanation for such behavior is that the local effect caused by the nodes as they switch their preferred upstream supplier results in a global effect of tier-wide competition. Given the nature of the model, having preferred upstream supplier status confers the advantage of receiving the first assignment of replenishment demand, thus guaranteeing a sales volume that would contribute to profits, which would then amplify the nodes attractiveness through its price and capacity adjustments. Not having the preferred upstream supplier status may lead the dampening of the node's attractiveness as it adapts negatively through its price and capacity adjustments. Over many periods and many nodes, the consequence of this advantage is a global effect that manifests itself in the form of a disparity between nodes that are attractive or unattractive as suppliers. As the disparity widens, it leads to the inevitable outcome that the concentration of power within the supply chain will fall only to a privileged few. With a lower rigidity threshold, or less habituation, switching decisions by nodes are faster, creating a more competitive environment, leading to a faster rate at which the disparity will grow and a shorter length of time for the high power concentration structure to emerge.

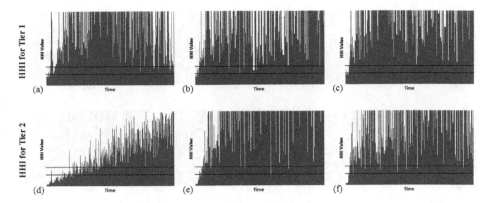

Fig. 1. The effect of varying values of ρ on the HHI for each tier is presented. For all cases, the following parameters are applied to all nodes: $\alpha = 5$, $\Delta p = 1$, and $\Delta c = 125$. Two lines are also included as visual indicators when the HHI is greater than 1500 and 2500[4]. In (a) & (d), $\rho = 0.1$; in (b) & (e), $\rho = 1.0$. (c) & (f), presents the results where nodes have perfect discovery of their upstream suppliers, i.e. all upstream suppliers are revealed. This extreme case was investigated because values of $\rho > 1.0$ produced results that were uncannily similar. From these results, it can be observed that with increasing values of ρ, the length of time needed for the structure of high power concentration to emerge is shorter[5].

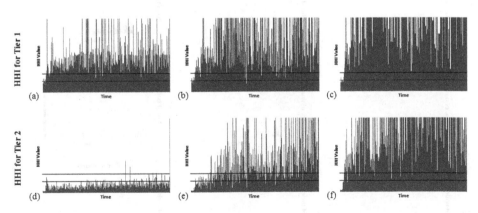

Fig. 2. The effect of varying values of Δc on the HHI for each tier is presented. For all cases, all nodes have perfect visibility of their upstream supplier nodes and the following parameters $\alpha = 5$ and $\Delta p = 1$ are applied. In (a) & (d), $\Delta c = 5$; in (b) & (e), $\Delta c = 25$; and in (c) & (f), $\Delta c = 125$. From these results, it can be observed that with increasing values of Δc, the length of time needed for the structure of high power concentration to emerge is shorter.

---

[4] The HHI values of 1500 and 2500 are dictated by the U.S. Department of Justice and Federal Trade Commission to indicate moderate and high concentration. See U.S. Department of Justice, Federal Trade Commission, *Horizontal Merger Guidelines*, 2010.

[5] This can be interpreted by observing the HHI values for each graph. Whenever the value of HHI exceeds 1500 or 2500, a structure of moderate or high concentration was reached.

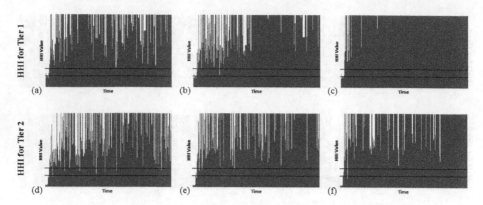

Fig. 3. The effect of varying values of Δp on the HHI for each tier is presented. For all cases, all nodes have perfect visibility of their upstream supplier nodes and the following parameters $\alpha = 5$ and $\Delta c = 125$ are applied. In (a) & (d), $\Delta p = 1$; in (b) & (e), $\Delta p = 5$; and, in (c) & (f), $\Delta p = 25$. From these results, it can be observed that with increasing values of Δp, the length of time needed for the structure of high power concentration to emerge is shorter.

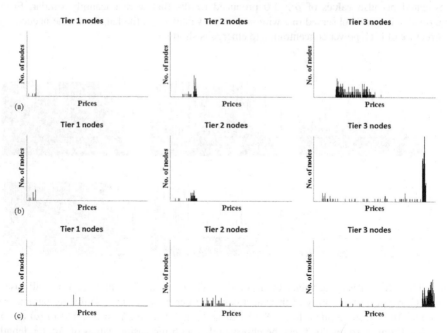

Fig. 4. The distributions for offered selling prices by each tier of nodes for the same three runs in Fig. 3. are presented. In (a), $\Delta p = 1$; in (b), $\Delta p = 5$; and, in (c), $\Delta p = 25$. From these results, it can be observed that with increasing values of Δp, there is an increase in spread of offered sell prices, thus showing a greater disparity between nodes since attractiveness is modelled to directly related to prices. Also, as evidenced by their positions on the x-axis, prices also evolve to become higher, leading to lower demand from the Sink economic agent as seen in Fig. 5.

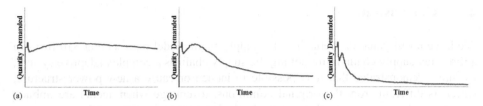

Fig. 5. The quantity of goods demanded over time by the Sink economic agents for varying values of Δp is presented. In (a), $\Delta p = 1$; in (b), $\Delta p = 5$; and, in (c), $\Delta p = 25$.

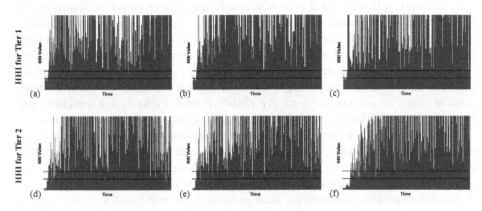

Fig. 6. The effect of varying values of α on the HHI for each tier is presented. For all cases, all nodes have perfect visibility of their upstream supplier nodes and the following parameters $\Delta p = 1$ and $\Delta c = 100$ are applied. Unlike the other experimental runs, a lower value Δc is applied to 'slow down' the process of evolution towards a structure of high power concentration power. In (a) & (d), $\alpha = 0$; in (b) & (e), $\alpha = 5$; and, in (c) & (f), $\alpha = 10$.

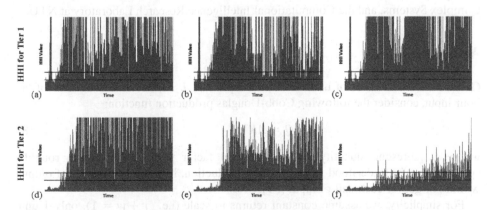

Fig. 7. These results are a continuation of the results presented in Fig. 6. In (a) & (d), $\alpha = 15$; in (b) & (e), $\alpha = 20$; and, in (c) & (f), $\alpha = 25$. From these results, it can be observed more obviously that with increasing values of α, the length of time needed for the structure of high power concentration to emerge is shorter.

4   Conclusion

We have investigated the behavior of a supply chain model representing 310 firms in a three-tier supply chain. In modelling the supply chain as a complex adaptive system, we have demonstrated that it is possible to induce or cause a new power structure, which is different from the original conditions, to emerge when agents are imbued with simple adaptive mechanisms. This form of emergence is different from previously published works in the area of modelling supply chains from a complex adaptive systems perspective.

In this work, it is shown that even with simple sets of micro-level mechanisms, it is possible to induce a change at the macro-level. This highlights the need to dedicate a greater depth of thought when examining the results of supply chain simulation runs. The extra effort is needed as it is possible for an unwitting modeler to implement mechanisms that alter the structure and dynamics of a supply chain during a simulation run. Therefore, in interpreting results, an explicit recognition or consideration must be given to account for a supply chain's propensity to evolve structures and dynamics that are different from its initial conditions and behavioral assumptions.

Based on this work, we see several possible opportunities for future research. Some promising directions include examining our work in the context of (1) its evolution when drastic changes, such as supply chain disruptions, are introduced; (2) when distributed values for the parameters of the nodes are used; and, (3) when more sophisticated adaptation mechanisms, such as mimicry or rule-based decision making, are modelled.

Acknowledgements. This work is partially supported under the A*Star-TSRP funding, Singapore Institute of Manufacturing Technology-Nanyang Technological University (SIMTech-NTU) Joint Laboratory and Collaborative research Programme on Complex Systems, and the Computational Intelligence Research Laboratory at NTU.

Appendix A

Given K as the number of units of capital input and L as the number of units of labour input, consider the following Cobb-Douglas production function:

$$Q = AK^r L^w \tag{8}$$

where Q represents quantity, and A the total factor productivity, which roughly speaking, can be understood as the scale of production, i.e. how many units of output would one get for each unit of input [31].

For simplicity, we assume constant returns to scale (i.e. $r + w = 1$), only 1 unit of capital is employed (i.e. $K = 1$), and capital and labour inputs contribute equally to the output (i.e. $r = w = 0.5$). Rearranging, Eq. 8,

$$L = \frac{Q^2}{A^2} \tag{9}$$

Now, given the price for each unit of labour P_L and price per unit of capital P_K, the total cost TC for a firm, which is the sum of its fixed costs and variable costs, can be expressed as

$$TC = P_K K + P_L L \tag{10}$$

where $P_K K$ represents the fixed cost, $P_L L$ the variable cost, and K and L retain the same conventions used in Eq 8.

Combining Eq 9 and Eq 10,

$$TC = P_K K + P_L \frac{Q^2}{A^2} \tag{11}$$

For a numerical example, using $K = 1, P_K = 1, P_L = 1, A = 100, r = w = 0.5,$

$$Q = 10 \cdot (1)^{0.5} \cdot L^{0.5} \Rightarrow Q = 10L^{0.5} \Rightarrow L = \frac{Q^2}{10000}$$

$$TC = (1)(1) + (10)\left(\frac{Q^2}{10000}\right) \Rightarrow TC = 1 + 0.001 \cdot Q^2$$

References

1. Choi, T.Y., Dooley, K.J., Rungtusanatham, M.: Supply networks and complex adaptive systems: control versus emergence. Journal of Operations Management 19(3), 351–366 (2001)
2. Surana, A., et al.: Supply-chain networks: a complex adaptive systems perspective. International Journal of Production Research 43(20), 4235–4265 (2005)
3. Pathak, S.D., et al.: Complexity and adaptivity in supply networks: Building supply network theory using a complex adaptive systems perspective. Decision Sciences 38(4), 547–580 (2007)
4. Ablowitz, R.: The theory of emergence. Philosophy of Science 6(1), 1–16 (1939)
5. Lovejoy, A.O.: The meanings of 'emergence' and its modes. Philosophy 2(06), 167–181 (1927)
6. Pepper, S.C.: Emergence. The Journal of Philosophy 23(9), 241–245 (1926)
7. Alexander, S.: Space, time, and deity: the Gifford lectures at Glasgow, 1916-1918, vol. 2. Macmillan (1920)
8. Morgan, C.L.: Emergent evolution (1923)
9. Martin, R., Sunley, P.: Forms of emergence and the evolution of economic landscapes. Journal of Economic Behavior & Organization 82(2), 338–351 (2012)
10. Bickhard, M.H., Campbell, D.T.: Emergence (2000)
11. Clayton, P., Davies, P.: The re-emergence of emergence. Oxford, New York (2006)
12. Cunningham, B.: The reemergence of emergence. Philosophy of Science, 62–75 (2001)
13. Kim, J.: Emergence: Core ideas and issues. Synthese 151(3), 547–559 (2006)
14. Lawson, T.: Ontology and the study of social reality: emergence, organisation, community, power, social relations, corporations, artefacts and money. Cambridge Journal of Economics 36(2), 345–385 (2012)
15. Sawyer, R.K.: Social emergence: Societies as complex systems. Cambridge University Press (2005)

16. Sawyer, R.K.: Emergence in Sociology: Contemporary Philosophy of Mind and Some Implications for Sociological Theory1. American Journal of Sociology 107(3), 551–585 (2001)
17. Stephan, A.: Emergence–a systematic view on its historical facets. Emergence or Reduction, 25–48 (1992)
18. Pathak, S., Dilts, D.: Simulation of supply chain networks using complex adaptive system theory. In: 2002 IEEE International Engineering Management Conference, IEMC 2002. IEEE (2002)
19. Pathak, S.D., Dilts, D.M., Biswas, G.: Next generation modeling III-agents: a multi-paradigm simulator for simulating complex adaptive supply chain networks. In: Proceedings of the 35th Conference on Winter Simulation: Driving Innovation, Winter Simulation Conference (2003)
20. Pathak, S.D., Dilts, D.M., Biswas, G.: On the evolutionary dynamics of supply network topologies. IEEE Transactions on Engineering Management 54(4), 662–672 (2007)
21. Li, G., et al.: Modeling and simulation of supply network evolution based on complex adaptive system and fitness landscape. Computers & Industrial Engineering 56(3), 839–853 (2009)
22. Li, G., et al.: The evolutionary complexity of complex adaptive supply networks: a simulation and case study. International Journal of Production Economics 124(2), 310–330 (2010)
23. Wilensky, U.: NetLogo. Center for Connected Learning and Computer-Based Modeling. Northwestern University, Evanston, IL (1999)
24. Cowling, K., Cubbin, J.: Price, quality and advertising competition: an econometric investigation of the United Kingdom car market. Economica, 378–394 (1971)
25. Blattberg, R.C., Levin, A.: Modelling the effectiveness and profitability of trade promotions. Marketing Science 6(2), 124–146 (1987)
26. Blattberg, R.C., Wisniewski, K.J.: Price-induced patterns of competition. Marketing Science 8(4), 291–309 (1989)
27. Krishnamurthi, L., Raj, S.: A model of brand choice and purchase quantity price sensitivities. Marketing Science 7(1), 1–20 (1988)
28. Bolton, R.N.: The robustness of retail-level price elasticity estimates. Journal of Retailing 65(2), 193–218 (1989)
29. Hanssens, D.M., Parsons, L.J.: Econometric and time-series market response models. Handbooks in Operations Research and Management Science 5, 409–464 (1993)
30. Varian, H.R.: Cost Curves, in Intermediate microeconomics: a modern approach, p. 379. WW Norton & Company, New York (2009)
31. Varian, H.R.: Technology, in Intermediate microeconomics: a modern approach, p. 335. WW Norton & Company, New York (2009)
32. Supply Chain Council, Supply Chain Operations Reference Model
33. U.S. Department of Justice and Federal Trade Commission, 2010 Horizontal Merger Guidelines (2010)
34. Hirschman, A.O.: The paternity of an index. The American Economic Review, 761–762 (1964)
35. Rhoades, S.A.: Herfindahl-Hirschman Index. The. Fed. Res. Bull. 79, 188 (1993)

Integrated Solid Waste Management System Design under Uncertainty

Jie Xiong, Tsan Sheng Adam Ng, and Shuming Wang

Department of Industrial and Systems Engineering, National University of Singapore,
1 Engineering Drive 2, Singapore 117576

Abstract. Solid waste management (SWM) has become one of the priority issues to decision makers in modern municipalities. This paper proposes a life cycle assessment-based (LCA-based) two-stage mixed-integer stochastic programming model for an integrated solid waste management (ISWM) system design. Different from previous models that commonly focus on improving the overall performance of an ISWM system, the proposed model considers the sustainability of each involved individual agent within the complex system. In addition, the proposed model simultaneously considers both internal and external uncertainties during the ISWM system design. The simulation results show that the proposed model can help decision makers identify what the optimal ISWM system design under uncertainties is and how uncertainties influence the optimal ISWM system design. The model comparison shows that considering individual agent sustainability during the ISWM system design process can help enhance the system stability under uncertain circumstances.

Keywords: Integrated solid waste management system, life cycle assessment, two-stage stochastic programming, individual agent sustainability.

1 Introduction

Industrialization, urbanization, population expansion, and economic growth result in continuous increase of solid waste generation in recent decades. Since generated solid wastes in municipalities pose a risk to contamination of surface and ground water, air pollution, soil pollution and spreading of diseases, properly managing them has become one of the priority issues to decision makers of modern municipalities.

The concept of ISWM was first proposed in the early 1990s, and has gradually become its current paradigm throughout the world [1, 2]. Compared to typical SWM approaches optimizing the operation of a certain SWM facility or the treatment and disposal of a certain solid waste type [e.g. 3, 4], the ISWM approach analyses, optimizes and manages the SWM practices from a holistic perspective with considering the full array of solid waste types and integration patterns among different SWM facilities throughout the entire life of SWM process [5]. Therefore, the ISWM approach shows a higher applicability in the real-world SWM practices.

This paper intends to apply the ISWM approach to design and evaluate the SWM system under uncertain circumstances. Specifically, a life cycle assessment (LCA-based) two-stage mixed-integer stochastic programming model with considering individual agent sustainability for the ISWM system design is established in this paper. The objective of this model is to maximize the joint success probability of achieving the profit target for each individual agent involved in the ISWM system. The contribution of the proposed model is to help decision makers answer: i) how to generate an appropriate ISWM system design under uncertainty, and ii) how uncertainty influences the ISWM system design.

The rest of this paper is organized as follows. Section 2 presents a literature review of current studies on the ISWM system design under uncertainties. Section 3 proposes the LCA-based two-stage mixed-integer stochastic programming model. Section 4 demonstrates a case study in Singapore by applying the proposed model. Section 5 concludes this paper.

2   Literature Review

2.1   ISWM System Design Based on LCA Method

The LCA method is initially applied to study the environmental impacts throughout a products (or services) life from raw material acquisition though production, use and final disposal (i.e. from cradle to grave) [6]. Although various LCA-based assessment models have been developed [e.g. 7, 8], the LCA-based optimization models for ISWM system design are still not commonly seen. [9] presented a LCA-based linear programming (LP) model to explore economically and environmentally efficient ISWM strategies that are defined by a set of waste treatment units and the amount of each waste processed by these units. [10] proposed a generic optimization model based on LCA to optimize ISWM system by developing waste collection and treatment schemes with objectives ranging from minimizing cost to minimizing greenhouse gas (GHG) emission while considering user-defined emissions and waste diversion constraints over multiple time stages. Besides, uncertainties have seldom been considered in LCA-based ISWM system design models. Therefore, this paper intends to propose a generic ISWM system design model with considering both internal and external uncertainties to support decision makings on SWM facility capacity planning and feedstock allocation. LCA approaches are applied to comprehensively assess inputs and outputs of each implemented SWM facility from economic, environmental and energy perspectives.

2.2   Uncertainty Design

Uncertainties existing inside and outside the real-world ISWM system have placed the system design problem beyond the capability of deterministic programming methods. The most prevailing techniques for addressing uncertainties in ISWM system design problem include stochastic programming, fuzzy programming, and interval programming [11, 12].

Two-stage stochastic programming has been proven as an effective method for handling optimization problems when the analysis of policy scenarios is desired while the related data are mostly uncertain [13]. [14] established an ISWM model with uncertainty for minimizing the total net cost by incorporating two-stage stochastic programming and interval parameter optimization. [15] improved this two-stage stochastic programming model by adding binary second-stage decision variables for facilities expansion, which is only allowable once during the whole planning horizon. [12] further improved this two-stage stochastic programming model by integrating robust optimization and chance-constrained programming to deal with the ambiguous coefficients and vague information of decision makers implicit knowledge for long-term waste flow allocation and facility capacity expansion.

Some shortcomings are commonly seen in previous works adopting two-stage stochastic programming for the ISWM system design. Firstly, current two-stage models commonly consider fixed solid waste and SWM facility portfolio. As a result, the designed ISWM system cannot flexibly accept emerging solid waste types and cannot be further improved by introducing new innovative facilities, as well as forming new integration patterns among these facilities. Secondly, current two-stage models do not comprehensively analyze the direct and indirect contributions of each facility to the overall performance of the entire ISWM system. Motivated by these issues, this paper incorporates the two-stage stochastic programming and the LCA approaches to comprehensively evaluate and optimize the ISWM system.

2.3   Target Achievement Approach

Current ISWM system design models commonly utilize weighted sum method to evaluate the overall system performance, and utilize it as the optimization measure to optimize the ISWM system design [e.g. 10, 14]. However, this method is usually inapplicable to the real-world ISWM system as SWM facilities are usually outsourced or contracted to different private agents. Therefore, improving the overall performance of the ISWM system might be at the expense of sacrificing their sustainability, which could result in significant difficulties in implementing the proposed system design. In this paper, a probability measure aiming at maximizing the joint success probability of achieving profit targets for each involved individual agent is introduced to emphasize on the significance of individual agent sustainability to the success of the designed ISWM system, which could better reflect the real-world SWM practices.

3   ISWM System Design Problem

3.1   Background Information

From a life cycle point of view, an all-inclusive real-world SWM process should include all essential operational units, i.e. collection, transportation, presorting, recycling, treatment, and final disposal. Generally speaking, once a type of solid waste being generated, it could be either collected separately (forming source-segregated solid waste) or with other type of solid waste (forming mixed solid waste). The mixed solid waste then needs to be presorted into several stable fractions first due to its uncertain physical and chemical character. The recycling process is usually aggregated into the collection and presorting process, and recyclables are then directly transported to re-manufacturing systems. Treatment and final disposal facilities are the cores of the ISWM system that deal with the incoming solid wastes, and recover energy and by-product.

It should be pointed out that treatment facilities mostly are not capable of fully converting incoming solid wastes into expected products. Undesired secondary solid residues that may bring negative impacts to human health and the environment are commonly generated after treatment processes. Therefore, these residues need to be further treated either by other treatment facilities or by final disposal facilities. And then, different SWM facilities are linked together and form the complete ISWM system. In this paper, the secondary solid residues from treatment facilities are called internally-generated feedstock as they are generated within the ISWM system. While the collected and presorted solid wastes are called externally-generated feedstock as they are originated from outside the ISWM system.

3.2   Problem Statement and Assumption

In this paper, we consider solid waste treatment and disposal facility capacity planning, and externally- and internally-generated feedstock allocation for a centralized large-scale ISWM system design with existing collection and presorting schemes. Hence, in the remaining parts of this paper, the term facility only refers to the solid waste treatment and disposal facility.

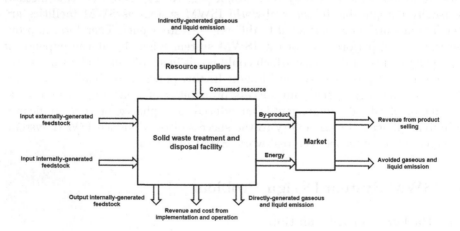

Fig. 1. Generic representation of a solid waste treatment and disposal facility based on the LCA approach

Fig. 1 demonstrates the generic representation of a solid waste treatment and disposal facility based on the LCA approach used in this paper. Each facility is featured by a series of input-specific and process-specific technical coefficients from both economic and environmental aspect. The impact of existing collection, presorting, recycling and transportation schemes can be accounted as parts of the unit operation cost of each facility. Besides, we assume that each facility is only operated by a single individual agent, and the time horizon for the ISWM system design is one year.

3.3   Model Formulation

The two-stage stochastic programming model for the ISWM system design problem is introduced in this section. In this model, facility capacity planning is treated as the first-stage decisions, while feedstock allocation is treated as the second-stage decision. The annual generation of each externally-generated feedstock is considered as the external uncertainty, and the energy generation coefficient and by-product generation coefficient are considered as the internal uncertainties. The proposed model is formulated below. Some notations in the proposed model are given in the Table 1, and the others are introduced as needed in this paper.

Table 1. Notations in the model

Notation	Definition
$\mathbb{F}$	Set of alternative facility type.
$\mathbb{H}$	Set of externally-generated feedstock type.
$\mathbb{I}$	Set of internally-generated feedstock type.
$\mathbb{J}$	Set of considered gaseous and liquid emission type.
$\mathbb{L}$	Set of generated by-product type.
$\mathcal{M}$	A significantly large positive constant.
a_f	Facility selection decision variable, where the variable has value 1 if the fth facility is selected and 0 otherwise, where $f \in \mathbb{F}$.
b_f	Annual designed capacity (ton/yr.) of the fth facility, where $f \in \mathbb{F}$.
x_{hf}	Annual amount (ton/yr.) of the hth externally-generated feedstock sent to the fth facility, where $h \in \mathbb{H}$, and $f \in \mathbb{F}$.
$y_{i'f'f}$	Annual amount (ton/yr.) of the $i'th$ internally-generated feedstock from the f'th facility and sent to the fth facility, where $i' \in \mathbb{I}$, and $f, f' \in \mathbb{F}$.
$\tilde{\varepsilon}_f$	Set of uncertain technical coefficients of the fth facility, where $f \in \mathbb{F}$.
$\tilde{w}_h$	Uncertain annual generation (ton/yr.) of the hth externally-generated feedstock, where $h \in \mathbb{H}$.
r_h	Recycling rate (%) of the hth externally-generated feedstock, where $h \in \mathbb{H}$.

$$\max \ \Pr\{\mathcal{C}_f(\mathbb{X}) + \mathcal{Q}_f(\mathbb{Y}, \tilde{\varepsilon}_f) \geq \lambda_f - \mathcal{M}(1 - a_f), \ \forall f\} \tag{1}$$

$$\text{s.t.} \ \sum_{f \in \mathbb{F}} x_{hf} = \tilde{w}_h(1 - r_h), \ \forall h \tag{2}$$

$$\sum_{f \in \mathbb{F}} y_{iff'} - \mathcal{S}_{fi}(\mathbb{Y}) = 0, \ \forall i, f, f' \tag{3}$$

$$\sum_{f \in \mathbb{F}} \mathcal{G}_{fj}(\mathbb{Y}, \tilde{\varepsilon}_f) \leq \overline{\mathcal{G}}_j, \ \forall j \tag{4}$$

$$\sum_{h \in \mathbb{H}} x_{hf} + \sum_{f' \in \mathbb{F}} \sum_{i' \in \mathbb{I}} y_{i'f'f} \leq b_f, \ \forall f \tag{5}$$

$$b_f \leq \overline{b}_f a_f, \ \forall f \tag{6}$$

$$y_{iff} = 0, \ \forall i, f \tag{7}$$

$$a_f \in \{0, 1\}, \ \forall f \tag{8}$$

$$b_f, x_{hf}, y_{i'f'f} \in [0, \infty), \ \forall h, i', f, f' \tag{9}$$

$$\text{where } \mathbb{X} := \{a_f, b_f, \forall f\}, \mathbb{Y} := \{x_{hf}, y_{i'f'f}, \forall h, i', f, f'\}$$

Eq.(1) represents the objective function aiming at maximizing the joint success probability of achieving profit targets (λ_f) for all selected facilities each year. Eq.(2) and Eq.(3) represent mass balance constraints for each type of externally-generated and internally-generated feedstock, respectively. Eq.(4) represents a series of strict constraints ensuring that the annual generation of each considered gaseous and liquid emission from the whole ISWM system ($\mathcal{G}_j$) will not exceed its upper bound ($\overline{\mathcal{G}}_j$) that might be regulated by decision makers. Eq.(5) represents a series of strict constraints that ensure the annual total amount of incoming feedstock to each selected facility does not surpass its annual designed capacity. Eq.(6) ensures that the annual designed capacity of each selected facility does not exceed its pre-regulated upper bound ($\overline{b}_f$). Eq.(7) indicates that each facility cannot treat its own secondary solid residues. Eq.(8) and Eq.(9) represent the value ranges of decision variables.

$$\mathcal{C}_f(\mathbb{X}) = -\alpha_f a_f - \beta_f b_f, \ \forall f \tag{10}$$

Eq.(10) represents the annual net profit related to the first-stage decision ($\mathcal{C}_f$). α_f and β_f refers to the fixed cost and unit capacity cost, respectively.

$$\mathcal{Q}_f(\mathbb{Y}, \tilde{\varepsilon}_f) = (\eta_f - \gamma_f)(\sum_{h \in \mathbb{H}} x_{hf} + \sum_{f' \in \mathbb{F}} \sum_{i' \in \mathbb{I}} y_{i'f'f}) + \nu \mathcal{R}_f(\mathbb{Y}, \tilde{\varepsilon}_f)$$
$$+ \sum_{l \in \mathbb{L}} u \mathcal{P}_{fl}(\mathbb{Y}, \tilde{\varepsilon}_f^2) - \sum_{f' \in \mathbb{F}} \eta_{f'} \sum_{i \in \mathbb{I}} y_{iff'}, \forall f \tag{11}$$

Eq.(11) represents the annual net profit related to the second-stage decision ($\mathcal{Q}_f$). η_f refers to the gate fee. γ_f refers to the unit operating cost. $\mathcal{R}_f$ refers to amount of energy recovered from each selected facility, and ν refers to the energy price. $\mathcal{P}_{fl}$ refers to the amount of each by-product generated from each selected facility, and ι_l refers to the price of each by-product. $\tilde{\varepsilon}_{\boldsymbol{f}}^2$ refers to the uncertain by-product generation coefficient set of each facility, where $\tilde{\varepsilon}_{\boldsymbol{f}}^2 \subset \tilde{\varepsilon}_{\boldsymbol{f}}, \forall f$.

$$\mathcal{S}_{fi}(\mathbb{Y}) = \sum_{h \in \mathbb{H}} \delta_{hfi} x_{hf} + \sum_{f' \in \mathbb{F}} \sum_{i' \in \mathbb{I}} \vartheta_{i'fi} y_{i'f'f}, \; \forall f, i \qquad (12)$$

Eq.(12) represents the annual generation of internally-generated feedstock in each facility ($\mathcal{S}_{fi}$) by treating all incoming feedstock. δ_{hfi} and $\vartheta_{i'fi}$ refer to the internally-generated feedstock generation coefficient of treating externally-generated feedstock and internally-generated feedstock in each facility, respectively.

$$\mathcal{G}_{fj}(\mathbb{Y}, \tilde{\varepsilon}_{\boldsymbol{f}}) = \sum_{h \in \mathbb{H}} \xi_{hfj} x_{hf} + \sum_{f' \in \mathbb{F}} \sum_{i' \in \mathbb{I}} \theta_{i'fj} y_{i'f'f} - \omega_j \mathcal{R}_f(\mathbb{Y}, \tilde{\varepsilon}_{\boldsymbol{f}}), \; \forall f, j \qquad (13)$$

Eq.(13) represents the annual net generation of considered gaseous and liquid emission from each selected facility, which includes the direct emission generated from treating incoming feedstock and the indirect emission avoidance by energy recovering. ξ_{hfj} and $\theta_{i'fj}$ refer to the emission generation coefficients of treating externally-generated and internally-generated feedstock in each facility, respectively. ω_j refers to the emission generation coefficients of generating energy in energy suppliers.

$$\mathcal{R}_f(\mathbb{Y}, \tilde{\varepsilon}_{\boldsymbol{f}}) = \sum_{h \in \mathbb{H}} (\mu_{hf} - \varphi_f) x_{hf} + \sum_{f' \in \mathbb{F}} \sum_{i' \in \mathbb{I}} (\rho_{i'f} - \varphi_f) y_{i'f'f} + \sum_{l \in \mathbb{L}} \tau_l \mathcal{P}_{fl}(\mathbb{Y}, \tilde{\varepsilon}_{\boldsymbol{f}}^2), \; \forall f$$
$$(14)$$

Eq.(14) represents the annual net energy generation from each selected facility, which includes the net energy generation from treating incoming feedstock and the avoided energy usage in typical by-product producers. μ_{hf} and $\rho_{i'f}$ refer to energy generation efficiency of treating externally-generated and internally-generated feedstock in each facility, respectively. φ_f refers to energy consumption efficiency of each facility. τ_l refers to energy consumption efficiency in by-product suppliers.

$$\mathcal{P}_{fl}(\mathbb{Y}, \tilde{\varepsilon}_{\boldsymbol{f}}^2) = \sum_{h \in \mathbb{H}} \psi_{hfl} x_{hf} + \sum_{f' \in \mathbb{F}} \sum_{i' \in \mathbb{I}} \sigma_{i'fl} y_{i'f'f}, \; \forall f, l \qquad (15)$$

Eq.(15) represents the annual by-product generation from each selected facility. ψ_{hfl} and $\sigma_{i'fl}$ refer to the by-product generation coefficient of treating externally-generated and internally-generated feedstock in each facility, respectively.

3.4   Solution Approach

Due to general difficulty in calculating the quantity of probability function, the sample average approximation (SAA) method is employed to replace the actual probability distribution of uncertainties in the model. In the original two-stage programming model, uncertainty is assumed to exist in the annual generation of externally-generated feedstock ($\tilde{w}_h$), the energy generation coefficients ($\tilde{\varepsilon}_f^1$) and the by-product generation coefficient ($\tilde{\varepsilon}_f^2$) of each alternative facility. We assume that K different samples are generated for SAA. In each sample, a set of deterministic values are given to uncertain parameters. These deterministic values in the kth sample are denoted by $\tilde{w}_h^k$ and $\varepsilon_f^k := \{\tilde{\varepsilon}_f^{1k}, \tilde{\varepsilon}_f^{2k}\}$, respectively. Then, the probability of the kth sample, which is denoted as p^k, should be equal to $\frac{1}{K}$ as all the samples are randomly selected from the sample space. Then, the SAA-based formulation of the original two-stage stochastic programming model is demonstrated by Eq.(16) to Eq.(24), which can be easily solved by mixed-integer linear programming solvers.

$$\max \quad \frac{1}{K} \sum_{k=1}^{K} \mathcal{I}^k \tag{16}$$

$$\text{s.t.} \quad (6), (8), \text{ and} \tag{17}$$

$$\mathcal{C}_f^k(\mathbb{X}, \tilde{\varepsilon}_f^k) + \mathcal{Q}_f^k(\mathbb{Y}^k, \tilde{\varepsilon}_f^k) - \lambda_f + \mathcal{M}(1 - a_f) \geq \mathcal{M}(\mathcal{I}^k - 1), \ \forall f, k \tag{18}$$

$$\sum_{f \in \mathbb{F}} x_{hf}^k = \tilde{w}_h^k(1 - r_h), \ \forall h, k \tag{19}$$

$$\sum_{f \in \mathbb{F}} y_{iff'}^k - \mathcal{S}_{fi}^k(\mathbb{Y}^k) = 0, \ \forall i, f, f', k \tag{20}$$

$$\sum_{f \in \mathbb{F}} \mathcal{G}_{fj}^k(\mathbb{Y}^k, \tilde{\varepsilon}_f^k) \leq \overline{\mathcal{G}}_j, \ \forall j, k \tag{21}$$

$$\sum_{h \in \mathbb{H}} x_{hf}^k + \sum_{f' \in \mathbb{F}} \sum_{i' \in \mathbb{I}} y_{i'f'f}^k \leq b_f, \ \forall f, k \tag{22}$$

$$y_{iff}^k = 0, \ \forall i, f, k \tag{23}$$

$$b_f, x_{hf}^k, y_{i'f'f}^k \in [0, \infty), \ \forall h, i', f, f', k \tag{24}$$

where $\mathbb{Y}^k := \{x_{hf}^k, y_{i'f'f}^k, \forall h, i', f, f', k\}$

4   Case Study

4.1   Background

This case study aims at supporting decision making on the ISWM system design in Singapore. SWM in Singapore begins at homes and businesses. Prior to collection process, recyclables are sorted and retrieved for processing to prolong the lifespan of recyclable materials. The remaining fractions are sent to four incineration plants. The incineration ash and other non-combustible wastes are then transported to Tuas Marine Transfer Station (TMTS) for barging operation to Semakau landfill plant.

However, current incineration plants generate a large amount of emissions and current landfill plant is estimated to be used up by 2035 due to the land scarcity in Singapore. Therefore, the National Environment Agency(NEA) determines to implement more innovative facilities to alleviate the SWM problem in Singapore. The facility portfolio in this case study contains anaerobic digestion (AD), aerobic composting (AC), gasification, incineration, and landfill. The externally-generated feedstock portfolio considers all solid waste categories generated in Singapore. The internally-generated feedstock includes the solid digested residues from AD, and ashes from gasification and incineration. Fig. 2 demonstrates the diagram of the ISWM system in the case study.

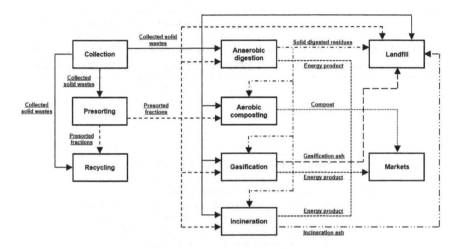

Fig. 2. ISWM system diagram in the case study

4.2 Experimental Setup

This experiment assumes that the profit target of each solid waste treatment facility is the net profit breakeven point, and there is no profit target for the solid waste disposal facility (landfill) as it is run by the local government. In addition, this experiment adopts 200,000 tons/yr. as the upper bound of annual designed capacity for landfill facility based on the current SWM practices in Singapore. The annual CO_2 emission is chosen to be the representative of environmental emission constraints and its upper bound is set as 2256 kg/cap/yr. [16]. The CPLEX solver is selected to conduct the experiment.

4.3 Result and Discussion

Basic Design. The basic design is obtained by running the proposed model with the initial data of each parameter. The sample size is chosen to be 1,000 by considering both simulation performance and computational time after conducting multiple sample size tests. The computational time is 134.513 seconds.

The maximum joint probability is 93.50%. Table 2 demonstrates the optimal annual designed capacity and average utilization of each implemented facility in the basic design. Fig. 3 shows the corresponding optimal annual average feedstock allocation. The basic design provides decision makers an insight into how different facilities collaborate within the ISWM system. Besides, it also provides a possibility to introduce the aerobic composting and gasification facility to improve the current SWM practices in Singapore.

Table 2. Basic design of the ISWM system

	AD	AC	Gasification	Incineration	Landfill
Designed capacity(tons/yr.)	0	1,450,740	1,828,460	1,329,840	200,000
Avg. utilization(tons/yr.)	0	1,334,800	1,174,770	1,213,300	199,508

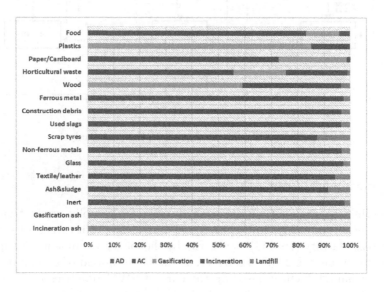

Fig. 3. Optimal annual average feedstock allocation in the basic design

Sensitivity Analysis under the Basic Design. The implication of the sensitivity analysis under the basic design is to provide decision makers managerial insights into how policies affect the performance of an existing ISWM system design. Electricity price is chosen as the representative to conduct this sensitivity analysis as it can be somehow utilized as a tool by the decision makers to adjust the performance of the ISWM system. The values of all the other parameters are fixed throughout the sensitivity analysis.

Fig. 4 shows the behavior of maximum joint success probability and the average utilization of each implemented facility with respect to different electricity prices under the basic design. The horizontal axis refers to the ratio of current electricity price compared to the initial one used in the basic design. Different colors in each bar refer to the annual designed capacity of each implemented facility. It can be seen that with the increase of the electricity price, the maximum joint success probability does not grow all the way. This is because the high electricity price enhances the marginal profit of Waste-To-Energy (WTE) facilities (i.e. AD, gasification, and incineration in the case study) but reduces the marginal profit of the non-WTE facilities (i.e. AC in the case study), and vice versa. Therefore, when the electricity price continuously decreases, more feedstock are sent to the incineration and gasification. However, less feedstock also causes a lower average annual net profit in AC, and thus leading to a lower probability of achieving profit target for AC. On the contrary, when the electricity price continuously increases, more feedstock are sent to AC. However, this could not always happen as AC can only treat the organic feedstock. Then, the average annual net profit of AC continuously drops, which negatively affect the maximum joint success probability.

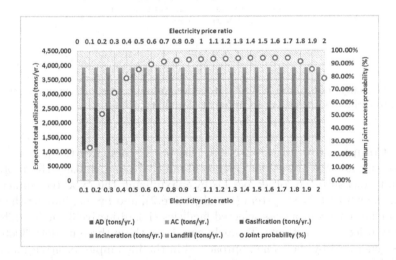

Fig. 4. The behavior of maximum joint success probability and current utilization of each implemented facility with respect to different electricity prices under the basic design

Sensitivity Analysis under the Re-Optimized Design. The implication of the sensitivity analysis under the re-optimized design is to provide decision makers managerial insights into how uncertainties influence optimal design of the ISWM system. Fig. 5 demonstrates the change of the optimal ISWM system design and corresponding maximum joint success probability with respect to different electricity prices. The horizontal axis is the same as the one in Fig. 4.

It can be seen that when the electricity price increases, the maximum joint success probability has the same trend as the one under the basic design, which can be explained the same way as the sensitivity analysis under the basic design. Besides, it is worthwhile to note that high electricity price promotes the implementation of AD facility and reduces the designed capacity of AC facility. Therefore, the WTE facilities should be implemented more under the high electricity price.

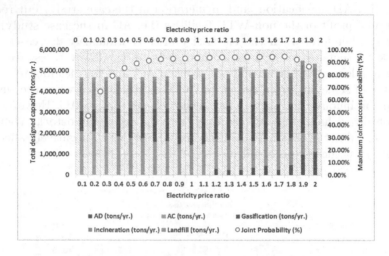

Fig. 5. The change of optimal ISWM system design and corresponding maximum joint success probability with respect to different electricity prices

Model Comparison. In this section, this paper compares the performances of three optimal ISWM system designs obtained by maximizing the proposed probability measure (i.e. objective 1), by maximizing the expected annual net profit of the whole ISWM system (i.e. objective 2), and by maximizing the minimum annual net profit across selected facilities (i.e. objective 3) under different electricity prices, respectively. To fairly compares these three models, 3000 new samples generated by the same approach as in the in-sample sensitivity analysis are utilized to conduct three out-of-sample tests. In each out-of-sample test, we apply the three optimal designs in each model to observe how the objective value changes under different electricity prices. In addition, expect for the objective functions, all the other assumptions, constraints, and data are same for all the models.

Fig. 6 demonstrates how the maximum joint success probability performs by applying the three optimal designs with respect to different electricity prices. Fig. 7 demonstrates how the maximum expected total net annual profit of the whole system performs by applying the three optimal designs with respect to different electricity prices. And Fig. 8 demonstrates how the maximum expected minimum net annual profit across selected facilities performances by applying the three optimal designs with respect to different electricity prices. It can be seen that optimal design under objective 1 and objective 3 performs more stable than the one under objective 2 in these three out-of-sample tests. But the optimal design under objective 3 performs better than the one under objective 1 in all three tests when the electricity price ratio is lower than a certain number, and vice versa. Besides, the optimal design under objective 2 sacrifices the individual agent economic sustainability to achieve a higher maximum expected total net annual profit for the whole ISWM system.

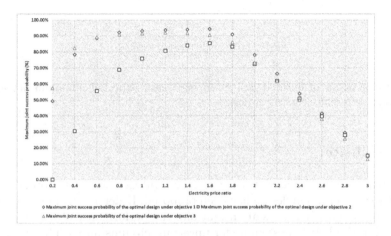

Fig. 6. Comparison of optimal designs in maximum joint success probability under different electricity prices

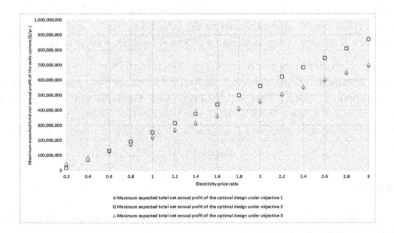

Fig. 7. Comparison of optimal designs in maximum expected total net annual profit of the whole system under different electricity prices

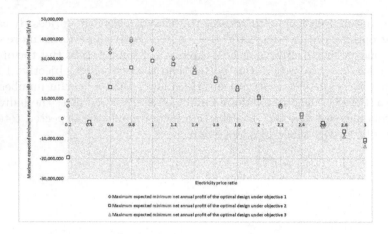

Fig. 8. Comparison of optimal designs in maximum expected minimum net annual profit under different electricity prices

5 Conclusion

In this paper, a LCA-based two-stage mixed-integer stochastic programming model for ISWM system design has been proposed. Different from previous models, the proposed model takes an insight into the performance of each individual agent within the ISWM system under uncertain circumstances. The experiment results show that the proposed model can help decision makers identify what the optimal ISWM system design under uncertainties is and how uncertainties influence the optimal ISWM system design.

Future work can be done to extend the current research. Firstly, the decisions on facility location could be introduced into the current model, which provide a possibility to design a decentralized ISWM system. Secondly, robustness optimization method could also be introduced into the current model to design an ISWM system that are not sensitive to internal and/or external uncertainties.

Acknowledgement. This research is funded by the Singapore National Research Foundation (NRF) Project 'Energy and Environmental Sustainability Solutions for Mega-Cities (E^2S^2)' and the publication is supported under the Campus for Research Excellence And Technological Enterprise (CREATE) programme.

References

1. Seadon, J.K.: Sustainable waste management systems. Journal of Cleaner Production 18(16), 1639–1651 (2010)
2. Marshall, R.E., Farahbakhsh, K.: Systems approaches to integrated solid waste management in developing countries. Waste Management 33(4), 988–1003 (2013)

3. Baldasano, J.M., Gasso, S., Perez, C.: Environmental performance review and cost analysis of msw landfilling by baling-wrapping technology versus conventional system. Waste Management 23(9), 795–806 (2003)
4. Louis, G., Shih, J.-S.: A flexible inventory model for municipal solid waste recycling. Socio-Economic Planning Sciences 41(1), 61–89 (2007)
5. McDougall, F.R., White, P.R., Franke, M., Hindle, P.: Integrated solid waste management: a life cycle inventory. John Wiley & Sons (2008)
6. ISO14040 ISO. 14040: Environmental management–life cycle assessment–principles and framework. London: British Standards Institution (2006)
7. Eriksson, O., Reich, M.C., Frostell, B., Björklund, A., Assefa, G., Sundqvist, J.-O., Granath, J., Baky, A., Thyselius, L.: Municipal solid waste management from a systems perspective. Journal of Cleaner Production 13(3), 241–252 (2005)
8. Banar, M., Cokaygil, Z., Ozkan, A.: Life cycle assessment of solid waste management options for eskisehir, turkey. Waste Management 29(1), 54–62 (2009)
9. Solano, E., Ranjithan, S.R., Barlaz, M.A., Brill, E.D.: Life-cycle-based solid waste management. i: Model development. Journal of Environmental Engineering 128(10), 981–992 (2002)
10. Levis, J.W., Barlaz, M.A., DeCarolis, J.F., Ranjithan, S.R.: A generalized multi-stage optimization modeling framework for life cycle assessment-based integrated solid waste management. Environmental Modelling & Software 50, 51–65 (2013)
11. Chang, N.-B., Wang, S.F.: A fuzzy goal programming approach for the optimal planning of metropolitan solid waste management systems. European Journal of Operational Research 99(2), 303–321 (1997)
12. Li, Y.P., Huang, G.H., Nie, X.H., Nie, S.L.: A two-stage fuzzy robust integer programming approach for capacity planning of environmental management systems. European Journal of Operational Research 189(2), 399–420 (2008)
13. Lv, Y., Huang, G.H., Li, Y.P., Yang, Z.F., Sun, W.: A two-stage inexact joint-probabilistic programming method for air quality management under uncertainty. Journal of Environmental Management 92(3), 813–826 (2011)
14. Maqsood, I., Huang, G.H.: A two-stage interval-stochastic programming model for waste management under uncertainty. Journal of the Air & Waste Management Association 53(5), 540–552 (2003)
15. Li, Y.P., Huang, G.H.: An inexact two-stage mixed integer linear programming method for solid waste management in the city of regina. Journal of Environmental Management 81(3), 188–209 (2006)
16. Couth, R., Trois, C., Vaughan-Jones, S.: Modelling of greenhouse gas emissions from municipal solid waste disposal in africa. International Journal of Greenhouse Gas Control 5(6), 1443–1453 (2011)

A Uniform Evolutionary Algorithm Based on Decomposition and Contraction for Many-Objective Optimization Problems

Cai Dai, Yuping Wang, and Lijuan Hu

School of Computer Science and Technology, Xidian University, Xi'an, 710071, China

Abstract. For many-objective optimization problems, how to get a set of solutions with good convergence and diversity is a difficult and challenging task. To achieve this goal, a new evolutionary algorithm based on decomposition and contraction is proposed. Moreover, a sub-population strategy is used to enhance the local search ability and improve the convergence. The proposed algorithm adopts a contraction scheme of the non-dominance area to determine the best solution of each sub-population. The comparison with the several existing well-known algorithms: NSGAII, MOEA/D and HypE, on two kinds of benchmark functions with 5 to 25 objectives is made, and the results indicate that the proposed algorithm is able to obtain more accurate Pareto front with better diversity.

Keywords: Many-objective optimization problems, Decomposition, Contraction Method.

1   Introduction

Multi-objective Evolutionary Algorithms (MOEAs) are a kind of effective methods for solving multi-objective problems [1-2]. Almost all well-known and frequently-used MOEAs, which have been proposed in the last twenty years [3-5], are based on Pareto dominance. Such Pareto dominance-based algorithms usually work well on problems with two or three objectives but their searching ability is often severely degraded by the increased number of objectives [6]. This is due to the fact that most solutions in a population could be non-dominated solutions even in the early stages of evolution for many objective optimization problems. When this happens, Pareto dominance-based fitness evaluation almost cannot generate any selection pressure toward the Pareto front (PF). Therefore, how to increase the selection pressure toward the PF is critical for many-objective optimization algorithms.

In the literature, there are mainly three categories which are used to MOEAs to deal with many-objective optimization problems. The first category uses an indicator function, such as the hypervolume [7-9], as the fitness function. This kind of algorithms is also referred to as IBEAs (indicator-based evolutionary algorithms), and their high search ability has been demonstrated in the literature [10]. Bader and Zitzler [11] proposed a fast hypervolume-based many-objective optimization algorithm which used Monte Carlo simulation to quickly approximate the exact hypervolume values. However, one of their main drawbacks is the computation time for the hypervolume calculation which exponentially increases with the number of objectives.

© Springer International Publishing Switzerland 2015

H. Handa et al. (eds.), *Proc. of the 18th Asia Pacific Symp. on Intell. & Evol. Systems – Vol. 2*,
Proceedings in Adaptation, Learning and Optimization 2, DOI: 10.1007/978-3-319-13356-0_14

The second category utilizes the scalarizing functions to deal with the many-objective problem. According to the literatures [12-14], scalarizing function-based algorithms could better deal with many-objective problems than the Pareto dominance-based algorithms. The main advantage of scalarizing function-based algorithms is the simplicity of their fitness evaluation which can be easily calculated even the number of objectives is large. The representative MOEA in this category is MOEA/D [15] (multi-objective evolutionary algorithm based on decomposition), which works well on a wide range of multi-objective problems with many objectives, discrete decision variables and complicated Pareto sets [16, 17]. In MOEA/D [15], the uniformity of the weighted vectors adopted determines the uniformity of the obtained non-dominated optimal solutions, however, the weighted vectors adopted in MOEA/D will be not very uniform for some problems and the size N of these weighted vectors should satisfy the restriction $N = C_{H+m-1}^m$. Thus N can not be freely assigned and it will increase exponentially with m, where m is the number of objectives and H is an integer.This restricts the application of MOEA/D to many-objective optimization problems (i.e., m is large). Therefore, for many-objective problems, how to set weight vectors is a very difficult but critical task.

The third category makes use of solution ranking methods. Specifically, solution ranking methods are used to discriminate among solutions in order to increase the selection pressure and enhance the convergence to PF. Bentley and Wakefield [18] proposed ranking composition methods which extract the separated fitness of every solution into a list of fitness values for each objective. Kokolo and Hajime [19] proposed a relaxed form of dominance (RFD) to deal with what they called dominance resistant solutions, i.e., solutions that are extremely inferior to others in at least one objective, but hardly-dominated. Farina and Amato [20] proposed a dominance relation which takes into account with how many objectives a solution is better, equal and worse than another solution. Sato et al. [21] proposed a method to strength or weaken the selection process by expanding or contracting the solutions' dominance area.

The ranking methods can improve the convergence and the decomposition can maintain the diversity, but, to the best of our knowledge, there exists no work to combine them to solve many-objective optimization problems. In this paper, an evolutionary algorithm based on decomposition and a ranking method (UREA/D) is proposed to solve the many-objective optimization problems. Firstly, a sup-population strategy is used to enhance the local search ability of the proposed algorithm. Each sub-problem generated by decomposition will have a sub-population and use the information provided by this sub-population to improve the convergence performance. Secondly, a new method based on contraction of the non-dominance area [22] is used to compare solutions, which will provide stronger selection pressure toward the PF than Pareto dominance. Thirdly, an update strategy based on decomposition is proposed to maintain the diversity. Moreover, the experiments demonstrate that UREA/D can significantly outperform MOEA/D, NSGAII-CDAS (NSGAII based on CDAS) and HypE on a set of test instances.

The rest of this paper is organized as follows: Section 2 introduces the main concepts of the multi-objective optimization; Section 3 presents a new many-objective evolutionary algorithm; while Section 4 shows the experiment results of the proposed algorithm and the related analysis; finally, Section 5 draws the conclusions and proposes the future work.

2 Multi-objective Optimization Problem

A multi-objective optimization problem (MOP) can be described as follows [23]:

$$
\begin{cases}
\min F(x) = (f_1(x), f_2(x), \cdots, f_m(x)) \\
\text{s.t.}\ g_i(x) \leq 0, i = 1,2, \cdots, q \\
\quad h_j(x) = 0, j = 1,2, \cdots p
\end{cases}
\tag{1}
$$

where $x = (x_1, \cdots, x_n) \in X \subset R^n$ is called decision variable and X is n-dimensional decision space. $f_i(x)(i = 1, \cdots, m)$ is the ith objective to be minimized, $g_i(x)(i = 1,2 \cdots q)$ defines ith inequality constraint and $h_j(x)(j = 1,2, \cdots p)$ defines jth equality constraint. Furthermore, all the constraints determine the set of feasible solutions which are denoted by Ω. To be specific, we try to find a feasible solution $x \in \Omega$ minimizing each objective function $f_i(x)(i = 1, \cdots, m)$ in F. In the following, four important definitions [24] for multi-objective problems are given.

Definition 1 (Pareto dominance): Pareto dominance between solutions $x, z \in \Omega$ is defined as follow. If

$$
\begin{aligned}
&\forall i \in \{1,2, \cdots m\} f_i(x) \leq f_i(z) \\
&\wedge \exists i \in \{1,2, \cdots m\} f_i(x) < f_i(z)
\end{aligned}
\tag{2}
$$

are satisfied, x is called to dominate (Pareto dominate) z (denoted $x \succ z$).

Definition 2 (Pareto optimal): A solution vector x is said to be Pareto optimal with respect to Ω, if $\nexists z \in \Omega: z \succ x$.

Definition 3 (Pareto optimal set (PS)): The set of Pareto optimal solutions (PS) is defined as:

$$
PS = \{x \in \Omega | \nexists z \in \Omega: z \succ x\}
\tag{3}
$$

Definition 4 (Pareto front): The Pareto optimal front (PF) is defined as:

$$
PF = \{F(x) | x \in PS\}
\tag{4}
$$

3 A New Evolutionary Algorithm Based on Decomposition and Contraction

The proposed algorithm (UREA/D) is specifically designed to obtain a set of solutions with good diversity and convergence. UREA/D consists four parts, i.e. space and population decomposition, crossover operator, selection strategy and update strategy. Space and population decomposition and update strategy are used to maintain the diversity of the obtained solutions. The crossover operator is used to search the decision space. The selection strategy can help the obtained solutions to converge to the true PF. In this section, these four parts are introduced in the following subsections. Moreover, the update strategy uses a new method based on contraction of the non-dominance area which is introduced in Subsection 3.2 to rank solutions of each sub-population.

3.1   Space and Population Decomposition

The objective space Ω of a MOP is decomposed into a set of sub-regions $\Omega_1, \Omega_2, \cdots, \Omega_N$ based on a set of weight vectors, and each sub-region will take K individuals from POP (randomly selecting some individuals from POP if POP has not K individuals in this sub-region) so that each sub-region owns K solutions. These K individuals constitute a sub-population in this sub-region. Specifically, for a given set of weight vectors $(\gamma^1, \gamma^2, \cdots, \gamma^N)$ which are generated by the uniform method [16] and a set of the current obtained solutions POP, where N is the number of the weight vectors. The decomposition of the objective space and the classification of these solutions to the sub-regions are carried out according to the following equations:

$$P^i = \{x \mid x \in POP, \Delta(F(x), \gamma^i) = \max_{1 \leq j \leq N}\{\Delta(F(x), \gamma^j)\}\}$$

$$\Delta(F(x), \gamma^i) = \frac{\gamma^i * (F(x) - Z)^T}{\|\gamma^i\| * \|F(x) - Z)\|}, i = 1, \cdots m$$

(5)

$$\Omega_i = \{F(x) \mid x \in \Omega, \Delta(F(x), \gamma^i) = \max_{1 \leq j \leq N}\{\Delta(F(x), \gamma^j)\}\}$$

(6)

where $Z = (Z_1, \cdots Z_m)$ is a reference point with $Z_i = \min\{f_i(x) \mid x \in \Omega\}$; $\Delta(F(x), \gamma^i)\}$ is the cosine of the angle between γ^i and $F(x) - Z$. The solution set POP is divided into N classes $P^1, P^2, \cdots, P^N$ by Eq. (5) and the objective space Ω is divided into N sub-regions $\Omega_1, \cdots, \Omega_N$ by Eq. (6). If the number of solutions in $P^i (1 \leq i \leq N)$ is smaller than K, some solutions are randomly selected from POP and put into P^i, which makes each sub-region have K solutions so that the diversity of obtained solutions is improved.

Note that above decomposition method has the following properties:

1)   The PFs of all these sub-regions constitute the PF of problem (1)

2)   The sub-PS of each sub-region is just a small part of the whole PS, and the obtained non-dominated solutions in this sub-region will be a good approximation to this sub-PS. So the union of these obtained non-dominated solutions will be a good approximation to whole PS even when the whole PS has a complex shape. Therefore, Eq. (5) and (6) make searching the PS of problem (1) become simpler.

3)   This classification (decomposition) method does not require any aggregation methods, and it just requires the user choose a set of weight vectors. Therefore, to some extent, it requires little computation.

3.2   New Method Based on Contraction of the Non-dominance Area

We have proposed a new method based on the contraction of the non-dominance area [22] to enhance the selection pressure. The proposed method controls the levels of contraction or expansion of the dominance area of solutions by modifying the fitness value of a solution for each objective function. After modifying the fitness values of each solution, the original dominance relation may change, e.g., some non-dominated solutions may become dominated ones. The fitness value modification of a solution x is defined by

$$f_i'(x) = r(max(sin(w_i), cos(w_i)))^H (i = 1, \cdots m) \tag{7}$$

where $r = \|F(x)\|$, $cos(w_i) = f_i(x)/\|F(x)\|$, $H > 0$. If $(f_1'(x), \cdots f_m'(x))$ dominates $(f_1'(z), \cdots f_m'(z))$, the solution x is better than the solution z; if $(f_1'(x), \cdots f_m'(x))$ and $(f_1'(z), \cdots f_m'(z))$ do not dominate each other, we can not determine which one of solutions x and z will be better.

H is a user-defined parameter which allows to control the dominance area of the solution x along the ith dimension. The size of the non-dominance area of the solution x will becomes large (small) when H becomes large (small). When the non-dominance area of a solution is contracted, the number of PFs will decrease and ranking of solutions by non-dominance will become coarser, which can enhance the selection pressure.

3.3   *POP* Update Strategy Based on Decomposition

In this section, a *POP* update strategy based on decomposition is proposed to maintain the diversity of obtained solutions and improve the convergence. After *POP* and the objective space Ω are classified, each weight vector corresponds a set of solutions (a sub-population) in the corresponding sub-objective space. Then the ranking method in Section 3 is used to rank the solutions of each sub-population. The detail is as follows. In each sub-region Ω_i, there are two cases:

1)   If the sub-region Ω_i contains no solution of P^i, the solution in P^i whose objective vector has the smallest angle to the weight vector γ^i is defined as the current best solution of P^i.

2)   If the sub-region contains solution(s) of P^i, there are two situations. If this sub-region contains only one solution of this sub-region, the solution is the current best solution of the P^i. If this sub-region contains more than one solutions of P^i, these solutions are firstly ranked by the ranking method. Then the solution of the first front whose objective vector has the smallest angle to the weight vector γ^i is defined as the current best solution of P^i.

Note that, if a sub-population does not have K solutions, some solutions are randomly selected from *POP* to put into the sub-population such that each sub-population has K solutions, and the best solution is stored as the K-th solution.

Thus, all $N \times K$ solutions can be denoted as $(x_1, \cdots x_K, \cdots, x_{(N-1)*K+1}, \cdots x_{N*K})$, where $(x_{(i-1)*K+1}, \cdots x_{i*K})$ are solutions of i-th sub-population and $x_{i \times K}$ is the best solution of P^i. The update strategy for the $N \times K$ solutions is introduced as follows. For a new solution y which belongs to the sub-region Ω_i, there are two cases:

1) If some solutions of P^i do not belong to the sub-region Ω_i, y is put into P^i and delete a solution of P^i. Update the best solution of P^i.

2) If all solutions of P^i belong to the sub-region Ω_i and the new solution y is better than the solution x_{K*i}, then a solution randomly selected from $\{x_{(i-1)*K+1}, \cdots x_{(i-1)*K+K-1}\}$ is removed, and y is put into P^i.

It can be seen that the update strategy has the following properties: 1) It makes each sub-region have K solutions. This can improve the diversity of the solutions of *POP* in objective space. 2). If both the non-dominated and dominated solutions exist in a sub-region, a non-dominated solution is chosen and kept. This can help solutions to converge to PF fast. 3) The objective vector of the best solution has the smallest angle to the weight vector of the corresponding sub-region. This can make the solutions in *POP* distribute relatively evenly in objective space. 4) The new solution only compares with the best solution of the corresponding sub-population, which reduce the consumption of computation. 5) After the best solution of a sub-population is updated, some worse solution of the sub-population be removed, which improve the quality of solutions in the sub-population.

3.4   Selection Strategy and Crossover Operator

The selection strategy has a great impact on the performance of local search and global search, thus an appropriate selection strategy can improve the performance of an algorithm. In this work, a selection strategy based on the decomposition is designed to achieve the goal. Firstly, compute the Euclidean distance between any two weight vectors and then find out T closet weight vectors to each weight vector [27]. For each $i = 1, \cdots N$, set $B(i) = \{K * i_1, \cdots, K * i_T\}$ where $\lambda^{i_1}, \cdots \lambda^{i_T}$ are the T closet weight vectors to λ^i and K is each sub-population size. Then set

$$P = \begin{cases} \{(i-1)*K+1, \cdots, i*K\}, if\ rand1 < p1 \\ \begin{cases} B(i), if\ rand2 < J \\ \{K*1, \cdots, K*N\}, otherwise \end{cases}, otherwise \end{cases} \quad (8)$$

where $rand1$ and $rand2$ are two random numbers in $[0,1]$, $p1$ and J are two parameters. J is set to 0.9 as the same in [21]. For the weight vector λ^i, the aggregation function value of x^{K*i} is the smallest among $x^{(i-1)*K+1}, \cdots, x^{K*i}$. When P is set, randomly select two indexes $r2$ and $r3$ from P, and generate a solution from x^{K*i}, x^{r2} and x^{r3} by the crossover operator by the following formula:

$$x_j^{new} = \begin{cases} x_j^{K*i} + L(x_j^{r2} - x_j^{r3}), if\ rand(0,1) < CR \\ x_j^{K*i}\ otherwise \end{cases} \quad (9)$$

where $L \in [0,2]$ is a scale factor which controls the length of the exploration vector $(x^{r2} - x^{r3})$; CR is a constant value (namely crossover rate); $j = 1, \cdots, n$ and x_j^{r2} indicates the j-th component of x^{r2}.

If P is set to $\begin{cases} B(i), if\ rand2 < J \\ \{K * 1, \cdots, K * N\}, otherwise \end{cases}$, the crossover helps to make the global search, If P is set to $\{(i - 1) * K + 1, \cdots, i * K\}$, the crossover helps to make the local search.

3.5   Proposed Algorithm

Based on all above, a uniform evolutionary algorithm based on decomposition and contraction (UEA/DC) is proposed and the steps of algorithm UEA/DC are as follows:

Input:
   N: the number of weight vectors (the sub-problems);
   K: the size of the sub-population;
   T: the number of weight vectors in the neighborhood of each weight vector,
   $0 < T < N$; $\lambda^1, \lambda^2, \cdots, \lambda^N$: a set of N uniformly distributed weight vectors;
Output: Obtained PS and PF: $\{F(x^K), F(x^{2*K}), \cdots, F(x^{N*K})\}$
Step 1. Initialization:
   Step 1.1. Generate an initial population $x^1, x^2, \cdots x^{N*K}$ randomly or by a problem-specific method.
   Step 1.2. Given an initial reference solution $Z = (z_1, \cdots z_m)$ by a problem-specific method.
   Step 1.3 Divide the initial population into N classes by Eq. (5) and each class has K solutions. Then determine the best solution of each class (sub-population) by the strategy of Subsection 3.3.
   Step 1.4. Compute the Euclidean distances between any two weight vectors and the find out the T closet weight vectors to each weight vector. For each $i = 1, \cdots, N$, set $B(i) = \{i_1, \cdots i_T\}$, where $\lambda^{i_1}, \cdots \lambda^{i_T}$ are the T closest weight vectors to λ^i.
Step 2. Update:
   For $i = 1, \cdots, N$, do
   Step 2.1. Crossover: a new solution y is generated by crossover in Subsection 3.4.
   Step 2.2. Mutation: Apply a mutation operator on y to produce y'.
   Step 2.3. Update of Z: For $k = 1, \cdots, m$, if $z_k < f_k(y')$, then set $z_k = f_k(y')$.
   Step 2.4. Determine the sub-region where the solution y' belongs to by the Eq. (5). The corresponding sub-population is updated by the update strategy of Subsection 3.3.
   End for.
Step 3. Stopping Criteria: If stopping criteria is satisfied, then stop and output $\{F(x^K), F(x^{2*K}), \cdots, F(x^{N*K})\}$; otherwise, go to Step 2.

4   Experimental Study

In this section, UEA/DC is fully compared with MOEA/D [17], NSGAII-CDAS [21] (NSGAII based on contracting or expanding the solutions' dominance area) and HypE [11] (fast hypervolume-based many-objective optimization algorithm) on DTLZ1 and DTLZ3 of the DTLZ family with different number of objectives [25]. Each of the problems is with 5-50 objectives. In this paper, the following three performance metrics are used to measure the performance of the different algorithm quantitatively: inverted generational distance (IGD) [26] and Wilcoxon Rank-Sum Test [27]. Wilcoxon Rank-Sum test [27] is used in the sense of statistics to compare the mean IGD of the compared algorithms. It tests whether the performance of UEA/DC on each test problem is better ("+"), same ("="), or worse ("-") than/as that of the compared algorithms at a significance level of 0.05 by a two-tailed test.

Table 1. IGD obtained by UEA/DC, MOEA/D, NSGAII-CDAS and HypE

Instance	UEA/DC		MOEA/D		NSGAII-CDAS		HypE	
	Mean	Std	Mean	Std	Mean	Std	Mean	Std
DTLZ1-5	0.0417	0.0009	0.0590(+)	0.0045	0.1359(+)	0.0190	0.1741(+)	0.0453
DTLZ1-10	0.0409	0.0005	0.0830(+)	0.0060	0.1209(+)	0.0221	0.5271(+)	0.0358
DTLZ1-15	0.0607	0.0008	0.1044(+)	0.0116	0.1379(+)	0.0332	0.3113(+)	0.0258
DTLZ1-20	0.0632	0.0028	0.0981(+)	0.0075	0.1652(+)	0.0374	0.5570(+)	0.0368
DTLZ1-25	0.0617	0.0053	0.0975(+)	0.0066	0.1610(+)	0.0234	0.3892(+)	0.0248
DTLZ3-5	0.1331	0.0011	0.1569(+)	0.0028	1.0058(+)	0.2559	0.5007(+)	0.0564
DTLZ3-10	0.2028	0.0020	0.3561(+)	0.0372	0.9322(+)	0.2828	0.7398(+)	0.0365
DTLZ3-15	0.2276	0.0032	0.4614(+)	0.0701	0.9794(+)	0.3951	0.7525(+)	0.1753
DTLZ3-20	0.2466	0.0043	0.7186(+)	0.0302	0.5750(+)	0.1143	0.7506(+)	0.0984
DTLZ3-25	0.2542	0.0102	0.7471(+)	0.0046	0.7672(+)	0.2920	0.7459(+)	0.1230

"+" means that UEA/DC outperforms its competitor algorithm, "-" means that UEA/DC is outperformed by its competitor algorithm, and "=" means that the competitor algorithm has the same performance as UEA/DC.

4.1   Experimental Setting

The experiments are carried out on a personal computer (Intel Xeon CPU 2.53GHz, 3.98G RAM). The solutions are all coded as real vectors. Polynomial mutation [28] operator and differential evolution (DE) [29] are applied directly to real vectors in three compared algorithms. The crossover rate and scaling factor in DE operator are set to 1.0 and 0.5, respectively. The aggregate function of MOEA/D is the Tchebycheff approach [17], and weight vectors are generated by using the uniform design method [16] for MOEA/D and UEA/DC. The number of the weight vectors in the neighborhood in MOEA/D and UEA/DC is set to 20 for all test problems. The parameter H of the ranking method is set to 0.75. The parameter of CDAS is set to 0.25 for NSGAII-CDAS. The parameters of HypE are same as [11]. For each algorithm, 30 independent runs are performed with population size of 200 for all these instances. The maximal number of function evaluations is set to 400000 for all test problems. The values of default parameters are the same as in the corresponding papers.

4.2   Comparisons of UEA/DC with MOEA/D, NSGAII-CDAS and HypE

Table 1 shows the mean and standard deviation of the IGD metric obtained by these four algorithms for these test problems with 5-25 objectives respectively, and Wilcoxon test for pair wise comparisons of algorithms on each test instance is also shown in Table I. DTLZ1-k represents that the number of objectives adopted in DTLZ1 is k.

It can be seen from Table I that, for the Wilcoxon test of the IGD metric, UEA/DC outperforms MOEA/D, NSGAII-CDAS and HypE on all these test problems. From the Table I, we also can see that the mean values of IGD obtained by MOEA/D are much smaller than those obtained by NSGAII-CDAS and HypE for all the test problems, which indicates that for these test problems, MOEA/D can obtain solutions with better diversity than NSGAII-CDAS and HypE. However, the mean values of IGD obtained by UEA/DC are much smaller than those obtained by MOEA/D, especially, e.g., for problem DTLZ3 with 10-25 objectives, the mean value of IGD obtained by UEA/DC is smaller at least 70% than that obtained by MOEA/D. These indicate that for these test problems, the solutions obtained by UEA/DC have better coverage than those obtained by other three algorithms and have good convergence. These also imply that the proposed update strategy based on decomposition is good at maintaining the diversity and the proposed algorithm has a good convergence.

5   Conclusions

In this paper, a new evolutionary algorithm based on decomposition and contraction is proposed to obtain a set of solutions with good convergence and diversity for many-objective optimization problems. In order to maintain the diversity of obtained solutions, an update strategy based on decomposition is designed. The objective space of a MOP is firstly divided into a set of sub-regions by a set of weight vectors which is generated by a uniform design. Secondly, this update strategy makes each sub-region have about same number of solutions. In addition, in order to obtain a set of solutions with good convergence, the new contraction method is used to enhance the selection pressure toward the true PF. Thirdly, a sub-population strategy is used to balance the local search and global search. The experimental results show that, with these schemes, the proposed algorithm is able to search well in continuous domain, and achieve accurate Pareto sets and wide Pareto fronts efficiently. Moreover, compared with three well known algorithms MOEA/D, NSGAII-CDAS and HypE by simulation, the results show that the proposed algorithm is able to find much better spread of solutions and these solutions have better convergence to the true PF. In the future, how to improve the efficiency of the algorithm is still to be studied.

Acknowledgments. This work was supported by the National Natural Science Foundation of China (No. 61472297).

References

1. Coello Coello, C.A., Lamont, G.B., Veldhuizen, D.A.: Evolutionary Algorithms for Solving Multi-Objective Problems (Genetic and Evolutionary Computation). Springer-Verlag, New York, Inc., Secaucus (2006)
2. Kokshenev, I., Braga, A.P.: An efficient multi-objective learning algorithm for RBF neural network. Neurocomputing 73(16-18), 2799–2808 (2010)
3. Deb, K., Agrawal, S., Pratap, A., Meyarivan, T.: A fast and elitist multiobjective genetic algorithm: NSGA-II. IEEE Transactions on Evolutionary Computation 6(2), 182–197 (2002)
4. Nebro, A., Durillo, J., et al.: SMPSO: a new PSO-based metaheuristic for multi-objective optimization. In: IEEE Symposium on Computational Intelligence in Multi-Criteria Decision-Making, MCDM 2009, 66–73 (2009)
5. Beume, N., Naujoks, B., Emmerich, M.: SMS-EMOA: multiobjective selection based on dominated hypervolume. European Journal of Operational Research 180(3), 1653–1669 (2007)
6. Purshouse, R.C., Fleming, P.J.: On the evolutionary optimization of many conflicting objectives. IEEE Trans. on Volutionary Computation 11(6), 770–784 (2007)
7. Friedrich, T., Horoba, C., Neumann, F.: Multiplicative approximations and the hypervolume indicator. In: Proc. of 2009 Genetic and Evolutionary Computation Conference, pp. 571–578 (2009)
8. Auger, A., Bader, J., Brockhoff, D., Zitzler, E.: Theory of the hypervolume indicator: Optimal μ-distributions and the choice of the reference point. In: Proc. of Foundations of Genetic Algorithm X, pp. 87–102 (2009)
9. Zitzler, E., Künzli, S.: Indicator-based selection in multiobjective search. In: Yao, X., et al. (eds.) PPSN 2004. LNCS, vol. 3242, pp. 832–842. Springer, Heidelberg (2004)
10. Wagner, T., Beume, N., Naujoks, B.: Pareto-, aggregation-, and indicator-based methods in many-objective optimization. In: Obayashi, S., Deb, K., Poloni, C., Hiroyasu, T., Murata, T. (eds.) EMO 2007. LNCS, vol. 4403, pp. 742–756. Springer, Heidelberg (2007)
11. Bader, J., Zitzler, E.: HypE: An algorithm for fast hypervolume- based many-objective optimization. Evolution Computation 19, 45–76 (2011)
12. Hughes, E.J.: Evolutionary many-objective optimization: Many once or one many? In: Proc. of 2005 IEEE Congress on Evolutionary Computation, pp. 222–227 (2005)
13. Hughes, E.J.: MSOPS-II: A general-purpose many-objective optimizer. In: Proc. of 2007 IEEE Congress on Evolutionary Computation, pp. 3944–3951 (2007)
14. Ishibuchi, H., Nojima, Y.: Optimization of scalarizing functions through evolutionary multiobjective optimization. In: Obayashi, S., Deb, K., Poloni, C., Hiroyasu, T., Murata, T. (eds.) EMO 2007. LNCS, vol. 4403, pp. 51–65. Springer, Heidelberg (2007)
15. Zhang, Q., Li, H.: MOEA/D: A multiobjective evolutionary algorithm based on decomposition. IEEE Trans. on Evolutionary Computation 11(6), 712–731 (2007)
16. Tan, Y., Jiao, C., Li, H., Wang, X.: MOEA/D +uniform design: A new version of MOEA/D for optimization problems with many objectives. Computers & Operations Research 40(6), 1648–1660 (2013)
17. Li, H., Zhang, Q.: Multiobjective optimization problems with complicated Pareto sets, MOEA/D and NSGA-II. IEEE Trans. on Evolutionary Computation. 13(2), 284–302 (2009)
18. Bentley, P.J., Wakefield, J.P.: Finding Acceptable Solutions in the Pareto-Optimal Range using Multiobjective Genetic Algorithms. In: World Conference on Soft Computing in Design and Manufacturing, pp. 231–240 (1997)
19. Kokolo, I., Hajime, K., Shigenobu, K.: Failure of Pareto-based MOEAs: Does Nondominated Really Mean Near to Optimal? In: Proceedings of the Congress on Evolutionary Computation, vol. 2, pp. 957–962 (2001)

20. Farina, M., Amato, P.: A fuzzy definition of "optimality" for many-criteria optimization problems. IEEE Transactions on Systems, Man, and Cybernetics Part A—Systems and Humans 34(3), 315–326 (2004)
21. Sato, H., Aguirre, H.E., Tanaka, K.: Controlling dominance area of solutions and its impact on the performance of MOEAs. In: Obayashi, S., Deb, K., Poloni, C., Hiroyasu, T., Murata, T. (eds.) EMO 2007. LNCS, vol. 4403, pp. 5–20. Springer, Heidelberg (2007)
22. Dai, C., Wang, Y.: A New Evolutionary Algorithm Based on Contraction Method for Many-objective Optimization Problems. Applied Mathematics and Computation, 192–205 (2014)
23. Van Veldhuizen, D.A.: Multiobjective evolutionary algorithms: Classifications, analyses, and new innovations. Ph.D. dissertation, Dept. Electr. Comput. Eng. Graduate School Eng., Air Force Instit. Technol., Wright- Patterson AFB, OH (May 1999)
24. Zitzler, E., Deb, K., Thie, L.: Comparison of multiobjective evolutionary algorithms: Empirical results. Evol. Comput. 8(2), 173–195 (2000)
25. Deb, K., Thiele, L., Laumanns, M., Zitzler, E.: Scalable multi-objective optimization test problems. In: Congress on Evolutionary Computation (CEC 2002), pp. 825–830 (2002)
26. Coello Coello, C.A., Cortés, N.C.: Solving multiobjective optimization problems using an artificial immune system. Genetic Programming and Evolvable Machines 6(2), 163–190 (2005)
27. Robert, S., Torrie, J., Dickey, D.: Principles and Procedures of Statistics: A Biometrical Approach. McGraw-Hill, New York (1997)
28. Deb, K.: Multiobjective Optimization Using Evolutionary Algorithms. Wiley, New York (2001)
29. Price, K., Storn, R.M., Lampinen, J.A.: Differential Evolution: A Practical Approach to Global Optimization. Natural Computing. Springer, Secaucus (2005)

Cell Segmentation Using Binarization and Growing Neural Gas

Tomoyuki Hiroyasu[1], Shunsuke Sekiya[2], Noriko Koizumi[1],
Naoki Okumura[1], and Utako Yamamoto[1]

[1] Faculty of Life and Medical Sciences, Doshisha University,
Tataramiyakodani, 1-3, Kyotanabe-shi, Kyoto, Japan
{tomo,utako}@mis.doshisha.ac.jp,
nkoizumi@mail.doshisha.ac.jp,
nokumura@koto.kpu-m.ac.jp
[2] Graduate School of Life and Medical Sciences, Doshisha University,
Tataramiyakodani, 1-3, Kyotanabe-shi, Kyoto, Japan
ssekiya@mis.doshisha.ac.jp

Abstract. In corneal endothelium tissue engineering, automatic judgment is vital to determine whether cultured cells are suitable for transplantation, which can be achieved by measuring indicators from cell images. Indicator measurement requires an accurate image processing method for cell segmentation. We previously propose the system that combines simple image-processing filters suitably by genetic programming. However, it is too difficult to obtain high accuracy because of overfitting. Therefore, achieving segmentation by an unsupervised learning method is an essential requirement for applying segmentation to several types of images. In this paper, we propose an unsupervised learning segmentation method using binarization and growing neural gas. This method segments the cells by performing vector quantization of cell borders and connecting units. The proposed method is comparable to a previously reported method. The results show that the proposed method has superior accuracy.

Keywords: Growing Neural Gas, Binarization, Genetic Programming, Cell Segmentation.

1 Introduction

The prevalent use of contact lenses and the increasing number of intraocular surgeries have been considered a common cause of corneal endothelial cell loss [1,2]. Thus, corneal endothelium tissue engineering has attracted significant attention as a treatment protocol[3,4]. In this treatment, corneal endothelial cells taken from donor eyes are cultured. Then, these cultured cells are transplanted into patients. When the cells are cultured, quality evaluation to determine their transplantability is important. Quality evaluation can be performed by measuring indicators, such as cell density, size dispersion, and shape, from the cell images taken by an optical microscope. In the present state of corneal endothelium tissue engineering, specialists evaluate these indicators by manual observations. Thus, the accuracy of the judgment depends on the experience and ability of the assessor. In addition, manual analysis is not realistic because numerous cell images are accumulated during long-term cultivation. Therefore, software that helps quantitatively evaluate the quality using image processing technique has been developed.

To measure the indicators from images, accurate image processing of the cell segmentation is important. Existing software, i.e., ImageJ [5] and Cell Profiler [6], can perform image processing of segmentation by combining pre-prepared image-processing filters. Thus, users are burdened with a complicated combination of filters for each image. To address this problem, a method that automatically optimizes a combination of filters by genetic programming (GP) [7], which is an evolutionary computing method, has been proposed [8]. This method is applied to some kinds of medical images, for example PET images [9], pathological images [10,11], and cell images [12]. Moreover, a system for measuring such cultivation indicators has been constructed [13]. In this system, the learning region (supervised image) is obtained from a part of the subject image. Then, the most suitable combination of multiple image-processing filters is searched by GP to perform accurate segmentation in a learning region. The combination of filters is applied to the other regions of the subject image (i.e., non-learning regions). To obtain higher accuracy for the non-learning regions, determining a suitable learning region is essential. This decision is very difficult and thus several types of cell images should be applied to the process. Therefore, a cell segmentation method that uses unsupervised learning is preferable.

In this paper, we propose a segmentation algorithm that uses binarization and growing neural gas (GNG) [14], which is an unsupervised learning method. The proposed method first divides the cell border and cell region using binarization. Then, it performs vector quantization of the cell border by GNG and connects the units to segment each cell. The proposed method is experimentally compared with a previous method. The effectiveness of the proposed method is examined by comparing the accuracy of cell segmentation images derived by the proposed and previous methods.

2   Previous Method

2.1   General Cell Image Analysis Software: ImageJ

Recently, software (ImageJ) for cell image analysis has been developed to evaluate the state of cultured cells quantitatively. ImageJ was developed by the National Institutes of Health to analyze medical images [5]. In ImageJ, the user performs the desired processing by combining multiple image-processing filters. To measure cultivation indicators, the combined filters must be constructed for segmentation. Thus, knowledge of image processing and parameter settings for each subject image is required. Therefore, existing software is labor-intensive. In particular, acquisition of image processing is a heavy burden for corneal endothelium tissue engineering researchers.

2.2 Cell Segmentation Method by Combinatorial Optimization of Image-Processing Filters Using GP

Generally, image processing involves the combination of simple image-processing filters. To use the cell image analysis software described in the previous section, the user must combine these filters arbitrarily; thus, user burden is problematic. Image processing can be regarded as a combinatorial optimization problem. Therefore, we employ a method that constructs a tree-structured filter (Fig. 1) by finding optimum solution of this problem.

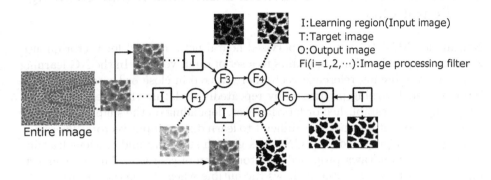

Fig. 1. Principle of tree-structured image-processing filter using GP

In this method, a learning region I is taken from the entire image. Then, a target image T, which is the ideal segmentation result of image I, is created manually using paint software. Image I is input to terminal nodes of the tree structure (i.e., a combination of image-processing filters F_i), and is processed in order by F_i. To approximate the output image O to the target image T, GP searches for the most suitable combination of F_i.

In this method, global consistency error (GCE) [15] is used to evaluate the GP function. Generally, GCE is used as a quantitative evaluation for segmentation, and it focuses on the same pixel p_i in images $S1$ and $S2$. The size of each cell $(R(T, p_i), R(O, p_i))$ that contains the chosen pixel is measured by counting the number of pixels. The local refinement error E is expressed by the difference between $R(S1, p_i)$ and $R(S2, p_i)$. Here, E is defined by formula (1). By measuring the average E of every pixel in the images, the calculated GCE can indicate segmentation precision. If the number of pixels composing the image is N, then GCE is defined by formula (2).

$$E(T, O, p_i) = \frac{|R(T, p_i) \setminus R(O, p_i)|}{|R(T, p_i)|} \tag{1}$$

$$GCE(T, O) = \frac{1}{N} min\left\{ \sum_i E(T, O, p_i), \sum_i E(O, T, p_i) \right\} \tag{2}$$

This method can obtain high accuracy for the learning region. The effectiveness of the method has been verified in a previous report [13]. However, when a searched combination is applied to the non-learning regions, accuracy is sometimes reduced because of over fitting. To address this problem, a method to determine the most suitable learning region is required; however, this is too difficult to implement. Therefore, a cell segmentation method that does not require learning is needed.

3   Segmentation Method with Binarization and Growing Neural Gas

Neural gas (NG) [16] is an unsupervised neural network used for vector quantization by approximating a few units to a set of input vectors. In the NG learning process, units having reference vectors are ranked in close order relative to the input vector. Learning progresses by repeatedly updating the reference vectors according to the established order. However, depending on the shape and density of input vectors, it is sometimes difficult to learn distant input vectors. Therefore, GNG has been proposed [14]. GNG has growth structure and efficient learning performance. Sunakawa proposed skeletonization of character data by connecting the GNG units [17]. Fig. 2 shows the outline when the method is applied to cell segmentation.

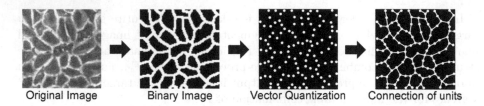

Original Image   Binary Image   Vector Quantization   Connection of units

Fig. 2. Outline of Cell Segmentation Using GNG

First, the original image is converted into a binary image to generate input vectors. When the binary image is created, k-means clustering [18] is performed for each 20×20 [pixel] window. k-Means clustering divides pixels in the image into two classes, i.e., cell region and cell border. The features used for clustering are the pixel value of the pixel of interest and its eight neighboring pixels.

Second, vector quantization of the cell border pixels is performed by GNG. Pixels that represent the cell border are the input vectors v. Each unit i on the image has a reference vector ω_i. The input vectors and the reference vectors are used to create a two-dimensional coordinate vector of the image. In addition, unit i has a counter C_i. The GNG learning algorithm is described as follows.

Step.1 Initialization

Repeat counter t is set to 0. The number of initial units $N(t)$ and reference vector $\omega_i(t)$ of each unit are set. The counter of each unit is $C_i(t)$, and the input vector in the repeat counter t is $v(t)$.

Step.2 Presentation of input vector

Input vectors $v(t)$ of the cell border pixels are given in uniform distribution.

Step.3 Calculation of distance from input vectors

The distance D_i between the given input vector and the reference vector of each unit is calculated using formula (3).

$$D_i = ||v(t) - \omega_i(t)|| \tag{3}$$

Step.4 Update of units

D_i of all units are compared and arranged in ascending order. When the order is k, D_k is represented by $D_k : D_0 < D_1 ... < D_{N-2} < D_{N-1}$. The reference vectors of all units are updated by formulas (4) and (5). Here, α is the learning rate.

$$\omega_i(t+1) = \omega_i(t) + h_k(v(t) - \omega_i(t)) \tag{4}$$

$$h_k = \alpha \exp(-k_i) \tag{5}$$

When the unit of $k = 0$ is a, and the unit that is the closest to a is b, the counter is updated by formula (6) when formula (7) is satisfied. The repeat counter maximum is T_{max}.

$$||v(t) - \omega_a(t)|| > (||\omega_a(t) - \omega_b(t)||)^{\frac{t}{T_{max}}} \tag{6}$$

$$C_a(t+1) = C_a(t) + 1 \tag{7}$$

Step.5 Addition of unit

The counters of all units are compared when repeat counter t reaches each time T_{int}. The unit with the maximum counter is c. If formula (8) is satisfied, a new unit is added.

$$C_c(t) > (t/N) \tag{8}$$

Here, the unit that is closest to c is d. A new unit r is added by formulas (9) to (12).

$$\omega_r(t+1) = \frac{\omega_c(t) + \omega_d(t)}{2} \tag{9}$$

$$N(t+1) = N(t) + 1 \tag{10}$$

$$C_r(t+1) = \frac{C_c(t)}{3} \tag{11}$$

$$C_c(t+1) = \frac{C_c(t)}{3} \tag{12}$$

Step.6 Completion judgment

When $t < T_{max}$ is satisfied, the repeat counter t is incremented by one, and the process returns to Step.2.

The GNG learning algorithm performs the quantization for the cell border. Finally, cell regions are divided by connecting the units. For all units created by GNG, the closest five units are searched. If the unit is not connected to any searched units, then they are connected. In addition, if formula (13) is satisfied, the connection is removed. The average length of connected units is d_{ave}, and the coefficient is ϵ_{cut}. After connection and removal, small regions created by the intersection of the connection are filled by closing processing. Closing processing performs dilation and erosion of the morphological operation. Then, the cell border is extracted by the line thinning algorithm [19].

$$||\omega_i - \omega_j|| = \epsilon_{cut} d_{ave} \tag{13}$$

4   Experiment

4.1   Experiment Outline

In this experiment, the cell segmentation method using GNG was compared with the combinatorial optimization method, i.e., image-processing filters using GP. The images used in the experiment were two 400×300 [pixel] cultured Macaca fascicularis corneal endothelial cell images (Fig. 3) photographed by a phase-contrast microscope (200 times magnification). In the GP method, an experimental image was divided into twelve 100×100 [pixel] images that were used as the learning regions. The accuracy of the proposed method was compared with the accuracy of the method that evaluates non-learning regions on the basis of data from each learning region. In addition, the accuracy of cell segmentation is defined as $1.0 - $GCE. The median from 10 trials was used for comparison. The GP image-processing filters used in the experiment are shown in Tables 1 and 2. The GP and GNG parameters are shown in Tables 3 and 4, respectively. These parameters were determined experimentally. The computing environment is shown in Table 5.

4.2   Experimental Results

The GNG learning process in Region7 of Image1 is shown in Fig. 4. We found that units approximate the pixels of a cell border as they increase.

In addition, the number of non-learning regions with GNG accuracy that was greater than GP are shown in Tables 6 and 7. For Image1, the accuracy of average seven non-learning regions out of eleven regions is higher than GP. For Image2, the accuracy of average eight non-learning regions is also higher than GP. Moreover, GNG took about 6 minutes per a trial, whereas, GP took about 40 minutes. These results indicate that the proposed method performs better than the previous method. As can been in Table 6, GNG is inferior to GP when Region1 of Image1 was learned. Comparative results are shown in Fig. 5(a). As can be observed in Fig. 5(a), the differences among Region4, Region6, and Region12 are small, and the accuracy of these regions is relatively high. The comparative results in which Region2 of Image2 was learned are shown in Fig. 5(b). The differences among Region9, Region10, and Region12 are also small.

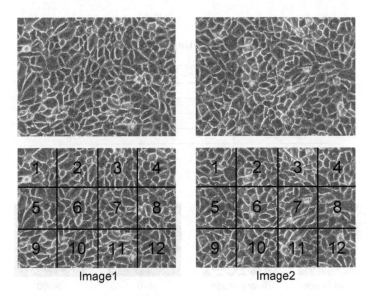

Image1   Image2

Fig. 3. Experimental Images

Table 1. One-Input Filters [12]

Number	Effect
f1	Min
f2	Max
f3	Light Pixel
f4	Dark Pixel
f5	Mean
f6	Median
f7	Gaussian
f8	Threshold(Otsu)
f9	Inversion
f10	Small Region
f11	Large Region
f12	Watershed [20]
f13	Watershed(distance transform)
f14	Watershed(binary)
f15	Watershed(distance transform+binary)

Table 2. Two-Input Filters [12]

Number	Effect
F1	Logical Sum
F2	Logical Prod
F3	Algebraic Sum
F4	Algebraic Prod
F5	Bounded Sum
F6	Bounded Prod

Table 3. GP Parameters

Parameter	Value
Number of Generations	300
Population Size	300
Selection	Tournament
Tournament Size	4
Crossover Rate	1.0
Mutation Rate	0.1
Penalty	1.0×10^{-3}

Table 4. GNG Parameters

Parameter	Value
Number of Max Repeat	12500
Number of Initial Unit	5
Interval of Addition	25
Learning Late	0.4
Removal Coefficient	2.0

Table 5. Computing Environment

OS	Windows Server 2012 Standard
CPU	Intel Core i7-3770
Clock Frequency	3.4 GHz
Core	4
Language	C#

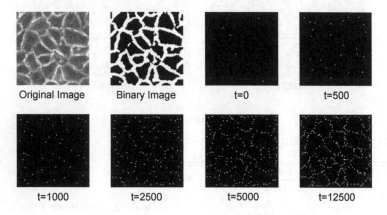

Original Image   Binary Image   t=0   t=500

t=1000   t=2500   t=5000   t=12500

Fig. 4. GNG Learning Results

Table 6. Comparative Results for Image1

Learning Region	Region1	Region2	Region3	Region4	Region5	Region6
Number of Region	5	6	8	6	8	8

Learning Region	Region7	Region8	Region9	Region10	Region11	Region12
Number of Region	6	6	8	10	10	6

Table 7. Comparative Results for Image2

Learning Region	Region1	Region2	Region3	Region4	Region5	Region6
Number of Region	8	5	8	9	8	7

Learning Region	Region7	Region8	Region9	Region10	Region11	Region12
Number of Region	10	9	10	7	7	9

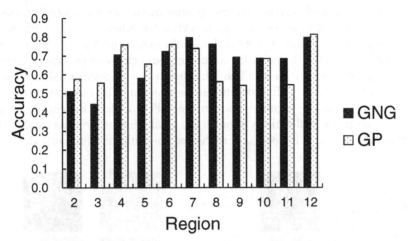

(a) Results when Region1 of Image1 was Learned

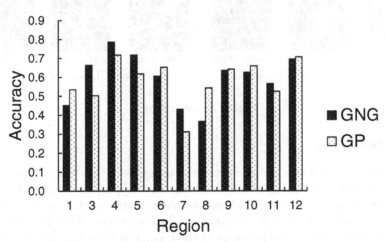

(b) Results when Region2 of Image2 was Learned

Fig. 5. Comparative Results (Accuracy)

5   Discussion

The results presented in the previous section indicate that accuracy of the proposed methods is inferior to GP when Region3 of Image1 and Region8 of Image2 were selected as the learning regions. The resulting images from the proposed method are shown in Fig. 6(a) and Fig. 6(b). The red, blue, and yellow regions are inaccurately segmented regions. The red regions are under-segmented regions. In these regions, the cell border could not be extracted when the binary image was created. In addition, sufficient connection could not be performed. Therefore, it is necessary to examine the rules for connecting and the post-processing method after creation of a binary image. The blue region is over-segmented regions due to over-connected units. This problem can be prevented by calculating the density of units around the segment line. In the yellow region, k-means clustering was performed in the cell regions because the local window size was smaller than the cell region. Thus, it is necessary to review the binarization method.

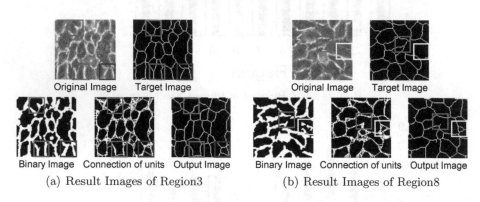

Original Image Target Image   Original Image Target Image

Binary Image Connection of units Output Image   Binary Image Connection of units Output Image
      (a) Result Images of Region3   (b) Result Images of Region8

Fig. 6. Result Images

6   Conclusion

In this paper, a cell segmentation method using GNG has been proposed for cell segmentation by unsupervised learning. In the proposed method, once a subject image has been converted into a binary image, vector quantization by GNG and connection of units are performed. The proposed method was compared experimentally with a previous method, i.e., combinatorial optimization of image-processing filters using GP. With the previous method, the experimental image was divided into twelve regions for learning. The accuracy of the proposed method was compared with the accuracy of accuracy of the method that evaluates non-learning regions on the basis of data from each learning region. It was found that the accuracy of the proposed method is higher than the previous method by an average of 7-8 non-learning regions out of every 11 regions. Moreover, the difference of the accuracy is small in the interior regions. In future, it will be necessary to examine the binary method, post-processing, and the rules for the required region.

References

1. Inaba, M.: Contact lens wear and corneal endothelial cell loss. Journal of the Eye 26(2), 187–192 (2009)
2. Tsubota, K., Hato, S.: Corneal disease and regenerative medicine. Trends in the Sciences 15(7), 7–13 (2010)
3. Koizumi, N., Nishida, K., Amano, S., Kinoshita, S.: Progress in the development of tissue engineering of the cornea in japan. Journal of Japanese Ophthalmological Society 111(7), 493–503 (2007)
4. Koizumi, N.: Cultivated corneal endothelial cell sheet transplantation in a primate model. Journal of Japanese Ophthalmological Society 113(11), 1050–1059 (2009)
5. Abrámoff, M.D., Magalhães, P.J., Ram, S.J.: Image processing with imagej. Biophotonics International 11(7), 36–42 (2004)
6. Carpenter, A.E., Jones, T.R., Lamprecht, M.R., Clarke, C., Kang, I.H., Friman, O., Guertin, D.A., Joo, C.H., Robert, L.A., Moffat, J., Golland, P., Sabatini, D.M.: Cellprofiler: image analysis software for identifying and quantifying cell phenotypes. Genome Biology 7(10), R100.1–R100.11 (2006)
7. Koza, J.R.: Genetic Programming on the Programming of Computers by Means of Natural Selection. MIT Press (1992)
8. Aoki, S., Nagao, T.: Automatic construction of tree-structural image transformations using genetic programming. In: Proceedings of the International Conference on Image Analysis and Processing, pp. 136–141 (1999)
9. Nakano, Y., Nagao, T.: 3d medical image processing using 3d-actit: Automatic construction of tree-structural image transformation (computer vision, medical applications and networked mm) (international workshop on advanced image technology (iwait2004)). IEICE technical report. Image Engineering 103(540), 49–53 (2004)
10. Hiroyasu, T., Fujita, S., Watanabe, A., Miki, M., Ogura, M., Fukumoto, M.: Comparison of gp and sap in the image-processing filter construction using pathology images. In: 2010 3rd International Congress on Image and Signal Processing (CISP), vol. 2, pp. 904–908 (2010)
11. Hiroyasu, T., Yamaguchi, H., Fujita, S., Miki, M., Yoshimi, M., Ogura, M., Fukumoto, M.: An algorithm for cancer nest feature extraction from pathological images. In: 2011 Annual International Conference of the IEEE Engineering in Medicine and Biology Society, EMBC, pp. 3423–3426 (2011)
12. Yamaguchi, H., Hiroyasu, T., Nunokawa, S., Koizumi, N., Okumura, N., Yokouchi, H., Miki, M., Yoshimi, M.: Comparison study of controlling bloat model of gp in constructing filter for cell image segmentation problems. In: 2012 IEEE Congress on Evolutionary Computation (CEC), pp. 1–8 (2012)
13. Hiroyasu, T., Nunokawa, S., Yamaguchi, H., Koizumi, N., Okumura, N., Yokouchi, H.: Algorithms for automatic extraction of feature values of corneal endothelial cells using genetic programming. In: 2012 Joint 6th International Conference on Soft Computing and Intelligent Systems (SCIS) and 13th International Symposium on Advanced Intelligent Systems (ISIS), pp. 1388–1392 (2012)
14. Bernd, F.: A growing neural gas network learns topologies. In: Advances in Neural Information Processing Systems 7, pp. 625–632 (1995)
15. Martin, D., Fowlkes, C., Tal, D., Malik, J.: A database of human segmented natural images and its application to evaluating segmentation algorithms and measuring ecological statistics. In: IEEE International Conference on Computer Vision, vol. 2, pp. 416–423 (2001)

16. Martinetz, T.M., Berkovich, S.G., Schulten, K.J.: 'neural-gas' network for vector quantization and its application to time-series prediction. IEEE Transactions on Neural Networks 4(4), 558–569 (1993)
17. Sunakawa, K., Saito, T.: A-2-29 growing neural gas and skeletonization. In: Proceedings of the Society Conference of IEICE, vol. 60 (2010)
18. Hartigan, J.A., Wong, M.A.: Algorithm as 136: A k-means clustering algorithm. Journal of the Royal Statistical Society. Series C 28(1), 100–108 (1979)
19. Tamura, H.: A comparison of line thinning algorithms from digital geometry viewpoint. In: Proc. 4th Int. Conf. Pattern Recognition, 715–719 (1978)
20. Vincent, L., Soille, P.: Watersheds in digital spaces: an efficient algorithm based on immersion simulations. IEEE Transactions on Pattern Analysis and Machine Intelligence 13(6), 583–598 (1991)

Self-adaptive Orthogonal Simplified Swarm Optimization for the Series-Parallel Redundancy Allocation Problem

Wei-Chang Yeh[1] and Chia-Ling Huang[2]

[1] Integration and Collaboration Laboratory,
Department of Industrial Engineering and Engineering Management,
National Tsing Hua University
yeh@ieee.org
[2] Department of Logistics and Shipping Management,
Kainan University
cl.hfirst@gmail.com

Abstract. This work presents a novel self-adaptive orthogonal simplified swarm optimization scheme (SR-SSO) that combines repetitive orthogonal array testing (ROAT), self-adaptive parameter control, and SSO to the series-parallel redundancy allocation problem (RAP) with a mix of components. The RAP is to decide a network structure to minimize the manufacturing cost under the reliability limitation by using redundant components in parallel. The results obtained in extensive experiments indicate that the proposed algorithm outperforms the previously-developed algorithms in the literature.

Keywords: Reliability, Series-parallel system, Redundancy allocation problem (RAP), Simplified swarm optimization (SSO), Orthogonal array (OA), Self-adaptive parameter control.

Acronym:

ACO	Ant Colony Optimization
GA	Genetic Algorithm
ISC	The Improved Surrogate Constraint Method
LP	Linear Programming Approach
OA/OAT	Orthogonal Array/OA testing
RAP	Redundancy Allocation Problem
ROAT	repetitive OAT
SSO/PSO	Simplified/Particle Swarm Optimization
SR-SSO	The proposed Self-adaptive Orthogonal SSO
TS	Tabu Search
VNS	Variable Neighborhood Search Algorithm

Notations:

n, N	The number of variables and solutions in SSO/SR-SSO in each generation.
T	The total number of independent generations in SSO/SR-SSO.
ρ	The random number uniformly distributed in [0,1].

© Springer International Publishing Switzerland 2015
H. Handa et al. (eds.), *Proc. of the 18th Asia Pacific Symp. on Intell. & Evol. Systems – Vol. 2*,
Proceedings in Adaptation, Learning and Optimization 2, DOI: 10.1007/978-3-319-13356-0_16

pBest, gBest	*pBest* represents the best one of a specified solution has achieved so far; *gBest* represents the best value of all solutions so far.
$c_w, c_p, c_g, c_r, C_w, C_p, C_g$	The parameters represent the probabilities of the new variable value generated from the current solution, *pBest, gBest* and a random number in SSO, respectively, where $c_w+c_p+c_g+c_r=1$. $C_w=c_w$, $C_p=C_w+c_p$, and $C_g=C_p+c_g$.
X_{ti}, x_{ij}^t	$X_{ti}=(x_{i1}^t, x_{i2}^t, ..., x_{in}^t)$ is the *i*th solution at generation *t*, where *i*=1,2,...,*N* and *t*=1,2,...,*T*.
P_i	$P_i=(p_{i1}, p_{i2}, ..., p_{in})$ is the current *pbest* w.r.t. the *i*th solution, where *i*=1,2,...,*N*.
G	$G=(g_1, g_2, ..., g_n)$ is the current *gBest*.
$F(\bullet)$	The fitness function value of •.
r_{ij}, c_{ij}, w_{ij}	The reliability, cost, and weight of the *j*th type of components in the *i*th subsystem, respectively.
$R(\bullet), C(\bullet), W(\bullet)$	The total reliability, cost, and weight of •, respectively.
C_{UB}, W_{UB}	The number of the required budget and weight, respectively.
$L_a(b^c)$	The general symbol for the *b*-level standard OAs.

Assumption: [15-23]

(1) The combination of the items for redundancy should satisfy the function at the system level. If an item is used, all its sibling items should be used or its function should be satisfied by the corresponding child items.
(2) The items for redundancy should be used in parallel at one combination.

1   Introduction

Soft computing methods have been utilized for obtaining optimal or good-quality solutions to difficult optimization problems in various fields. Among soft computing techniques including Neural Network Approach [1], Genetic Algorithms (GAs) [2], Simulated Annealing Methods [3,4], Tabu Search Methods (TSs) [5], Ant Colony Optimizations (ACOs) [6], Immune System Algorithm [7], Estimation Distribution of Algorithm [7], Particle Swarm Optimizations (PSOs) [8-11], and Simplified Swarm Optimization (SSOs) [12-14]. Compared with SA, TS, GA, and ACO, PSO has some appealing features, including easy implementation, few parameters to tune and a fast convergence rate [11]. However, PSO is not suitable for the problems with discrete variables and sequencing problems [11].

To overcome the drawback of PSOs for discrete problems, Yeh proposed SSO, called discrete PSO (DPSO) originally, in 2009 [12]. SSO is simple, efficient, and flexible. Simulation results reveal that SSO has better convergence with quality solutions than PSO [12-14]. Hence, SSO has attracted attention in a range of applications such as the multiple multi-level redundancy allocation [12], the disassembly sequencing [13], and the data mining for breast cancer [14]. Thus, this work presents a novel improved SSO by integrating the novel repetitive orthogonal array testing (ROAT) and self-adaptive parameter control and is applied to the series-parallel redundancy allocation problem (RAP).

RAP is perhaps the most common problem in design-for-reliability. Onishi *et al.* proposed the improved surrogate constraint method (ISC) and reported that ISC can obtain more exact optimal solutions to a set of 14-subsystem 33-variation benchmark problem efficiently and effectively [22]. ISC may be the best-known method for solving the RAP. However, ISC is a mathematical-based algorithm and there is no evidence that ISC can solve all larger-size RAPs easily without highly computational complexity. Hence, there is a need to develop a heuristic-based algorithm.

2   The SSO

The exploration effectively and efficiency of soft computing techniques depends on parameter values [24], e.g., c_w, c_p, and c_g in SSO. These parameters play important roles in the exchange of information among solutions in SSOs/SR-SSOs, particles in PSOs, and chromosomes in GAs. However, specifying and adjusting suitable parameters is difficult and challenging [24]. In SSO/SR-SSO, the values of c_r, c_w, c_g, and c_p are in the range of 0–1, and represent the probability of a random number, the corresponding variable of the current solution, *gBest*, and *pBest* generates a value of a variable in a new candidate solution. In this section, a self-adaptive parameter control is proposed to adjust c_w, c_p, c_g, and c_r, automatically.

The current *gBest* maybe a premature solution if it unchanged for a certain number of generations, say τ, with large number of left generation, i.e., $T-t$. The premature solution is a local optimizer and cause all solutions trapped in the local optimizer forever without improving solution quality. The bigger the value of τ and the smaller value of $T-t$, the greater probability the current *gBest* is a premature solution. Hence, the proposed self-adaptive c_r depends on $\dfrac{\tau}{T-t}$ to enhance the capacity to escape from the local optimum.

The role of c_r is to prevent unwanted premature solutions and to reguide the search towards unexplored regions in the solution space. If c_r is larger, the next solution is more likely to be generated randomly and not as a premature solution, and vice versa. To maintain a balance between the portion of random-generated variables and non-random-generated variables, the minimal and maximal values of c_r are set to .15 and .85, *i.e.*, in generation t, c_r is defined as follows:

$$c_r = Min\{Max\{.15, \frac{\tau}{T-t}\},.85\} \cdot \tag{1}$$

The following figure demonstrates the trends of c_r (the z-axis) under different $T-t=0,10,\ldots,100$ (the y-axis), and $\tau=0,10,\ldots,90$ (the x-axis). From Fig. 1, the slope increases rapidly as τ decreases for fixed $T-t$, and c_r is varied over the generations in the proposed SR-SSO to achieve a trade-off between exploration and exploitation from .15 at the beginning of the search to .85 at the end.

In the proposed self-adaptive parameter control, to each solution, say X_{ij}, the corresponding values of c_w, c_p, and c_g are based on the reciprocal of the difference between the upper-bound, e.g. 1 in RAP, of the system reliability and $F(X_{ti})$, $F(P_i)$, and $F(G)$, respectively. Hence, these solutions with superior fitness values are likely to propagate their variable values to produce offspring. Based on this concept, the following equations are proposed to ensure that c_w, c_g, and c_p are proportional to the values of $F(X_{ti})$, $F(P_i)$, and $F(G)$, respectively.

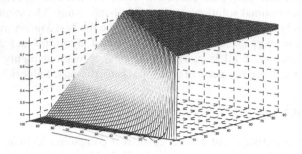

Fig. 1. The trend of c_r (z-axis) under different $T-t$ (y-axis), and τ (x-axis)

$$F^*(\bullet)=\frac{1}{1-F(\bullet)} \tag{2}$$

$$C_g =1-Min\{Max\{.15,\ \frac{\tau}{T-t}\},85\} \tag{3}$$

$$c_w=\frac{C_g \cdot F^*(X_{ti})}{F^*(X_{ti})+F^*(P_i)+F^*(G)} \tag{4}$$

$$c_p=\frac{C_g \cdot F^*(P_i)}{F^*(X_{ti})+F^*(P_i)+F^*(G)} \tag{5}$$

Note that the term C_g is multiplied in the numerator in Eqs.(4) and (5) so that $C_{gi}+c_{tr}$ would be equal to 1.

The proposed self-adaptive scheme is simple and can automatically adapt c_w, c_p, and c_g simultaneously to provide the necessary momentum for a solution to move across the search space. Parameters tune instead of predefining parameters according to fitness values throughout the entire search space.

3   Repetitive Orthogonal Array Testing (ROAT)

This work proposes ROAT as a local search algorithm to improve solutions. The purpose of using OAT is to produce a potentially good approximation systematically and efficiently [3]. Rather than exploring all possible combinations of assignments, OAT is able to prune some solutions based the orthogonal array (OA) during a search. The details of OAT are given as follows. The class of the three-level OAT $L_9(3^4)$ is used in this work. The proposed ROAT integrated the traditional OAT (part of STEPs R3-R6) is given below with the new design to set up the new values in the selected variables (STEP R4).

STEP R1. Let $d=1$.
STEP R2. Let $A=\{c$ distinctive variables selected randomly$\}$, and $X^{(h)}=X$ for
$\qquad h=1,2,\ldots, c+1$.

STEP R3. Construct $L_a(3^{(a-1)/2})$.

STEP R4. Update $X^{(h)}=(x_{h1},x_{h2},\ldots,x_{hn})$ and calculate $F(X^{(h)})$, for $h=1,2,\ldots,c$,

$$x_{hj}=\begin{cases} x_{1j}-1 & \text{if } a_{hi}=2 \text{ and } x_{1j}>0 \\ u_j & \text{if } a_{hi}=2 \text{ and } x_{1j}=0 \\ x_{1j} & \text{if } a_{hi}=1 \\ x_{1j}+1 & \text{if } a_{hi}=3 \text{ and } x_{1j}<u_j \\ 0 & \text{if } a_{hi}=3 \text{ and } x_{1j}=u_j \end{cases} \quad \text{for } j \text{ is the } i^{th} \text{ element in } A. \quad (6)$$

STEP R5. Calculate $S_{ij}=\sum_{k\in K}\dfrac{F(X^{(k)})}{3}$, where $K=\{\kappa \mid a_{\kappa i}=j\}$ for all $i\in A$.

STEP R6. Choose the best one in $\{S_{ij}\mid j=1,2,3\}$, say $S_{ik(i)}$, for all $i\in A$.

STEP R7. Update $x_{(c+1),i}$ based on the value corresponding to the level $k(i)$ of factor i and Eq.(6) and calculate $F(X^{(c+1)})$.

STEP R8. Let $X=X^*$, where $F(X^*)$ is the best one in $\{F(X^{(h)}) \mid h=1,2,\ldots,c+1\}$ (break arbitrarily if there is a tie).

STEP R9. If $d<D$, then let $d=d+1$ and go to STEP R2.

4   The Proposed SR-SSO for RAP

The proposed SR-SSO is based on the fundamental concept of the standard SSO together with self-adaptive parameter control and ROAT. The details of solution representations, the objective function, initial population, stop criterion, and overall procedure of the proposed SR-SSO for solve RAPs are presented below.

To guide a search toward unexplored regions in the solution space, a penalty function is implemented [23,25]. For solution X with the total system cost $C(X)$ and/or weight $W(X)$ that exceed C_{UB} and W_{UB}, respectively, a penalized reliability is calculated as

$$F(X)=\begin{cases} R(X) & \text{if } C(X)\leq C_{UB} \text{ and } W(X)\leq W_{UB} \\ R(X)\cdot Min\left\{\dfrac{C_{UB}}{C(X)},\left[\dfrac{W_{UB}}{W(X)}\right]^3\right\} & \text{otherwise} \end{cases} \quad (7)$$

where exponent 3 is preset amplification parameters,

$$R(X)=\prod_{i=1}^{\alpha}\left[1-\prod_{j=1}^{\beta_\alpha}[1-R(x_{ij})]\right], \quad (8)$$

$$C(X)=\sum_{i=1}^{\alpha}\sum_{j=1}^{\beta_\alpha}C(x_{ij}), \quad (9)$$

$$W(X)=\sum_{i=1}^{\alpha}\sum_{j=1}^{\beta_\alpha}W(x_{ij})\cdot \quad (10)$$

Eq.(7) encourages solutions to explore the feasible region and infeasible region near the border of the feasible area such that the search does not go too far into the infeasible region. Thus, the promising feasible and infeasible regions in the search space are explored efficiently and effectively to identify an optimal or near-optimal solution.

According to the discussion in the previous subsection, Sections 4 and 5, the steps of overall proposed SR-SSO are described as follows.

STEP 1. Generate X_{0i} with n variables randomly, calculate $F(X_{ti})$, and let $X_{0i}=P_i$, where $i=1,2,\ldots,N$.

STEP 2. Improve X_{ti} using the proposed ROAT discussed in Section 5, where $i=1,2,\ldots,N$.

STEP 3. Let $t=1$.

STEP 4. Calculate c_r, c_w, c_p, and c_g based on Eqs.(2)-(5) listed in Section 3.

STEP 5. Update X_{ti} based and calculate $F(X_{ti})$, where $i=1,2,\ldots,N$.

STEP 6. Improve X_{ti} using the proposed ROAT discussed in Section 4.

STEP 7. If $F(X_{ti})<F(P_i)$, then let $P_i=X_{ti}$ for $i=1,2,\ldots,N$. Otherwise, go to STEP 7.

STEP 8. If $F(P_i)<F(G)$, then let $G=P_i$.

STEP 9. If $t=T$ and/or CPU time are met, then halt; otherwise let $t=t+1$ and go to STEP 3.

5   Numerical Examples

To evaluate the quality and performance of the proposed SR-SSO for RAPs, the proposed SR-SSO is applied to the most famous benchmark problem in RAP, which was originally proposed by Fyffe et al. [15] and revised by Nikagawa [16]. This benchmark RAP has 33 variations of the original Fyffe problem. It is a well-known test problem and has been extensively used in literature to evaluate alternative approaches for RAP such as LP [19], GA [2], ACO [6], TS [5], VNS [23], and ISC [22].

In this 33-variation benchmark problem, the series-parallel system is connected by 14 parallel subsystems; each subsystem has three or four alternative component choices, the cost constraint is fixed to $C=130$, and the weight constraint is 159–191. In this work, component mixing is considered without restrictions.

The proposed SR-SSO is implemented in C programming language and run on an Intel Pentium 2.6 GHz PC with 256 MB of memory. The runtime unit of was CPU seconds. For fair comparison, the number of generations (1000), number of independent runs (10), and populations (100) for the proposed SR-SSO are taken directly from [13]. Fig. 2a–2c summarize the experimental results.

Fig. 2a and Fig. 2b revealed that the proposed SR-SSO is a robust optimization algorithm for RAP in running time and solution quality. Figure 4c showed that the average numbers of the current generation starts to converge is more than 500. Hence, the proposed self-adaptive is able to prevent the occurrence of premature solutions in the early stage. An interesting observation from Figs. 2a and 2b is that reliability and CPU running time are lower upon increasing as the weight increases.

Table 1 shows the final reliability (the fitness function value of the *gBest*) obtained from the SR-SSO, LP [19], GA [2], ACO [6], TS [5], VNS [23] and ISC [22]. The shaded parts in Table 1 shows the best result. The bottom row in Table 1 indicates that the fraction of obtained solutions equal to solutions obtained from ISC.

Experimental results indicate that the proposed SR-SSO is superior to LP [19], GA [2], ACO [6], TS [5], and VNS [23], and can efficiently solve 33-variation benchmark RAP in Table 1. Additionally, SR-SSO perform as well as ISC [22], which may be the best-known method for solving the RAP. Moreover, computational time for solving each benchmark RAP using the proposed SR-SSO is <0.2 sec, which is almost that by ISC (implemented on a Pentium 4 3.0 GHz running Windows XP).

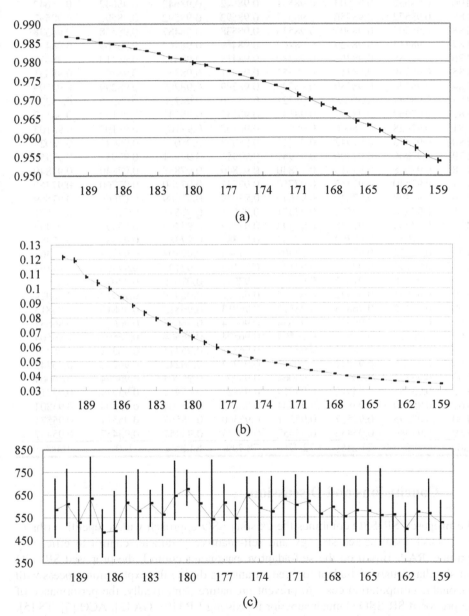

Fig. 2. Range of performance for 10 runs with means represented by horizontal dashes and weight represented the horizontal axis: (a) reliability (b) CPU seconds (c) the generation that started to converge.

Table 1. Comparison of the best solutions among heuristics

W	LP [19]	GA [2]	ACO [6]	TS [5]	VNS [23]	ISC [22]	SR-SSO
191	0.98671	0.98675	0.98675	0.98681	0.98681	0.98681	0.98681
190	0.98632	0.98603	0.98591	0.98642	0.98642	0.98642	0.98642
189	0.98572	0.98556	0.98577	0.98592	0.98592	0.98592	0.98592
188	0.98503	0.98503	0.98533	0.98538	0.98487	0.98538	0.98538
187	0.98415	0.98429	0.98469	0.98469	0.98467	0.98469	0.98469
186	0.98388	0.98362	0.98380	0.98418	0.98418	0.98418	0.98418
185	0.98339	0.98311	0.98351	0.98351	0.98351	0.98351	0.98351
184	0.98220	0.98239	0.98299	0.98299	0.98299	0.98299	0.98299
183	0.98147	0.9819	0.98221	0.98226	0.98226	0.98226	0.98226
182	0.97969	0.98102	0.98147	0.98152	0.98147	0.98152	0.98152
181	0.97928	0.98006	0.98068	0.98103	0.98103	0.98103	0.98103
180	0.97833	0.97942	0.98029	0.98029	0.98029	0.98029	0.98029
179	0.97806	0.97906	0.97951	0.97951	0.97951	0.97951	0.97951
178	0.97688	0.97810	0.97840	0.97840	0.97838	0.97840	0.97840
177	0.97540	0.97715	0.97760	0.97747	0.97760	0.97760	0.97760
176	0.97498	0.97642	0.97649	0.97669	0.97669	0.97669	0.97669
175	0.97350	0.97552	0.97571	0.97571	0.97571	0.97571	0.97571
174	0.97233	0.97435	0.97493	0.97479	0.97493	0.97493	0.97493
173	0.97053	0.97362	0.97383	0.97383	0.97381	0.97383	0.97383
172	0.96923	0.97266	0.97303	0.97303	0.97303	0.97303	0.97303
171	0.96790	0.97186	0.97193	0.97193	0.97193	0.97193	0.97193
170	0.96678	0.97076	0.97076	0.97076	0.97076	0.97076	0.97076
169	0.96561	0.96922	0.96929	0.96929	0.96929	0.96929	0.96929
168	0.96415	0.96813	0.96813	0.96813	0.96813	0.96813	0.96813
167	0.96299	0.96634	0.96634	0.96634	0.96634	0.96634	0.96634
166	0.96121	0.96504	0.96504	0.96504	0.96504	0.96504	0.96504
165	0.95992	0.96371	0.96371	0.96371	0.96371	0.96371	0.96371
164	0.95860	0.96242	0.96242	0.96242	0.96242	0.96242	0.96242
163	0.95732	0.96064	0.96064	0.95998	0.96064	0.96064	0.96064
162	0.95555	0.95912	0.95919	0.95821	0.95919	0.95919	0.95919
161	0.95410	0.95804	0.95804	0.95692	0.95804	0.95804	0.95804
160	0.95295	0.95567	0.95571	0.95560	0.95567	0.95571	0.95571
159	0.95080	0.95432	0.95457	0.95433	0.95457	0.95457	0.95457
ratio	0%	21.21%	72.72%	78.79%	84.85%	100%	100%

6   Conclusions and Discussions

The proposed SR-SSO combines the existing SSO approach with a repetitive (three-level) OAT local search strategy and self-adaptive parameter control to efficiently optimize RAP. Based on the self-adaptive parameter control, the proposed SR-SSO can simultaneously self-adapt to three parameters during the exploration process with reasonable computation costs to prevent premature. Empirically, the performance of the proposed SR-SSO is much superior to existing LP [19], GA [2], ACO [6], TS [5], and VNS [23] in solving the 33-variation benchmark RAP. Moreover, the proposed SR-SSO has proven to be effective in terms of all computational speeds were less than 0.5 second.

References

1. Yeh, W.C., Lin, C.H., Lin, Y.C.: A MCS-Based Neural Network Approach to Extract Network Approximate Reliability Function. In: Park, J.-W., Kim, T.-G., Kim, Y.-B. (eds.) AsiaSim 2007. CCIS, vol. 5, pp. 287–297. Springer, Heidelberg (2007)
2. Coit, D.W., Smith, A.E.: Reliability optimization of series-parallel systems using a genetic algorithm. IEEE Transactions on Reliability 45, 254–260 (1996)
3. Ho, S.J., Ho, S.Y., Shu, L.S.: OSA: Orthogonal Simulated Annealing Algorithm and Its Application to Designing Mixed H2/H∞ Optimal Controllers. IEEE Transactions on Systems, Man and Cybernetics, Part A: Systems and Humans 34(5), 588–600 (2004)
4. Kim, H.G., Bae, C.O., Park, D.J.: Reliability-redundancy optimization using simulated annealing algorithms. Journal of Quality in Maintenance Engineering 12(4), 354–363 (2006)
5. Kulturel-Konak, S., Smith, A.E., Coit, D.W.: Efficiently solving the redun-dancy allocation problem using Tabu search. IIE Transactions 35(6), 515–526 (2003)
6. Liang, Y.H., Smith, A.E.: An ant colony optimization algorithm for the re-dundancy allocation problem (RAP). IEEE Transactions on Reliability 53(3), 417–423 (2004)
7. Chang, W.W., Yeh, W.C., Huang, P.C.: A Hybrid Immune-Estimation Distribution of Algorithm for Mining Thyroid Gland Data. Expert Systems with Applications 37(3), 2066–2071 (2010)
8. Kennedy, J., Eberhard, R.C.: Particle swarm optimization. In: Proceedings of IEEE International Conference on Neural Networks, Piscataway, NJ, USA, pp. 1942–1948 (1995)
9. Kennedy, J., Eberhard, R.C., Shi, Y.: Swarm intelligence. Morgan Kaufmann, San Francisco (2001)
10. Yeh, W.C., Lin, Y.C., Chung, Y.Y., Chih, M.C.: A Particle Swarm OptimizationApproach Based on Monte Carlo Simulation for Solving the Complex Network Reliability Problem. To appear in IEEE Transactions on Reliability (TR2008-176) (April 09, 2007)
11. Shi, X.H., Lianga, Y.C., Leeb, H.P., Lub, C., Wanga, Q.X.: Particle swarm optimization-based algorithms for TSP and generalized TSP. Information Processing Letters 103, 169–176 (2007)
12. Yeh, W.C.: A Two-Stage Discrete Particle Swarm Optimization for the Problem of Multiple Multi-Level Redundancy Allocation in Series Systems. Expert Systems with Applications 36(5), 9192–9200 (2009)
13. Yeh, W.C., Chang, W.W., Chung, Y.Y.: A new hybrid approach for mining breast cancer pattern using Discrete Particle Swarm Optimization and Statistical method. Expert Systems with Applications 36(4), 8204–8211 (2009)
14. Yeh, W.C., Lin, H.Y.: A Soft Computing Algorithm for Disassembly Sequencing. In: International Conference on Engineering and Computational Mathematics (ECM 2009), Hong Kong, May 27-29 (2009)
15. Fyffe, D.E., Hines, W.W., Lee, N.K.: System reliability allocation and a computation algorithm. IEEE Transactions on Reliability R-17, 64–69 (1968)
16. Nakagawa, Y., Miyazaki, S.: Surrogate constraints algorithm for reliability optimization problems with two constraints. IEEE Transactions on Reliability 30(2), 175–180 (1981)
17. Chern, M.S.: On the computational complexity of reliability redundancy allocation in a series system. Operations Research Letters 11, 309–315 (1992)
18. Kuo, W., Prasad, V.R.: An annotated overview of system-reliability optimization. IEEE Transactions on Reliability 49(2), 176–187 (2000)
19. Hsieh, Y.C.: A linear approximation for redundant reliability problems with multiple component choices. Computers & Industrial Engineering 44, 91–103 (2002)

20. Yeh, W.C.: A MCS-RSM Approach for the Network Reliability to Minimize the Total Cost. International Journal of Advanced Manufacturing Technology 22(9-10), 681–688 (2003)
21. Kuo, W., Wan, R.: Recent Advances in Optimal Reliability Allocation. IEEE Transactions on Systems, Man and Cybernetics, Part A 37(2), 143–156 (2007)
22. Onishi, J., Kimura, S., James, R.J.W., Nakagawa, Y.: Solving the Redundancy Allocation Problem With a Mix of Components Using the Improved Surrogate Constraint Method. IEEE Transactions on Reliability 56(1), 94–101 (2007)
23. Liang, Y.C., Chen, Y.C.: Redundancy allocation of series-parallel systems using a variable neighborhood search algorithm. Reliability Engineering and System Safety 92, 323–331 (2007)
24. Brest, J., Greiner, S., Boskovic, B.: Self-adapting control parameters in differential evolution: a comparative study on numerical benchmark problems. IEEE Transactions on Evolutionary Computation 10, 646–657 (2006)
25. Agarwal, M., Gupta, R.: Penalty function approach in heuristic algorithms for constrained. IEEE Transaction on Reliability 54(3), 549–558 (2005)

Impact of the Length of Optical Flow Vectors in Estimating Time-to-Contact an Obstacle

Willson Amalraj Arokiasami*, Tan Kay Chen,
Dipti Srinivasan, and Prahlad Vadakkepat

National University of Singapore,
Department of Electrical and Computer Engineering,
4 Engineering Drive 3,
Singapore 117583
{willson,eletankc,dipti,prahlad}@nus.edu.sg

Abstract. Obstacle detection and avoidance for autonomous aerial vehicles is one of the most researched problem in the field of robotics. Vision-based obstacle detection and avoidance in aerial vehicles has received significant attention recently because of their inherent advantage of low power consumption and less weight. In this work an optical flow based time-to-contact method is used for obstacle avoidance. Though the accuracy of this method is good at close range, it cannot be used when obstacles are far away. This work investigates the reason for its less accurate performance for distant obstacles. It helps to identify a boundary within which the performance of the optical flow based time-to-contact method will be more accurate. Using Robot Operating System(ROS) and Gazebo - a 3D simulator, simulations are performed to estimate the time-to-contact an obstacle for an unmanned aerial vehicle (Parrot AR Drone). The results obtained and the analysis performed helps to identify the reason for the poor performance and suggests a proper boundary within which its performance will be more accurate.

Keywords: optical flow, time-to-contact, obstacle detection, unmanned aerial vehicle.

1   Introduction

Vision-based navigation, guidance and control has been one of the most focused research topic in autonomous operation of Unmanned Aerial Vehicles (UAVs) [1], [2], [3], [4] and [5]. Obstacle detection methods using sonar and laser sensors are not suitable for all UAVs as these sensors consume more power and are generally heavy to carry. It can be found from nature that birds and insects use vision as an exclusive sensor for object detection and navigation. Camera is a powerful vision sensor providing numerous types of information that can be explored in many contexts for obstacle detection. Furthermore, it is efficient to use a vision sensor in micro aerial vehicles (MAVs) as it is compact, light-weight and low cost.

---

* Corresponding author.

© Springer International Publishing Switzerland 2015   201
H. Handa et al. (eds.), *Proc. of the 18th Asia Pacific Symp. on Intell. & Evol. Systems – Vol. 2*,
Proceedings in Adaptation, Learning and Optimization 2, DOI: 10.1007/978-3-319-13356-0_17

Vision-based navigation algorithms were used initially for ground mobile robots and later imported to UAV systems. Identifying ground plane is an important process for detecting obstacles in the flight path of an aerial vehicle. Color and texture recognition have been used to separate the ground plane in images obtained from the on-board camera [6]. Vision-based algorithms are highly dependent on the type of environment. For corridor type environments, due to many straight line components, edge-based obstacle detection methods have been implemented [7]. If the ground plane is considered to be free of obstacles, then the same edge-based detection can be used for detecting obstacles [8]. The performance of these above mentioned techniques is good only in static lighting conditions and the performance decreases when applied to general conditions. For example, texture recognition assumes that the ground texture remains constant while color recognition assumes that objects are not of the same color as the ground. It has been proved in nature that spatial vision systems operate in conjunction with visual motion system [9].

Optical flow is an vector field generated by the changes in light patterns across optic array which is the fundamental stimulus of vision. Optical flow fields contain spatio-temporal information about the environment that have been shown to be associated with control and timing actions within nature, such as bee flight control, human balance control, etc. Optical flow can be defined by a set of vectors with each vector describing the motion of individual features in the image space [9], [13]. These optical flow vectors can provide a 2D representation of the robot's motion and the 3D structure and motion of the environment under correct conditions. Optical flow has been used in many different ways for navigation and obstacle avoidance [11], [12]. [10] has shown that optical flow templates can be successfully used to identify flat traversable surfaces for common ground plane segmentation. [14] and [15] show that the divergence maps obtained from optical flow fields can be used to manoeuvre around the obstacles successfully, while divergence methods for frontal collision detection have been successfully implemented in [16].

It is possible to estimate successfully the time-to-contact an obstacle without any knowledge of velocity or distance from the surface of the obstacle using only optical measurements. This is possible if the transitional component of the optical flow field is known [17]. The theory of time-to-contact was first introduced in 1985 by [18]. This work uses optical flow based time-to-contact information to avoid obstacles. It is a commonly accepted fact that the accuracy of time-to-contact estimation using optical flow techniques is good only if the obstacles are close to the robot. This work helps to identify the reason for this poor performance and suggests a safe distance within which the accuracy of this method will be better. The main contributions of this paper are

- Identification of the reason why time-to-contact is less accurate for far away obstacles and
- Suggestion of a safe limit within which the performance of this optical flow based time-to-contact method will be fairly accurate.

This paper is organized as follow. Section 2 explains the theory behind time-to-contact estimation using optical flow. Section 3 explains about the simulation set-up and the procedure to conduct experiments. Section 4 provides a discussion and analysis about the results obtained and conclusion is presented in section 5.

2   Time-to-Contact an Obstacle Using Optical Flow

Determination of the time-to-contact or time-to-collision is one of the most practical uses of an optical flow field. Time-to-contact is mathematically defined as the time until an object crosses an infinite plane defined by the image plane. The first detailed discussion of time-to-contact from visual information can be found in [19]. It showed that time-to-contact is a ratio of the visual angle of an object over the rate of change of the visual angle.

$$\tau = \frac{\theta}{\dot{\theta}} \tag{1}$$

where τ represents the time-to-contact and θ represent the visual angle. Time-to-contact estimated using this method is imperfect and is valid only for small visual angles. Based on (1) time-to-contact can also be estimated using motion vectors such as optical flow field. It is possible to estimate the time-to-contact a surface only using the velocity of the robot and the translational component of the optical flow vectors. Let us assume that a robot is moving with a constant velocity and the camera on the robot is facing the direction of motion.

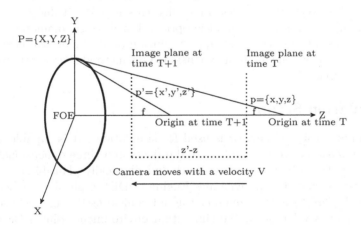

Fig. 1. The projections of a point P onto the image of a moving plane [20]

In Figure 1, $P = (X, Y, Z)$ is a point on the obstacle which projects through the image plane at time T, to point $p = (x, y, z)$. The image plane is positioned at distance f from origin. The camera moves along z-axis with a velocity $V = \frac{\delta Z}{\delta t}$ over a distance $\Delta z = z' - z$, approaching the focus of expansion (FOE). At time $T+1$, the point p now projects to a new point on the image plane, $p' = (x', y', z')$. From the properties of similar triangles we know that,

$$\frac{y}{f} = \frac{Y}{Z} \tag{2}$$

$$y = f\frac{Y}{Z} \tag{3}$$

on differentiating y with respect to time we get

$$\frac{\delta y}{\delta t} = f\left(\frac{\frac{\delta Y}{\delta t}}{Z}\right) - fY\left(\frac{\frac{\delta Z}{\delta t}}{Z^2}\right) \tag{4}$$

$\frac{\delta Y}{\delta t} = 0$, as Y does not change with time. Substituting (3) in (4) and $\frac{\delta Z}{\delta t} = V$, we get

$$\frac{\delta y}{\delta t} = -y\left(\frac{V}{Z}\right) \tag{5}$$

by rearranging we get,

$$\frac{y}{\frac{\delta y}{\delta t}} = -\frac{Z}{V} = TTC \tag{6}$$

where $\frac{dy}{dt}$ is the length of the optical flow vector and y is the distance from the pixel of the obstacle to the focus of expansion. From (6) it can be seen that in order to compute the time-to-contact the obstacle, it is necessary to estimate the focus of expansion. Focus of expansion is the single point in space where all the optical flow vectors should originate [21]. Theoretically it is the point at which two optical flow vectors intersect. In reality this is affected due to noise and other errors involved in computing the optical flow field. A commonly accepted solution by [22] is to use the least square solution of all the optical flow vectors computed to estimate the focus of expansion. This method is used here to obtain the focus of expansion.

3   Simulation Setup

Gazebo, a multi-robot simulator is used for simulations. It is capable of simulating a population of robots, sensors and objects in a three dimensional world. It generates both realistic sensor feedback and physically plausible interactions between objects using Open Dynamics Engine (ODE), a physics engine. Parrot AR Drone (Figure 2) is the unmanned aerial vehicle used. Figure 2 shows the visualization of AR Drone in the simulation environment using ROS Groovy Galapagos [23]. It has a forward facing camera and a downward facing camera. Images from the forward facing camera are used for time-to-contact computation. Location of the UAV is estimated using IMU. The simulation environment contains a pillar, which is the obstacle to be detected and time-to-contact this pillar is to be estimated. Figure 4 shows the top view and front view of the pillar. Figure 3 shows the different starting positions for the simulation. Positions A, B, C, D and E are 2m apart from each other and position A is 10m away from the obstacle. The aerial vehicle moves with a constant velocity of 0.1m/s throughout the simulation.

Fig. 2. Parrot AR Drone used in simulation

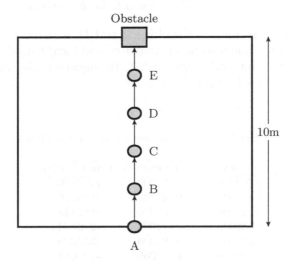

Fig. 3. starting positions of robot for simulation

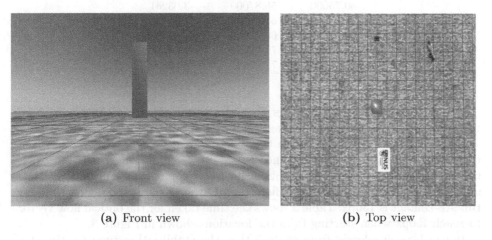

(a) Front view   (b) Top view

Fig. 4. Simulation environment with single obstacle

4   Simulation Results

Figures 5, 6, 7, and 8 show the simulation results when the aerial vehicle starts from position A, B, D and E shown in Fig 3. In all the figures, error in distance to the obstacle (pillar), computed using time-to-contact estimated, change in time-to-contact with time and change in length of the optical flow vector used for computing time-to-contact are shown. Tables 1, 2, 3 and 4 show the mean error in distance to the obstacle computed over an interval of 5s. Mean error to the obstacle is computed using,

$$Mean\ error = \frac{\sum_{t=i}^{t=k}(D_t - \hat{D}_t)}{No.\ of\ elements\ in\ the\ interval} \tag{7}$$

where, D_t is the true distance to the obstacle and $\hat{D}_t$ is the distance to the obstacle estimated using time-to-contact computed and i and k denote the starting and ending time of the interval between which the mean error is computed. Time interval between i and k is approximately 5s.

Table 1. Mean error of the distance to the obstacle estimated when starting 10m from the obstacle

Start Time (s)	End Time (s)	Mean Error(m)
8.75000	14.11000	14.77505
14.11000	19.22000	9.83226
19.22000	24.39000	0.62443
24.39000	29.71000	-2.43871
29.71000	34.92000	-3.75778
34.92000	40.33000	-3.89701
40.33000	45.39000	-4.31287
45.39000	50.75000	-3.74661
50.75000	55.87000	-3.03261
55.87000	61.12000	-2.29101
61.12000	66.43000	-1.61453
66.43000	71.79000	-1.03233
71.79000	71.99000	-0.69896

It can be seen clearly from Figs 5b and 6b that time-to-contact computed will have a lot of fluctuations during the initial period and it settles down after some point of time. If we analyze the plots, we can find that the time taken to stabilize decreases as the distance to the obstacle decreases. If we look at the corresponding length of the optical flow vector plots, we can find that the approximate length of the optical flow vector around the stabilization time for time-to-contact is 150px. Table 5 shows the time taken for the optical flow vector to reach 150px when starting from the locations shown in Figure 3.

It can be seen clearly from table 5 that the stabilization time for time-to-contact decreases as the distance to the obstacle decreases. The reason for this

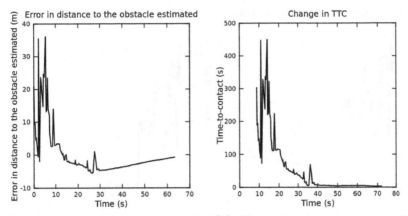

(a) Error in distance computed using time-to-contact estimated

(b) Change in time-to-contact with time

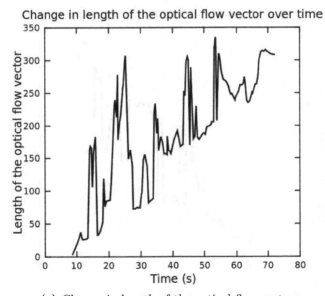

(c) Change in length of the optical flow vector

Fig. 5. Simulation results when starting 10m from the obstacle (Position A in Figure 3). Figure 5a shows the change error in the distance to the obstacle computed from time-to-contact method used. Figure 5b shows the change in time-to-contact computed with time and Figure 5c shows the change in length of the optical flow vector computed, in pixels

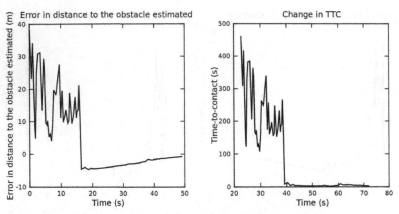

(a) Error in distance computed using time-to-contact estimated

(b) Change in time-to-contact with time

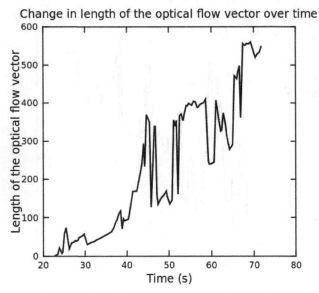

(c) Change in length of the optical flow vector

Fig. 6. Simulation results when starting 8m from the obstacle (Position B in Figure 3). Figure 6a shows the change error in the distance to the obstacle computed from time-to-contact method used. Figure 6b shows the change in time-to-contact computed with time and Figure 6c shows the change in length of the optical flow vector computed, in pixels

Table 2. Mean error of the distance to the obstacle estimated when starting 8m from the obstacle

Start Time (s)	End Time (s)	Mean Error(m)
22.79000	27.95000	22.87588
27.95000	33.16000	14.50319
33.16000	38.23000	13.17127
38.23000	43.48000	-0.73166
43.48000	48.55000	-4.17175
48.55000	53.71000	-3.52019
53.71000	58.71000	-2.79364
58.71000	64.27000	-1.82093
64.27000	69.29000	-1.17550
69.29000	71.79000	-0.82194

Table 3. Mean error of the distance to the obstacle estimated when starting 4m from the obstacle

Start Time (s)	End Time (s)	Mean Error(m)
51.91000	56.91000	15.80010
56.91000	62.23000	4.26219
62.23000	67.23000	0.48042
67.23000	71.79000	-0.69162

Table 4. Mean error of the distance to the obstacle estimated when starting 2m from the obstacle

Start Time (s)	End Time (s)	Mean Error(m)
66.88000	71.99000	6.23388
71.99000	71.99000	-0.71799

Table 5. Approximate time taken for the length of the optical flow vector to reach 150px

Starting distance from the obstacle (m)	Time (s)
10	30
8	20
4	10
2	2

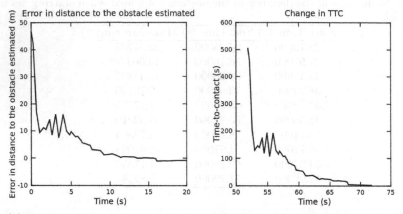

(a) Error in distance computed using time-to-contact estimated

(b) Change in time-to-contact with time

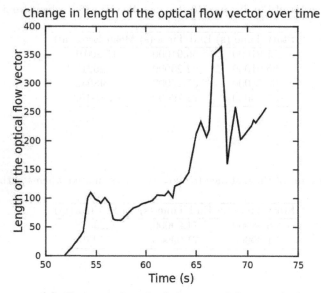

(c) Change in length of the optical flow vector

Fig. 7. Simulation results when starting 4m from the obstacle (Position D in Figure 3). Figure 7a shows the change error in the distance to the obstacle computed from time-to-contact method used. Figure 7b shows the change in time-to-contact computed with time and Figure 7c shows the change in length of the optical flow vector computed, in pixels

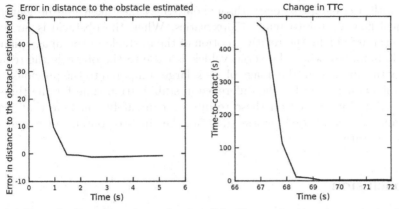

(a) Error in distance computed using time-to-contact estimated

(b) Change in time-to-contact with time

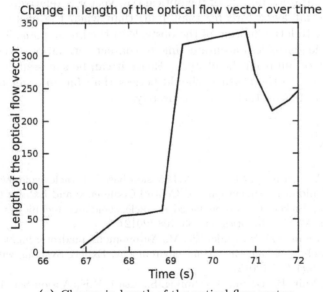

(c) Change in length of the optical flow vector

Fig. 8. Simulation results when starting 2m from the obstacle (Position E in Figure 3). Figure 8a shows the change error in the distance to the obstacle computed from time-to-contact method used. Figure 8b shows the change in time-to-contact computed with time and Figure 8c shows the change in length of the optical flow vector computed, in pixels

can be understood by observing (6) closely. From (6) we can find that the length of the optical flow vector is in the denominator and when the length of the optical flow vector is small, time-to-contact increases. Even though the aerial vehicle is moving with a constant velocity throughout, the time taken for the length of the optical flow vector to reach 150px varies. When the obstacle is far away from the aerial vehicle, the relative motion of the obstacle in the image plane is too small. Similarly when the aerial vehicle is close to the obstacle the relative motion of the obstacle in the image plane is large compared to the previous case. This is the main reason for the difference in stabilization time for the time-to-contact estimation. Based on these results we can safely conclude that, when the distance to the obstacle is less than 5m the time-to-contact estimated will be more accurate.

5   Conclusion

Based on the simulations carried out it can be found that length of the optical flow vector plays in an important role in estimating the time-to-contact an obstacle more accurately. Length of the optical flow vector is relatively small when the obstacle is far away (greater than 10m) than when the obstacle is closer (less than 5m). Relative motion of the obstacle in the image plane is the reason for this. This leads to less accurate time-to-contact estimations when the obstacle is less than 5m from the obstacle. Hence it can be successfully concluded that the distance to the obstacle should be less than 5m in-order to estimate time-to-contact that obstacle more accurately.

References

1. Watanabe, Y., Anthony, J.C., Eric, N.J.: Vision-based obstacle avoidance for UAVs. In: AIAA Guidance, Navigation and Control Conference and Exhibit (2007)
2. Guzel, M.S., Bicker, R.: Vision based obstacle avoidance techniques. In: Recent Advances in Mobile Robotics, pp. 83–108 (2012)
3. Eresen, A., İmamoğlu, N., Önder Efe, M.: Autonomous quadrotor flight with vision-based obstacle avoidance in virtual environment. Expert Systems with Applications 39(1), 894–905 (2012)
4. Diskin, Y., Nair, B., Braun, A., Duning, S., Asari, V.K.: Vision-based navigation system for obstacle avoidance in complex environments. In: 2013 IEEE (AIPR) Applied Imagery Pattern Recognition Workshop: Sensing for Control and Augmentation, pp. 1–8. IEEE (2013)
5. Heng, L., Meier, L., Tanskanen, P., Fraundorfer, F., Pollefeys, M.: Autonomous obstacle avoidance and maneuvering on a vision-guided MAV using on-board processing. In: 2011 IEEE International Conference on Robotics and Automation (ICRA), pp. 2472–2477. IEEE (2011)
6. Cheng, G., Alexander, Z.: Goal-oriented behaviour-based visual navigation. In: Proceedings of the 1998 IEEE International Conference on Robotics and Automation, vol. 4. IEEE (1998)

7. Ohya, A., Akio, K., Avi, K.: Vision-based navigation by a mobile robot with obstacle avoidance using single-camera vision and ultrasonic sensing. IEEE Transactions on Robotics and Automation 14(6), 969–978 (1998)
8. Lorigo, L.M., Rodney, A.B., Grimsou, W.E.L.: Visually-guided obstacle avoidance in unstructured environments. In: Proceedings of the 1997 IEEE/RSJ International Conference on Intelligent Robots and Systems, IROS 1997, vol. 1. IEEE (1997)
9. Toby, L., Gordon, W.: Obstacle detection using optical flow. In: Proceedings of the 2005 Australasian Conf. on Robotics and Automation (2005)
10. Ilic, M., Masciangelo, S., Pianigiani, E.: Ground plane obstacle detection from optical flow anomalies: a robust and efficient implementation. In: Proceedings of the Intelligent Vehicles 1994 Symposium (1994)
11. Zingg, S., Scaramuzza, D., Weiss, S., Siegwart, R.: MAV navigation through indoor corridors using optical flow. In: IEEE International Conference on Robotics and Automation (ICRA), pp. 3361–3368. IEEE (2010)
12. Chessa, M., Solari, F., Sabatini, S.P.: Adjustable linear models for optic flow based obstacle avoidance. Computer Vision and Image Understanding 117(6), 603–619 (2013)
13. Kuiaski, J.R., Lazzaretti, A.E., Neto, H.V.: Focus of Expansion Estimation for Motion Segmentation from a Single Camera. In: Anais do VII Workshop de Visao Computacional (WVC 2011), Curitiba, Brazil, pp. 272–277 (2011)
14. Camus, T., Coombs, D., Herman, M., Hong, T.H.: Real-time single workstation obstacle avoidance using only wide-field flow divergence. In: Proceedings of the 13th International Conference on Pattern Recognition, vol. 3, pp. 323–330. IEEE (1996)
15. Nelson, R.C., Jhon, A.: Obstacle avoidance using flow field divergence. IEEE Transactions on Pattern Analysis and Machine Intelligence 11(10), 1102–1106 (1989)
16. Coombs, D., Herman, M., Hong, T.H., Nashman, M.: Real-time obstacle avoidance using central flow divergence, and peripheral flow. IEEE Transactions on Robotics and Automation 14(1), 49–59 (1998)
17. Camus, T.: Real-time optical flow: Society of Manufacturing Engineers (1994)
18. David, N.L., David, S.Y.: Brain mechanisms and spatial vision. Dordrecht and Boston and Nijhoff (1985)
19. Hoyle, F.: The black cloud. Penguin, London (1957)
20. Donovan, P.O.: Optical flow: Techniques and applications. The University of Saskatchewan, TR, Vol. 502425 (2005)
21. Kuiaski, J.R., Lazzaretti, A.E., Neto, H.V.: Focus of Expansion Estimation for Motion Segmentation from a Single Camera. In: Anais do VII Workshop de Visao Computational (WVC 2011), Curitiba, Brazil, pp. 272–277 (2011)
22. Tistarelli, M., Grosso, E., Sandini, G.: Dynamic stero in visual navigation. In: Proceedings of the IEEE Computer Society Conference on Computer Vision and Pattern Recognition, CVPR 1991, pp. 186–193 (1991)
23. Quigley, M., Conley, K., Gerkey, B., Faust, J., Foote, T., Leibs, J., Ng, A.Y.: ROS: an open-source Robot Operating System. In: ICRA Workshop on Open Source Software, vol. 3(3.2), p. 5 (May 2009)

Evolving Emotion Recognition Module for Intelligent Agent

Rahadian Yusuf, Ivan Tanev, and Katsunori Shimohara

Doshisha University, Kyoto, Japan
{yusuf2013,itanev,kshimoha}@sil.doshisha.ac.jp

Abstract. An emotion recognition module is crucial in designing a computer agent that is capable of interacting with emotional expressions. Under-standing user's current emotion can be achieved by several methods, but current researches are either using still images, or sensors that are not pervasive. Usual approach is using a generalized classifier to recognize pattern of emotion features captured by sensors. Unlike most researches, this research focuses on pervasive sensors and a single user, using evolution algorithm. This research also discusses about the classifier evolutions using Genetic Programming, and comparing several directed evolutions in evolving the emotion recognition module.

Keywords: Emotion, Computer Agent, Evolutionary Algorithm, Genetic Programming.

1   Introduction

1.1   Background

Computer Agents that are capable of recognizing user's emotion will improve the usability of computers. Such agent would be able to understand user's intention better, thus would be able to give a better and proper feedback and reply.

Common approach of recognition using classifier is generalization, which relies a large datasets consisted of numerous person's data to train the classifier (or in this case the computer agent). The classifier would then select a certain dominant feature to distinguish between each input.

Another method of recognition is by using customization, which uses a single user's input and focuses on their unique features. The main difference between generalization and customization can be described by a simple real life analogy: a psychologist and a family member. A psychologist is capable of understanding a client's psychological conditions due to their meticulous study and previous learning on general pattern. A family member is capable of understanding another member's psychological conditions due to the routine and massive interaction with the person, without a vast collection of knowledge on psychology.

H. Handa et al. (eds.), *Proc. of the 18th Asia Pacific Symp. on Intell. & Evol. Systems – Vol. 2*,
Proceedings in Adaptation, Learning and Optimization 2, DOI: 10.1007/978-3-319-13356-0_18

The main ongoing research is about designing computer agent capable of recognizing user's emotion through interaction, applying customized approach focusing on a single user. In order to be able to recognize user's emotion, a classifier is evolved during the construction of emotion recognition module.

1.2   Objectives

There are several research objectives on this paper. The first is to examine the possibility of customization approach, as well as to explore the feasibility of employing genetic programming to automatically select and combine the features in mathematical expressions that can be used to recognize emotion. In addition, we should also discuss the efficiency of applied evolutionary approach in terms of both the computational effort and computational performance.

2   Related Researches

2.1   Similar Works

There are currently numerous researches on emotion recognition. Common method is using still images or photographs and/or non-pervasive sensors (EEG, ECG, etc.).

Many researches are using datasets consisted of photographs of numerous human's facial expressions, however these photographs might be not expressing the appropriate emotion or not representing a unique individual well enough.

Non-pervasive sensors are also commonly found on similar researches, as they give a specific data regarding user's physiological information, which might relate to user's current state of emotion. However using non-pervasive sensors might cause some inconvenience on user, thus might give some levels of noise due to the anxiety. This research uses a pervasive Microsoft Kinect (acting similar to a webcam) to capture user's facial expression.

Another point about similar researches, a generic classifier is usually constructed using other AI methods and not using evolution; therefore this paper would like to take the approach of using Genetic Programming to evolve the emotion recognition module.

2.2   Emotion Classifications

There are several emotion classifications. Paul Ekman proposed several types of emotion that is recognizable regardless of human culture. Many researches on emotion recognition based their classifications on Paul Ekman's.

Another classification is used for this research, and it is based on James Russel's circumplex model of emotion. Circumplex model of emotion describes that there are two dimensions of emotion: Valence (pleasure-ness) and Arousal. If plotted into a graph, Arousal is on the vertical axis and Valence is on the horizontal axis. This model can also mean at least five emotions can be classified:

- Happy (positive valence, positive arousal)
- Relaxed (positive valence, negative arousal)
- Sad (negative valence, negative arousal)
- Angry (negative valence, positive arousal)
- Neutral (no detected valence nor arousal)

2.3   Features of Human's Facial Expressions

This research is using CANDIDE-3 for the changes on facial expression as selected features. There are several AUs (action units) selected:

- Upper Lip Raiser
- Jaw Lowerer
- Lip Stretcher
- Brow Lowerer
- Lip Corner Depressor
- Outer Brow Raiser
- Head Tilt Poses (yaw, pitch, roll)

These features are extracted using Microsoft Kinect as non-pervasive sensor.

As an addition, Alex Pentland stated that there are other signals that might relate to emotions, and not represented on facial expressions. One of them is Activity or the movement and stillness of the person. This Activity will be used in this research, by extracting the average and standard deviation of each AU values in a specific time frame.

3   Proposed System

The functionality of the proposed system includes interacting with the user and learning about his (or her) emotional state through an interaction. The proposed system consists of several modules, as elaborated below.

A Microsoft Kinect sensor will act similar as webcam in capturing videos of user during operating a PC. Then from the data, several features are extracted and are used by the agent to evolve and to be used on current classifier. The result of the evolution will be used to adapt the classifier.

Detailed information and explanations regarding the proposed system can be found on another paper [1].

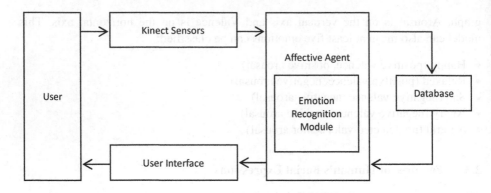

Fig. 1. A diagram of the proposed system. Emotion Recognition Module is one part of the Affective Agent.

Short explanations on the components of the system are shown below:

Table 1. Short Explanation of Proposed System

Component	Details
User	A single user who interacts with the system
Kinect	The main sensor of the system, consisted of several cameras and microphones
Affective Agent	The main module to manage and synchronize flows of data, including the module for machine learning
Database	Storage of features
U.I.	To interact with the user, the system designed will use system tray application (to help the user in focusing on their task and made the system pervasive)

The focus of this paper would be Emotion Recognition Module, one part of the Affective Agent component. Meanwhile, the Affective Agent has the perceptions, actions, and behaviors shown on the following table.

Table 2. Perceptions, Actions, and Behaviors of Affective Agent Component

Category	Descriptions
Perceptions	Features extracted
	Collected features from Database
Actions	Features to be stored in Database
	Features to be deleted from Database
	Proper response to user's current situation to UI
Behaviors	Synchronization of feature extractions and time
	Recognition (classifier)
	Self-evolution
	Analyzing important feature / irrelevant ones

3.1   Genetic Programming

In order to evolve the emotion recognition module, this research is using an in-house Genetic Programming engine called XML-based Genetic Programming (XGP).

One of the reason of using XGP is the evolved expressions of features is represented in human-readable form (XML-text), and therefore, they could be analyzed by human in order to allow the latter to understand the dominant features for a specific person that contribute to a specific emotion. Out assumption is that, due to various personal and cultural differences, not all persons express their emotions in an identical way. Rather, there might be differences in the way of expressing emotions, with different set of the dominant features.

4   Emotion Recognition Module

4.1   Designing Emotion Recognition Module

Emotion Recognition Module works as the classifier of the system. This paper is focusing on designing the framework of this Emotion Recognition Module.

XGP. This paper will focus on evolution of the Emotion Recognition Module using XGP. The XGP is an in-house engine of Genetic Programming that uses XML-based genotypic representations of candidate solutions (genetic programs), XML-schema to determining the allowed syntax of the genotypes, and UDP channel to communicate between fitness evaluator and the XGP manager (which manages the population of genetic programs and performs the genetic operations – selection, crossover and mutation - on them) to perform evolution on the individuals. Figure 2 illustrates how XGP works and interacts with fitness evaluator.

Genes. There are 9 basic AU values gathered from Microsoft Kinect sensor, and each of them is normalized into a scale of -1000 to 1000. The system then extracts 3 values (average, standard deviation, power) from each AU for a specific time frame, which results into 27 extracted features. These features, together with random constant within the range (1..10) comprise the set of terminal symbols in XGP. The set of non-terminal (functional) symbols consist of arithmetical operations (+, -, *, /) and threshold evaluator at the root of tree-representation of candidate solution. This evaluator determines the condition of each Valence and Arousal of the inputs.

Table 3. Features for XGP

Feature Name	Values
27 Extracted Features (terminal)	-1,000 – 1,000
Random Constants (terminal)	1 – 10
Arithmetical Operators (non-terminal)	+, -, *, /
Comparison Operators (non-terminal)	<, >

Fitness Function. In order to evaluate the fitness of an individual, a fitness function is needed. This research uses a variant of accuracy and precision: Matthew's Correlation Coefficient. It considers True Positive, True Negative, False Positive, and False Negative to represent the 'truth-ness' of a result. This research uses MCC and converts it into a percentage (instead of -1 to 1, it is converted into 0% to 100%).

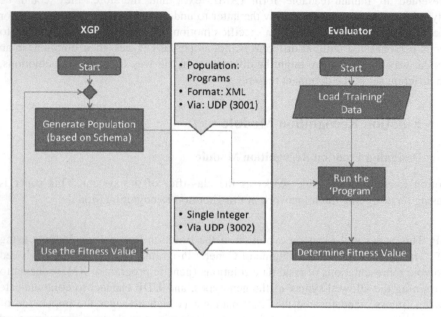

Fig. 2. How XGP (in-house Genetic Programming engine developed by Socioinformatics lab of Doshisha University) interacts with evaluator

Tree Structure. The XGP would produce a schema in the form of XML, to represent a tree structure of the 'program'. The tree structure consisted of multiple node, each nodes is a multiple variety of genes. Each gene (or node or branch) can be used by XGP for crossover, mutations, and anything else deemed needed for evolution. The root of the tree is separated into two distinct trees. One of them is the Valence tree; the other is the Arousal tree. Each tree branched into a comparison between two of the branches below them. The value of the tree (also the Valence and Arousal, respective to their tree) is set according to the condition of the comparison (to True or False). The branches below them are a combination between arithmetical genes, constant genes, and variable genes.

The syntax of the genes can be represented by the following BNF:

```
IF "F" "Comp" "F" THEN True
Comp::= "<" | ">"
F::= "Const"|"Var"|"F", "Op", "Const"|"Var"|"F"
Const::= "0"… "100"
```

```
Var::=  "v_0"…"v_29"
Op::=  "+"  |  "-"  |  "/"  |  ":"
```

The tree structure can be represented as shown in Figure 3 below.

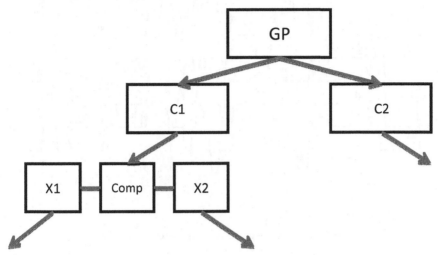

Fig. 3. Tree representation of genetic programs (GP) in XGP. C1 and C2 are Valence and Arousal, respectively. Comp is a comparison operation (<, >). X1 and X2 are sub-trees that could be either random constant, variable (a feature), or another operational branch.

5   Experiments and Results

5.1   Feature Extraction

One of the first experiments conducted is comparing the data of two persons who express the same emotion for around 30 seconds. The persons then express another emotion for around 1 minute.

The extracted features then compared and analyzed. The result was that each person does not have a similar graph.

These data then put into an evolution of several runs. The result was by separating the person into a different classifier (customized) would improve the accuracy of the classifier, compared with putting both data on a single evolution.

Raw graph of extracted features between two persons expressing the same emotion can be seen on Figure 4 below.

The graph shows that, the LS (lip stretcher) is one of the positive factor for the first person in expressing one emotion, while it is a quite negative factor for the second person in expressing the same emotion. In the data acquisition for emotions, there are cases where a person do not smile a lot while happy (despite popular belief), and not all person frowns when they are sad.

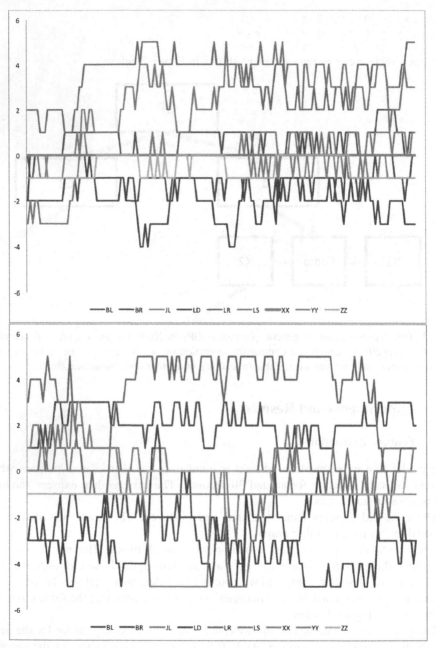

Fig. 4. Comparison between two persons' simplified data (same emotion). It can be seen that some lines are contradictory.

5.2   Time Frame

The features extracted also included an Activity, which is dependent on the changes of value in one time frame. An experiment is performed to test which time frame is feasible.

The experiment uses a single person data, expressing two kinds of emotion. The data is separated into training set and test set. A same number of 48 independent XGP session runs were performed for two training events, first for 30 frames data (around 1 second) of time frame and then for 100 frames data (around 3 seconds) of time frame.

The settings for the XGP are shown below:

Table 4. XGP Settings for Experiments on Time Frame

XGP Setting	Value
Termination	Generation > 200	Fitness Value < 2	Stagnation
# of Individuals	100
Elite Individuals	2
Selection Rate	10%
Mutation Rate	2%

The experiments then yielded result as follows:

Table 5. Time Frame Experiment, comparing 30 frames and 100 frames

Category	30 frames	100 frames
Average Fitness (training)	85.60%	86.84%
Average Fitness (test)	61.25%	82.10%
Best Fitness (training)	93.18%	92.10%
Best Fitness (test)	84.18%	97.49%
Best Individual (average)	85.23%	86.13%
Average Generations	81	127
Average Nodes	202	405

Average Fitness is the average of all result. Best Fitness is the best result from the whole sessions. Best Individual is the individual with good enough result for both training set and test set (by averaging them). Average Generations is the average generations needed to have the result. Average Nodes is the average number of Nodes of the best individuals from the whole sessions.

From the result shown on Table 5 above, it is safe to conclude that using 100 frames instead of 30 frames would give better result.

5.3   Bloat Penalty

As a common sense, a smaller tree would represent a better and faster operation. However, when the system is complex, sometimes it is better to have a bit larger tree, to put into considerations multiple variables/features at once.

An experiment is set to test this hypothesis. The controlled environment is achieved by setting a similar setup for data and XGP settings. The differences are just Bloat Penalty, maximum stagnation, and initial depth of the tree.

The experiment resulted in an increase of accuracy, but also increasing the number of generations needed to achieve the result.

5.4   Different Test Data

During different days, people might change their behavior. Environmental conditions during the different day of data acquisition might also changes, in example the position of seating, the lightings of the room, etc.

An experiment was performed to test the effect of environmental changes on the recognition. This experiment uses a data set from one day for training, and two test sets: one of the same day and the other from a totally different day (with different environmental setting to add noise). None of the test data set is used for training.

The training set consisted of 120 rows of data, while each test sets consisted of 20 rows of data.

The settings for the XGP are shown below:

Table 6. XGP Settings for Experiment on Different Test Data

XGP Setting	Value
Termination	Generation > 300	Fitness Value < 2	Stagnation
# of Individuals	100
Elite Individuals	2
Selection Rate	10%
Mutation Rate	3%

The results are as follows:

Table 7. Comparison between same-day data and different-day data

Category	Same-day	Different-day
Average Fitness	82.10%	66.24%
Standard Deviation Fitness	1.74%	7.17%
Best Fitness	97.49%	80.02%

It is also worth to notice that the average fitness of the training sets was 86.84% with a standard deviation of 2.98%. The best fitness for the training sets was 92.10%.

From the results shown on Table 7 above we can safely conclude that the change of environment condition significantly lowers the accuracy.

5.5   Calculating Time Needed For Evolutions

The time needed for evolution was calculated by putting a time-stamp on each generation. The result was more or less homogenous, despite other changes in XGP setting or the evaluator setting. The number of individuals per generation stays at 100 individuals.

The time for each generation needed using XGP and evaluator is around 36 seconds.

However it should be noted that the time also includes the UDP transfer.

5.6   Separating Valence and Arousal Tree

Previous experiments were using a single tree to be evolved using XGP. However, upon short analysis on the tree results, it can be seen that there were cases where the GP would evolve the 'wrong' tree.

This is the result of the fitness function that calculates both tree at once, thus the XGP does not receive information about which tree needs improvement. The XGP only receives information regarding the general accuracy of both trees.

An experiment was then prepared to investigate the possibility of splitting trees, and the comparing the computational effectiveness and computational efficiency. Computational effectiveness can be seen by the fitness result, while computational efficiency can be seen from the time and generations needed to achieve result.

In order to perform the experiments, two instances of XGP are prepared, as the tree schema must be changed. One XGP is similar to the XGP used on previous experiments, while the other XGP is using a different schema. The schema is edited by removing one branch altogether, and run separate sessions for Valence tree and Arousal tree.

The training data used for all session runs are the same. The other settings for XGP are using the settings shown on Table 6.

The result of the experiments can be seen from the Table 8, below:

Table 8. Raw Comparisons of Valence, Arousal, and Both Tree

Category	Valence	Arousal	Both
Average Fitness	88.51%	87.89%	84.46%
Stdev Fitness	2.72%	2.99%	4.09%
Best Fitness	93.69%	92.18%	89.32%
Worst Fitness	82.31%	81.93%	77.99%
Average Generation	90.75	92.8	150.45
Stdev Generation	32.49	35.79	63.68
Average Node	307.1	334.75	463.09
Stdev Node	263.29	244.76	276.37
Average Total Minute	58.25	62.85	99.27
Stdev Total Minute	20.96	30.44	43.63
Average Second/Generation	38.59	39.8	39.49
Stdev Second/Generation	2.05	3.73	0.90

From Table 8, it can be seen that the computational effectiveness increases when separating the trees and evolve them on their own. Not only the average fitness is increased, the best fitness also increased and the worst fitness value also decreased quite significantly.

Unfortunately, the generations and time needed to evolve separately is higher than to evolve them using a single XGP. Therefore it can be said that separating the tree will reduce the computational efficiency of the system.

Another interesting point is the standard deviation of the result. The variance of fitness value and number of generations from separated tree is much lower than the single tree. Meanwhile, the standard deviation of average time needed to evaluate a single generations is lower when processing a single tree.

6   Conclusions

The experiments have shown that Genetic Programming can be used to evolve classifiers that can recognize emotions. Focusing on evolving a classifier for a single user only improves accuracy, underlining that each user might have unique ways in expressing emotions. Separating valence and arousal and evolving them separately would also give a better computational effectiveness, but it cannot be concluded that this method would give a better computational efficiency.

Emotion Recognition also needs a complex program (or GP tree) to evolve a good classifier, and it might show the complexity of how human expresses their emotions. Also a slightly longer time frame gives a better result compared to a narrow time frame.

7   Future Works

In the future, we are planning to reduce the effect of environmental noise on the quality of recognition. Also, we should implement recognition of emotion using the offline-evolved classifiers on real-time test data.

References

1. Yusuf, R., Wang, S., Tanev, I., Shimohara, K.: Designing Evolving Computer Agent Capable of Emotion Recognition and Expression. In: AAAI Spring Symposium, Palo Alto, California (2014)
2. Pentland, A.: Honest Signals. MIT Press (2008)
3. Ekman, P.: Emotions Revealed. Times Books (2003)
4. Picard, R.W.: Affective Computing. MIT Media Laboratory Perceptual Computing Section Technical Report no. 321
5. Picard, R.W.: Affective Computing: From Laughter to IEEE. IEEE Transactions on Affective Computing 1(1) (2010)
6. Ledoux, J.: The Emotional Brain. Touchstone (1996)
7. Alfaro-Cid, E., Sharman, K., Espercia-Alcazar, A.I.: Evolving a Learning Machine by Genetic Programming. In: IEEE Congress on Evolutionary Computation, Canada (2006)
8. Sun, Y., Li, Z., Tang, C.: An Evolving Neural Network for Authentic Emotion Classification. In: Fifth International Conference on Natural Computation (2009)
9. Takahashi, K.: Remarks on Computational Emotion Recognition from Vital Information. In: 6th International Symposium on Image and Signal Processing and Analysis (2009)

A Novel Multi-objective Optimization Framework Combining NSGA-II and MOEA/D

Xin Qiu[1], Ye Huang[2], and Kay Chen Tan[2]

[1] NUS Graduate School for Integrative Sciences and Engineering,
National University of Singapore, 28 Medical Drive, Singapore 117456
[2] Department of Electrical and Computer Engineering,
National University of Singapore, 4 Engineering Drive 3, Singapore 117576
{qiuxin,huangye,eletankc}@nus.edu.sg

Abstract. Multi-objective Evolutionary Algorithms (MOEAs) are efficient tools for solving multi-objective problems (MOPs). Current existing algorithms such as Multi-Objective Evolutionary Algorithms based on Decomposition (MOEA/D) and Non-dominated Genetic Algorithm II (NSGA-II) have achieved great success in the field by introducing important concept such as decomposition and non-dominated sorting. It would be interesting to employ these crucial ideas of the two algorithms in a hybrid manner. This paper proposes a new framework combining the key features from MOEA/D and NSGA-II. The new framework is a grouping approach aiming to further improve the performance of the current existing algorithms in terms of overall diversity maintenance. In the new framework, original MOP is decomposed into several scalar subproblems and every group is assigned with two scalar subproblems as their new objectives in the searching process. Non-dominated sorting is conducted within each group respectively at every generation. Experimental results demonstrate that the overall performance of the new framework is competitive when dealing with 2-objective problems.

Keywords: Multi-objective evolutionary algorithm, hybrid, decomposition.

1 Introduction

Nowadays Multi-Objective Evolutionary Algorithms (MOEAs) are popular approaches in searching for Pareto optimal solutions to a multi-objective optimization problem. A Multi-Objective Optimization Problem (MOP) can be defined as follows:

$$\max/\min F(x) = \left(f_1(x), \dots, f_m(x)\right)^T \tag{1}$$

$$subject\ to\ x \in \Omega$$

where x refers to the decision variables which lie in the decision (variable) space Ω. The MOP consists of m objective functions and it maps the decision space Ω into an m-dimensional objective space R^m, i.e. $F: \Omega \rightarrow R^m$.

Objectives of a MOP are often conflicting with each other, meaning that the opti-

© Springer International Publishing Switzerland 2015
H. Handa et al. (eds.), *Proc. of the 18th Asia Pacific Symp. on Intell. & Evol. Systems – Vol. 2*,
Proceedings in Adaptation, Learning and Optimization 2, DOI: 10.1007/978-3-319-13356-0_19

mized solution in one objective does not produce optimal result for the other objectives. Thus, there are many or even infinite Pareto Optimal solutions for a MOP instead of a single solution to optimize all the objectives simultaneously. The best tradeoff among objectives is defined as the Pareto Front (PF).

Current existing algorithms such as Multi-Objective Evolutionary Algorithms based on Decomposition (MOEA/D) [1] and Non-dominated Genetic Algorithm II (NSGA-II) [2] have achieved great success in the field by introducing important concepts of decomposition and non-dominated sorting [3-8]. Inspired by the two existing algorithms, it is ideal to have an algorithm that could combine the non-dominated sorting technique and the decomposition concept. This paper proposes a new framework of MOEA combining the key features from the current existing algorithms. In the proposed group-based approach, original MOP is decomposed into several scalar subproblems and every group is assigned with two scalar subproblems as their new objectives during optimization. Non-dominated sorting is then conducted within each group respectively for survival selection. Promising results are observed from the empirical study.

The rest of the paper is organized as follows. Section 2 reviews some related work. Section 3 provides the details of the proposed framework. Section 4 shows the empirical results compared with those of MOEA/D and NSGA-II with brief discussions. Conclusions are drawn in Section 5.

2   Related Work

2.1   NSGA-II

NSGA-II makes use of the important techniques of non-dominated sorting and density estimation in the survival selection process [2].

The concept of domination can be explained as follows.

In a maximization problem, let u and v be two points in an m-dimensional objective space R^m, i.e. u, v $\in R^m$, u is said to dominate v if and only if:

1. $u_i > v_i$ for at least one index i $\in$ {1, 2, ... , m}
2. $u_j \geq v_j$ for every index j $\in$ {1, 2, ..., m}

This is to say, performance of u must be better than v in at least one objective (i.e. condition 1) and cannot be worse than v in any of the m objectives (i.e. condition 2) in order for us to say u dominates v.

NSGA-II makes use of the non-dominated sorting technique to give every individual solution a rank. Individuals which are not dominated by any other solution (i.e. non-dominated solutions) are ranked with zero, while all those which are dominated by others (i.e. dominated solutions) are ranked with the number of solutions that are able to dominate them.

Density estimation computes the distance between individual solutions. In the case that there are more than required number of individuals with the same rank in a selection process, the algorithm considers the contribution of the individuals in diversity maintenance as well. Density estimation ensures that solution set that presents best diversity in objective space is selected for the next generation.

Evaluation of the overall fitness in NSGA-II is based on the rank as well as density estimation results. From the second generation onwards, each individual is to generate its own offspring through crossover and mutation. Survival selection is conducted among parents as well as offspring. Population for the next generation is selected through non-dominated sorting as well as density estimation.

2.2   MOEA/D

Essentially, MOEA/D decomposes a MOP into a set of scalar subproblems using uniformly distributed aggregation weight vectors and optimizes all of them simultaneously. Throughout the searching process, each individual solution is assigned with a scalar subproblem as its new objective. By doing so, individual solutions are in fact assigned with specified searching directions in the objective space. Uniformly distributed aggregation weight vectors are utilized to ensure the searching directions are evenly distributed in the objective space. Thus, a good approximation of PF with individual solutions evenly distributed along the real PF can be expected. The most common decomposition approaches that have been adopted are weight sum approach and Tchebycheff approach.

Weighted Sum Approach [9]. Weighted sum approach involves a convex combination of the different objectives in a MOP. In this case, each scalar subproblem is in fact a linear combination of the original objectives in the MOP with all the coefficients to be non-negative and sum to 1. Let $\lambda = (\lambda_1, ..., \lambda_m)^T$ be a weight vector of a MOP with m objectives, $\lambda_i \geq 0$ for $i = 0, 1, ..., m$; and $\sum_{i=1}^{m} \lambda_i = 1$. The corresponding scalar function g(x) produced with this weight vector λ would be:

$$\text{max/min } g(x|\lambda) = \sum_{i=1}^{m} \lambda_i \cdot f_i(x) \tag{2}$$

where $f_i(x)$ is the real objective value obtained on objective i.

Tchebycheff Approach [10]. In the Tchebycheff approach, the scalar optimization problem with weight vector λ is the difference between the current performance on objective i and the optimal result obtained on the same objective, while i is decided to be the objective producing the maximum value of such difference.

$$\min g(x|\lambda) = \max_{1 \leq i \leq m} \{\lambda_i | f_i(x) - z_i^* |\} \tag{3}$$

where z_i^* is the reference point storing the optimal value found so far for objective i and $|f_i(x) - z_i^*|$ gives the absolute difference between the performance of decision variable x on objective i and the optimal result on objective i stored in the reference point.

3   Methodology

3.1   General Idea

The new framework proposed in this paper is a grouping approach combining the non-dominated sorting technique from NSGA-II and the decomposition concept from MOEA/D. After the very first initialization of the population, the overall population is immediately grouped into N groups. The original objectives from the MOP are decomposed into a fixed number of scalar subproblems and every group will be assigned with two of the scalar subproblems. Non-dominated sorting will thus be conducted within each group with the 2 scalar functions as the new objectives.

The overall flow of the algorithm is summarized as follows.

Input:
- MOP with m objectives
- A stopping criterion
- n: population size
- N: number of groups

Output:
- Current Population (CP) to store the survived individuals after each generation

Step 1: Initialization (1st generation)
1.1   Set CP = $\emptyset$
1.2   Generate an initial population x^1, x^2, ..., x^n randomly; where x^i denotes the ith individual containing decision variables set $\{x_1^i, x_2^i, ..., x_k^i\}$
1.3   Generate a uniform spread of (N+1) weight vectors: $\lambda_1, \lambda_2, ..., \lambda_{N+1}$

Step 2:   Update (from 2nd generation onwards)
2.1   Group the current population into N groups
2.2   For each group i = 1, 2, ... , N,
    2.2.1   Update the group population
    2.2.2   Reproduction: every individual is to generate its own offspring through crossover and mutation
    2.2.3   Assign 2 scalar functions as the 2 new objectives for the individuals in the group
    2.2.4   Update the objective values for the individuals in the group
    2.2.5   Rank the overall population in the group (i.e. parents together with offspring) using non-dominated sorting
    2.2.6   Survival selection based on non-dominated ranking, if needed, conduct density estimation for further selection
    2.2.7   Update CP with the selected individual solution set from the current group

Step 3: Check for stopping criteria
If stopping criteria is satisfied, stop and output CP.
Otherwise, go to **Step 2.2**.

3.2   Implementation

The essence of the new approach is to assign groups of individuals to look for different sections of the real PF while expecting the overall coverage by all the sections is a good approximation of the real PF.

For a specific scalar function given, there always exists a corresponding point in the objective space representing the intersection of the specified searching direction by the scalar function with the real PF. This point is referred as the optimal solution for the specific scalar function [11]. Taking benchmark problem UF2 in Fig. 1 as an example, the green line is the real PF of UF2 that we are looking for. For scalar function $0.5 \times f_1 + 0.5 \times f_2$, the corresponding searching direction is indicted as the orange arrow while point A is the point of intersection of the searching direction (i.e. orange arrow) with the real PF (i.e. green line). In this case, both the weight vectors assigned to f_1 and f_2 are 0.5 (i.e. $\lambda_1 = \lambda_2 = 0.5$) implying that both of the two objectives are equally important. Thus the searching direction defined by this scalar function is 45° away from both objectives. Another scalar function of $0.8 \times f_1 + 0.2 \times f_2$ has its searching direction indicated by the blue arrow. In this case, the weight vector assigned to f_1 is 0.8 while the weight vector assigned to f_2 is only 0.2 (i.e. $\lambda_1 = 0.8, \lambda_2 = 0.2$). It implies that f_1 is relatively 4 times more important compare to f_2. The searching direction is thus shifted towards f_2 because the problem is a minimization problem. Point B is the optimal solution on the real PF corresponding to scalar function $0.8 \times f_1 + 0.2 \times f_2$.

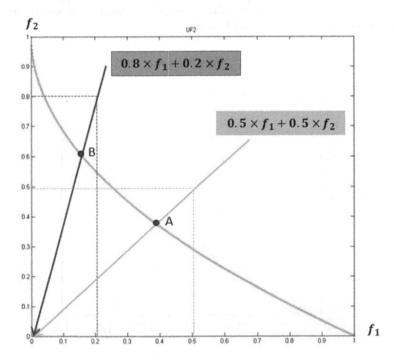

Fig. 1. Optimal Points produced by Scalar Functions

In the case where a MOP has two objectives, the corresponding PF is a line in the 2-Dimensional (2D) objective space. Thus the optimization task is simplified to assign different groups to look for different sections of the line.

The new framework is first implemented targeting at solving 2-objective MOPs. At this stage, it is free to decide the group number N on our own. However, special considerations must be given to the tradeoff between the diversity maintenance and the convergence rate of the solutions. Intuitively, it is desired to have as many groups as possible to cover all the possible searching directions in the objective space. However, as the number of groups increases, the group size for each group decreases since the population size must be fixed. Having too few individuals within a group affects the convergence rate because selection of parents is within each group to avoid large perturbation in the searching process. It is in fact a tradeoff between the explorative ability and exploitative ability of the framework. Results from test runs show that the new framework gives the optimal result when N is taken to be 5.

N+1 scalar functions must be identified when the overall population is to be divided into N groups. Thus there would be N+1 optimal points existing in the objective space corresponding to the N+1 scalar functions. The real PF is evenly divided into N sections by these scalar objective functions since uniformly distributed aggregation weight vectors are used in the decomposition process. Each group is able to focus on searching one of the sections of the real PF. Therefore, diversity maintenance can be expected by this grouping approach. Fig. 2 illustrated the idea when group number is taken to be 5 (i.e. N=5). The original two objectives f_1 and f_2 from the problem are decomposed into 6 scalar functions using uniformly distributed aggregation weight vectors. The corresponding searching directions are indicated as the arrows in Fig. 2, which divide the objective space into 5 sections.

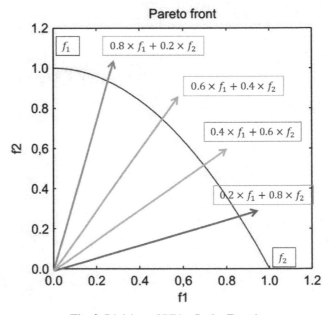

Fig. 2. Division of PF by Scalar Functions

3.3   Selection of Decomposition Scheme

Initially, Weighted Sum approach was adopted in decomposing the original objectives into (N+1) scalar functions. However, the performance of the new framework was much worse than expected and clusters of individuals can be observed in the objective space. Such phenomenon occurs frequently in WFG problems as well as in the benchmark problem UF4 where the real PFs are concave in nature. In these cases, the scalar functions as the new objectives are no longer conflicting with each other, causing some solutions to be missed out in the selection process.

Taking the UF4 problem in Fig. 3 as an example, the real PF is the concave green line. The final solutions cluster in groups instead of the uniformly distributed along the real PF. To further illustrate such phenomenon, consider the simplest case when the scalar function is $0.5f_1 + 0.5f_2$. Weighted sum approach computes the scalar function values for all the solution points as their fitness values. In this case both points A and B share the same scalar objective value, while point C has a larger fitness value and thus less fitter since UF4 is a minimization problem. If the real PF is a straight line, both point A and B have equal opportunity in survival selection. However, when the real PF is concave in nature, points around point A always outperform those around point C, thus would be selected to survive for the next generation. Nevertheless, those around point C are in fact desired values to be kept as well in order to better approximate the real PF. Therefore, weighted sum approach is not suitable to be used in problems of which real PFs are concave in nature.

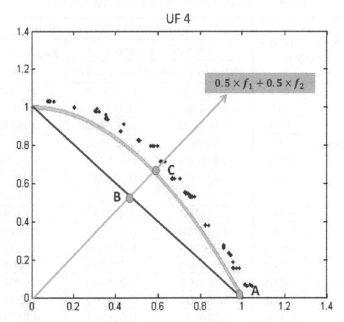

Fig. 3. Performance with weighted sum decomposition scheme

Tchebycheff approach is adopted instead in the new framework targeting at 2-objective MOPs. For each individual solution, Tchebycheff approach computes the differences between its performance (i.e. objective value) on objective i and the reference point, and assigns weight vector λ_i to this difference. The weighted difference with max value is selected to be the scalar function value (i.e. fitness value) for this individual solution. Tchebycheff approach prevents the clustering of solutions in the objective space when real PF of the MOP is concave in nature.

4   Preliminary Results and Discussion

Performance of the proposed algorithm is evaluated with various representative benchmark problems [12][13]. The experimental setting is fixed with 50,000 evaluation times for all the 2-objective problems. Population size is taken to be 100 for all the simulation runs. The Inverted Generational Distance (IGD) [14][15] is utilized to provide a quantitative evaluation of the performance. Table 1 shows the mean of IGD values of the proposed framework, traditional MOEA/D algorithm and traditional NSGA-II algorithm over 30 independent runs. The problems that the new algorithm could provide best performance are marked in boldface. Same offspring reproducing mechanisms (i.e. simulated binary crossover (SBX) [16]) were adopted by all the algorithms in this evaluation of their performance. This is to avoid the effects from different offspring reproducing mechanisms on the performance, thus to ensure the performances are solely based on the frameworks.

Table 1. Average IGD values over 30 independent runs

Benchmark Problems (2-objective)		Average IGD values for 30 runs		
		Existing Algorithms		New framework (SBX)
		MOEA/D (SBX)	NSGA-II (SBX)	
UF cases	UF1	0.156800	0.123028	**0.097386**
	UF2	0.064000	0.048061	0.049881
	UF3	0.306400	0.217866	**0.157220**
	UF4	0.056000	0.053263	0.058972
	UF5	0.431800	0.325709	0.557963
	UF6	0.437400	0.230237	0.240605
	UF7	0.353600	0.235911	**0.216400**
WFG cases	WFG1	1.048300	1.079039	1.144853
	WFG2	0.187100	0.160381	0.211946
	WFG3	0.020300	0.021076	0.067391
	WFG4	0.016700	0.018876	0.019551
	WFG5	0.069100	0.070546	0.070735
	WFG6	0.082000	0.064002	0.081076
	WFG7	0.020500	0.016961	**0.015987**
	WFG8	0.127000	0.137067	0.129444
	WFG9	0.060600	0.084358	0.083656

According to the experimental results, the proposed framework performs best among all the algorithms in 4 out 16 benchmark problems, and gives a comparable performance in most other problems. The reasons that the proposed algorithm performs poorly in some benchmark functions is explained as: 1. The Pareto Front for some problems are discontinues, thus some groups may search within the area without true optimal solutions. 2. All the WFG problems have disparately scaled objectives, so diversity problem would be encountered in the decomposition process. Traditionally, uniformly distributed aggregation weight vectors have been used to decompose the original objectives into scalar subproblems. However, such weight vector assignment scheme in fact distributes uneven number of individuals to different parts of the asymmetrical PF.

In dealing with 2-objective MOP, grouping approach helps to prevent the situation in which the entire solution set is trapped when there is a local optimal. This is illustrated with benchmark problem WFG9 in Fig. 4. The final solutions obtained by NSGA-II are plotted using the red dots whereas solutions found by the new framework are plotted using blue dots. Green dots sketch the true Pareto Optimal Front. It can be seen that the performance of original NSGA-II is not good in terms of convergence because final solutions are trapped into a local optimal. The new framework adopted the grouping approach and parents' selection process is within each group. It in turn prevents such situation in which entire solution set being trapped into local optimal simultaneously. It can be seen that in this case only the last group is trapped into a local optimal.

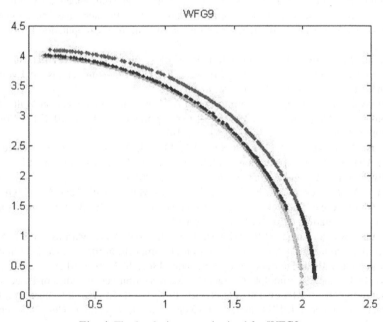

Fig. 4. Final solution sets obtained for WFG9

5   Conclusion

The new framework proposed in this paper combines the key features from MOEA/D and NSGA-II. Experimental results prove that the new approach is promising in solving MOPs with 2 objectives. The grouping approach enables the new framework to maintain the overall diversity easily and prevents the undesired situation in which all the individual solutions are trapped into the local optimal. Future work could be working on MOPs with disparately scaled objectives. Diversity problem are encountered in the decomposition process for problems with asymmetrical PF. One possible solution could be tracking the reference points after a few generations, and then weight vectors can be generated accordingly. More even distribution of the searching directions with respect to the PF is expected. Another future direction is to extend the framework into 3-objective or many-objective problems. Additional mechanism for handling high-dimensional objective space needs to be designed.

References

1. Zhang, Q., Li, H.: MOEA/D: A multiobjective evolutionary algorithm based on decomposition. IEEE Transactions on Evolutionary Computation 11(6), 712–731 (2007)
2. Deb, K., Pratap, A., Agarwal, S., Meyarivan, T.: A fast and elitist multiobjec-tive genetic algorithm: NSGA-II. IEEE Transactions on Evolutionary Computation 6(2), 182–197 (2002)
3. Antonin, P., Antonio, L.J., Carlos, A.C.C.: A Survey on Multiobjective Evolutionary Algorithms for the Solution of the Portfolio Optimization Problem and Other Finance and Economics Applications. IEEE Transactions on Evolutionary Computation 17(3), 321–344 (2013)
4. Anirban, M., Ujjwal, M., Sanghamitra, B., Carlos, A.C.C.: A Survey of Multiobjective Evolutionary Algorithms for Data Mining: Part I. IEEE Transactions on Evolutionary Computation 18(1), 4–19 (2014)
5. Pindoriya, N.M., Singh, S.N., Kwang, Y.L.: A Comprehensive Survey on Multi-objective Evolutionary Optimization in Power System Applications. In: 2010 IEEE Power and Energy Society General Meeting, pp. 1–8 (2010)
6. Liu, B., Fernandez, F.V., Zhang, Q., Pak, M., Sipahi, S., Gielen, G.: An enhanced MOEA/D-DE and its application to multiobjective analog cell sizing. In: IEEE Congress on Evolutionary Computation, CEC 2010, pp. 1–7 (2010)
7. Kafafy, A., Bounekkar, A., Bonnevay, S.: Hybrid Metaheuristics based on MOEA/D for 0/1 multiobjective knapsack problems: A comparative study. In: IEEE Congress on Evolutionary Computation, CEC 2012, pp. 1–8 (2012)
8. Carvalho, R., Saldanha, R.R., Gomes, B.N., Lisboa, A.C., Martins, A.X.: A Multi-Objective Evolutionary Algorithm Based on Decomposition for Optimal Design of Yagi-Uda Antennas. IEEE Transactions on Magnetics 48(2), 803–806 (2012)
9. Voss, T., Beume, N., Rudolph, G., Igel, C.: Scalarization versus indicator-based selection in multi-objective CMA evolution strategies. In: Proc. (IEEE World Congress on Computational Intelligence). IEEE Congress on Evolutionary Computation, CEC 2008, pp. 3036–3043 (2008)
10. Miettinen, K.: Nonlinear Multiobjective Optimization. Kluwer, Norwell (1999)

11. Qi, Y.T., Ma, X.L., Liu, F., Jiao, L.C., Sun, J.Y., Wu, J.S.: MOEA/D with Adaptive Weight Adjustment. Evolutionary Computation 22(2), 231–264 (2014)
12. Zhang, Q., Zhou, A., Zhao, S., Suganthan, P.N., Liu, W., Tiwari, S.: Multiobjective optimization test instances for the CEC 2009 special session and competition. Tech. Rep. CES-487, University of Essex and Nanyang Technological University (2008)
13. Bradstreet, L., Barone, L., While, L., Huband, S., Hingston, P.: Use of the WFG toolkit and PISA for comparison of MOEAs. In: IEEE Symposium on Computational Intelligence in Multicriteria Decision Making, pp. 382–389 (2007)
14. Veldhuizen, D.A.V., Lamont, G.B.: On measuring multiobjective evolutionary algorithm performance. In: 2000 Congress on Evolutionary Compuation, vol. 1. IEEE Service Center, Piscataway (2000)
15. Veldhuizen, D.A.V., Lamont, G.B.: Multiobjective evolutionary algorithm research: A history and analysis. Technical Report TR-98-03, Department of Electrical and Computer Engineering, Graduate School of Engineering, Air Force Institute of Technology, Wright-Patterson AFB, OH (1998)
16. Deb, K., Agrawal, R.B.: Simulated binary crossover for continuous search space. Complex Syst. 9, 115–148 (1995)

A Study on Multi-level Robust Solution Search for Noisy Multi-objective Optimization Problems

Tomohisa Hashimoto and Hiroyuki Sato

The University of Electro-Communications,
1-5-1 Chofugaoka, Chofu, Tokyo 182-8585 JAPAN
hashimoto@hs.hc.uec.ac.jp, sato@hc.uec.ac.jp

Abstract. For noisy multi-objective optimization problems involving multiple noisy objective functions, we aim to develop a two-stage multi-criteria decision-making system considering not only the objective values but also the noise level of each solution. In the first stage, the decision maker selects a solution with a preferred balance of objective values from the obtained Pareto optimal solutions without considering the noise level. In the second stage, for the preferred balance of objective values, this system shows several solutions with different levels of the noise and guides the decision-making considering the noise level of solutions. For the two-stage multi-criteria decision-making system, in this work we propose an algorithm to simultaneously find multi-level robust solutions with different noise levels for each search direction in the objective space. The experimental results using noisy DTLZ2 and multi-objective knapsack problems shows that the proposed algorithm is able to obtain multi-level robust solutions with different noise levels for each search direction in a single run of the algorithm.

Keywords: noisy multi-objective optimization, evolutionary algorithms, multi-level robust solutions

1 Introduction

Evolutionary algorithms are particularly suited to solve multi-objective optimization problems (MOPs) involving multiple objective functions since Pareto optimal solutions (POS) approximating the optimal trade-off among objectives can be obtained from the population in a single run [1,2]. So far, multi-objective evolutionary algorithms (MOEAs) have been intensively studied for solving noise-free MOPs that objective values of each solution are uniquely determined by one-time evaluation. However, in some MOPs such as real-world engineering optimization problems, objective values of each solution are not uniquely determined, and they are varied in every evaluation due to the influence of noise [3]. This is because, several unknown elements not considered as decision variables affect the objective values. For these noisy multi-objective optimization problems (NMOPs), several algorithms optimizing solutions based on the average objective

H. Handa et al. (eds.), *Proc. of the 18th Asia Pacific Symp. on Intell. & Evol. Systems – Vol. 2*,
Proceedings in Adaptation, Learning and Optimization 2, DOI: 10.1007/978-3-319-13356-0_20

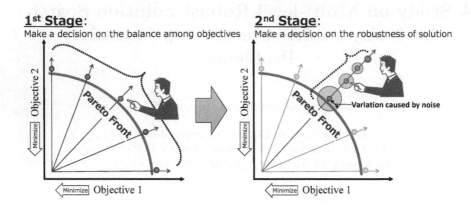

Fig. 1. Two-stage multi-criteria decision-making considering the balance among objectives and the noise levels of solutions

values have been studied so far [4,5,6,7]. However, in these approach, the noise level of each solution cannot be considered in the decision-making process when a decision maker tries to select the final solution from the obtained solutions. If the noise level of each solution can be considered in the decision-making process, the number of valuable choices is increased for the decision maker. Meanwhile, the decision maker will be able to select a solution by considering not only objective values but also the noise level.

For NMOPs, we aim to develop a multi-criteria decision-making (MCDM) system considering not only the objective values but also the noise level of solutions. Fig. 1 shows an overview of the MCDM system. In this system, the decision-making process is divided into two stages. In the first stage, the decision maker selects a solution with a preferred balance of objective values from the obtained POS without considering the noise level in the objective space. In the second stage, the decision maker selects a solution based on the noise levels of solutions. For the preferred balance of objective values, this system shows solutions with different levels of the noise and guides the decision-making considering the noise level of solutions. In Fig. 1, the size of circle around each solution indicates the noise level. Thus, the two-stage MCDM system allows the decision maker to select the final solution by considering not only the objective values but also the noise levels.

For the two-stage MCDM system considering the noise, in this work we propose an algorithm to simultaneously find multi-level robust solutions with different noise levels for each search direction in the multi-dimensional objective space. The proposed algorithm is designed as a variant of the conventional MOEA/D [8] which is a representative MOEA for solving MOPs without the noise. The proposed algorithm tries to find multi-level robust solutions for a number of search directions in a single run of the algorithm. In this work we extend the conventional DTLZ2 problem [9] and multi-objective knapsack problem [10] to noisy problems, and use them to verify the effectiveness of the proposed algorithm.

2 Noisy Multi-objective Optimization Problem

A noisy multi-objective optimization problem (NMOP) is defined by

$$\text{Minimize (or Maximize)} \quad \boldsymbol{f}^z(\boldsymbol{x}) = \{f_1^z(\boldsymbol{x}), f_2^z(\boldsymbol{x}), \ldots, f_m^z(\boldsymbol{x})\}. \tag{1}$$

The task is to find solution(s) $\boldsymbol{x}$ minimizing (or maximizing) m kinds of noisy objective functions. Each noisy objective function f_j^z ($j = 1, 2, \ldots, m$) is defined by

$$f_j^z(\boldsymbol{x}) = f_j(\boldsymbol{x}) + z \quad (j = 1, 2, \ldots, m), \tag{2}$$

where, $f_j(\boldsymbol{x})$ is the original objective function value, and z is a probabilistic noise value. Each of noisy objective function values $f_j^z(\boldsymbol{x})$ ($j = 1, 2, \ldots, m$) is varied in every evaluation of $\boldsymbol{x}$ due to the influence of the probabilistic noise z.

To solve NMOPs by using evolutionary algorithms, several noise-handling algorithms have been proposed so far [3]. Several algorithms tries to optimize solutions based on the average objective function values calculated by repeatedly evaluating a solution, and other algorithms optimize solutions based on a balance between the average objective function values and its variance [4,5,6,7]. However, in these approaches, the noise levels of solutions cannot be considered in the decision-making process when the decision maker tries to select the final solution from the obtained solutions. If the noise levels of solutions can be considered in the decision-making process, the number of valuable choices is increased for the decision maker. Meanwhile, the decision maker will be able to select a solution by considering not only the objective values but also the robustness.

3 MOEA/D

In this work, we utilize the algorithm framework of MOEA/D [8]. MOEA/D is a representative MOEA for solving MOPs without the noise. In this section, we introduce the conventional MOEA/D. Fig. 2 (a) shows a conceptual figure of the conventional MOEA/D in a $m = 2$ dimensional minimization problem.

3.1 Algorithm

MOEA/D decomposes a MOP into a number of single-objective optimization problems. The single-objective optimization problems are defined by scalarizing functions s using uniformly distributed weight vectors $\mathcal{L} = \{\boldsymbol{\lambda}^1, \boldsymbol{\lambda}^2, \ldots, \boldsymbol{\lambda}^N\}$. Each weight vector $\boldsymbol{\lambda}^i$ determines a search direction in the m dimensional objective space. Each element λ_j^i ($j = 1, 2, \ldots, m$) is one of $\{0/H, 1/H, \ldots, H/H\}$ based on the decomposition parameter H, and $N = C_{H+m-1}^{m-1}$ kinds of weight vectors satisfying $\sum_{j=1}^m \lambda_j^i = 1.0$ are used for the solution search. In the following, the algorithm of MOEA/D [8] is briefly described.

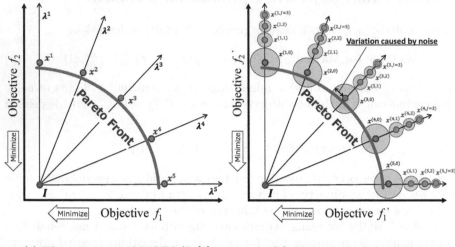

(a) The conventional MOEA/D [8]   (b) The proposed algorithm

Fig. 2. Conceptual figures of the conventional MOEA/D and the proposed algorithm in a $m = 2$ dimensional minimization problem

Step 1) Initialization:

Step 1-1) Compute the Euclidean distances between any two weight vectors and find the T-nearest weight vectors to each weight vector. For each $i \in \{1, 2, \ldots, N\}$, set $B(i) = \{i_1, i_2, \ldots, i_T\}$, where $\boldsymbol{\lambda}^{i_1}, \boldsymbol{\lambda}^{i_2}, \ldots, \boldsymbol{\lambda}^{i_T}$ are the T-nearest weight vectors to $\boldsymbol{\lambda}^i$.

Step 1-2) Randomly generate the population $\{\boldsymbol{x}^1, \boldsymbol{x}^2, \ldots, \boldsymbol{x}^N\}$.

Step 2) Solution Search:

For each $i \in \{1, 2, \ldots, N\}$, perform the following procedure.

Step 2-1) Randomly choose two indices k and l from $B(i)$, and then generate an offspring $\boldsymbol{y}$ from parents $\boldsymbol{x}^k$ and $\boldsymbol{x}^l$ by applying genetic operators.

Step 2-2) For each index $j \in B(i)$, if $s(\boldsymbol{y}|\boldsymbol{\lambda}^j)$ is better than $s(\boldsymbol{x}^j|\boldsymbol{\lambda}^j)$, then the current solution $\boldsymbol{x}^j$ is replaced by the generated offspring $\boldsymbol{y}$ $(\boldsymbol{x}^j = \boldsymbol{y})$.

Step 3) Stopping Criteria:

If the termination criterion is satisfied, then stop and pick POS from the population $\{\boldsymbol{x}^1, \boldsymbol{x}^2, \ldots, \boldsymbol{x}^N\}$ as the output of the optimization. Otherwise, go to **Step 2**.

3.2 Scalarizing Function

In MOEA/D, there are several scalarizing approaches to aggregate m kinds of objective function values [8]. In this work, we employ the weighted Tchebycheff scalarizing function. The scalar optimization problem of the weighted Tchebycheff function s [11] is defined by

$$\text{Minimize } s(\boldsymbol{x}|\boldsymbol{\lambda}) = \max_{1 \le j \le m} \{\lambda_j \cdot |f_j(\boldsymbol{x}) - I_j|\}, \tag{3}$$

where, $\boldsymbol{I}$ is the obtained ideal point. In this work, each element I_j $(j = 1, 2, \ldots, m)$ is set to the best[1] objective function value f_j in the population. The weighted Tchebycheff approach searches a solution minimizing s toward $\boldsymbol{I}$. The weighted Tchebycheff has an advantage that both convex and concave Pareto front can be approximated.

4 Proposal: Multi-level Robust Solution Search

4.1 Concept

For the two-stage MCDM considering the noise level, in this work we propose an algorithm to simultaneously find multi-level robust solutions with different noise levels for each search direction in the objective space. The proposed algorithm is designed as a variant of MOEA/D. Fig. 2 (b) shows a conceptual figure of the proposed algorithm. In this figure, the size of circle around each solution indicates the noise level. As shown in Fig. 2 (a), the conventional MOEA/D only obtain blue solutions since the noise level is not considered. For the two-stage MCDM considering the noise level, as shown in Fig. 2 (b), the proposed algorithm tries to simultaneously obtain multiple red solutions with different levels of the noise for each search direction.

4.2 $m + 1$ Fitness Values Considering Noise

The proposed algorithm solves a noisy m objective optimization problem as a $m + 1$ objective optimization problem considering not only the optimization of m objectives but also the minimization of the noise. The proposed algorithm evaluates a solution based on $m+1$ kinds of fitness values f'_j $(j = 1, 2, \ldots, m+1)$. The first m fitness values f'_j $(j = 1, 2, \ldots, m)$ are the average objective function values calculated by repeatedly evaluating a solution r times. For a solution $\boldsymbol{x}$, the fitness values $f'_j(\boldsymbol{x})$ $(j = 1, 2, \ldots, m)$ are calculated by

$$f'_j(\boldsymbol{x}) = \frac{1}{r} \cdot \sum_{i=1}^{r} f^z_j(\boldsymbol{x}) \quad (j = 1, 2, \ldots, m). \tag{4}$$

---

[1] In minimization problems, the best indicates the minimum. In the maximization problems, the best indicates the maximum.

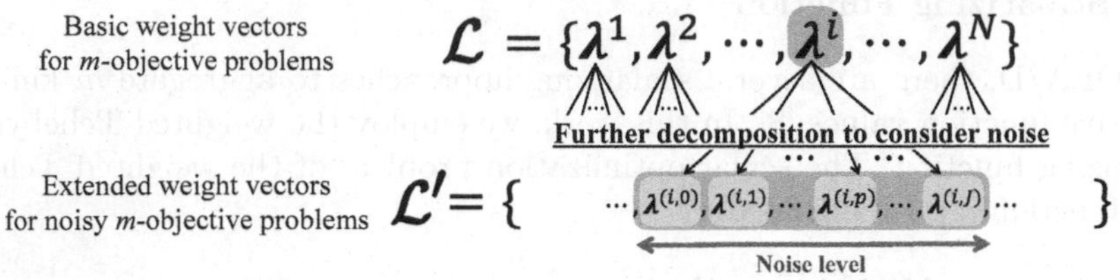

Fig. 3. The basic weight vectors $\mathcal{L}$ and the extended weight vectors $\mathcal{L}'$ in the proposed algorithm

The last fitness value f'_{m+1} is the noise level. In this work, we use the total standard deviation. For a solution x, the fitness value $f'_{m+1}(x)$ is calculated by

$$f'_{m+1}(x) = \sum_{j=1}^{m} \sqrt{\frac{1}{r} \cdot \sum_{i=1}^{r} \{f_j^z(x) - f'_j(x)\}^2}. \tag{5}$$

4.3 $m+1$ Dimensional Weight Vectors Considering Noise

The proposed algorithm needs $m+1$ dimensional weight vectors since a NMOP with m objectives is solved as a MOP with $m+1$ objectives. The proposed algorithm generates $m+1$ dimensional weight vectors in a different way from the conventional MOEA/D [8].

For a NMOP with m objectives, first we generate m dimensional basic weight vectors $\mathcal{L}$ in the same manner as the conventional MOEA/D [8]. The number of elements in each of basic weight vectors $\boldsymbol{\lambda}^i$ $(i = 1, 2, \ldots, N)$ is equivalent to the number of objectives m. Next, we generate the extended weight vectors $\mathcal{L}'$ from the basic weight vectors $\mathcal{L}$ as shown in Fig. 3. Each extended weight vector in $\mathcal{L}'$ is represented as $\boldsymbol{\lambda}^{(i,p)}$, where i is the basic weight vector index, and p is the extended weight vector index. $\boldsymbol{\lambda}^{(i,p)}$ $(p = 0, 1, \ldots, J)$ are weight vectors extended from the basic weight vector $\boldsymbol{\lambda}^i$ to consider the noise in J levels, and the number of elements in $\boldsymbol{\lambda}^{(i,p)}$ becomes $m+1$. We generate $\boldsymbol{\lambda}^{(i,p)}$ $(p = 0, 1, \ldots, J)$ from $\boldsymbol{\lambda}^i$ by the following equation.

$$\lambda_j^{(i,p)} = \begin{cases} \lambda_j^i \cdot (1.0 - p/J) & \text{for } j = 1, 2, \ldots, m \\ p/J & \text{for } j = m+1 \end{cases} \quad (j = 1, 2, \ldots, m, m+1) \ (6)$$

In this way, the extended weight vectors $\boldsymbol{\lambda}^{(i,p)}$ $(p = 0, 1, \ldots, J)$ consider the additional dimension for the noise while maintaining the balance of m kinds of elements in the basic weight vector $\boldsymbol{\lambda}^i$. The number of the basic weight vectors in $\mathcal{L}$ is $N = C_{H+m-1}^{m-1}$ as described in Section 3.1. The number of the extended weight vectors in $\mathcal{L}'$ becomes $N' = N \times (J+1)$ since each of the basic weight vectors is further decomposed into $J + 1$ kinds of the extended weight vectors.

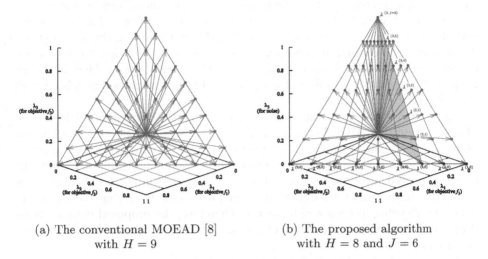

(a) The conventional MOEAD [8]   (b) The proposed algorithm
with $H = 9$ with $H = 8$ and $J = 6$

Fig. 4. Difference of the weights distribution between the conventional MOEA/D considering $m = 3$ objectives and the proposed method considering $m = 2$ objectives plus 1 noise

Although $m + 1$ dimensional weight vectors can also be generated by the same way of the conventional MOEA/D, the proposed method generates $m + 1$ dimensional weight vectors by using the above procedure for the two-stage MCDM system shown in Fig. 1. Fig. 4 shows examples of weight vector distributions in the weight space. $m + 1 = 3$ dimensional weights generated by the conventional MOEA/D are shown in Fig. 4 **(a)**. Also, $m + 1 = 3$ dimensional weights generated by the proposed method is shown in Fig. 4 **(b)**. Thus, weight distributions of the conventional MOEA/D and the proposed method are different. For the two-stage MCDM, the proposed method varies the extended weight λ_3 for the noise while maintaining the balance of the elements in the basic weight vector. For example, in Fig. 4 **(b)**, after the balance of $\lambda^{(3,0)}$ is selected in the first stage of the decision-making process, in the second stage the decision maker can select a solution from solutions obtained with $\lambda^{(3,0)}, \lambda^{(3,1)}, \ldots, \lambda^{(3,J)}$ while maintaining the balance of the objectives selected in the first stage of the decision-making process.

4.4 Algorithm

The proposed algorithm is designed as a variant of MOEA/D. The proposed algorithm has two main differences from the conventional MOEA/D.

The first difference is the way to determine neighborhood weight vectors. MOEA/D selects parents from solutions of neighborhood weight vectors and generates offspring. As shown in Fig. 4 **(a)**, the weight vectors of the conventional MOEA/D are uniformly distributed in the weight space. Therefore, in the the conventional MOEA/D, neighborhood weight vectors are T-nearest neighbors to each weight vector. That is, the number of neighborhood weight vectors is T for any weight vectors. On the other hand, as shown in Fig. 4 **(b)**, the weight vectors of the proposed method is not uniformly distributed in the weight space. Therefore, the proposed method determines neighborhood weight vectors based on an user-defined Euclidean distance D in the weight space. That is, in the proposed method, the number of neighborhood weight vectors depends on each weight vector.

The second difference is the way to generate offspring. Since the weight vectors of the conventional MOEA/D are uniformly distributed in the weight space, the objective space can be uniformly searched by equally giving the chance to generate offspring to each weight vector. On the other hand, since the weight vectors of the proposed algorithm is not uniformly distributed in the weight space, the objective space cannot be uniformly searched by equally giving the chance to generate offspring to each weight vector. Therefore, the proposed method introduces a counter $c^{(i,p)}$ for each weight vector and controls the number of offspring generations.

The detailed algorithm is described in the following.

Step 1) Initialization:

Step 1-1) Compute Euclidean distances between any two weight vectors in $\mathcal{L}'$ and find the nearest weight vectors within the user-defined distance D to each weight vector. For each $\boldsymbol{\lambda}^{(i,p)}$ in $\mathcal{L}'$, set $B(i,p) = \{(i_1, p_1), (i_2, p_2), \dots\}$, where $\boldsymbol{\lambda}^{(i_1,p_1)}$, $\boldsymbol{\lambda}^{(i_2,p_2)}, \dots$ are the weight vectors within the user-defined distance D from $\boldsymbol{\lambda}^{(i,p)}$. [2]

Step 1-2) For each $i \in \{1, 2, \dots, N\}$, randomly generate one solution and initialize all $\boldsymbol{x}^{(i,p)}$ ($p = 0, 1, \dots, J$) by the generated solution, i.e. $\boldsymbol{x}^{(i,0)} = \boldsymbol{x}^{(i,1)} = \dots = \boldsymbol{x}^{(i,J)}$.

Step 1-3) Initialize each counter $c^{(i,p)}$ of the weight vector $\boldsymbol{\lambda}^{(i,p)}$ by zero.

Step 2) Solution Search:

Perform the following procedure N times.

Step 2-1) Find the index $(i^{\min}, p^{\min})$ of the weight vector $\boldsymbol{\lambda}^{(i^{\min}, p^{\min})}$ which has the minimum counter value $c^{(i^{\min}, p^{\min})}$ among all weight vectors in $\mathcal{L}'$. Then, select $\boldsymbol{x}^{(i^{\min}, p^{\min})}$ as the first parent.

Step 2-2) Randomly choose an index $(i_q^{\min}, p_q^{\min})$ from $B(i^{\min}, p^{\min})$ and select $\boldsymbol{x}^{(i_q^{\min}, p_q^{\min})}$ as the second parent. Then, generate an offspring $\boldsymbol{y}$ from two parents $\boldsymbol{x}^{(i^{\min}, p^{\min})}$ and $\boldsymbol{x}^{(i_q^{\min}, p_q^{\min})}$ by applying genetic operators.

---

[2] The proposed algorithm determines the nearest weight vectors based on the user-defined distance D. Therefore, the number of nearest weight vectors is not fixed value like T in the conventional MOEA/D, and it depends of each weight vector.

Step 2-3) For each index $(i_j^{\min}, p_j^{\min}) \in B(i^{\min}, p^{\min})$, if $s(\boldsymbol{y} \mid \boldsymbol{\lambda}^{(i_j^{\min}, p_j^{\min})})$ is better than $s(\boldsymbol{x}^{(i_j^{\min}, p_j^{\min})} \mid \boldsymbol{\lambda}^{(i_j^{\min}, p_j^{\min})})$, then the current solution $\boldsymbol{x}^{(i_j^{\min}, p_j^{\min})}$ is replaced by the generated offspring $\boldsymbol{y}$, i.e. $\boldsymbol{x}^{(i_j^{\min}, p_j^{\min})} = \boldsymbol{y}$.

Step 2-4) For each index $(i_j^{\min}, p_j^{\min}) \in B(i^{\min}, p^{\min})$, $c^{(i_j^{\min}, p_j^{\min})}$ is incremented by one.

Step 3) Stopping Criteria:

If the termination criterion is satisfied, then stop and pick non-dominated solutions on fitness vector $\boldsymbol{f}'$ from $\{\boldsymbol{x}^{(i,0)}, \boldsymbol{x}^{(i,1)}, \ldots, \boldsymbol{x}^{(i,J)}\}$ for each $i \in \{1, 2, \ldots, N\}$ as the output of the optimization. Otherwise, go to **Step 2**.

5   Experimental Setup

5.1   Continuous Test Problem: nDTLZ2

In this work, we extend DTLZ2 problem [9] to a noisy continuous test problem. The noisy DTLZ2 (nDTLZ2) is defined by

Minimize

$$f_1^z(\boldsymbol{x}) = (1 + g(\boldsymbol{x}^M)) \cos(x_1 \tfrac{\pi}{2}) \cos(x_2 \tfrac{\pi}{2}) \cdots \cos(x_{m-2} \tfrac{\pi}{2}) \cos(x_{m-1} \tfrac{\pi}{2}) + g^z(\boldsymbol{x}^M),$$

$$f_2^z(\boldsymbol{x}) = (1 + g(\boldsymbol{x}^M)) \cos(x_1 \tfrac{\pi}{2}) \cos(x_2 \tfrac{\pi}{2}) \cdots \cos(x_{m-2} \tfrac{\pi}{2}) \sin(x_{m-1} \tfrac{\pi}{2}) + g^z(\boldsymbol{x}^M),$$

$$f_3^z(\boldsymbol{x}) = (1 + g(\boldsymbol{x}^M)) \cos(x_1 \tfrac{\pi}{2}) \cos(x_2 \tfrac{\pi}{2}) \cdots \sin(x_{m-2} \tfrac{\pi}{2}) + g^z(\boldsymbol{x}^M),$$

$$\vdots$$

$$f_m^z(\boldsymbol{x}) = (1 + g(\boldsymbol{x}^M)) \sin(x_1 \tfrac{\pi}{2}) + g^z(\boldsymbol{x}^M), \tag{7}$$

where g is the distance function, and g^z is the noise function. g^z is the only difference from the original DTLZ2. A solution $\boldsymbol{x}$ consists of n variables $\{x_1, x_2, \ldots, x_{m-1}, x_m^M, x_{m+1}^M, \ldots, x_n^M\}$, and all the elements are real values in the range $[0, 1]$. The last part $\boldsymbol{x}^M = \{x_m^M, x_{m+1}^M, \ldots, x_n^M\}$ in $\boldsymbol{x}$ determines the distance $g(\boldsymbol{x}^M)$ between $\boldsymbol{x}$ and the true Pareto front in the objective space. $g(\boldsymbol{x}^M)$ is defined by the following equation.

$$g(\boldsymbol{x}^M) = \sum_{x_i^M \in \boldsymbol{x}^M} (x_i^M - 0.5)^2. \tag{8}$$

A solution with $\boldsymbol{x}^M = \{0.5, 0.5, \ldots, 0.5\}$ achieves the minimum $g(\boldsymbol{x}^M) = 0$ and becomes a true Pareto optimal solution.

In the nDTLZ2, the last part $\boldsymbol{x}^M$ also determines the noise level $g^z(\boldsymbol{x}^M)$. $g^z(\boldsymbol{x}^M)$ is defined by the following equation.

$$g^z(\boldsymbol{x}^M) = R \cdot N(0, 1) \cdot \sum_{x_i^M \in \boldsymbol{x}^M} (x_i^M - 0.7)^2, \tag{9}$$

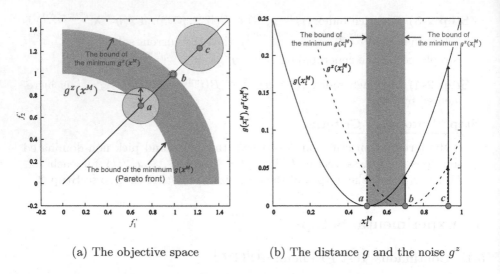

(a) The objective space   (b) The distance g and the noise g^z

Fig. 5. The noisy DTLZ2 problem (nDTLZ2)

where R is the problem parameter to control the noise level, and N is the zero-mean Gaussian probability distribution function with the standard deviation 1. In the nDTLZ2, a solution with $x^M = \{0.7, 0.7, \ldots, 0.7\}$ has the smallest noise.

Fig. 5 **(a)** shows the objective space of the nDTLZ2 with $m = 2$ objectives, and Fig. 5 **(b)** shows g and g^z functions for a variable x_i^M ($\in x^M$). In the nDTLZ2, the distance to the true Pareto front in the objective space is determined by g. For example, the solution a with the smallest g in Fig. 5 **(b)** is the closest to the true Pareto front among three solutions in Fig. 5 **(a)**. Contrary, the solution c with the largest g in Fig. 5 **(b)** is the farthest to the true Pareto front among three solutions in Fig. 5 **(a)**. Next, in the nDTLZ2, the noise level is determined by g^z. In Fig. 5 **(a)**, the size of circle around each solution indicates the noise level. Although the noise of b is the smallest, its distance to the true Pareto front is larger than a. Contrary, although a is closest to the true Pareto front in the objective space, its noise is larger than b. Additionally, c is the farthest from the true Pareto front and its noise is the largest among three solutions. That is, there is the optimal trade-off between the distance g and the noise g^z in the nDTLZ2, and the task of the proposed algorithm in the nDTLZ2 is to find solutions distributed in the range $0.5 \le x_i^M \le 0.7$ (gray region) by the solution search with the extended weight vectors.

5.2 Discrete Test Problem: nMOKP

We extend multi-objective knapsack problem (MOKP) [10] to a noisy discrete test problem. The noisy MOKP (nMOKP) is defined as follows.

$$\begin{cases} \text{Maximize} \;\; f_j^z(x) = \sum_{l=1}^{n}(p_{l,j} + z(l)) \cdot x_l \\ \text{Subject to} \; \sum_{l=1}^{n} w_{l,j} \cdot x_l \le c_j \end{cases} \quad (j = 1, 2, \ldots, m). \quad (10)$$

The noise function $z(l)$ is the only difference from the original MOKP. In the nMOKP, there are n items and m knapsacks (objectives). Each item l has m kinds of profits $p_{l,j}$ ($j = 1, 2, \ldots, m$) and m kinds of weights $w_{l,j}$ ($j = 1, 2, \ldots, m$). The task is to find combinations of items $x = \{x_1, x_2, \ldots, x_n\} \in \{0, 1\}^n$ which maximizes the total of profits on m kinds of objectives subject to the total of weights does not exceed m kinds of knapsack capacities c_j. The capacities of knapsacks c_j are defined as

$$c_j = \phi \cdot \sum_{l=1}^{n} w_{l,j} \quad (j = 1, 2, \ldots, m), \tag{11}$$

where ϕ is the feasibility ratio for each knapsack (constraint). In this work, infeasible solutions are repaired by repeatedly removing a randomly chosen item from the solution until all constraints are satisfied.

The noise function $z(l)$ for each item l is defined by

$$z(l) = R \cdot N(0, l/3n) \quad (l = 1, 2, \ldots, n), \tag{12}$$

where R is the problem parameter to control the noise level, and N is the zero-mean Gaussian probability distribution function with the standard deviation $l/3n$. Note that the noise value $z(l)$ is increased by increasing item number l.

All profits $p_{l,j}$ and weights $w_{l,j}$ are generated by random integers in the range of $[10, 100]$ in the same manner as the original MOKP [10]. However, for analysis of the experimental result, a special reordering of pairs of profit $p_{l,1}$ and weight $w_{l,1}$ ($l = 1, 2, \ldots, n$) is performed in this work. We calculate ratios $p_{l,1}/w_{l,1}$ ($l = 1, 2, \ldots, n$) and reorder pairs of profit $p_{l,1}$ and weight $w_{l,1}$ to satisfy $p_{1,1}/w_{1,1} > p_{n,1}/w_{n,1} > p_{2,1}/w_{2,1} > p_{n-1,1}/w_{n-1,1} \cdots$. That is, the ratio $p_{l,1}/w_{l,1}$ is increased by increasing the difference $|l - n/2|$ between item number l and the intermediate item number $n/2$.

5.3   Parameters

In this work, we use the nDTLZ2 problem with $m = 2$ objectives, $n = 11$ variables and the noise parameter $R = 0.125$. For solving the nDTLZ2, we use SBX crossover with the crossover ratio 0.8 and the distribution parameter $\eta_c = 15$ and the polynomial mutation with the mutation ratio $1/n$ and the distribution parameter $\eta_m = 20$ [12]. The termination criterion to stop the algorithm is set to $8,000$ generations.

Also, we use the nMOKP with $m = 2$ objectives, $n = 500$ items (bits) and the noise parameter $R = 15$. For solving the nMOKP, we use the uniform crossover with the crossover ratio 0.8 and the bit-flip mutation with the mutation ratio $4/n$. The termination criterion is set to $10,000$ generations.

For both problems, as the common parameters, the basic decomposition parameter and the extended decomposition parameter are set to $H = 200$ and $J = 40$, respectively. The distance to determine the neighborhood weight vectors is set to $D = 0.06$. To calculate the average objective function values f'_j ($j = 1, 2, \ldots, m$) and the total standard deviation f'_{m+1}, the number of evaluation for each generated solution is set to $r = 100$.

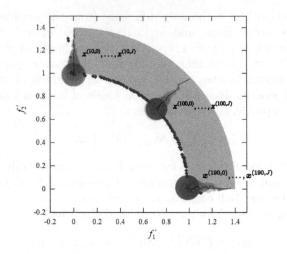

Fig. 6. The obtained solutions on the nDTLZ2 problem

6 Experimental Results and Discussion

6.1 Results on nDTLZ2 Problem

First, we verify the effectiveness of the proposed algorithm on the nDTLZ2 problem. Fig. 6 shows the obtained solutions in the objective space. Note that the both objective functions of the nDTLZ2 should be minimized. All solutions are plotted based on their average objective function values f_1' and f_2'. First, the blue points are the obtained solutions $\boldsymbol{x}^{(1,0)}, \boldsymbol{x}^{(2,0)}, \ldots, \boldsymbol{x}^{(N,0)}$ without considering the noise. Next, we focus only on three basic weight vector indices $i = \{10, 100, 190\}$, and solutions $\boldsymbol{x}^{(10,p)}, \boldsymbol{x}^{(100,p)}, \boldsymbol{x}^{(190,p)}$ ($p = 0, 1, \ldots, J$) considering the noise are plotted with red circles. The size of circle indicates the noise level.

From the result, first we can see that the blue solutions $\boldsymbol{x}^{(1,0)}, \boldsymbol{x}^{(2,0)}, \ldots, \boldsymbol{x}^{(N,0)}$ without considering the noise approximate the trade-off between two objective functions. These blue solutions are used for the first stage of the decision-making shown in Fig. 1. Next, for each of three basic weight vector indices $i = \{10, 100, 190\}$, we can see that the proposed algorithm can simultaneously obtain multi-level robust solutions with different levels of noise in a single run of the algorithm. These red solutions are used for the second stage of the decision-making shown in Fig. 1.

Next, we analyze the obtained solutions in the variable space. Here, we focus on the obtained solutions $\boldsymbol{x}^{(190,p)}$ ($p = 0, 1, \ldots, J$) for the basic weight vector index $i = 190$. Fig. 7 shows the distance function g and the noise function g^z for the variable x_2^M ($\in \boldsymbol{x}^M$), and the obtained solutions $\boldsymbol{x}^{(190,p)}$ ($p = 0, 1, \ldots, J$) are plotted on the horizontal axis. The solution $\boldsymbol{x}^{(190,0)}$ without considering the noise is plotted as the blue point, and the solutions $\boldsymbol{x}^{(190,p)}$ ($p = 1, 2, \ldots, J$) considering the noise are plotted as the red points. From the result, we can see that the

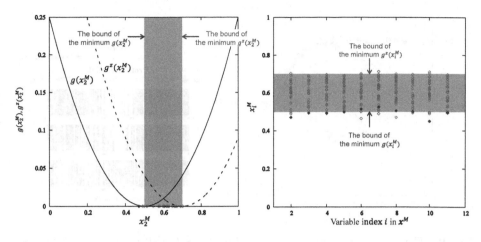

Fig. 7. The obtained solutions $\boldsymbol{x}^{(190,p)}$ $(p = 0, 1, \ldots, J)$ on the variable x_2^M

Fig. 8. The obtained of solutions $\boldsymbol{x}^{(190,p)}$ $(p = 0, 1, \ldots, J)$ on all variables in $\boldsymbol{x}^M$

obtained solutions are distributed in or near the gray region ($0.5 \leq x_2^M \leq 0.7$) showing the optimal trade-off between the distance g and the noise g^z. Also, Fig. 8 shows the obtained solutions $\boldsymbol{x}^{(190,p)}$ $(p = 0, 1, \ldots, J)$ on all variables $\boldsymbol{x}^M = \{x_2^M, x_3^M, \ldots, x_{n=11}^M\}$. From the result, we can see that the obtained solutions are distributed in or near the optimal gray regions of all variables in $\boldsymbol{x}^M$.

6.2   Results on nMOKP Problem

Next, we verify the effectiveness of the proposed algorithm on the nMOKP. Fig. 9 shows the obtained solutions in the objective space. Note that the both objective functions of the nMOKP should be maximized. The blue points are the obtained solutions $\boldsymbol{x}^{(1,0)}, \boldsymbol{x}^{(2,0)}, \ldots, \boldsymbol{x}^{(N,0)}$ without considering the noise, and the red points are the obtained solutions $\boldsymbol{x}^{(10,p)}, \boldsymbol{x}^{(100,p)}, \boldsymbol{x}^{(190,p)}$ $(p = 0, 1, \ldots, J)$ considering the noise for three basic weight vector indices $i = \{10, 100, 190\}$, respectively.

From the result, as a general tendency, we can see that solutions with large objective values f_1' and f_2' have large noises with large circles. This result in the nMOKP also reveals that the proposed algorithm is able to simultaneously find multi-level robust solutions with different levels of the noise for each search direction in the objective space in the single run of the algorithm.

Next, we focus on the basic weight vector index $i = 10$ and observe the selected items of the obtained solutions in the nMOKP. The selected items of three solutions $\boldsymbol{x}^{(10,0)}$, $\boldsymbol{x}^{(10,22)}$ and $\boldsymbol{x}^{(10,J=40)}$ are shown in Fig. 10. For each item (bit) l, selected item (1) is shown in black, and unselected item (0) is shown in white in this figure. As described in Section 5.2, the noise level is increased

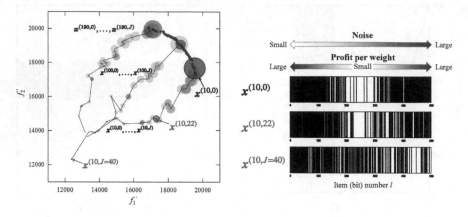

Fig. 9. The obtained solutions on the nMOKP problem

Fig. 10. Selected items (bits) of three solutions with different noise levels

by increasing the item (bit) number l. Also, the ratio $p_{l,1}/w_{l,1}$ is increased by increasing the difference $|l - 250|$ between item number l and the intermediate item number 250. From this result, we can see that the solution $x^{(10,0)}$ with a large noise selects many items with large item number l. That is, $x^{(10,0)}$ selects items with high $p_{l,1}/w_{l,1}$ even if their noises are high. On the other hand, the robust solution $x^{(10,J=40)}$ with a small noise tends to not select items with large l. That is, $x^{(10,J=40)}$ avoids to select items with high noise even if their $p_{l,1}/w_{l,1}$ are high. This result reveals that the proposed algorithm is able to extract items (variables) with low influences of noise.

7 Conclusions

For the two-stage MCDM considering the noise level, in this work we proposed an algorithm to simultaneously find multi-level robust solutions with different noise levels for each search direction in the multi-dimensional objective space of NMOPs.

To verify the effectiveness of the proposed algorithm, we used continuous nDTLZ2 and discrete nMOKP problems extended from the conventional DTLZ2 and MOKP, respectively. The experimental results showed that the proposed algorithm was able to obtain multi-level robust solutions with different noise levels for each search direction in a single run of the algorithm. In the nDTLZ2 problem, we showed that the proposed algorithm obtained solutions in the optimal region showing the optimal trade-off between the distance to the true Pareto front and noise levels. Also, in the nMOKP, we showed that the proposed algorithm was able to extract items (variables) with low influences of noise.

As future work, we will try to decrease the total number of evaluations in the proposed algorithm since the current algorithm needs many evaluations to assign fitness values for each solution.

Acknowledgment. This work was supported by JSPS KAKENHI Grant Number 26730129.

References

1. Deb, K.: Multi-Objective Optimization using Evolutionary Algorithms. John Wiley & Sons (2001)
2. Coello, C.A.C., Van Veldhuizen, D.A., Lamont, G.B.: Evolutionary @ Algorithms for Solving Multi-Objective Problems. Kluwer Academic Publishers, Boston (2002)
3. Goh, C., Tan, K.C.: Evolutionary Multi-objective Optimization in Uncertain Environments. SCI, vol. 186. Springer, Heidelberg (2009)
4. Park, T., Ryu, K.: Accumulative Sampling for Noisy Evolutionary Multi-Objective Optimization. In: Proc. of the 2011 Genetic and Evolutionary Computation Conference, pp. 793–800 (2011)
5. Babbar, M., Lakshmikantha, M., Goldberg, D.E.: Modified NSGA-II to solve Noisy Multi-objective Problems. In: Proc. of the 2003 Genetic and Evolutionary Computation Conference, Late-Breaking Papers, pp. 21–27 (2003)
6. Teich, J.: Pareto-Front Exploration with Uncertain Objectives. In: Zitzler, E., Deb, K., Thiele, L., Coello Coello, C.A., Corne, D.W. (eds.) EMO 2001. LNCS, vol. 1993, pp. 314–328. Springer, Heidelberg (2001)
7. Thompson, A.: Evolutionary Techniques for Fault Tolerance. In: Proc. of the UKACC International Conference on Control, pp. 693–698 (1996)
8. Zhang, Q., Li, H.: MOEA/D: A Multiobjective Evolutionaly Algorithm Based on Decomposition. IEEE Trans. on Evolutionary Computation 11(6), 712–713 (2007)
9. Deb, K., Thiele, L., Laumanns, M., Zitzler, E.: Scalable Multi-Objective Optimization Test Problems. In: Proc. of CEC 2002, pp. 825–830 (2002)
10. Zitzler, E., Thiele, L.: Multiobjective Evolutionary Algorithms: A Comparative Case Study and the Strength Pareto Approach. IEEE Trans. on Evolutionary Computation 3(4), 257–271 (1999)
11. Miettinen, K.: Nonlinear Multiobjective Optimization. Kluwer, Norwell (1999)
12. Deb, K., Goyal, M.: A Combined Genetic Adaptive Search (GeneAS) for Engineering Design. Computer Science and Informatics 26(4), 30–45 (1996)

Directional Differential Evolution: Approaching to the Superior, Departing from the Inferior, or Both?

Xinchao Zhao[1], Junling Hao[2], and Xingquan Zuo[3]

[1] BJ10086uk School of Science, Beijing University of Posts and Telecommunications,
Beijing 100876, China
[2] School of Statistics, University of International Business and Economics,
Beijing 100029, China
[3] School of Computer Science, Beijing University of Posts and Telecommunications,
Beijing 100876, China

Abstract. Differential evolution (DE) is an efficient and powerful stochastic population-based optimization algorithm. The DE-variants obtain new solutions through the scaled differences of randomly selected and distinct individuals. Many generation strategies and improved variants have been proposed. However, the directional information for difference evolution attracts little consideration until now. In this paper, several kinds of directional difference information (DDI), approaching to the superior solutions, departing from the inferior solutions, and both, are considered and incorporated into DE-mutation strategy. Analytic and experimental results indicate that DE/rand/1 is significantly improved by any one of DDI, especially for the negative DDI. Furthermore, this idea does not introduce any new items and operations.

Keywords: Differential evolution, direction information, negative information, DDI.

1   Introduction

Differential evolution (DE) is a simple yet powerful evolutionary algorithm for global optimization, which was introduced by Price and Storn [13]. DE algorithm has gradually become more and more popular and has been used in many practical problems [4], mainly because it has many attractive characteristics, such as compact structure, ease to use, speediness, and robustness. It uses mutation, crossover, and selection operators at each generation to drive its population toward the global optimum.

The core operator of DE algorithm is differential mutation, and generally, the parents in the mutation for both base vector and difference vector are usually randomly selected from the current population. However, both base vector and difference vector usually ignore the beneficial direction information, approaching to the superior or departing from the inferior individuals, especially for the latter. It is possible to make algorithm being trapped in local optimum more

H. Handa et al. (eds.), *Proc. of the 18th Asia Pacific Symp. on Intell. & Evol. Systems – Vol. 2*,
Proceedings in Adaptation, Learning and Optimization 2, DOI: 10.1007/978-3-319-13356-0_21

frequently if the best individual is far away from the global optimum or the current position is close to a local trap, especially for multimodal optimization [11,2]. Therefore, the directional difference information (DDI), approaching to the superior solutions, departing from the inferior solutions, and both, is exploited in this paper for even more efficient DE-mutation strategy. By the way, a stochastic neighborhood structure of population is simultaneously composed while constructing base vector and DDI. It is special to say that the proposed idea of DDI does not introduce any new items, new operations or generation strategies.

The major contributions of this paper are as follows.

- The directional difference information, approaching to the superior solutions, departing from the inferior solutions, and both, are constructed for even more efficient search.
- A stochastic neighborhood structure of population is simultaneously constructed while selecting base vector and DDI.
- A simple yet effective way of synergizing DDI and a simultaneous yet effortless neighborhood structure is presented to enhance DE.

The rest of this paper is organized as follows. Differential evolution and related works are presented in Section 2. DDI and the proposed algorithm is presented in Section 3. Experimental comparisons are given in Section 4. This paper is concluded in Section 5.

2   Differential Evolution

Four main steps in DE are initialization, mutation, recombination and selection of parents for next generation from the current parent and offspring.

2.1   Initialization

The individuals of DE take the form as follows: $\mathbf{x}_i^t = (x_{i1}^t, x_{i2}^t, \ldots, x_{in}^t)$, $i \in \{1, 2, \ldots, NP\}$, where x_{ij}^t denotes the j-th component of the i-th individual in the t-th generation. NP is the population size. n is the dimension size of domain space. The initialization phase for all individuals as follows,

$$x_{ij}^0 = x_{min}^j + rand(0,1) \cdot (x_{max}^j - x_{min}^j) \tag{1}$$

where x_{min}^j and x_{max}^j are the minimum and maximum bounds at j-th dimension.

2.2   Mutation Operation

After initialization, DE employs mutation operation to produce a mutant vector v_i^t with respect to each individual x_i^t, so-called target vector, in the current population. Although different strategies have been suggested [4,10], this paper uses DE/rand/1 only as an analytic model strategy as Eq.(2).

$$\mathbf{v}_i^t = \mathbf{x}_{r_1}^t + F_1 \cdot (\mathbf{x}_{r_2}^t - \mathbf{x}_{r_3}^t) \tag{2}$$

where $r_1, r_2, r_3 \in [1, NP]$ are different random integers, and they are also different from vector index i. Scale factor $F > 0$ is a constant and the effective range of F is usually between 0.4 and 1 in classic DE.

2.3 Crossover Operation

After mutation operation, crossover operation is applied to each pair of target vector $\mathbf{x}_i^t$ and its corresponding mutant vector $\mathbf{v}_i^t$ to generate a trial vector $\mathbf{u}_i^t$.

$$u_{ij}^t = \begin{cases} v_{ij}^t, & rand \leq \text{CR or j=}j_{rand} \\ x_{ij}^t, & \text{Otherwise} \end{cases} \tag{3}$$

where $rand$ represents a uniform random number between 0 and 1. CR is crossover factor in $[0, 1]$. $j_{rand} \in [1, 2, \cdots, n]$ is a randomly chosen index, which ensures that u_i^t gets at least one component from v_i^t.

2.4 Selection Operation

The child candidate $\mathbf{u}_i^t$ and the parent individual $\mathbf{x}_i^t$ take a competition mechanism in their fitness and the winner has the chance enter into next generation.

$$\mathbf{x}_i^{t+1} = \begin{cases} \mathbf{u}_i^t, & f(\mathbf{u}_i^t) \leq f(\mathbf{x}_i^t) \\ \mathbf{x}_i^t, & \text{Otherwise} \end{cases} \tag{4}$$

The above three basic steps repeat until the termination condition is met and a final candidate solution to $f(\mathbf{x})$ is output.

2.5 Related Works

DE has been more and more widely used by many researchers [7,14,9] due to its evident features and excellent performance. However, it is known that DE is impossible to be free from stagnation and premature convergence due to the limited amount of exploratory moves [4]. Many improved DE-variants are proposed to overcome the existing problems. Either an extra component (e.g.,[15,7]) or a modified structure (e.g., [5,8]) is integrated into DE. Here, a little wellknown DE variants are introduced.

Brest et al. [1] proposed a new version DE algorithm with self-adaptive control parameter settings. It showed good performance on numerical benchmarks. Rahnamayan et al. [12] presented an opposition-based DE with opposition-based learning for population initialization and for generation jumping. Zhang and Sanderson [15] implemented a new mutation strategy "DE/current-to-$p$best" with optional external archive and updating control parameters in an adaptive manner. Qin et al. [11] proposed a self-adaptive DE. Both trial vector generation strategies and their associated control parameters are gradually self-adapted by their experiences when generating promising solutions. Wang et al. [14] studied whether DE can be improved by combining several vector generation strategies with suitable control parameter settings. A novel method (CoDE) has been proposed which uses three trial vector generation strategies and three control parameter settings. Cai and Wang [2] introduced two novel operators, namely, the neighbor guided selection scheme for parents involved in mutation and the direction induced mutation strategy, to fully exploit the neighborhood and direction information of the population, respectively. Gong and Cai [7] proposed a ranking-based mutation operator for DE algorithm, where some of the parents in mutation are proportionally selected according to their rankings in the current population. Das. et al. [3] proposed a dynamic DE variant with Brownian and quantum individuals (DDEBQ), which uses a neighborhood-driven double mutation strategy to control the perturbation and then prevent the algorithm from converging too quickly.

3 Directional Difference Information

3.1 Motivations

As natural and social rules indicate, successive/positive experience is learned and utilized to help us solve the challenging problems. At the same time, failing/negative experience is often intentionally or unintentionally ignored. In fact, they are also our valuable fortunes and should be paid more attention. As far as DE is concerned, the positive heuristic information, such as the successful experience from the best or the superior individuals, has been exploited by many scholars [11,15,7]. Feoktistov [6] also discussed the similar technique on the direction of differential evolution. However, the negative heuristic information or the experienced penalties from the worse or the inferior individuals are far from consideration.

This paper aims at considering several kinds of directional heuristic information, i.e., approaching directions to the superior solutions, departing directions from the inferior solutions, and both.

3.2 Directional Difference and Base Vectors Construction

The usually and widely used DE mutation, DE/rand/1 in Eq.(2), is adopted as an analytic model strategy in this paper. Parental individuals for both base vector and difference vector are selected in a completely random manner. Recently, Cai and Wang [2] employed the directional information to the original DE mutation. However, they introduced some additional Convergence/Attraction/Repulsion direction vectors with a few computing costs based on the original DE mutation.

This paper will utilize the directional heuristic information through adjusting the sequence of three selected individuals, which is very different from and much easier than those of [2].

Approaching to the Superior. DE randomly select three individuals from DE population to produce a mutant vector $\mathbf{v}_i^t$ with respect to each individual $\mathbf{x}_i^t$. In order to learn the positive experience from the superior individuals, the best one of three selected individuals is taken as base vector and the other two individuals produce a difference vector as Eq.(5).

$$\mathbf{v}_i^t = \mathbf{x}_{r_1'}^t + F_1 \cdot (\mathbf{x}_{r_2}^t - \mathbf{x}_{r_3}^t) \tag{5}$$

where $\mathbf{x}_{r_1'}^t$ is the best individual of three randomly selected individuals.

Departing from the Inferior. In order to depart the negative experience from the inferior individuals, the worst one of three selected individuals is taken as the subtrahend during the difference vector generating process. The other two individuals remain untouched.

$$\mathbf{v}_i^t = \mathbf{x}_{r_1}^t + F_1 \cdot (\mathbf{x}_{r_2}^t - \mathbf{x}_{r_3'}^t) \tag{6}$$

where $\mathbf{x}_{r_3'}^t$ is the worst individual of three randomly selected individuals.

Both. In order to learn the positive information from the superior individuals and depart the negative experience from the inferior individuals simultaneously, the best one of three selected individuals is taken as base vector and the worst one of three is taken as the subtrahend during the difference vector generating process. It is indicated as Eq.(7).

$$\mathbf{v}_i^t = \mathbf{x}_{r_1'}^t + F_1 \cdot (\mathbf{x}_{r_2}^t - \mathbf{x}_{r_3'}^t) \tag{7}$$

where $\mathbf{x}_{r_1'}^t$ is the best one of three randomly selected individuals, $\mathbf{x}_{r_3'}^t$ is the worst one of three randomly selected individuals.

4   Experimental Comparisons

4.1   Benchmarks, Competitors and Parameters

Ten classical 30-dimensional benchmarks [16] and four competitors are chosen to verify the enhancement and cooperation of two ideas, namely the positive and the negative heuristics. Benchmarks refer to Table 1.

Four competitors are original DE with mutation Eq.(2), DEpos with positive mutation Eq.(5), DEneg with negative mutation Eq.(6), DEboth with mutation Eq.(7). Other operations of these four algorithms are all the same.

Four competitors take the same parameters setting in the experiment. Maximal function evaluation is 100 000; population size is 100; dimension of benchmark is 30; all algorithms for every function are independently run 50 times; scale factor F = 0.5; crossover probability CR = 0.95.

Table 1. Benchmark functions, where $f_1 - f_5$ are unimodal and $f_6 - f_{10}$ are multimodal, whose known optimal values are $f_{min} = 0$.

Benchmark functions	Domain	x_{min}
$f_1 = \sum\limits_{i=1}^{n} x_i^2$	$[-100, 100]^n$	$\{0\}^n$
$f_2 = \sum\limits_{i=1}^{n} |x_i| + \prod\limits_{i=1}^{n} |x_i|$	$[-10, 10]^n$	$\{0\}^n$
$f_3 = \sum\limits_{i=1}^{n} \left(\sum_{j=1}^{i} x_j \right)^2$	$[-100, 100]^n$	$\{1\}^n$
$f_4 = \max\limits_{i}\{|x_i|, 1 \le i \le n\}$	$[-100, 100]^n$	$\{0\}^n$
$f_5 = \sum\limits_{i=1}^{n} (\lfloor x_i + 0.5 \rfloor)^2$	$[-100, 100]^n$	$\{0\}^n$
$f_6 = 418.983333 * n - \sum\limits_{i=1}^{n} (x_i \sin(\sqrt{|x_i|}))$	$[-500, 500]^n$	$\{420.9687\}^n$
$f_7 = \sum\limits_{i=1}^{n} [x_i^2 - 10\cos(2\pi x_i) + 10]$	$[-5.12, 5.12]^n$	$\{0\}^n$
$f_8 = -20\exp\left[-0.2\sqrt{\frac{1}{n}\sum\limits_{i=1}^{n} x_i^2} \right] -$ $\exp\left(\frac{1}{n}\sum_{i=1}^{n}\cos(2\pi x_i) \right) + 20 + e$	$[-32, 32]^n$	$\{0\}^n$
$f_9 = \frac{\pi}{n}\left\{ 10\sin^2(\pi y_1) + \sum\limits_{i=1}^{n-1}(y_i - 1)^2 \cdot [1 + 10\sin^2(\pi y_{i+1})] + (y_n - 1)^2 \right\}$ $+ \sum\limits_{i=1}^{n} u(x_i, 10, 100, 4),\ y_i = 1 + \frac{1}{4}(x_i + 1),$	$[-50, 50]^n$	$\{1\}^n$
$f_{10} = 0.1\left\{ 10\sin^2(3\pi x_1) + \sum\limits_{i=1}^{n-1}(x_i - 1)^2 \cdot [1 + 10\sin^2(3\pi x_{i+1})] + (x_n - 1)^2 \cdot [1 + \sin^2(2\pi x_n)] \right\} + \sum\limits_{i=1}^{n} u(x_i, 5, 100, 4),$ where $u(x_i, a, r, s) = \begin{cases} r(x_i - a)^s, & \text{if } x_i > a \\ 0, & \text{if } x_i \in [-a, a] \\ r(-x_i - a)^s, & \text{if } x_i < -a. \end{cases}$	$[-50, 50]^n$	$\{1\}^n$

4.2   Numerical Comparison

The final results of four algorithms in 50 runs are presented in Table 2. A two-sided rank sum test is performed with the proposed directional DE to the original DE, in which the hypothesis is that two independent samples come from distributions with equal medians, and returns the p-value from the test. p-value is the probability of two independent samples having equal medians, namely, no significant difference. "H" returns the result of the hypothesis test, performed at the 0.05 significance level. H $= 0/1$ indicates that the null hypothesis cannot/can be rejected at the 5% level.

It is clear to see that whatever the positive DDI from the superior individuals (DEpos) or the negative DDI from the inferior individuals (DEneg) can enhance the performance of DE. DEboth with two kinds of heuristic information generally performs best for most functions. All four algorithms have the same performance for function f_5. However, the next figure, Fig.1, indicates that they have different converging speed, i.e., DEboth converges fastest and De converges slowest.

DEpos outperforms DE statistically on seven out ten functions except for functions f_5, f_9 and f_{10}. Table 2 shows that DEpos has much better "Best" items than those of DE except for f_4, f_5, f_6, which indicates that the excellent enhancement for exploitation of positive DDI. Table 2 shows that DEneg significantly outperforms DE on nine out ten functions except for function f_5. It indicates even more excellent enhancement of negative DDI than positive DDI. This experimental results remind us that more attention should be paid to the negative heuristic information for optimization. DEboth has statistically best results on five in ten functions except for function f_5. It also has the best "Best" items for functions f_7, f_9 and f_{10}. These results indicate the cooperative property of the positive and negative DDI. However, comparatively speaking, DEboth with both positive and negative DDI has much greater advantages than other three algorithms on unimodal functions over multimodal functions.

Interestingly, "Mean" items of functions f_9 and f_{10} of algorithms DEpos and DEboth are worse than those of DE. However, "Mean" items of DEneg are much better than those of its competitors. It is possible that positive information executes superfluous drives to evolve these two functions, which are non-separable and have huge local optima. In contrast, negative heuristic information does not show this tendency.

Furthermore, simulation results empirically tell us that neither positive DDI from the superior individuals nor negative DDI from the inferior individuals are ignored to enhance the performance of DE.

4.3   Evolving Behaviors Comparison

In order to further verify the evolutionary performance of the proposed positive and negative DDI, the average best function values found-until-now of DE, DEpos, DEneg and DEboth at every iteration in 50 runs versus function evaluations are plotted in Fig. 1.

Table 2. Result comparisons between DE, DEpos, DEneg and DEboth. Data are the statistical results in 50 independent runs, where "Best", "Mean" and "STD" are the best, average results and standard deviation of the final results in 50 runs. [†] is a two-sided rank sum test of two independent samples, which is significant at $\alpha = 0.05$. p-value in "[†]" is the probability that two samples have no significant difference. "[‡]" means that rank sum test does not present p-value for the same results.

Function	Items	DE	DEpos	DEneg	DEboth
f_1	Best	1.79e-10	4.94e-22	4.86e-16	**2.12e-23**
	Mean	7.71e-10	4.01e-21	4.94e-15	**8.98e-22**
	STD	2.32e-09	1.88e-20	1.56e-14	**9.36e-21**
[†]	p-value(H)		7.07e-18(1)	7.07e-18(1)	7.07e-18(1)
f_2	Best	1.23e-5	7.99e-12	2.31e-08	**1.02e-11**
	Mean	4.65e-5	6.31e-11	1.04e-07	**3.03e-11**
	STD	1.24e-4	1.71e-10	1.94e-07	**6.89e-11**
[†]	p-value(H)		7.07e-18(1)	7.07e-18(1)	7.07e-18(1)
f_3	Best	7.34e-02	2.64e-05	2.79e-04	**1.58e-06**
	Mean	4.29e-01	5.72e-04	2.96e-03	**5.46e-05**
	STD	2.32e+00	2.48e-03	1.03e-02	**4.17e-04**
[†]	p-value(H)		7.07e-18(1)	7.07e-18(1)	7.07e-18(1)
f_4	Best	1.02e-01	1.02e-01	5.21e-03	**4.82e-04**
	Mean	1.95e+00	8.69e-01	5.77e-02	**1.05e-02**
	STD	7.40e+00	5.25e+00	1.85e-01	**4.61e-02**
[†]	p-value(H)		3.60e-04(1)	1.28e-17(1)	7.06e-18(1)
f_5	Best	0	0	0	0
	Mean	0	0	0	0
	STD	0	0	0	0
[†]	p-value(H)		‡ (0)	‡ (0)	‡ (0)
f_6	Best	7.89e-01	1.18e+02	**1.34e-02**	1.18e+02
	Mean	2.90e+03	4.62e+02	**3.39e+02**	4.80e+02
	STD	6.68e+03	**8.29e+02**	9.05e+02	**8.29e+02**
[†]	p-value(H)		2.79e-08(1)	5.14e-10(1)	8.56e-08(1)
f_7	Best	1.55e+02	1.89e+01	9.51e+01	**1.29e+01**
	Mean	1.83e+02	**1.27e+02**	1.70e+02	1.42e+02
	STD	2.07e+02	**1.87e+02**	1.99e+02	1.94e+02
[†]	p-value(H)		4.65e-13(1)	8.88e-05(1)	4.49e-11(1)
f_8	Best	3.49e-06	2.83e-12	5.47e-09	**2.46e-12**
	Mean	9.14e-06	1.74e-11	1.99e-08	**8.17e-12**
	STD	1.98e-05	3.98e-11	5.22e-08	**2.31e-11**
[†]	p-value(H)		7.06e-18(1)	7.06e-18(1)	7.06e-18(1)
f_9	Best	4.76e-12	6.04e-24	2.17e-17	**8.64e-25**
	Mean	2.07e-03	4.15e-03	**5.29e-16**	4.15e-03
	STD	1.03e-01	1.04e-01	**3.51e-15**	1.04e-01
[†]	p-value(H)		2.05e-15(1)	7.06e-18(1)	2.05e-15(1)
f_{10}	Best	8.29e-11	1.64e-22	1.22e-16	**1.54e-23**
	Mean	5.01e-10	4.39e-04	**1.07e-14**	2.19e-04
	STD	2.89e-09	1.09e-02	**2.19e-13**	1.09e-02
[†]	p-value(H)		2.29e-15(1)	7.07e-18(1)	1.35e-16(1)

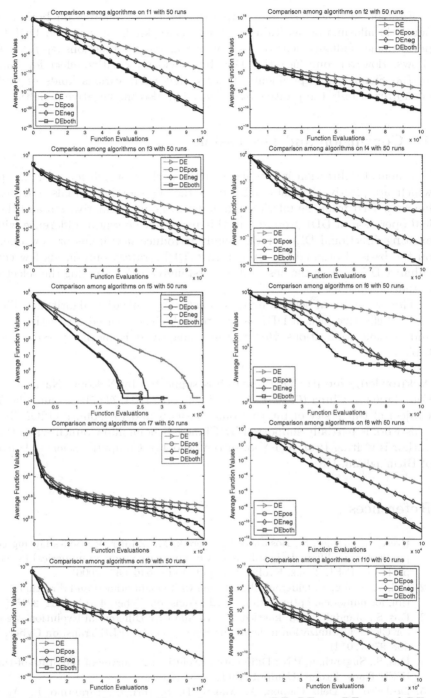

Fig. 1. Evolutionary Performance Comparisons among DE, DEpos, DEneg and DEboth

First of all, Fig. 1 strongly supports the above analysis on both kinds of heuristic information as Table 2 shows. Both kinds of heuristic DDI benefit performance enhancement of DE for the first eight functions f_1-f_8 as Fig. 1 shows. However, only negative DDI has obviously positive effect for functions f_9-f_{10}. Simulation experiments indicate that what various kinds of problems need are not only the positive experience but also the negative penalties.

5   Conclusion and Discussion

The heuristic direction information in DE is the research focus of this paper, which are positive heuristic DDI from the superior individuals and negative heuristic DDI from the inferior individuals. The initial motivation and the critical operation of DDI proposed in this paper have essential difference with the existing directional DEs, which usually introduce new items or new operation on the basis of original DE. In contrast, DDI strategy only needs few costs to enhance DE, picking the best or the worst individual, or sorting the selected few parental solutions.

This research can benefit many population optimization algorithms. The simple and effective idea of DDI can be refined and enhanced for further research and possible applications. How to use it adaptively is another possible research topic.

Acknowledgement: This research is supported by National Natural Science Foundation of China (61105127, 61375066, 61374204, 71171053) and the Youth Project of University of International Business and Economics (12QNGLX02). This work is partially supported by Chinese Scholarship Council when the first author is visiting Prof. Xin Yao, in the School of Computer Science, University of Birmingham, UK.

References

1. Brest, J., Greiner, S., Boskovic, B., Mernik, M., Zumer, V.: Self-adapting control parameters in differential evolution: A comparative study on numerical benchmark problems. IEEE Trans. Evolut. Comput. 10(6), 646–657 (2006)
2. Cai, Y., Wang, J.: Differential evolution with neighborhood and direction information for numerical optimization. IEEE Trans. on Cyber. 43(6), 2202–2215 (2013)
3. Das, S., Mandal, A., Mukherjee, R.: An Adaptive Differential Evolution Algorithm for Global Optimization in Dynamic Environments. IEEE Trans. on Cyber. 44(6), 966–978 (2014)
4. Das, S., Suganthan, P.N.: Differential Evolution: A Survey of the State-of-the-Art. IEEE Trans. Evolut. Comput. 15(1), 4–31 (2011)
5. De Falco, I., Della Cioppa, A., Maisto, D., Scafuri, U., Tarantino, E.: An adaptive invasion-based model for distributed Differential Evolution. Information Sciences 278, 653–672 (2014)
6. Feoktistov V., Differential Evolution - In Search of Solutions. Springer (2006)

7. Gong, W.Y., Cai, Z.H.: Differential evolution with ranking-based mutation operators. IEEE Trans. on Cyber. 43(6), 2066–2081 (2013)
8. Maio, F.D., Baronchelli, S., Zio, E.: Hierarchical differential evolution for minimal cut sets identification: Application to nuclear safety systems. European Journal of Operational Research 238(2), 645–652 (2014)
9. Mezura-Montes, E., Coello Coello, C.A., Velazquez-Reyes, Munoz-Davila, J.L.: Multiple trial vectors in differential evolution for engineering design. Engineering Optimization 39(5), 567–589 (2007)
10. Price, K., Storn, R., Lampinen, J.: Differential evolution: A practical approach to global optimization. Springer, Berlin (2008)
11. Qin, A.K., Huang, V.L., Suganthan, P.N.: Differential evolution algorithm with strategy adaptation for global numerical optimization. IEEE Trans. Evolut. Comput. 13(2), 398–417 (2009)
12. Rahnamayan, S., Tizhoosh, H.R., Salama, M.M.A.: Opposition-based differential evolution. IEEE Trans. Evolut. Comput. 12(1), 64–79 (2008)
13. Storn, R., Price, K.: Differential evolution: A simple and efficient heuristic for global optimization over continuous spaces. J. Global Optimization 11(4), 341–359 (1997)
14. Wang, Y., Cai, Z.: Constrained evolutionary optimization by means of $(\mu+\lambda)$-differential evolution and improved adaptive trade-off model. Evolutionary Computation 19(2), 249–285 (2011)
15. Zhang, J., Sanderson, A.C.: JADE: adaptive differential evolution with optional external archive. IEEE Trans. Evolut. Comput. 13(5), 945–958 (2009)
16. Zhao, X., Liu, Z., Yang, X.: A multi-swarm cooperative multistage perturbation guiding particle swarm optimizer. Appl. Soft Comput. 22, 77–93 (2014)

Working Memory Training Strategies and Their Influence on Changes in Brain Activity and White Matter

Tomoyuki Hiroyasu[1], Shogo Obuchi[2], Misato Tanaka[3],
Tatsuya Okamura[2], and Utako Yamamoto[1]

[1] Faculty of Life and Medical Sciences, Doshisha University,
Tataramiyakodani, 1-3, Kyotanabe-shi, Kyoto, Japan
{tomo,utako}@mis.doshisha.ac.jp
[2] Graduate School of Life and Medical Sciences, Doshisha University,
Tataramiyakodani, 1-3, Kyotanabe-shi, Kyoto, Japan
{sobuchi,tokamura}@mis.doshisha.ac.jp
[3] Faculty of Science and Engineering, Doshisha University,
Tataramiyakodani, 1-3, Kyotanabe-shi, Kyoto, Japan
mtanaka@mikilab.doshisha.ac.jp

Abstract. In this study, we investigated whether different working memory training tasks influence brain activity and white matter changes. Thirteen participants were involved in our interventional study over a period of one month. During pre- and post-training, brain activity and structural integrity were measured using functional magnetic resonance imaging and diffusion tensor imaging. The reading span task was used to measure working memory capacity in participants performing different strategies. Participants were classified into a training group (10 participants) and a control group (4 participants). The training group was further divided into the imagery strategy group and rehearsal strategy group. Only the imagery strategy group improved working memory capacity, showing significantly increased activation in the anterior cingulate cortex and fractional anisotropy adjacent to the right temporal gyrus. Consequently, adopting the appropriate strategy is important for improving working memory capacity as different strategies affect brain activity and white matter to different degrees.

Keywords: functional Magnetic Resonance Imaging, Diffusion Tensor Imaging, Working Memory, Reading Span Task, Myelination, Strategy, Training.

1 Introduction

Working memory is the limited capacity storage system involved in the maintenance and manipulation of information over short periods of time, facilitating various forms of complex thinking [1]D According to the multicomponent model proposed by Baddeley et al. (2000), working memory comprises three storage

H. Handa et al. (eds.), *Proc. of the 18th Asia Pacific Symp. on Intell. & Evol. Systems – Vol. 2,*
Proceedings in Adaptation, Learning and Optimization 2, DOI: 10.1007/978-3-319-13356-0_22

buffers and the central executive function [2]D These three buffers are defined as follows: 1) the phonological loop for storing verbal information, 2) the visuo-spatial sketchpad for the storage of visual information, and 3) the episodic buffer for binding information from subsidiary systems and long-term memory. The central executive function controls these three subsystems. Because the subsystems are involved in short-term storage, working memory has capacity, and the maximum information that can be stored is called the working memory capacity [1]D Individual working memory capacity is correlated with a wide range of cognitive functions such as reading comprehension abilities [3], and reasoning tasks [4]. The reading span task (RST) [5] proposed by Daneman & Carpenter is well known as a task that can measure individual differences in working memory capacity [6,7]. In the RST, a series of sentences are read aloud by the subject at their own pace and subjects are asked to recall a target word in each sentence. Osaka and Nishizaki (2000), and Endo (2012) examined individual differences in strategies used to memorize target words in the Japanese RST [8,9]. Two common strategies are the imagery strategy and the rehearsal strategy. Subjects who obtained high scores on the RST used the imagery strategy, whereas subjects who obtained low scores used the rehearsal strategy [8]. Therefore, strategies can affect the working memory capacity.

Previous studies have indicated that we can improve our working memory capacity [10]. Moreover, it has become clear that training on a working memory task and a cognitive task improves not only the results of trained tasks but also other working memory tasks [11–13]. Improvement in working memory capacity, for example, enhanced results on reasoning tasks [4] and reaction inhibition tasks [10]. It is also expected that improving working memory can relieve the symptoms of attention deficit/hyperactive disorder [14,15].

Previous studies using functional magnetic resonance imaging (fMRI) have shown that an increase in working memory capacity changes brain activity [10, 16]. Olesen et al. (2004) reported that brain activation in the prefrontal region increased after working memory task training [10]. According to Osaka et al. (2012), training using an imagery strategy in the RST improved working memory capacity, and brain activation in the anterior cingulate cortex (ACC) increased significantly. Thus, it is clear that training strategies can change brain activity [16].

In contrast, it has been shown that improvements in working memory capacity not only affect brain activity, but also white matter as measured by diffusion tensor imaging (DTI) [17]. Takeuchi et al. (2010) examined the effect of working memory training on structural connectivity using voxel-based analysis (VBA) of fractional anisotropy (FA) measures of fiber tracts. After training, they identified increases in FA in white matter regions adjacent to the left intraparietal sulcus and the corpus callosum, and the degree of change correlated with the amount of working memory training. Thus, it is clear that working memory training induced plasticity in regions that are considered to be critical for cognitive processing.

Nevertheless, no previously reported study has examined effective methods for improving working memory capacity, and the variations in brain activity and white matter related to such methods. Here, we report the impact on brain activity and white matter caused by different working memory training strategies. To explore this, we undertook an interventional study of different training strategies within the RST over a period of one month and determined whether or not such strategies changed brain activity and the structural integrity of the white matter as measured by fMRI and DTI. As training strategies, we select the imagery strategy and the rehearsal strategy because a previous study [8] showed that high-RST performers tended to use a mental imagery strategy more often than low-RST performers, who appear to restrict themselves to verbal rehearsal. We hypothesized that there are different effects on working memory capacity, brain activity, and the structural integrity between strategies.

2 Materials and Methods

2.1 Participants

Thirteen healthy right-handed individuals (11 men, 2 women) participated. The mean age was 22.3 years (SD, 0.95). This study was approved by the Ethics Committee of Doshisha University Faculty of Life Medical Sciences. Written informed consent was obtained from each participant. We classified participants into training and control groups. Nine participants constituted the training group and performed the working memory task for one month. Four participants constituted the control group. The control group was arranged because participants were university students who studied and played sports in their daily lives, possibly affecting brain structure [18,19]. We instructed five participants to use the imagery strategy and four participants to use the rehearsal strategy. One participant who failed to undergo the required training over the month was omitted from the analysis.

2.2 Procedure

We measured the results of RST, brain activity during the RST, and the integrity of white matter pre- and post-training using an MR scanner. The scanning order was fMRI scan, then DTI scan, and lastly T1 anatomical scan. The working memory training programs were developed in-house, and participants trained on their personal computer. Participants performed the working memory training task five days a week. RST was used as the working memory task using the following sequence.

1. Each of the five sentences was displayed. Participants read a series of sentences aloud at their own pace and memorized a target word in each sentence.
2. When "Recognition Time" was displayed, participants were asked to recall the target words and record them on paper.
3. The correct answers were then revealed.

Stages 1–3 were repeated five times each day. Participants were instructed to read a series of sentences aloud at their own pace, and go to the next sentence as soon as they had finished reading the previous sentence. Five participants in the training group used an imagery strategy, which involved making a visual image of each target word to memorize it, and four participants used a rehearsal strategy, which involved reciting the target word repeatedly. If for any reason a participant was unable to undergo training, we allowed them to skip a session; however, at least 20 training sessions had to be completed within the month for their results to be included. To confirm that they had undergone the required number of training sessions, they were asked to submit a log book.

2.3   Experimental Design

Each participant undertook the RST during both pre- and post-training experimental sessions. Fig. 1 shows an experimental block (modified RST fitted to an fMRI experiment). We used the RST created by Osaka (2002) [20]. One session followed this sequence:

1. RST
   Six sentences were displayed, one at a time, for 5 s. Participants read silently and attempted to memorize the target word in each sentence simultaneously. To confirm they had finished reading each sentence, participants pressed a button.
2. Recognition
   Participants recalled the target words in order and identified each of the target words from a closed set of three words. If the target word appeared among the set of three words, participants pressed the button pertaining to the target word, and if it did not appear, they pressed a button with a cross.
3. Control
   "Left" or "Right" was shown for 2.5 s. Participants pressed the button for the corresponding words.
4. Read
   Six sentences were displayed for 5 s each. Participants only read silently. To confirm they had finished reading each sentence, participants pressed a button.
5. Control
   "Left" or "Right" was shown for 2.5 s. Participants pressed the button for the corresponding words.

Stages 1–5 were repeated four times during each session. Participants performed this experiment for two sessions in the MR scanner. In the post-training experiment, we instructed five participants in the imagery strategy training group to make a visual image of each target word to memorize it, and four participants in the rehearsal strategy training group to recite the target word repeatedly. To enable us to differentiate between proper brain activity and activity derived from the RST, participants underwent the Read condition. We then subtracted brain activity measured in the Read condition from the Recognition condition to ascertain a true reading of brain activity during RST.

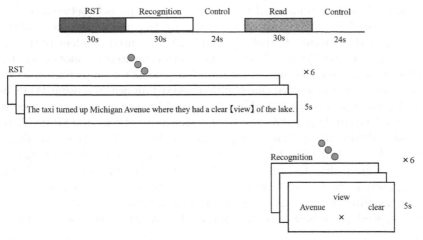

Fig. 1. Design of the fMRI experiment

2.4   Image Acquisition

Whole brain imaging data were acquired on a 1.5 T MR scanner (Hitachi Medico Echelon Vega). The experiment was controlled by Presentation software (Neurobehavioral System Inc.), and the fORP 932 Subject Response Package (Cambridge Research System) was used to acquire subject responses. For functional imaging, a gradient-echo echo planar (GE-EPI) imaging sequence was used (TR = 2500 ms, TE = 50 ms, FA = 90°, FOV = 240×240 mm, matrix = 64×64, thickness = 6.0 mm, slice number= 20). For DTI, a diffusion-weighted EPI (DWI) was used (TR = 2317 ms, TE = 74.3 ms, FA = 90°, FOV = 240 × 240 mm, matrix = 256 × 256, thickness = 3.0mm, slice number = 50, b value = 1000 s/mm^2, direction = 21). For T1 anatomical imaging, rf-spoiled steady state gradient echo (RSSG) was used (TR = 9.4 ms, TE = 4.0 ms, FA = 8°, FOV = 256 × 256 mm, matrix = 256 × 256, thickness = 1.0 mm, slice number= 192).

2.5   Analysis

An analysis of fMRI data was initially performed for each individual participant for each pre- and post-training session. Data were analyzed using SPM8 (Wellcome Department of Cognitive Neurology) on Matlab (MathWorks). Six initial images from each scanning session were discarded from the analysis in order to eliminate non-equilibrium effects of magnetization, leaving 448 images in total for analysis. All functional images were realigned to correct for head movement. After realignment, the anatomical images were co-registered to mean functional images. Functional images were then normalized with the anatomical image and spatially smoothed using a Gaussian filter (8 mm full width-half maximum). The box-car reference function was adopted for individual analysis to identify voxels under each task condition. Single participant data were analyzed using a fixed-effects model, whereas group data from pre- and post-training sessions were analyzed using a random-effects model. As a comparison between post- and pre-training sessions, we performed a paired t-test to examine the effects of training. Automated anatomical labeling (AAL) [21] was used to identify the brain regions that were activated. All DWI data were corrected for motion artifacts, and then diffusion tensors and scalar diffusion parameter maps were calculated for each participant in native space with standard algorithms using DTIStudio (John Hopkins University) [22]. The anatomical images were co-registered to b value = 0 images and then DWI images were normalized with the SPM8 T1 template using SPM8. By applying the parameters derived from the normalization of the b value = 0 images, FA maps were also normalized and then smoothed using a Gaussian filter (10 mm full width-half maximum). In the whole brain analysis, using a paired t-test, we investigated regions that showed increased FA following working memory training. In this computation, we included only voxels that showed FA values > 0.2 to extract the "white matter only" structural changes.

3   Results

3.1   Behavioral Performance

Recognition accuracy during the experimental RST was calculated for both pre- and post-training phases. Table 1 shows the group results and Fig. 2 shows the individual results. The mean percentage recognition accuracies were higher only in the imagery strategy group in the post- than in the pre-training session (effect size = 0.84). In contrast, in the rehearsal strategy group and control group, there was little difference between pre- and post-training session results. One participant in the imagery strategy group showed a marked decrease in recognition accuracy post-training and his data was omitted from the fMRI and DTI analysis; he was judged as an outlier using the interquartile method.

Table 1. Effect of training strategies

Strategy	Pre [%]	Post [%]	t value	effect size
Imagery	82.81 ± 7.09	90.63 ± 4.96	-4.39^*	0.84
Rehearsal	77.08 ± 2.95	77.08 ± 4.50	0.00	0.00
Control	82.29 ± 4.77	82.81 ± 4.75	-0.10	0.11

3.2   fMRI Results

The fMRI results indicated that working memory affects brain activity. Increased activation after training was compared using the paired t-test. Comparisons were made between the pre- and post-training sessions based on a comparison of the data captured during the RST condition and the Read condition. Results are shown in Table 2 (voxel-level threshold uncorrected for multiple comparison $p < 0.001$, cluster size > 10) and Fig. 3 ($x = 6$). There were significantly more activation regions in the imagery strategy group and the control group. There was no significant activation in the rehearsal strategy group.

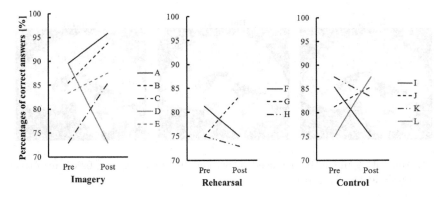

Fig. 2. Effect of training strategies in individual participants

Table 2. Significant clusters of activation in the three groups during the RST condition comparing the Read condition ($p < 0.001$, uncorrected; extent threshold voxels = 10)

Cortical region	x	y	z	z score	cluster size
Image					
Left Inferior Parietal	−32	−58	40	4.61	53
Right Frontal Middle	36	10	62	4.59	14
Left Precuneus	−4	−50	38	4.53	54
Left Inferior Frontal	−48	30	22	4.43	40
Right Superior Occipital	28	−82	24	4.41	23
Right Cuneus	18	−74	32	4.40	24
Right Anterior Cingulate	6	48	16	4.23	13
Left Caudate	−20	−12	28	4.17	21
Right Middle Cingulate	0	−18	26	3.99	22
Left Paracentral Lobule	−14	−26	80	3.91	11
Rehearsal					
Control					
Right Inferior Frontal	40	28	28	4.90	64
Right Middle Cingulate	4	−12	36	4.25	12
Right Angular	54	−54	34	4.19	26
Right Superior Temporal	48	−36	8	4.09	31
Left Superior Parietal	−18	−60	60	4.03	19
Right Middle Occipita	48	−74	30	4.00	18
Left Middle Frontal	−44	−28	46	3.99	21
Right Superior Frontal	20	48	18	3.96	16
Right SupraMarginal	60	−38	36	3.87	28
Left Superior Temporal	−66	−32	18	3.78	18

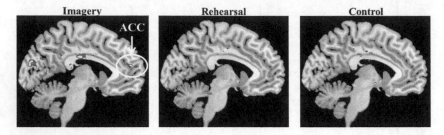

Fig. 3. Areas of significantly increased activation after training ($p < 0.001$, uncorrected; extent threshold voxels $= 10$)

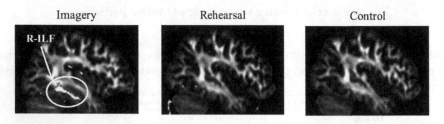

Fig. 4. Areas of significantly increased FA after training ($p < 0.001$, uncorrected; extent threshold voxels $= 10$)

3.3 DTI Results

The DTI results, using a paired t-test, indicated that working memory training affected the structural integrity. In the VBA, we used a paired t-test to assess whether there were group changes in FA after training. A significant increase in FA was found only in the imagery strategy group in an anatomical cluster in a white matter region adjacent to the inferior temporal gyrus (ITG). However, in the rehearsal strategy group and the control group, there was no significant increase in FA (x, y z $= 46, -42, -16$; paired t-test, $t = 21.10$; df $= 3$, $p < 0.001$, uncorrected, cluster size > 10) (Fig. 4). In contrast, there was no significant decrease in FA.

4 Discussion of the Variations in Working Memory Capacity, Brain Activity, and White Matter Induced by Training Strategies

4.1 Changes in Working Memory Capacity by Training Strategy

Since the mean percentage recognition accuracies were higher only in the imagery strategy group in the post- than in the pre-training session (effect size $= 0.84$), only the imagery strategy group improved their working memory capacity (Table 1). Therefore, the imagery strategy is an effective method to improve working memory capacity. These results are consistent with the experiment conducted by Osaka et al. (2012) [16]. Based on the model of working memory proposed by Baddeley et al. (2000), participants adopting the imagery strategy use the visuo-spatial sketchpad, and thus they memorize target words accurately without interfering with reading sentences. In contrast, the rehearsal strategy users used the phonological loop to memorize target words and read sentences simultaneously. This indicates that the phonological loop burden on brain activity is significant, and because of this, the participants in the Rehearsal condition were unable to use working memory effectively.

4.2   Impacts on Brain Activity

The ACC activation significantly increased only in the imagery strategy group post-training (Fig. 3). The ACC has an attention coordination function, including inhibiting attention directed toward inappropriate stimuli [23, 24]. An increase in ACC activity and control of attention to target words are both related to improvements in recognition accuracy on the RST. In the imagery strategy group, significant increased activations in the left precuneus, right superior occipital lobule, and right cuneus were also observed. We concluded that these activations were due to visual imageries, because Granis et al. (2004) [25] and Knauff et al. (2000) [26] reported that the visual cortex is also activated when people create a visual image. In contrast, there were no significant activations in the rehearsal strategy group. The rehearsal strategy was to recite target words mentally and read sentences silently simultaneously. Participants also read sentences silently in the Read condition; therefore, this eliminated potential differences in brain activity. In the control group, it was possible that participants changed their strategy without training. In this experiment, we did not ask about pre-training strategies and therefore we could not control strategies to a certain extent. We need to control this in further experiments.

4.3   Impacts on White Matter

Only in the imagery strategy group was working memory training associated with FA increases in the white matter region close to the ITG. Using DTI analysis, it has been well established that myelination continues even during adulthood [27, 28]. Furthermore, human studies have shown practice-induced white matter plasticity, even in adulthood [29, 30]. We concluded that increased myelination caused by neuronal activity in fiber tracts during working memory training is one possible mechanism underlying the observed FA increase. The region in which FA significantly increased was the inferior longitudinal fasciculus (ILF). As FA increased significantly only in the imagery strategy group, the ILF plays an important role in transferring information to the visuo-spatial sketchpad. The ILF is involved in short-term memory of visual stimuli and visual imagery [31]. Therefore, the increase in neuronal activities caused by the training imagery strategy increased myelination. This myelination improves working memory capacity by increasing the effectiveness of the connection of the working memory.

5   Conclusion

It has not been clearly demonstrated that variations in brain activity and white matter can be induced by training strategies designed to improve working memory capacity. In this paper, with RST as the training task, we examined whether different training strategies designed to improve working memory affects working memory capacity, brain activity, and the structural integrity of white matter. Participants trained their working memory over a month and the brain activity and structural integrity of their white matter were measured using fMRI and DTI. Working memory capacity improved significantly only in the imagery strategy; significant increased activation in the ACC and FA in the ILF were also observed. We concluded that different strategies affect the working memory capacity, brain activity, and white matter in different degrees.

References

1. Baddeley, A.D., Hitch, G.: Working memory. The Psychology of Learning and Motivation 8, 47–89 (1974)
2. Baddeley, A.: The episodic buffer: a new component of working memory. Trends in Cognitive Sciences 4, 417–423 (2000)
3. Just, M.A., Carpenter, P.A.: A capacity theory of comprehension: Individual differences in working memory. Psychological Review 99, 122–149 (1992)
4. Kyllonen, P.C., Christal, R.E.: Reasoning ability is (little more than) working-memory capacity. Intelligence 14, 389–433 (1990)
5. Daneman, M., Carpenter, P.A.: Individual differences in working memory and reading. Journal of Verbal Learning and Verbal Behavior 19, 450–466 (1980)
6. Friedman, N.P., Miyake, A.: The reading span test and its predictive power for reading comprehension ability. Journal of Memory and Language 51, 136–158 (2004)
7. Buchweitz, A., Keller, T.A., Meyler, A., Just, M.A.: Brain Activation for Language Dual-Tasking: Listening to Two People Speak at the Same Time and a Change in Network Timing. Human Brain Mapping 33, 1868–1882 (2012)
8. Osaka, M., Nishizakki, Y.: Processing Characteristics in the Central Executive of Working Memory–Understanding and Integration in RST Performance– (In Japanese). In: Brain and Working Memory, 1st edn., pp. 203–223. Kyoto Daigaku Gakujutsu Shuppan Kai (2000) (in Japanese)
9. Endo, K., Osaka, M.: Individual differences in strategy use in the Japanese version of reading span test. Shinrigaku Kenkyu 82, 554–559 (2012) (in Japanese)
10. Olesen, P.J., Weserberg, H., Klingberg, T.: Increased prefrontal and parietal activity after training of working memory. Nature Neuroscience 7, 75–79 (2004)

11. Perrig, W.J., Hollenstein, M., Oelhafen, S.: Can We Improve Fluid Intelligence with Training on Working Memory in Persons with Intellectual Disabilities? Journal of Cognitive Education and Psychology 8, 148–164 (2009)
12. Kawashima, R., Okita, K., Yamazaki, R., Tajima, N., Yoshida, H., Taira, M., Iwata, K., Sasaki, T., Maeyama, K., Usui, N.: Reading aloud and arithmetic calculation improve frontal function of people with dementia. Biological and Medical Sciences 60, 380–384 (2005)
13. Uchida, S., Kawashima, R.: Reading and solving arithmetic problems improves cognitive functions of normal aged people: a randomized controlled study. Age 30, 21–29 (2008)
14. Barkel, R.A.: Behavioral inhibition, sustained attention, and executive functions: Constructing a unifying theory of ADHD. Psychological Bulletin 121, 65–94 (1997)
15. Klingberg, T., Fernell, E., Olesen, P.J., Johnson, M., Gustafsson, P., Dahlstrom, K., Gillberg, C.G., Forssberg, H., Westerberg, H.: Computerized training of working memory in children with ADHD-a randomized, controlled trial. Journal of American Academy of Child and Adolescent Psychiatry 44, 177–186 (2005)
16. Osaka, M., Otsuka, Y., Osaka, N.: Verbal to visual code switching improves working memory in older adults: an fMRI study. Frontiers in Human Neuroscience 6, 24 (2012)
17. Takeuchi, H., Sekiguchi, A., Taki, Y.: Training of working memory impacts structural connectivity. The Journal of Neuroscience 30, 3297–3303 (2010)
18. Draganski, B., Gaser, C., Kempermann, G., Kuhn, H.G., Winkler, J., Buchel, C., May, A.: Temporal and spatial dynamics of brain structure changes during extensive learning. Journal of Neuroscience 26, 6314–6317 (2006)
19. Jancke, L., Koeneke, S., Hoppe, A., Rominger, C., Hanggi, J., Earley, R.L.: The Architecture of the Golfer's Brain. PLoS One 4, e4785 (2009)
20. Osaka, M.: The memo pad of brain: working memory, 1st edn. Shinyo-sha (2002) (in Japanese)
21. Mazoyer, T.N., Landeau, N., Papathanassiou, D., Crivello, F., Etard, O., Delcroix, N., Mazoyerand, B., Joliot, M.: Automated Anatomical Labeling of Activations in SPM Using a Macroscopic Anatomical Parcellation of the MNI MRI Single-Subject Brain. NeuroImage 15, 273–289 (2002)
22. Jiang, H., van Zijl, P., Kim, J., Pearlson, G.D., Mori, S.: DtiStudio: Resource program for diffusion tensor computation and fiber bundle tracking. Computer Methods and Programs in Biomedicine 81, 106–116 (2006)
23. Smith, E.E., Jonides, J.: Storage and executive processes in the frontal lobes. Science 283, 1657–1661 (1999)
24. Braver, T.S., Barch, D.M., Gray, J.R., Molfese, D.L., Snyder, A.: Anterior cingulate cortex and response conflict: Effects of frequency, inhibition and errors. Cerebral Cortex 11, 825–836 (2001)
25. Granis, G., Thompson, W.L., Kosslyn, S.M.: Brain areas underlying visual mental imagery and visual perception: an fMRI study. Cognitive Brain Research 20, 226–241 (2004)
26. Knauff, M., Kassubek, J., Mulack, T., Greenlee, M.W.: Cortical activation evoked by visual mental imagery as measured by fMRI. NeuroReport 11, 3957–3962 (2000)
27. Benes, F.M., Turtle, M., Khan, Y., Farol, P.: Myelinationin of a key relay zone in the hippocampal formation occurs in the human brain during childhood. Archives of General Psychiatry 51, 477–484 (1994)

28. Yakovlev, P.I., Lecours, A.R.: Regional development of the brain in early life. Blackwell Scientific, Oxford (1967)
29. Bengtsson, S.L., Nagy, Z., Skare, S., Forsman, L., Forssberg, H., Ulen, F.: Extensive piano practicing has regionally specific effects on white matter development. Nature Neuroscience 8, 1148–1150 (2005)
30. Scholz, S., Klein, M.C., Beherens, T.E.J., Johansen-Berg, H.: Training induces changes in white-matter architecture. Nature Neuroscience 12, 1370–1371 (2009)
31. Shinoura, N., Suzuki, Y., Tsukada, M., Katsuki, S., Yamada, R., Tabei, Y., Saito, K., Yagi, K.: Impairment of Inferior Longitudinal Fasciculus plays a Role in Visual Memory Disturbance. Neurocase 13, 127–130 (2007)

A Decision Support System for Laptop Procurement Using Mixed Text and Image Analytics

Wei-Yi Tay[1] and Chee Khiang Pang[1,*]

Department of Electrical and Computer Engineering,
National University of Singapore, 4 Engineering Drive 3, Singapore 117576

Abstract. The amount of data in today's world is exploding and the heterogeneous nature of the data both on the internet and offline makes it very difficult to integrate and analyze in order to generate useful insights to aid decision making. In this paper, a Decision Support System (DSS) is designed for intelligent decision making using the large amount of heterogeneous data available online. Text analytics techniques such as the Latent Semantic Analysis (LSA) and Latent Dirichlet Allocation (LDA) are combined with image analytics techniques to extract useful information from huge volume of data. The DSS is implemented as a Graphical User Interface (GUI) to extract unsupervised topics from the textual data and identify documents that are semantically similar to the topics or image text, and relevant documents are then identified and presented to the user in aid of laptop procurement.

1 Introduction

In today's information technology era, there has been phenomenal growth in multimedia data such as images, audio, text, animations, video and other sensory data [1]. In the future, every other physical object on the planet will be connected to the internet in a giant Internet of Things (IoT) [2]. Information obtained from different sources such as sensors in home appliances and cars, mobile phones and internet comes in many different formats such as digital, sensory, imagery and audio. This immense growth of data has led to traditional data technology not being able to keep up and analyze the huge wave of new heterogeneous information to extract useful information. Furthermore, due to the heterogeneous nature of the data online and offline, this makes it even more difficult to integrate all these data and generate useful insights.

The proposed framework in this paper aims to develop a decision support system that is capable of extraction and integration of the heterogeneous data using artificial intelligence and data-mining principles. The data analytics system proposed in this paper addresses the problem by analyzing a large volume of text and image data that are available from different sources. Useful insights

---

* Corresponding author.

H. Handa et al. (eds.), *Proc. of the 18th Asia Pacific Symp. on Intell. & Evol. Systems – Vol. 2,*
Proceedings in Adaptation, Learning and Optimization 2, DOI: 10.1007/978-3-319-13356-0_23

are obtained by identifying latent topics that are present inside the entire text corpus, using techniques that reduces the dimensionality of the data and identification of topics by probabilistic topic modelling. Together with image data which are processed and combined with the text data, insights can be achieved from the topics that are obtained as it helps to extract the salient features from the large amount of information into a summarized form.

The rest of this paper is organized as follows, Section II discusses the literature review about the analytics techniques that are being applied in this paper. Section III discusses the text analytics techniques that are used in this paper. Section IV discusses the image analytics techniques. Section V showcases the implementation of the data analytics system, applied to an example to highlight how the different analytics techniques can be put together to support a decision making process. Discussions are presented in Section VI, future work in Section VIII and conclusion in Section VIII.

2   Text Analytics

Text is the most prevalent type of data present in unstructured data, as well as dates, numbers and facts present. By representing textual data in a matrix form, powerful mathematical techniques can be applied to it in order to extract useful information and insights from the unstructured data.

2.1   Textual Data Representation

A standard model that is used for text representation is known as the "vector space model". In this model, the document space consist of documents which are identified by one or more weighted index terms [3] which Term Frequency-Inverse Document Frequency (TF-IDF) metric is commonly used. Similarity between two term vectors can be found by their inner product, which the normalized scalar product is often used [4]

$$sim(v_q, v_d) = \frac{v_q^T v_d}{\|v_q\|_2 \|v_d\|_2}, \tag{1}$$

where v_q is the query vector and $\|v\|_2$ is the L_2 norm of v.

The principle behind TF-IDF is that rarer words make better keywords, as they uniquely identify each document, and the more frequently those words appear in the document, they represent the document better [4]. This is calculated by

$$t_i = n_w^d \log_2 \frac{N}{N_w}, \tag{2}$$

where N is the number of total documents, N_w is the number of documents containing the word and n_w^d is the frequency of the word in the document.

2.2 Latent Semantic Analysis (LSA)

The limitation of basic corpus representation techniques is the difficulty in representing semantic information between different words and documents. LSA is a "statistical model of language learning and representation that uses vector averages in semantic space to assess the similarity between words or text" [5].

For this paper, reduced Singular Value Decomposition (SVD) will be used to perform document retrieval and word similarity calculation. By performing the reduction in dimension, each word or passage is now represented by a vector in a *semantic space* defined by hundreds of abstract dimensions [6].

The main value of LSA is its power in representing the complex relationships in language. LSA has been successful in modelling semantic similarity between words and it has been used in human language experiments. Gagne *et al.* have used LSA scores to measure the semantic similarity between words for their experiment on relational priming of words [7].

In LSA, the vector space matrix is decomposed into a product of three separate matrices, where $\mathbf{H}$ is the original vector space matrix

$$\mathbf{H} = \mathbf{USV^T},\tag{3}$$

where $\mathbf{U}$ and $\mathbf{V}$ are unitary matrices of dimensions $n \times n$ and $m \times m$, respectively, and $\mathbf{S}$ is a $n \times m$ diagonal matrix.

Since many of the singular values tend to be small, they can be removed and the dimensionality of the data can be reduced. By choosing the largest k singular values of matrix $\mathbf{S}$ and keeping the corresponding columns in $\mathbf{U}$ or $\mathbf{V}$, the resulting matrix

$$\mathbf{\hat{H}} = \mathbf{U^T H} = \mathbf{SV^T}\tag{4}$$

is a projection of the original term document matrix in the least-square sense.

By applying LSA on a text corpus, it can be used to identify semantic information between terms by calculating the cosine score of all the terms in the corpus with respect to a reference word. Cosine scores in the reduced space are higher than those in the full space due to the smaller dimensionality of the data after performing SVD.

2.3 Latent Semantic Indexing (LSI)

LSI is an information retrieval technique based on the spectral analysis of the term-document matrix [8], the vectors representing the documents are projected onto a low dimensional space using SVD of the term-document matrix. By performing a similar dimension reduction process as LSA, LSI represents the original text corpus in a low dimension representation, where it is small enough to enable fast retrieval but large enough to capture he structure of the corpus [8]. The retrieval system works by predicting what terms or documents are really implied by a query based on the latent semantics between them [9].

However, a limitation of LSI is that it will usually only work well in a collection of meaningful documents [8]. Large data sets are often required to allow LSI or

LSA to extract meaningful results from them. Another common critique of LSI is the high computation and memory requirement to compute the results [10], but modern processors have allowed LSI to become more feasible today.

In LSI, a query vector is used to query the term document matrix that has been reduced by SVD. Documents that are related are documents with a high similarity score to the query vector, where the query vector can be a vector that has not appeared in the text data before.

In a normal vector space retrieval process, documents are retrieved by mapping $\mathbf{q}$, the query vector in to the document space of the term document matrix $\mathbf{H}$

$$\mathbf{w} = \mathbf{q}^{\mathbf{T}}\mathbf{H}, \tag{5}$$

where $\mathbf{w}$ is the vector which contains the score of how relevant each document vector is to the query vector.

By applying SVD on the vector space representation of the text and keeping the k largest singular values, $\mathbf{H}$ is then approximated by

$$\mathbf{H} \approx \mathbf{H_k} = \mathbf{U_k}\mathbf{S_k}\mathbf{V_k^T}. \tag{6}$$

$\mathbf{H}_k$ is then used as the matrix for the vector space retrieval process [11].

2.4   Latent Dirichlet Allocation (LDA)

LDA is a common technique for topic modelling in text documents. Topic modelling provides methods to organize, understand and search large electronic archives [12]. LDA has been shown to pick out meaningful aspects of the structure of the documents and reveal some of the relationships between the different documents [13]. Since the introduction of LDA modelling, it has been applied in data mining, in text analysis and computer vision. LDA has been shown to explain word association concepts like word frequency and asymmetry, and has been used to further the understanding of how language is used [14].

LDA is a generative probabilistic model of a corpus, where it assumes that documents are represented as a random mixture over latent topics, where each topic is defined by a probabilistic distribution over a vocabulary of words [12]. Words are modelled as observed random variables and the topics are observed as latent random variables. For each document in a corpus, it has a K-dimensional topic weight vector θ_m from the distribution

$$p(\theta|\alpha) = Dirichlet(\alpha). \tag{7}$$

Each word in the document is generated by choosing a topic z_n from the multinomial distribution,

$$p(z_n = k|\theta_m) = \theta_m^k, \tag{8}$$

and choosing a word from the chosen topic z_n from the probability distribution

$$p(w_n = i|z_n = j, \beta) = \beta_{ij}, \tag{9}$$

where

$$\beta_{ij} = p(w^j = 1 | z^i = 1), \tag{10}$$

where β_{ij} is a $k \times V$ matrix to be estimated.

The parameter α is a K-dimensional parameter that remains constant over all of the documents within a corpus. This allows LDA to analyze unseen documents in relation to the rest of the existing documents in the corpus, overcoming the limitation of LSA. β parameterizes the word probabilities in the documents according to the topics.

Using this generative model, LDA seeks to infer the underlying topic structure by observing the documents themselves. Given the parameters α and β, the joint distribution of a topic mixture θ, a set of N topics **z**, and a set of N words **w** is given by

$$p(\theta, z, w | \alpha, \beta) = p(\theta | \alpha) \prod_{n=1}^{N} p(z_n | \theta) p(w_n | z_n, \beta). \tag{11}$$

The probability of a corpus is the product of the marginal probabilities of single documents

$$p(D | \alpha, \beta) = \prod_{d=1}^{M} \int p(\theta_d | \alpha) \left(\prod_{n=1}^{N_d} p(z_{dn} | \theta_d) p(w_{dn} | z_{dn}, \beta) \right) d\theta_d. \tag{12}$$

Gibbs Sampling algorithm is used to infer the probabilities of ϕ and θ, where $\phi_w^{(j)} = P(w | z = j)$. With the addition of the Dirichlet prior on ϕ

$$p(\phi | \beta) = Dirichlet(\beta), \tag{13}$$

this probability model can be used for the estimation of the ϕ and θ. A Markov chain Monte Carlo is used where the Markov chain is constructed to converge to the target distribution, then which samples are extracted from that Markov chain. Each state of the chain is an assignment of values to the variable z, which represents the topics present in the corpus.

Transitions between states uses Gibbs Sampling, whereby the next state is achieved by sequentially sampling all variables from their distribution when conditioned on the current values of all other variables and the data [13]. For this algorithm to work, the full conditional distribution

$$P(z_i = j | z_{-i}, w) \propto \frac{n_{-i,j}^{(w_i)} + \beta}{n_{-i,j}^{(\cdot)} + W\beta} \frac{n_{-i,j}^{(d_i)} + \alpha}{n_{-i,\cdot}^{(d_i)} + T\alpha} \tag{14}$$

is needed. $n_{-i}^{(\cdot)}$ is a count that does not include the current assignment of z_i.

With this conditional distribution, the Monte Carlo algorithm is implemented by initializing the z_i variables to values in $(1, 2, \cdots, T)$, where T is the number of topics that is defined beforehand. The chain is ran for a number of iterations and a new state is obtained each time by sampling each z_i from the distribution given in (14). Once enough iterations are performed such that the chain can approach the target distribution, the current values of the z_i variables are used to estimate ϕ and θ by

$$\phi_j^{(w)} = \frac{n_{-i,j}^{(w_i)} + \beta}{n_{-i,j}^{(\cdot)} + W\beta}, \tag{15}$$

$$\hat{\theta}_j^{(d)} = \frac{n_{-i,j}^{(d_i)} + \alpha}{n_{-i,\cdot}^{(d_i)} + T\alpha}. \tag{16}$$

3   Image Analytics

Image data plays an important part in the large amount of data available on the web. Image contains information about the object in the image and can be used for many analytics purposes. In this paper, the text that are present in the images are extracted in order to integrate with the textual information. The merged data will be analyzed together using the text analytics techniques described above.

3.1   Template Matching and Correlation Techniques

A simple OCR implementation is used, based on a template matching and correlation technique for performing image analytics. A set of prototype characters are used, in order to represent the different letters of the alphabet and the image of each segmented input character is then directly matched with the prototype character. The distance between the pattern and each prototype is compared, and it is used to find the best match to the input pattern [15].

3.2   OCR Post Processing

Once the raw text has been extracted from the image, post processing of the raw text will take place in order to clean up the text. In order to correct non-word errors, where words that are present in the raw text is not a real word, which may be due to an incorrect matching by the OCR of a character, like 'house' being read as 'hause'. The way to correct these errors is to compare the raw text with a text corpus that contains a large amount of accurate words. A similarity measure has to be used in order to select the word that is closest to the erroneous word. For this paper, the Levenshtein-Distance is used as this similarity measure and it is defined by

$$lev_{a,b}(i,j) = \begin{cases} \max(i,j) & \text{if } \min(i,j) = 0, \\ \min \begin{cases} lev_{a,b}(i-1,j)+1 \\ lev_{a,b}(i,j-1)+1 \\ lev_{a,b}(i-1,j-1)+1_{(a_i \neq b_j)} \end{cases} & \text{otherwise,} \end{cases} \tag{17}$$

where $1_{(a_i \neq b_j)}$ is an indicator function equal to 0 when $a_i = b_j$ and 1 otherwise.

For the algorithm, two metrics are used for the correction of non-word errors, first is the Levenshtein-Distance and second is the Term Frequency metric. For each word from the raw text, the algorithm search through the entire corpus and return the two scores. The Levenshtein-Distance metric takes priority over the term frequency metric so a corrected word that is closer to the raw word will be used even if there is another word that has a high term frequency but further from the raw word.

4   GUI Implementation

For this paper, the scenario used for illustration of this decision support system will be a decision making process for the purchase of a laptop computer. By using the data analytics system as seen in Fig. 1, this paper will show how it can help a consumer make a decision amongst the huge amount of information available on the web.

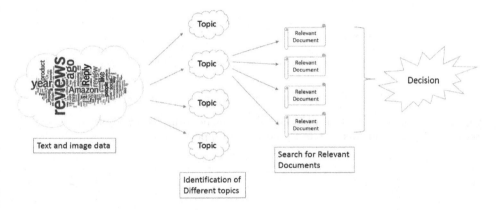

Fig. 1. How the data analytics system can support a decision making process. Topics from the mashed text and image data can be extracted and used to search for relevant documents using LSI. The relevant documents can then help the user make better decisions.

In this paper, the data analytics implementation is in the form of a Graphical User Interface (GUI) in MATLAB. The GUI allows the user to analyze a particular collection of documents and then derive useful information from the collection. The analytics system can also analyze text documents and retrieve similar words that are closest to a word input by the user.

From Fig. 2, it can be seen that the input file is the base file for the collection of documents that will be processed in the data analytics system. A stop list consisting of common English words can also be applied to the collection, removing common words such "the," "there," or "these". Removal of stop words will often lead to better results when calculating similarity scores between the words and documents.

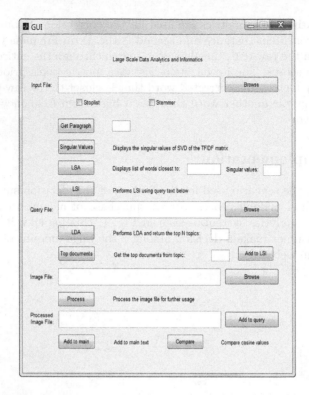

Fig. 2. GUI of data analytics system. User can perform the functions discussed in Sections II and III in this system as a decision support system

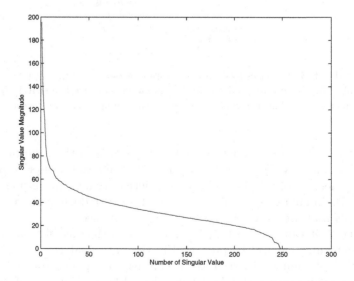

Fig. 3. Plot of singular values of the text corpus. This can be used to decide on the number of singular values to be used for the dimension reduction of LSA

Once the input file has been selected, text analytics techniques described in the previous chapters can be applied to the input text. The singular values button performs SVD on the input text and returns the plot of the singular values of the text corpus as seen in Fig. 3.

5   Industrial Application

A consumer in the market for a laptop computer today may face many choices from the wide variety of brands offering laptops to the huge number of models that each brand manufacturer may produce. Hence, the consumer may go onto the internet to obtain more information about the different kinds of laptops available on the market. For example, on Amazon.com, reviews are written by users who have bought the products and they help potential buyers to learn more about the product itself.

On Amazon.com, thousands of product reviews may be available for a single product. Coupled with the problem that there are a huge number of laptops, and even different specifications for one laptop, the consumer truly face a problem to make a decision amongst the wave of information.

For this scenario, Amazon.com reviews will be used as the text data for the data analytics system. More than 100 reviews of Apple MacBook Air are collected from the Amazon.com website. The reviews are split into good and bad reviews, according to the number of stars that are given by the customers who have bought the MacBook Air.

Amazon.com provides a 5-star rating system where consumers who have bought and used the product before can give their reviews and assign a score to the product. However, the star value may be less useful to a consumer because it does not give information about the features of the product. Rather, it is an average score that aggregates the good and bad features of the product.

Hence, LDA becomes very useful as it extracts the main features about the product and reviews can then be sorted according to the features. The user will then be able to find the reviews that discusses about a certain topic in order to understand more about the feature that is most important to that user. Using the reviews obtained from Amazon.com, it is analyzed using the data analytics system.

From the topics that were obtained by LDA, it can be seen there are some topics that describe the features of the MacBook Air. Some of the topics describes the battery life, the RAM and about the track pad of the MacBook Air. These are important features of the MacBook Air that a user may be interested in.

By using these topics, related reviews can be extracted from the system using LSA so that the user can obtain more information about the features. From these reviews, the user will be able to understand more about a feature of the MacBook (such as battery life) without having to look through all of the reviews. As LDA is an unsupervised topic modelling process, different topics can be generated from all the different reviews for different products in Amazon.

Table 1. TOPICS EXTRACTED USING LDA

Topic 1	Topic 2	Topic 3	Topic 4	Topic 5
laptop	air	windows	computer	ram
macbook	light	battery	hours	hard
need	computer	thing	battery	price
lot	user	days	don	feel
perfect	running	browsing	heavy	going
ipad	ssd	video	find	solid
portable	store	internet	laptops	external
looking	bought	nice	left	perform
want	doesn	storage	hour	pretty
office	faster	wifi	entire	4gb

Topic 6	Topic 7	Topic 8	Topic 9	Topic 10
drive	mac	apple	machine	quality
fast	macbook	love	makes	pro
buy	couple	screen	amazing	web
think	retina	keyboard	trackpad	desktop
design	happy	things	working	purchase
difference	fine	model	easy	say
system	gaming	run	full	128GB
programs	cool	amazon	word	built
haswell	extremely	consider	easily	compared
reason	previous	laptops	software	college

In addition to the discovery of topics for the user to support the decision making process, the user can use pictures containing information about the laptop and search for reviews that are relevant to it.

The pictures are processed using the image analytics system and then added to the main text. By integrating the image data with the text data, the cosine score distribution of the image to the rest of the data can be seen in Fig. 5.

With LSI, the related reviews can be found and extracted, so that the user can read these related reviews and obtain more relevant information about the product. The following reviews were obtained by analyzing Fig. 4.

Paragraph 218:
"Power: 8/10. The 2013 Macbook Air packs the latest Haswell processors from Intel which promises improved graphics and better power management (more details in battery life review). This is the perfect laptop I've wanted for everyday use. If you are looking for video editing intense gaming, I believe the MacBook Pro would be more suitable. For everyday use such as web browsing, Office computing and video chatting, the Air is great."

Paragraph 236:
"+ Haswell cpu/Intel HD5000 gpu
I am very impressed with the performance of this chipset. I haven't done much in the way of heavy gaming but it can handle Borderlands 2 without a hiccup. I

All-new Intel Core processors and graphics

 Fourth-generation Intel Core processors and Intel HD Graphics 5000 make even the most demanding computing tasks easier. Whether you're editing photos, perfecting a presentation, or just browsing the web, everything moves faster.

Fig. 4. An image obtained from Apple online store that describes a feature of the MacBook Air. These images will be analyzed using the data analytics systems and mashed into the Amazon reviews. Documents related to this image can be extracted for the user to learn more about this feature.

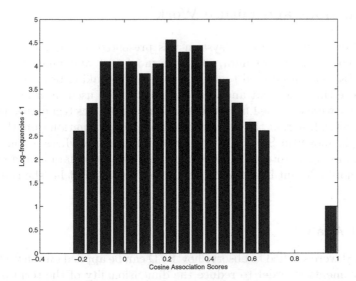

Fig. 5. Distribution of cosine scores of image shown in Fig.4. IT can be seen that there are many documents that are not relevant to this image (scores below 0.5). Hence, an extraction of relevant articles will help the user to find the most related reviews.

have yet to tap the full potential of the i7, the only part of this entire system that causes any sort of lag at all (only when im gaming or filesharing) is the RAM (I would suggest you go with 8gb)."

Paragraph 49:
"Haswell architecture has enabled an unheard of real-world 10-12 hour wireless browsing battery life while packing a fully featured quad core i5 chip with HD5000 graphics. Unprecedented for this size and weight. HD5000 tackles Skyrim at comfortable mid-level graphics and playable frames, at the expense of extreme system heat."

From these results, it can be seen that very relevant reviews are obtained from the large amount of text. Even though the image itself does not contain the word 'Haswell', the results obtained are all able to link to Haswell as the processor being used for the MacBook Air. With these information, the user will be able to understand more about the processor, and hence be able to make a better decision whether to purchase the laptop or not.

By using the data analytics system, the user is better able to find more relevant information amongst the large amount of data available from the internet. By filtering the information the user has to browse through, the system helps to improve the decision making process, supporting the user with relevant information that truly matters.

6   Discussions and Future Work

In this paper, a data analytics system was presented in order to process and analyze a large amount of data, in the form of text and image, so that the insights obtained can be used to support a decision making process.

To achieve this goal, text and image analytics were used in order to analyze the data that were obtained from the web. These analytics techniques were integrated into a GUI, so that the user can easily obtain information from the system. The example shown in Section V highlighted a scenario where a consumer interested in buying a laptop computer can make use of the data analytics system to obtain more relevant information about the features that he/she is interested in.

6.1   Text Analytics

LSA and LSI were studied to observe how SVD can be applied on the vector space of text documents in order to reduce the dimensionality of the term document matrix which resulted in a lower dimension semantic space. The semantic space was useful in order to obtain relationship between the terms and the documents which is better represented in this semantic space.

LDA was studied as a topic modelling technique which generated topics from the text corpus using the probabilistic distribution of the words in the corpus. By applying a Gibbs Sampling Markov Chain Monte Carlo algorithm to the observed distribution of words, the hidden topics from the text corpus can be inferred. These topics allowed an extraction of salient features of the text corpus which produced a good summarized form of the corpus.

6.2   Image Analytics

In this paper, a simple OCR implementation was used together with a post processing algorithm that took the corpus information into account in order to apply a correction of non-word errors of the text obtained from OCR. A metric consisting of the Levenshtein-Distance and the term frequency values was used to identify the words that are most related to the erroneous word. The result was an extraction of useful and related words from the image so that it can be mashed together with the text data.

6.3   Future Work

In this paper, the data analytics system was able to improve the decision making process by supplying the user with more relevant information. However, that was still a very manual process where the user will have to look through the filtered information before making the decision.

Future work may include the implementation of sentiment analysis techniques and ontology in order to provide an automatic recommendation system. By combining semantic information that is available from this paper together with sentiment analysis techniques, the reviews can be automatically classified as good or bad categories. By mapping these reviews onto an ontology, recommendation scores can be calculated that takes into account the comments about a particular topic and hence generate a positive or negative score about a feature of many similar products. The scores can then be used to generate recommendation scores of different laptops based on the important features that the user has indicated.

Another scope for future work is the implementation of LDA for image analysis. With a topic model for images, large amount of images can be analyzed with a topic assigned to the images. Combined with the topics from the text analytics, users will be able to search for relevant images in addition to searching for relevant documents. As the data on the web are often very heterogeneous, like photo reviews for products, or product reviews video, this will allow a search for a greater variety of information that may further improve the decision making process.

7   Conclusion

In this paper, a Decision Support System (DSS) is designed for intelligent decision making using the large amount of heterogeneous data available online. By combining text and image analytics techniques, the DSS derives important topics from the large amount of text and image data, and search for relevant documents present in the data. This allows the user to filter out irrelevant and unneeded information from the huge amount of heterogeneous data available, saving previous time and making the decision making process faster. The DSS was implemented as a Graphical User Interface (GUI) to aid the laptop procurement process considering the huge volume of online reviews.

Acknowledgement. This work was supported in part by Singapore MOE AcRF Tier 1 Grants R-263-000-A44-112 and R-263-000-A52-112.

References

1. Bhatt, C., Kankanhalli, M.: Probabilistic Temporal Multimedia Data Mining. ACM Trans. Intell. Syst. Technol. 2, 17:1–17:19 (2011)
2. Kawamoto, Y., Nishiyama, H., Fadlullah, Z.M., Kato, N.: Effective Data Collection via Satellite-Routed Sensor System (SRSS) to Realize Global-Scaled Internet of Things. IEEE Sensors Journal 13, 3645–3654 (2013)
3. Salton, G., Wong, A., Yang, C.S.: A Vector Space Model for Automatic Indexing 18, 613–620 (1975)
4. Sivic, J., Zisserman, A.: Efficient Visual Search of Videos Cast as Text Retrieval. IEEE Transactions on Pattern Analysis and Machine Intelligence 31, 591–606 (2009)
5. Simmons, S., Estes, Z.: Using Latent Semantic Analysis to Estimate Similarity. In: Proceedings of the Cognitive Science Society, pp. 2169–2173 (2006)
6. Landauer, T.K.: Latent Semantic Analysis. Encyclopedia of Cognitive Science (2006)
7. Gagné, C.L., Spalding, C.L., Ji, H.: Re-examining Evidence for the Use of Independent Relational Representations during Conceptual Combination. Journal of Memory and Language 53, 445–455 (2005)
8. Papadimitriou, C.H., Raghavan, P., Tamaki, H., Vempala, S.: Latent Semantic Indexing: A Probabilistic Analysis 61, 217–235 (2000)
9. Deerwester, S., Dumais, S.T., Furnas, G.W., Landauer, T.K., Harshman, R.: Indexing by Latent Semantic Analysis. Journal of the American Society for Information Science 41, 391–407 (1990)
10. Zukas, A., Price, R.J.: Document Categorization Using Latent Semantic Indexing. In: Proceedings, Symposium on Document Image Understanding Technology, pp. 87–91 (2003)
11. Garron, A., Kontostathis, A.: Applying Latent Semantic Indexing on the TREC 2010 Legal Dataset. In: TREC (2010)
12. Blei, D.M., Ng, A.Y., Jordan, M.I.: Latent Dirichlet Allocation. The Journal of Machine Learning Research 3, 993–1022 (2003)
13. Griffiths, T.L., Steyvers, M.: Finding Scientific Topics. Proceedings of the National Academy of Sciences of the United States of America 101, 5228–5235 (2004)
14. Griffiths, T.L., Tenenbaum, J.B., Steyvers, M.: Topics in Semantic Representation. Psychological Review 114, 211–244 (2007)
15. Eikvil, L.: Optical Character Recognition, http://www.nr.no/~eikvil/OCR.pdf

Numerical Simulations for Hybrid Electromagnetism-Like Mechanism Optimization Algorithms with Descent Methods

Hirofumi Miyajima[1], Noritaka Shigei[2],
Hiroki Taketatu[2], and Hiromi Miyajima[2]

[1] Graduate School of Science and Engineering, Kagoshima University, 1-21-40,
Korimoto, Kagoshima, Japan
[2] Kagoshima University
1-21-40 Korimoto, Kagoshima, Japan
k3768085@kadai.jp,
{shigei,miya}@eee.kagoshima-u.ac.jp

Abstract. Electromagnetism-like Mechanism (EM) method is known as one of metaheuristics. The basic idea is that the position of a charged particle represents a solution for the optimization problem, and the charge amount of the particle corresponds to the evaluation value of the solution. Starting with a population of particles whose positions are randomly initialized, the population converges to a neighborhood of the optimal or semi-optimal solution. Like other metaheuristics, one of its drawbacks is that it takes too much time to converge to a solution. In this paper, we will perform numerical simulations to investigate how to use and combine EM and steepest descent methods. Further, we will show that hybrid EM methods are superior in accuracy to conventional methods.

Keywords: Electromagnetism-like Mechanism method, descent method, k-means, Backpropagation.

1   Introduction

Metaheuristics are generally applied to problems for which there is no effective problem-oriented algorithm to solve them. They are widely used to solve complex problems in science and engineering[1]. Well-known metaheuristics are random search (RS) by Matyas (1965), the simulated annealing (SA) by Kirkpatrick (1982), genetic algorithm (GA) by Goldberg (1989), bee colonies (BC) by Walker (1993), particle swarm optimization (PSO) by Kennedy (1995), electromagnetism-like mechanism (EM) (2005) and so on [1,2,3,5]. Any method of them is not always universal, so problem-oriented algorithm must be selected for each problem. The difficult points are to search vast space to find the optimal or semi-optimal solution and complex distribution of fitness of the objective function[1]. The criterion to select any method of metaheuristics is how effective the search is performed by using global and local searches. It is well known that local search takes too much time[1,4]. In order to construct the effective algorithm, a fast local search with high

© Springer International Publishing Switzerland 2015
H. Handa et al. (eds.), *Proc. of the 18th Asia Pacific Symp. on Intell. & Evol. Systems – Vol. 2*,
Proceedings in Adaptation, Learning and Optimization 2, DOI: 10.1007/978-3-319-13356-0_24

accuracy is required[6,16,18,19]. Therefore, though some hybrid methods combining metaheuristics and back propagation (BP) method are introduced, the satisfactory result is not always obtained [15,17]. RS, PSO and EM methods are well known ones and their convergence to a solution for algorithms is guaranteed[2,5,7]. EM method is more complicate than RS, but is simpler than PSO. Further, it is known that PSO and EM methods have the same ability in accuracy and they are superior in accuracy to RS method[6,19]. However, there is few result on how to combine EM and BP and how to use hybrid EM with k-means method.

In this paper, we will perform numerical simulations to investigate how to use and combine EM and steepest descent methods. Electromagnetism-like Mechanism is known as one of random search algorithms[5,6,10,11]. The basic idea is that the position of a charged particle represents a solution for the optimization problem, and the charge amount of the particle corresponds to the evaluation value of the solution. Like other metaheuristics, one of its drawbacks is that it takes too much time to converge to a solution. In order to improve them, hybrid EM methods combining EM method with BP and k-means methods as local search are presented and are compared with the conventional methods. Though we assume to use BP and k-means methods, we can also consider hybrid EM method based on other steepest descent methods. In section 2, the conventional method of metaheuristics as EM and RS, BP and k-means are introduced. In section 3, some hybrid EM methods are proposed. In section 4, some numerical simulation results are shown and the effectiveness of the proposed methods is demonstrated.

2   Preliminaries

Let $\mathbb{R}$ be the set of all real numbers. Let $\mathbb{Z}_i = \{1, \cdots, i\}$ for positive integer i. Let [0,1] be the set of all real numbers between 0 and 1. The problem is finding an optimal or semi-optimal solution of a non-linear optimization problem with boundary variables in the form;

$$\min_{L \leq x \leq U} f(x) \tag{1}$$

where $f(x)$ is the evaluation function to be minimized, $x \in \mathbb{R}^n$ is the variable vector (parameter), and $L = (l_1, \cdots, l_n)$ and $U = (u_1, \cdots, u_n)$ are the lower and upper bounds of x, respectively.

For example, let us consider a neural network. Given learning data, let us determine the weights of neural network identifying the learning data. In the case, the mean square error between learning data and output of the network corresponds to evaluation function and the weights of the network corresponds to parameters.

In general, metaheuristics provide how to find the optimum or semi-optimum solution effectively. They consist of global and local searches. In the following, some methods used for local and global searches are introduced.

2.1   EM Method

EM algorithm simulates the interaction, attraction and repulsion, caused by electromagnetic force between electrically charged particles. The general scheme of EM method is shown as Fig.1[5]. It consists of four phases. In step 1, a set of particles is initialized. In step 4, local search to find local optimum is performed. In step 5, the force worked on each particle is calculated. In step 6, each particle moves in the direction of the compounded force.

Algorithm EM(m, MAX, LS, δ)
m: number of sample points
MAX: maximum number of iterations
LS: maximum number of iterations for local search
δ: local search parameter, $\delta \in [0, 1]$
1: Initialize()
2: $iteration \leftarrow 1$
3: while $iteration < MAX$ do
4:        Local(LS, δ)
5:        $F \leftarrow$ Calc F()
6:        Move(F)
7:        $iteration \leftarrow iteration + 1$
8: end while

Fig. 1. Outline of EM method

In Initialization step, the sample points, MAX, LS, and δ are given. Fig.2 shows the algorithm of local search of step 4 in Fig.1[5].

In the algorithm Local(LS, δ), a neighbor point $\boldsymbol{y}$ for the parameter $\boldsymbol{x}$ is given and $f(\boldsymbol{x})$ is compared with $f(\boldsymbol{y})$. In the case of $f(\boldsymbol{y}) < f(\boldsymbol{x})$, $\boldsymbol{x}$ is set to $\boldsymbol{y}$. In other case, the same process for other neighbor $\boldsymbol{y}'$ is repeated. After local searches for all parameters are performed, $\boldsymbol{x}^{best}$ is updated.

Figs.3 and 4 show the algorithm of global search that the new position of each particle in the electromagnetic theory is computed[5].

2.2   Three-Layered Neural Network and BP Method

Let us consider the case of n inputs and one output without loss of generality. Let $\mathbb{J} = [0, 1]$. Let $\boldsymbol{z}^i \in \mathbb{J}^n$ for $i = 1, \cdots, l$ and $d : \mathbb{J}^n \to \mathbb{J}$. Giving learning data $\{(\boldsymbol{z}^i, d(\boldsymbol{z}^i)) | i \in \mathbb{Z}_l\}$, let us determine the three-layered neural network identifying the learning data by BP method[12]. Let $h = g \circ e : \mathbb{J}^n \to \mathbb{J}$ be the function defined by a neural network. Let the number of elements in second layer be p. Let $\boldsymbol{w}$ and $\boldsymbol{v}$ be weights for the second and output layers, respectively. Then g and e are defined as follows(See Fig.5):

$$y_j = e_j(\boldsymbol{z}) = \tau \left(\sum_{i=0}^{n} w_{ij} z_i \right),$$

Algorithm Local(LS, δ)
u_k : the k-th element of upper bound
l_k : the k-th element of lower bound
$U(0,1)$: standard uniform distribution
1: $counter \leftarrow 1$
2: $Length \leftarrow \delta\ (\max_k\{u_k - l_k\})$
3: for $i = 1$ to m do
4:    for $k = 1$ to n do
5:        $\lambda_1 \leftarrow U\,(0,\,1)$
6:        while $counter < LS$ do
7:            $\boldsymbol{y} \leftarrow \boldsymbol{x}^i$
8:            $\lambda_2 \leftarrow U\,(0,\,1)$
9:            if $\lambda_1 > 0.5$ then
10:               $y_k \leftarrow y_k + \lambda_2(Length)$
11:           else
12:               $y_k \leftarrow y_k - \lambda_2(Length)$
13:           end if
14:           if $f(\boldsymbol{y}) < f(\boldsymbol{x}^i)$ then
15:               $\boldsymbol{x}^i \leftarrow \boldsymbol{y}$
16:               $counter \leftarrow LS$ - 1
17:           end if
18:           $counter \leftarrow counter + 1$
19:       end while
20:    end for
21: end for
22: $\boldsymbol{x}^{best} \leftarrow \arg\min_{\boldsymbol{x}^i}\{f(\boldsymbol{x}^i)\}$

Fig. 2. Local search of EM algorithm

Algorithm Calc F()
$||\boldsymbol{a} - \boldsymbol{b}||$: *the distance between* $\boldsymbol{a}$ *and* $\boldsymbol{b}$
1: for $i=1$ to m
2:    $q_i \leftarrow \exp\left(-n\frac{f(\boldsymbol{x}^i - f(\boldsymbol{x}^{best}))}{\sum_{k=1}^{n}(f(\boldsymbol{x}^k) - f(\boldsymbol{x}^{best}))}\right)$
3:    $F^i \leftarrow 0$
4: end for
5: for $i=1$ to m
6:    for $j=0$ to m
7:        if $f(\boldsymbol{x}^i) < f(\boldsymbol{x}^j)$ then
8:            $F^i \leftarrow F^i + (\boldsymbol{x}^i - \boldsymbol{x}^j)\frac{q^i q^j}{||\boldsymbol{x}^i - \boldsymbol{x}^j||^2}$
9:        else
10:           $F^i \leftarrow F^i$ - $(\boldsymbol{x}^i - \boldsymbol{x}^j)\frac{q^i q^j}{||\boldsymbol{x}^i - \boldsymbol{x}^j||^2}$
11:       end if
12:   end for
13: end for

Fig. 3. The force of each particle in EM algorithm

Algorithm Move (F)
1: for $i=1$ to m
2:   if $i \neq best$ then
3:     $\alpha \leftarrow U(0,1)$
4:     $F^i \leftarrow \frac{F^i}{||F^i||}$
5:     for $k=1$ to N
6:       if $F_k^i > 0$ then
7:         $x_i = x_i + \alpha F_k^i(u_k - x_k^i)$
8:       else
9:         $x_i = x_i + \alpha F_k^i(x_k^i - l_k)$
10:      end if
11:    end for
12: end if
13: end for

Fig. 4. The movement of each particle in EM algorithm

$$z_0 = 1,$$
$$\tau(u) = \frac{1}{1 + \exp(-u)}$$

where

$$\boldsymbol{z} = (z_1, \cdots, z_n) \in \mathbb{J}^n$$
$$\boldsymbol{y} = (y_1, \cdots, y_p) \in \mathbb{J}^p$$

and w_{0j} means the threshold value.
Further,

$$g(\boldsymbol{y}) = \tau \left(\sum_{j=0}^{p} v_j y_j \right),$$

$$y_0 = 1,$$

where v_0 means the threshold value.

Then, the evaluation function is defined as follow:

$$E = \frac{1}{2l} \sum_{i=1}^{l} ||h(\boldsymbol{z}^i) - d(\boldsymbol{z}^i)||^2 \qquad (2)$$

The weights $\boldsymbol{w}$ and $\boldsymbol{v}$ are updated based on BP method as follow[12]:

$$\triangle v_i = -\frac{\alpha_1}{l} \sum_{k=1}^{l} \delta_{2k}(\boldsymbol{z}^k) e_i(\boldsymbol{z}^k) \qquad (3)$$

$$\triangle w_{ij} = -\frac{\alpha_2}{l} \sum_{k=1}^{l} \delta_{1j}(\boldsymbol{z}^k) z_i^k \qquad (4)$$

$$(i = 0, \cdots, p, j = 1, \cdots, n)$$

298   H. Miyajima et al.

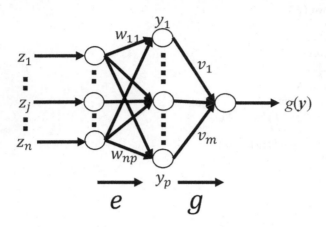

Fig. 5. Three-layered neural network

where α_1 and α_2 are learning coefficients,

$$\delta_{2j}(\boldsymbol{z}) = (h(\boldsymbol{z}) - d(\boldsymbol{z}))h(\boldsymbol{z})(1 - h(\boldsymbol{z}))e_j(\boldsymbol{z}) \tag{5}$$

and

$$\delta_{1j}(\boldsymbol{z}) = \delta_{2j}v_j e_j(\boldsymbol{z})(1 - e_j(\boldsymbol{z})). \tag{6}$$

2.3   Clustering by k-means Method

Vector quantization techniques encode a data space, e.g., a subspace $\boldsymbol{U} \subseteq \mathbb{R}^n$, utilizing only a finite set $\boldsymbol{W} = \{\boldsymbol{w}_i | i \in \mathbb{Z}_r\}$ of reference vectors (also called cluster centers), which n and r are positive integers. Let us introduce k-means method as one of vector quantization techniques[13].

Let the winner vector $\boldsymbol{w}_{i(\boldsymbol{v})}$ be defined for any vector $\boldsymbol{v} \in V$ as follows:

$$i(\boldsymbol{v}) = \arg\min_{i \in \mathbb{Z}_r} ||\boldsymbol{v} - \boldsymbol{w}_i|| \tag{7}$$

From the finite set $\boldsymbol{W}, \boldsymbol{V}$ is partioned as follows:

$$\boldsymbol{V}_i = \{\boldsymbol{v} \in V | ||\boldsymbol{v} - \boldsymbol{w}_i|| \leq ||\boldsymbol{v} - \boldsymbol{w}_j|| \text{ for } j \in \mathbb{Z}_r\} \tag{8}$$

The evaluation function for the partition is defined as follows:

$$E = \sum_{i=1}^{r} \sum_{\boldsymbol{v} \in \boldsymbol{V}_i} ||\boldsymbol{v} - \boldsymbol{w}_{i(\boldsymbol{v})}||^2 \tag{9}$$

Each parameter $\boldsymbol{w}$ is updated based on the steepest descent method as follow[13]:

$$\triangle\boldsymbol{w}_i = \varepsilon\delta_{ij(\boldsymbol{v}(t))}(\boldsymbol{v}(t) - \boldsymbol{w}_i) \tag{10}$$

where t is the step, ε is learning constant and δ_{ij} is Kronecker Delta. $\boldsymbol{v}(t)$ means data selected from $\boldsymbol{V}$ randomly at step t. The k-means method is denoted by KM.

3   Hybrid EM Algorithms

EM method[1] is an effective method of metaheuristics to solve the optimization problem. On the other hand, it takes too much time to converge to an optimum or semi-optimum solution . Therefore, we propose hybrid algorithms combining EM method with the steepest decent methods such as k-means and BP methods.

Why is the hybrid algorithm needed? Let us explain the reason using EM and BP methods. As shown in Fig.6, the evaluation value $E(\boldsymbol{w}, \boldsymbol{v})$ is determined based on the parameters $\boldsymbol{w}$ and $\boldsymbol{v}$. The initial vectors are selected randomly and the vectors are updated by local search. Since BP method is based on the steepest descent method, $E(\boldsymbol{w}, \boldsymbol{v})$ always decreases. On the other hand, local search in the conventional EM and other methods needs to find a new vector which decreases $E(\boldsymbol{w}, \boldsymbol{v})$ and it takes too much time. Further, if $E(\boldsymbol{w}, \boldsymbol{v})$ does not decrease so much, then the vector moves to a new position by using the force F. Then, the new evaluation value $E(\boldsymbol{w}, \boldsymbol{v})$ is lower than or equal to the old evaluation value. Continuing the processes, the vector is updated so as to improve the function $E(\boldsymbol{w}, \boldsymbol{v})$. In the case of BP algorithm, it is possible only to search the vectors (parameters) locally and is difficult to search them globally. Hence, there exists global search such as moving by F in the proposed method.

In this section, hybrid EM, hybrid random search and hybrid PSO methods are introduced[8,9,19]. The basic algorithms about EM, random search and PSO are used[1,2,3],

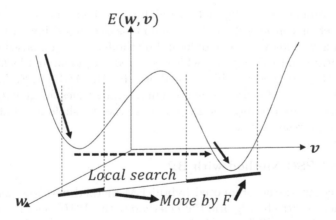

Fig. 6. The figure to explain hybrid EM method

3.1   Hybrid EM Method with BP

A hybrid EM method with BP (HEM-BP) employs BP method[12] which is one of the steepest descent ones. In HEM-BP, the parameter x of EM method represents the weights w and v of three layered neural network. For the algorithm of HEM-BP, step 4 in Fig.1, Local(LS, δ), is replaced with BP procedure BP(LS, ε) whose flowchart is shown in fig.7, where LS and ε are the number of learning and the threshold for local search, respectively. The BP procedure receives the current values of the parameters in $X = \{x^i|i\in\mathbb{Z}_m\}$ and the current best of the parameters x^{best} from the calling side, updates them based on the steepest descent method with the learning data $\{(z_i, d(z_i))|i\in\mathbb{Z}_l\}$ and returns the updated values to the calling side. HEM-BP involves additional parameters MAX, LS and ε, and it will be denoted as HEM-BP(MAX, LS, ε).

3.2   Hybrid EM Method for Clustering

A hybrid EM method with k-means (HEM-KM) employs k-means method for clustering of data. In HEM-KM, the parameter x of EM method represents the reference vectors $W = \{w_i|i\in\mathbb{Z}_r\}$, where r is the number of reference vectors. For the algorithm of HEM-KM, step 4 in Fig.2, Local(LS, δ), is replaced with a k-means procedure KM(LS, ε), where LS and ε are the number of learning and the the threshold for local search, respectively. The k-means procedure receives the current values of the parameters in $X = \{x^i|i\in\mathbb{Z}_m\}$ and the current best of the parameters x^{best} from the calling side, updates them based on the steepest descent method with the learning data $\{(v(t)\in V\subseteq\mathbb{R}^n|t = 1, 2, \cdots, LS\}$ and returns the updated values to the calling side. HEM-KM involves additional parameters MAX, LS and ε, and it will be denoted as HEM-KM(MAX, LS, ε).

3.3   Hybrid Random Search Algorithm with BP

In order to compare the proposed hybrid EM method with the conventional hybrid one, let us consider the hybrid random search algorithm with BP[4],[8]. The method is well known as one of hybrid methods and is composed of random search algorithm and BP method, which are used for global and local searches, respectively. The method will be denoted as HRS-BP(MAX, LS, RS, ε_0, ε_1), where MAX, LS and RS are the maximum learning times for global, local and random searches, respectively, and ε_0 and ε_1 are the threshold for random search and BP method, respectively.

3.4   Hybrid PSO Method with BP

In order to compare the prooposed hybrid EM method with the conventional hybrid one, let us introduce hybrid PSO method with BP(HPSO-BP) as the same method as section 3.1. PSO is well known as the popular one of metaheuristic methods. Figs. 8 and 9 show the basic PSO algorithm and the outline of HPSO-BP with the parameters MAX, LS, T_{max} and δ[9].

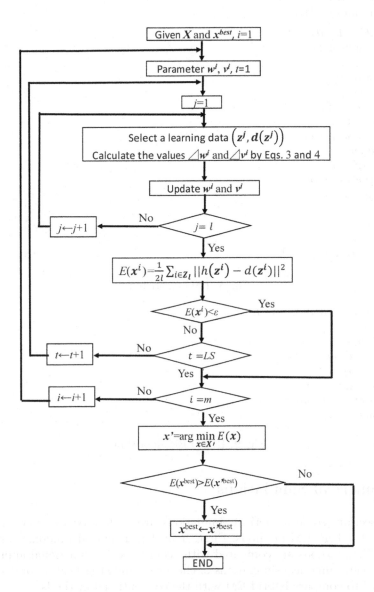

Fig. 7. The flowchart of BP(LS, ε)

Algorithm PSO(T_{max})
Initialize():
$\boldsymbol{x}^i \leftarrow$ U($\boldsymbol{L},\boldsymbol{U}$) for $i \in Z_m$
$\boldsymbol{p} \leftarrow \boldsymbol{x}^i$ for $i \in Z_m$
$\boldsymbol{x}^{best} \leftarrow \arg \min\limits_i \{f(\boldsymbol{x}^i)\}$
$\boldsymbol{v}^i \leftarrow$ U($-|\boldsymbol{U} - \boldsymbol{L}|, |\boldsymbol{U} - \boldsymbol{L}|$)
w, c_1, c_2;parameters
1: *counter* $\leftarrow$ 1
2: while *counter* $< T_{max}$
3:    for i=1 to m
4:        $r_1, r_2 \leftarrow$ U(0, 1)
5:        for k=1 to n
6:            $v_k^i \leftarrow wv_k^i + c_1 r_1 (p_k^i - x_k^i) + c_2 r_2 (x_k^{best} - x_k^i)$
7:        end for
8:        $\boldsymbol{x}^i \leftarrow \boldsymbol{x} + \boldsymbol{v}^i$
9:        if $f(\boldsymbol{x}^i) < f(\boldsymbol{p}^i)$
10:           $\boldsymbol{p}^i \leftarrow \boldsymbol{x}^i$
11:       end if
12:   end for
13: end while
14: $\boldsymbol{x}^{best} \leftarrow \arg \min\limits_i \{f(\boldsymbol{p}^i)\}$

Fig. 8. PSO algorithm

Algorithm HPSO-BP(MAX, LS, T_{max}, δ)
1: Initialize
2: *Iteration* $\leftarrow$ 1
3: while *iteration* $< MAX$
4:    BP(LS, δ)
5:    PSO(T_{max})
6:    *iteration* $\leftarrow$ *iteration* $+ 1$
7: end while

Fig. 9. HPSO-BP algorithm

4   Numerical Simulations

In this section, some numerical simulations are performed to show the effectiveness of hybrid EM methods. First, hybrid EM, hybrid random search and hybrid PSO methods are compared with accuracy using functional approximation. Further, numerical simulations of clustering and classification problems are performed to compare hybrid EM with the conventional methods.

Table 1. The conditions for function approximation

	EM	HEM	RS	HRS	PSO	HPSO
♯ parameters	73	73	73	73	73	73
♯ neurons for input layer	4	4	4	4	4	4
♯ neurons for second layer	12	12	12	12	12	12
m	20	20			20	20
MAX	5000	5000	100000	5000	-	5000
LS	20	20	-	20	-	20
T_{max}	-	-	-	-	100000	1
w	-	-	-	-	1.0	1.0
c_1	-	-	-	-	0.5	0.1
c_2	-	-	-	-	0.5	0.1

Table 2. Result of function approximation

	Eq.(11)		Eq.(12)		Eq.(13)	
	Training	Test	Training	Test	Training	Test
EM	8.29	9.95	4.39	5.77	31.66	40.72
HEM-BP	0.32	0.90	0.44	0.86	13.64	28.26
RS	4.13	5.73	14.05	16.25	56.79	64.36
HRS-BP	0.82	1.64	0.74	1.21	12.95	29.78
PSO	1.72	2.82	2.24	2.78	25.90	38.09
HPSO-BP	0.19	0.69	0.31	0.79	6.23	22.54

4.1   Function approximation by HEM-BP

This simulation uses four systems specified by the following fonctions with $[0, 1] \times [0, 1] \times [0, 1] \times [0, 1]$.

$$y = \frac{(2x_1 + 4x_2^2 + 0.1)^2}{37.21} \cdot \frac{(4\sin(\pi x_3) + 2\cos(\pi x_4) + 6.0)}{12.0} \tag{11}$$

$$y = \frac{(\sin(\pi x_1) + \cos(\pi x_2) + \sin(\pi x_3) + x_4 + 3.0)}{7.0} \tag{12}$$

$$y = \frac{(\sin(2\pi x_1) \cdot \cos(\pi x_2) \cdot \sin(\pi x_3) \cdot x_4 + 1.0)}{2.0} \tag{13}$$

The conditions of the simulation are shown in Table 1. The numbers of learning and test data selected randomly are 225 and 5000, respectively. Table 2 shows the results on comparison among EM, HEM-BP, RS, HRS, PSO and HPSO. In Table 2, MSE of training($\times 10^{-4}$) and MSE of test($\times 10^{-4}$) are shown. The result of simulation is the average value from twenty trials.

The result shows that hybrid EM and hybrid PSO methods are superior in accuracy to other methods and hybrid PSO needs more parameters than hybrid EM to determine before learning.

Table 3. Data for numerical simulation

	The number of input	The number of clusters	The number of data
iris	4	3	150
wine	13	3	178
sonar	60	2	208
BCW	9	2	683

Table 4. Result for classification by the conventional and proposed methods

	iris	wine	sonar	BCW
(1)	4.0	13.7	18.5	4.3
	[0.0,10.0]	[11.4,20.0]	[14.6,24.4]	[2.2,8.1]
(2)	4.0	16.6	19.5	4.3
	[0.0,6.7]	[8.6,25.7]	[12.2,29.3]	[2.2,6.6]
(3)	4.7	9.7	19.5	4.3
	[0.0,13.3]	[5.7,14.3]	[17.1,24.4]	[2.9,5.1]
(4)	3.3	4.0	15.1	4.0
	[0.0,6.7]	[2.2,5.1]	[9.1,24.4]	[2.2,5.1]

4.2 Classification by HEM-BP

Iris, Wine, Sonar and BCW data from USI database shown in Table 3 are used for numerical simulation[14].

In this simulation, 5-fold cross validation is used.

Then, the following algorithms are compared with each other:

(1) BP (50000, 10^{-5})

(2) EM(500, 5)

(3) HRS-BP(500, 1000, 10000, 10^{-5}, 10^{-5})

(4) HEM-BP(500, 1000, 10^{-5})

Table 4 shows the result of classification for the above four algorithms (1), (2), (3) and (4), where the upper and lower values in each box show the rate of misclassification (%) and [minimum value, maximum value] of misclassification for trials.

Table 4 shows that HEM-BP is most effective and the variance of result for HEM-BP is small. It means that dependency of initial value for parameter is low.

4.3 Clustering by HEM-KM

Giving the weights with the number of clusters shown in Table 3, clustering for each data in Table 3 is performed. Then, the following algorithms are compared with each other.

(1) KM(50000, 10^{-4})

(2) EM(500, 1000)

(3) HRS(500, 1000, 10000, 10^{-4}, 10^{-4})

(4) HEM-KM(500, 1000, 10^{-4})

Table 5 shows the result of classification for the above four algorithms (1), (2), (3) and (4), where each value in each box is the same meaning as one in 4.2. Each value is the average value from twenty trials.

Table 5. Result for clustering by the conventional and proposed methods

	iris	wine	sonar	BCW
(1)	11.3	19.7	45.6	3.9
	[4.0,33.3]	[5.6,33.9]	[43.8,46.2]	[3.8,4.0]
(2)	4.5	39.7	46.6	4.4
	[4.0,6.7]	[36.5,40.0]	[46.6,46.6]	[4.0,5.0]
(3)	4.0	23.1	45.7	3.9
	[4.0,4.0]	[5.6,41.6]	[45.7,45.7]	[3.8,4.0]
(4)	4.0	6.8	45.0	4.0
	[4.0,4.0]	[6.2,7.3]	[44.7,45.7]	[3.9,4.0]

As shown in Table 3, the proposed method is superior in accuracy to the other methods. Further, the variance of result is smallest compared with the other methods.

5   Conclusion

In this paper, we performed numerical simulations to investigate how to use and combine EM and steepest descent methods. It is shown that they are superior in accuracy to the conventional EM, BP and k-means methods in numerical simulations. Further, they showed better performance than the hybrid random search methods with BP and k-means methods. As a result, it is shown that many and few learning times of global and local searches for hybrid EM are needed for clustering and classification problems, respectively and reverse is true for functional approximation. Though we assumed to use k-means and BP methods as local search techniques, the other methods based on the steepest descent methods can be also applied. As the future works, we will consider to propose hybrid EM methods with the other steepest descent methods and to prove the convergence property of the proposed hybrid EM methods.

References

1. Boussaid, I., Lepagnot, J., Siarry, P.: A Survey on Optimization Metaheuristics. Information & Sciences 237, 82–117 (2013)
2. Matyas, J.: Random Optimization. Automation & Remote Contr. 26, 246–253 (1965)

3. Kennedy, J., Eberhart, R.: Particle Swarm Optimization. In: IEEE Inf. Conf. on Neural Networks, vol. 4, pp. 1942–1948 (1995)
4. Baba, N.: A new Approach for Finding the Global Minimum of Error Function of Neural Networks. Neural Networks 2, 367–373 (1989)
5. Birbil, S.I., Fang, S.C.: An Electromagnetism-like Mechanism for Global Optimization. Journal of Global Optimization 25, 263–282 (2003)
6. Jiang, J., Shang, H., Liu, K., Su, Q., Zhang, L.: A Clustering Method Using Electromagnetism-like Mechanism Algorithm. Journal of Computational Information Systems 9(10), 3985–3991 (2013)
7. Clerc, M., Kennedy, J.: The Particle Swarm-Explosion, Stability, and Convergence in a Multidimensional Complex Space. IEEE Trans. on Evolutionary Computation 6(1) (2002)
8. Baba, N.: A Hybrid Algorithm for Finding the Global Minimum of Error Function of Neural Networks and its applications. Neural Networks 7(8), 1253–1265 (1994)
9. Yuan, H., Ahi, J., Liu, J.: Application of Particle Swarm Optimization Algorithms based Fuzzy BP Neural Network for Target Damage Accesment. Scientific Research and Essays 6(15), 3109–3121 (2011)
10. Lin, J.L., Wa, C.H., Chung, H.Y.: Performance Comparison of Electromagnetism-like Algorithms for Global Optimization. Applied Mathematics 3, 1265–1275 (2012)
11. Lee, C.H., Li, C.T., Chang, F.Y.: A Species-based improved Electromagnetism-like Mechanism Algorithm for TSK-type integral-valued Neural Fuzzy System Optimization. Fuzzy Set and Systems 171, 22–43 (2011)
12. Gupta, M.M., Jin, L., Honma, N.: Static and Dynamic Neural Networks. IEEE Press, Wiley-Interscience (2003)
13. Martinetz, T.M., Berkovich, S.G., Schulten, K.J.: Neural Gas Network for Vector Quantization and its Application to Time-series Prediction. IEEE Trans. Neural Network 4(4), 558–569 (1993)
14. Repository, U.C.I.: of Machine Learning Databases and Domain Theories, ftp://ftp.ics.uci.edu/pub/machinelearning-Databases
15. Lee, C.H., Chang, F.Y., Lee, C.T.: A hybrid of electromagnetism-like mechanism and back-propagation algorithms for recurrent neural fuzzy systems design. Journal of Systems Science 43(2), 231–247 (2012)
16. Lee, C.H., Chang, F.K., Lee, Y.C.: An Improved Electromagnetism-like Algorithm for Recurrent Neural Fuzzy Controller Design, J. Fuzzy Systems 12(4), 280–290 (2010)
17. Chang, H.H., Huang, T.Y.: Mixture Experiment Design Using Artificial Neural Networks and Electromagnetism-like Mechanism Algorithm. In: Proc. Second International Conference on Innovative Computing, Information and Control, p. 397 (2007)
18. Wang, X.-J., Gao, L., Zhang, C.-Y.: Electromagnetism-Like Mechanism Based Algorithm for Neural Network Training. In: Huang, D.-S., Wunsch II, D.C., Levine, D.S., Jo, K.-H. (eds.) ICIC 2008. LNCS (LNAI), vol. 5227, pp. 40–45. Springer, Heidelberg (2008)
19. Lee, C.H., Chang, F.K., Lee, Y.C.: Nonlinear systems design by a novel fuzzy neural system via hybridization of electromagnetism-like mechanism and particle swarm optimization algorithms. Information Sciences 186(1), 59–72 (2012)

Feature Extraction for Classification of Welding Quality in RSW Systems

Xing-Jue Wang[1,2], Jun-Hong Zhou[2], and Chee Khiang Pang[1,*]

[1] Department of Electrical and Computer Engineering, National University of
Singapore, 4 Engineering Drive 3, 117576 Singapore
[2] A*STAR Singapore Institute of Manufacturing Technology, Singapore
{a0120725,justinpang}@nus.edu.sg
jzhou@simtech.a-star.edu.sg

Abstract. Resistance spot welding is one of the most important welding
techniques widely used in manufacturing, and monitoring the welding
quality draws close attention in industries. The common on-line monitoring systems apply multiple sensors to acquire various signals to predict the welding quality. In this paper, only the voltage and current data
are acquired for classification of welding quality. In this way, the cost of
data acquisition system is significantly reduced while acceptable accuracy
is maintained. Furthermore, the past works are sensitive to the random
perturbation from various sources. By extracting the key features such as
RMS values of each half cycles of voltage and current, resilience against
signal disorder due to noise from the environment is enhanced. By feeding various features into SOM neural network, classification of the welds
with 92.9% accuracy is achieved, showing great potential to deal with various conditions and different materials because of the fast training speed
compared with BP neural networks.

Keywords: current, dynamic resistance, self-organizing map, spot welding, voltage.

1 Introduction

Resistance spot welding (RSW) as an important welding technique invented
in 1877 is extensively used in industry nowadays, especially the automotive industry. Traditionally, the main quality control tests are the off-line destructive
chisel test and peel test examining the weld nugget obtained from the production line [1]. These traditional quality tests are very time-consuming and raise
the labor costs as well as product wastage. To solve these problems, many online monitoring schemes have been proposed by making use of various combination of associate parameters such as electrode displacement, weld current,
weld voltage, dynamic resistance, ultrasound inspection and electrode force.
Among them, dynamic resistance and electrode displacement provide the most

---

* Corresponding author.

© Springer International Publishing Switzerland 2015
H. Handa et al. (eds.), *Proc. of the 18th Asia Pacific Symp. on Intell. & Evol. Systems – Vol. 2*,
Proceedings in Adaptation, Learning and Optimization 2, DOI: 10.1007/978-3-319-13356-0_25

significant information about nugget formation proving to be excellent indicators of the quality of the nugget [2]. In the work of Dickinson et al. [3], how dynamic resistance changes with nugget formation in welds with various materials and qualities is discussed in detail and it shows a tight correlation between nugget formation and dynamic resistance.

Neural network has the advantages of pattern recognition and classification. Thus several on-line monitoring systems with neural networks are proposed. In 1991, Javed and Sanders [4] proposed an on-line monitoring model using a multi-layer neural network as a self-organisational structure. Dynamic resistance was used as the input for its monitoring. Brown et al. [5] used multiple parameters like weld current, dynamic resistance and electrode tip diameters as input for their multi-layer neural network to estimate the size of weld nugget. Cho et al. [6] employed Hopfield neural network to classify the weld quality based on dynamic resistance. Podrzaj et al. [7] made use of linear vector quantization (LVQ) network and chose signals of dynamic resistance, electrode displacement and welding force as the input to identify the expulsion. El-Banna et al. [8] employed LVQ and used dynamic resistance profile to do nugget quality classification that could distinguish normal welds, cold welds, and expulsion welds. Besides neural network, the utility of genetic theory [9], wavelet transform [10], SVM regression [11], k-means clustering [12] and linear regression methods [13] for resistance spot welding have been explored.

The traditional methods usually target on one specific material and should be under ideal conditions. The disorder of the signals can ruin the classification. Thus in this paper, only the key features are extracted. By this way, the whole data can be transferred into lower dimensional space. The features will be fed into SOM network. Self-organizing map neural network is a very efficient unsupervised artificial neural network that can classify signals based on the likelihood with past experience. It has faster training speed compared with back-propagation neural network. Thus we can do classification under various experiment conditions dealing with more data. It is more flexible and generic.

The rest of paper is organised as follows. First of all, a brief introduction will be given about the welding quality in Section 2. In Section 3, the signal processing method which includes noise attenuation and feature extraction will be discussed. In Section 4, SOM neural network used in this paper will be described in detail. Section 5 and Section 6 will cover the experimental setup, the experiment results and discussions.

2   Welding Quality

Based on nugget size, welding quality can be generally classified into four groups, no weld, undersized weld, good weld and expulsion. No weld, undersized weld, and expulsion are considered as welding faults and they should be avoided during welding. When input power is insufficient, no weld and undersized weld will occur. Under this condition, the nugget size will be too small or there is even no nugget. The consequence is that the welded material does not obtain sufficient mechanical strength. Expulsion is a common welding fault. It is a phenomena that melted metal is expelled from the weld pool by the excessive heat. Expulsion can induce lots of problems such as waste of energy, loss of liquid metal from the nugget, deterioration of electrode tip and porosity. Many factors can result in expulsion, such as over-high current, contaminant on the faying surface or lack of pressure on the parent metal.

In this paper, only two welding faults, no weld and expulsion are studied as only these two faults occur in the samples. For no weld case, large amount of current passes through the corners of the samples rather than the welding areas and creates weld marks on the corner as shown in Fig. 1. Dynamic resistance is very useful in indication of these two faults. Expulsion is reported to be accompanied with rapid decay of dynamic resistance. In the case of no weld, the dynamic resistance is slightly reduced because of current leaking on the edges and corners. To enhance the accuracy, other features other than dynamic resistance are used.

Fig. 1. Sample with no weld.

3   Signal Processing

Electrical signals carry the key information of nugget formation. Many monitoring systems utilise electrical signals as important indicators of the nugget quality. Typically, dynamic resistance is tightly related to the formation of nugget and expulsion is found to reduce the dynamic resistance significantly.

In this paper, both current and voltage signals are used. The acquired current and voltage signals have to be preprocessed technically before classifying the samples. The first step is to filter the signals to reduce their noise. Then each half cycles are identified from the sinusoidal electrical signals based on peak and valley positions. In the next step, essential features are extracted from each half cycles for analysis. The features will be combined to form a group of inputs for classification.

3.1 Noise Attenuation

There is noise observed in the welding signals, especially the experimental voltage signals. The noise comes from the disturbance from the circuit and it causes great difficulty in data processing. In order to suppress the effect of the noise on classification, a second order low pass digital butterworth filter from MATLAB toolbox with 5000 Hz cutoff frequency was used to attenuate the noise in voltage. It allows low-frequency signals to pass the filter and attenuates signals with frequencies higher than the cutoff frequency. Fig. 2 shows an example of original voltage signal and the filtered voltage signal and it indicates that the huge noise of the voltage signal is attenuated efficiently with the low pass filter.

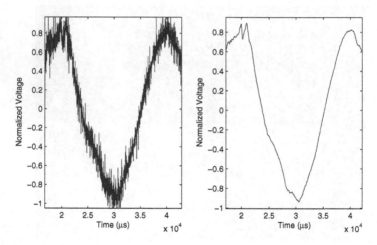

Fig. 2. Noise of voltage is attenuated by low-pass filter. Left: Original. Right: Filtered.

3.2 Features Extraction

Using statistical features is a popular mathematical approach to signal processing in time domain. Feature extraction can reduce the signal dimension effectively retaining the key information. The statistical features used in this paper are dynamic resistance, RMS voltage and RMS current. Dynamic resistance is an important indicator of welding quality and it was derived from the RMS current and RMS voltage of each half cycles. RMS voltage and RMS current carry important information about the energy transmitted to the welded samples within each half cycles and they are calculated by expression $\sqrt{\frac{1}{T}\int_{t_0}^{t_0+T} f^2(t)dt}$.

The current tends to increase in the beginning due to the melting of the metal in the faying surface and the increase slows down corresponding to the cooling of the nugget. In the case of expulsion, the current will continue to increase leading to sharp drop of dynamic resistance.

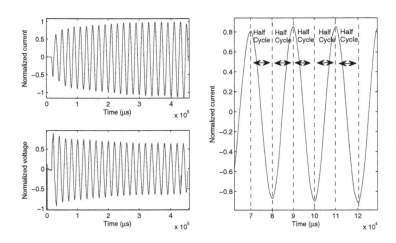

Fig. 3. Typical examples of current and voltage curves and half cycles where the features are extracted

Features were extracted from each half cycles with interval of around 0.01s to form a time series. A half cycle was chosen to be the interval from one peak to the adjacent valley or vice versa. Typical examples of half cycles are shown in Fig. 3. There are thirty-nine half cycles in total that are considered for feature extraction respectively. RMS voltage, RMS current and dynamic resistance were normalized respectively before passing through the neural network for the purpose of speeding up the training. As each features include equal number of inputs, the normalized features were combined together sequentially to form the input vector for the neural network.

Fig. 4a, Fig. 4b, and Fig. 4c show how RMS current, RMS voltage and dynamic resistance of each samples change with cycles, respectively. In general, the figures show that RMS voltage decreases with cycles and RMS current rises gradually with cycles resulting a drop in dynamic resistance. However, samples of various qualities have clear difference in their RMS current, RMS voltage and dynamic resistance curves and they can be classified by the SOM neural network introduced in the next section.

4   Self-Organizing Maps (SOMs)

Self-Organising Map neural network is a very useful unsupervised neural network proposed by Kohonen in 1982 [14]. It uses competitive learning process

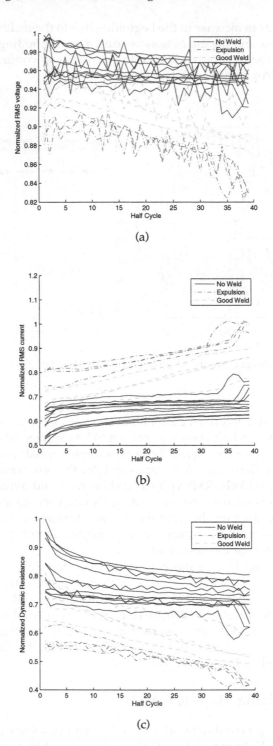

(a)

(b)

(c)

Fig. 4. RMS voltage, RMS current, and dynamic resistance curves of all samples

to do classification according to likelihood of the inputs. As an unsupervised neural network, the speed of SOM is far more rapid than the conventional BP neural network. The SOM neural network used in this paper has four hundred neurons arranged into 20×20 square shape.

According to the algorithm of SOM [14], each neurons will be initially assigned a random vector of weights (w) with the same dimensionality as the inputs (w_p =[$w_{1,p}$, $w_{2,p}$, . . . , $w_{n,p}$]). The input vector x of each welds is presented to the network sequentially. A best matching neuron will be selected according to the Euclidean distance with the input vector. The weight of the winner and the neighbourhood neurons will be updated by expression

$$w(k+1) = w(k) + \alpha(k)\,\Omega(k)(x - w(k)). \tag{1}$$

Here α is the learning rate with expression

$$\alpha(k) = \alpha_0 e^{-\frac{k}{T}}, \tag{2}$$

and Ω (k) is the neighbourhood function related to the Euclidean distance $d_{x,y}$ to the winner neuron. The neighborhood function was chosen to be Gaussian function with expression

$$\Omega(k) = e^{-\frac{d_{x,y}^2}{2\sigma(k)^2}}. \tag{3}$$

T is set to one thousand in our paper. $\sigma(k)$ controls the effective width of the topological neighborhood and only the neurons within the neighborhood width are updated [15]. Both learning rate and neighbourhood width shrink with iterations k. The neighbourhood function initially includes almost all neurons with initial width of the 'radius' of the lattice, then it exponentially shrinks to only affect immediate neighbors after T iterations. The expression of neighborhood width is

$$\sigma(k) = \sigma_0 e^{-\frac{k}{T}}. \tag{4}$$

The learning rate can be adjusted [16,17] to simplify the neural network and further improve the speed. The expression of the learning rate is

$$\alpha(k) = \alpha_0(1 - \frac{k}{T}), \tag{5}$$

where α_0 is set to be 0.3. Similar adjustment can be done to the neighborhood width [16]. However, it results in slightly slowing down of the neural network and thus it is not applied in our paper.

5   Experimental Setup

The whole experiment setup includes a welding machine, a voltage and a current sensor, a data acquisition system and a PC computer and it is schematically shown in Fig.5. Our experiment was performed on Miller LMSW series portable welding machine. This welding machine has a drawback that the input

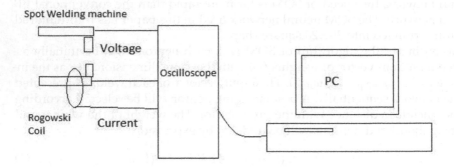

Fig. 5. Experiment setup scheme

current cannot be adjusted. The only parameter that can be tuned is the welding time. Samples were selected as 3 mm thick, 2.5 cm wide and long stainless steel plate. Welding time was approximately 0.3s. Twenty welds were conducted. After the welds, the quality of the samples were examined by destructive test. Two types of faults occurred and all the samples were classified into no weld, good weld and weld with expulsion.

The measurement setup is shown in Fig. 6a. The current data and voltage data were collected for monitoring of the welding quality at sampling rate of 10^6 Hz. The voltage was measured by connecting two tongs of the welding machine to the data acquisition system which is Picoscope as shown in Fig. 6b. The range of measurement of the RSW voltage signal was between -5V and +5V. The current was measured by implementation of Rogowski coil. The output voltage was also directed to the Picoscope.

The voltage signal obtained from the Rogowski coil is proportional to the derivative of current in the welding. In this paper, the current was calculated by integrating the voltage signal from Rogowski coil with the help of MATLAB. A linear downward drift of current was observed after the integration. To solve this problem, compensation was made to correct the current signal.

6   Results and Discussions

The classification by destructive tests are listed in Table. 1 for reference. Even through the same material, the same welding machine and the same setting of the welding machine were applied, the quality of the samples varied due to difference in pressure applied to the welded samples, difference in the surface condition of the samples and perturbations from the environment. The voltage and current during the welds were measured and fed into the SOM neural network for evaluation of their quality. The classification results based on voltage and current data were compared with their true classes by the destructive tests for evaluation of the accuracy of our SOM neural network.

(a)

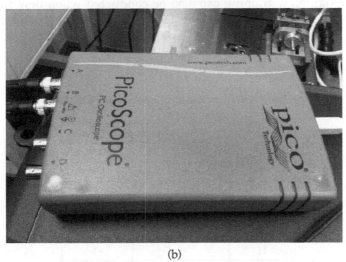

(b)

Fig. 6. Measurement setup

Table 1. Quality of samples. Label 0: Sample with good quality; Label 1: Sample with expulsion; Label -1: Sample with no weld (current passing through the corners)

Sample No.	1	2	3	4	5	6	7	8	9	10
Label	-1	-1	0	-1	-1	0	-1	-1	-1	1
Sample No.	11	12	13	14	15	16	17	18	19	20
Label	-1	-1	-1	0	0	-1	-1	1	1	1

Two samples with no weld (Label -1), two samples with good quality (Label 0) and two samples with expulsion (Label 1) were selected for training of the SOM neural network and the rest welds were used for testing. In the training part, input vectors of Sample 1, 2, 3, 6, 10, 18 were used to train the network repeatedly. The training lasted one thousand iterations. In each iteration, the input vectors of Sample 1, 2, 3, 6, 10, 18 were presented to the neural network in random sequence. This process took approximately 17.8s. In the testing part, the input vectors from the rest welds were fed into the neural network sequentially for evaluation of their quality.

The classification results by SOM are shown in Table. 2. From the table we can see that only sample 17 with no weld (Label -1) is misclassified into good weld (Label 0). All the samples with expulsion fault were classified correctly. The accuracy of the classification is approximately 92.9%. The results show that SOM neural network is fast maintaining satisfying accuracy. With fast speed of neural network, more neurons can be included for more precise classification and fault detection. Thus a single or a few SOM neural networks can cover various experiment conditions such as different sample thickness and different sample material. New training can also be conducted when new situation is introduced and it is very flexible. However, more experiments and more data are needed to proof the effectiveness of the scheme.

Table 2. Quality classification results by SOM

Sample No.	Predicted Quality	Real Quality
4	-1	-1
5	-1	-1
7	-1	-1
8	-1	-1
9	-1	-1
11	-1	-1
12	-1	-1
13	-1	-1
14	0	0
15	0	0
16	-1	-1
17	0	-1
19	1	1
20	1	1

7   Conclusion

In this paper, easily obtained electrical signals are made use of to monitor the quality of the spot welding. The electrical signals prove to have direct association with the quality of the weld nugget and they can be utilised to distinguish various qualities of the welding. Compared with displacement sensor and force sensor, collection of electrical signals has the advantage of avoiding touch with samples thus resulting in longer life time of sensing equipment. Dynamic resistance is very useful for classification. Moreover, more statistical features such as RMS voltage and RMS current were used to further improve the accuracy. In this paper, samples with no weld, good weld and expulsion are classified with 92.9% accuracy by making use of a SOM neural network. The training time of the neural network is around 18 seconds. The speed is faster than conventional BP neural network that shows great potential to include more neurons and input data covering more experiment conditions. However, more experiments are needed with various sample size, input current and voltage to show that the system is flexible and able to deal with multiple welding conditions.

Acknowledgment. The Authors would like to thank Da-You Pan from A*STAR Singapore Institute of Manufacturing Technology for his help on the welding experiment, and Xiao Teng from Department of Electrical and Computer Engineering, National University of Singapore, for his help on feature extraction. This work was supported in part by Singapore MOE AcRF Tier 1 Grants R-263-000-A44-112 and R-263-000-A52-112.

References

1. Al-Jader, M.A., Cullen, J.D., Athi, N., Al-Shamma'a, A.I.: Spot welding theoretical and practical investigations of the expulsion occurrence in joining metal for the automotive industry. In: Proceedings of the 2nd International Conference on Developments in eSystems Engineering, Abu Dhabi, pp. 425–430 (December 2009)
2. Subramanian, D., Dongarkar, G.K., Das, M., Fernandez, V., Grzadzinsky, G.: Real time monitoring and quality control of resistance spot welds using voltage, current, and force Data. In: Proceedings of International Conference on Electro/Information Technology, Milwaukee, WI, pp. 211–220 (August 2004)
3. Dickison, D.W., Franklin, J.E., Stanya, A.: Characterization of spot welding behavior by dynamic electrical parameter monitoring. Welding Journal 59(6), 170s–176s (1980)
4. Javed, M.A., Sanders, S.A.C.: Neural networks based learning and adaptive control for manufacturing systems. In: Proceedings of the IEEE/RSJ International Workshop on Intelligent Robots and Systems-IROS 1991, Osaka, Japan, pp. 242–246 (1991)
5. Brown, J.D., Rodd, M.G., Williams, N.T.: Application of artificial intelligence techniques to resistance spot welding. Ironmaking and Steelmaking 25(3), 199–204 (1998)
6. Cho, Y., Rhee, S.: Quality estimation of resistance spot welding by using pattern recognition with neural networks. IEEE Transactions on Instrumentation and Measurement 53(2), 330–334 (2004)

7. Podrzaj, P., Polajnar, I., Diaci, J., Kariz, Z.: Expulsion detection system for resistance spot welding based on a neural network. Measurement Science and Technology 15, 592–598 (2004)
8. El-Banna, M., Filev, D., Chinnam, R.B.: Online qualitative nugget classification by using a linear vector quantization neural network for resistance spot welding. The International Journal of Advanced Manufacturing 36, 237–249 (2008)
9. Panchakshari, A.S., Kadam, M.S.: Optimization of the process parameters in resistance spot welding using genetic algorithm. International Journal of Multidiciplinary Science and Engineering 4(3) (March 2013)
10. Pan, C., Zhao, P., Du, S., Wang, J.: Quality assessment of aluminum alloy resistance spot welding based on wavelet and statistic analysis. In: Proceedings of International Conference on Information and Automation, Zhuhai, Macau, pp. 1438–1442 (June 2009)
11. Zhang, J., Zhang, P.-X.: A SVM regression predicting model for indentation depth of welding spot based on digital image processing. In: Proceedings of the 4th International Conference on Digital Image Processing, vol. 8334, Kuala Lumpur, Malaysia (April 2012)
12. Zhang, H., Hou, Y.: Quality estimation of the resistance spot welding based on genetic k-means cluster analysis. In: Proceedings of International Conference on Control, Automation and Systems Engineering, Singapore, pp. 1–4 (July 2011)
13. Hao, M., Osman, K.A., Boomer, D.R., Newton, C.J.: Developments in characterization of resistance spot welding of aluminum. Welding Journal 75(1), 1–8 (1996)
14. Zurita-Milla, R., van Gijsel, J.A.E., Hamm, N.A.S., Augustijn, P.W.M., Vrieling, A.: Exploring spatiotemporal phenological patterns and trajectories using self-organizing maps. IEEE Transactions on Geoscience and Remote Sensing 51, 1914–1921 (2012)
15. Lo, Z.-P., Bavarian, B.: On the rate of convergence in topology preserving neural networks. Biological Cybernetics 65, 55–63 (1991)
16. Wang, Y., Peyls, A., Pan, Y., Claesen, L., Yan, X.: A fast self-organizing map algorithm for handwritten digit recognition. In: Park, J.J.(J.H.), Ng, J.K.-Y., Jeong, H.-Y., Waluyo, B. (eds.) Multimedia and Ubiquitous Engineering. LNEE, vol. 240, pp. 177–183. Springer, Heidelberg (2013)
17. Salem, A.-B.M., Syiam, M.M., Ayad, A.F.: Improving self-organizing feature map (SOFM) training algorithm using K-means initialization. In: IEEE Proceedings of International Conference on Intelligent Engineering Systems, INES, vol. 40, pp. 41–46 (2003)

Cross-Layer Secured IoT Network and Devices

Paul Loh Ruen Chze, Kan Siew Leong, Ang Khoon Wee, Elizabeth Sim,
Kan Ee May, and Hing Siew Wing

Communications & Networks Group, School of Engineering,
Nanyang Polytechnic, Singapore
{Paul_LOH,KAN_Siew_Leong,ANG_Khoon_Wee,Elizabeth_SIM,
KAN_Ee_May,HING_Siew_Wing}@nyp.edu.sg

Abstract. This paper introduces a scalable secured network and device architecture that enhances IoT devices' security to reduce the risk of them from becoming a target of attacks. The proposed network and device architecture combines user-definable parameters, information from the physical, network and application layers to generate an encrypted unique identification key to be embedded in the IoT devices. The network architecture uses a WiFi ad-hoc mode that enables the IoT devices to authenticate among themselves during the initial network formation. The IoT devices are only allowed to join an autonomous IoT network if the devices are running the users' pre-agreed application(s), in the list of permitted devices, and have an unique identification. Both indoor and outdoor field tests were conducted. The results of the field tests showed that the proposed network and device architecture is scalable and is suitable to be deployed for IoT.

Keywords: cross-layer, secured network, secured devices, IoT, security.

1   Introduction

As sensors' and machine-to-machines' (M2M) communication technologies advance, more objects are embedded with sensors that processed the abilities to communicate and collaborate [1, 2]. Internet of Things (IoT) enable these objects to be uniquely recognizable, locatable and at the same time allow the sharing of information with other *"things"* in the network using the Internet Protocol that unites the Internet. The resulting networks promise to enable new business models, new user experiences, optimize processes, optimize services and reduce costs in various industry segments, ranging from industrial, building, transportation, environmental, healthcare, retail, security, smart grid, *etc.*[3].

IoT is projected to be one of the fastest growing technology segments in the next 3 to 5 years [4]. The IoT market in 2011 was worth $44 billion, and is expected to grow to $290 billion by 2017 with a compounded annual growth rate of 30.1% from 2012 to 2017 [4].   However, as indicated by Frost & Sullivan [5] and [6, 7], security is a

© Springer International Publishing Switzerland 2015   319
H. Handa et al. (eds.), *Proc. of the 18th Asia Pacific Symp. on Intell. & Evol. Systems – Vol. 2,*
Proceedings in Adaptation, Learning and Optimization 2, DOI: 10.1007/978-3-319-13356-0_26

major hindrance for wide scale adoption of IoT. Constant sharing of information between *"things"* and users can occur without proper authentication and authorization. Currently, there are no secured platforms that provide access control and personalized security policy based on users' needs and context across different types of *"things"* [5].

The *"things"* in any IoT network are often unattended most of the time; therefore, they are vulnerable to attacks. Moreover, most IoT communications are wireless which make eavesdropping easy [3, 6]. In addition, most *"things"* in any IoT network are with low computing capacity and thus, are not able to implement complex schemes to support security and privacy [3, 6, 7].

This paper addresses the security issues of IoT network and the *"things"* in IoT. By the term "security", we mean the *"things"* in the IoT network are protected from unintended or unauthorized access. The contribution of this paper is threefold. Firstly, we introduce a Cross-Layer Secured Network and Device Architecture (CSND) that embeds the authentication of the *"things"* in the routing protocol. The CSND implements three layers of security parameters (*i.e.*, time-sensitive user-controllable variable, network and application information). In addition, the CSND offers an efficient and effective authentication method for the *"things"* in the IoT that uses three identifications: (1) Matching User-Controllable Identification (UCID), (2) Similar users' pre-agreed application(s) and (3) Permitted devices list based on the network and data-link addresses. Using this methodology, the CSND disregards those *"things"* that do not conform to any of the three identifications. The CSND is adopted from the User-Controllable Multi-Layer Secure Algorithm for mobile ad-hoc networks [8]. This methodology has been patented in Singapore [9, 10]. Secondly, we offer two *"things"*, namely, the *WiSeMAN-Sense* and the *WiSeMAN-Range* that were created for the secured IoT network. Both the *WiSeMAN-Sense* and the *WiSeMAN-Range* were embedded with the CSND routing protocol. Thirdly, we present field tests results for both indoor and outdoor environments with *"things"* executing the CSND routing protocol.

This paper is organized as follows. Section 2 discusses the related work. Section 3 introduces the CSND architecture. The *WiSeMAN-Sense* and *WiSeMAN-Range* are presented in Section 4. Followed by the descriptions of the field tests and conclusion in Section 5 and Section 6, respectively.

2   Related Work

IoT devices carrying RFID tags with a unique Electronic Product Code (EPC) and with the help of Object Naming Service (ONS) has been proposed to address the security issues for IoT [7, 11]. However, these solutions inherit the traditionally weakness of DNS and also subjected to single point of failure.

Interconnectivity amongst the IoT devices requires new peer discovery methods, physical and MAC layer procedures that are different from traditional wired and wireless architectures [6]. Reference [12] proposed a Service-Oriented-Architecture (SOA) to discover devices and services provided by each IoT device. On the other hand, [13] described a Perci framework based on web services to make connectivity to the neighboring devices. However, not all IoT devices will or can provide a service because of resource limitations within each devices.

Reference [14] proposed a security-critical multimedia service architecture that requires traffic classification and analysis. However, both the classification and analysis of traffic needs additional computational resources and time.

3   Cross-Layer Secured Network and Device Architecture (CSND)

The cross-layer secured network architecture (CSND) consists of two elements; the network architecture and the device architecture. The network architecture of CSND is adopted from [15] that presented a secured multi-hop routing algorithm for IoT communications. This paper enhances the security of the network architecture with cross-layer security authentication device architecture. Figure 1 shows the overview of the cross-layer security authentication device architecture.

The Key Management System (which can be an IoT node or any smart wireless device) defines and distributes the key to a variety of wireless systems or devices (WSs) which include but not limited to any IoT nodes, wireless sensor nodes, Access Points (AP), base stations, laptop computers, tablets, and many other future mobile devices.

The Key Management System comprises of the following modules:

- Selection and Sequencing module for security parameters,
- Random or User-defined Entity Generator,
- Scheduling Mechanism module,
- Database module, and
- Communications module.

Through the Selection and Sequencing module, user can first define the security parameters of applications, network and system level of interest to be used. The security parameters of application level include login ID, password, encryption/license key, *etc.*, while the network and systems level security parameters include WPA2 key, encryption key, SSID, channel number, frequency, MAC address, physical system identification no, *etc.* These selected parameters will then be sequenced according to user's preference or based on a randomly generated sequence.

The Entity Generator will generate a set of entities covering the Count Number, Length, Dummy characters and Timestamp either automatically or in user-defined manner. These entities, together with the selected sequence of parameters will be fed to the Scheduling mechanism. The data structure used by the Entity Generator is shown in Fig. 2.

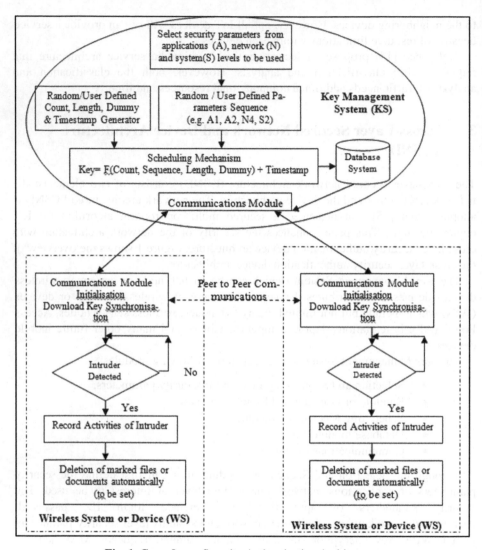

Fig. 1. Cross-Layer Security Authentication Architecture

The Scheduling Mechanism functions are as such:

  Security parameters of application level as A1, A2, …An

        where A1 being application security parameter
       1

  Security parameters of network level as N1, N2, ….Nn

      where N1 being network security parameter 1

  Security parameters of system level as S1, S2, … Sn

      where S1 being system security parameter 1

Selected Sequence of parameter as Seq;
Count number as C; Length as L; Timestamp as T;
Key as K, a function of (C, Seq, L, Dummy)

Then, the steps are as follows:

- Seq = any combination of selected A, N and S with any preferred sequence
- Repeat Seq in for a number of C time for as long as L time. L must be large enough to cover at least one period of Seq repeating C time. The remaining time if any will be filled up with randomly generated dummy characters. These dummy characters will serve as noisy character to "fool" the intruder if any and can be sufficiently long.
- K = F(C, Seq, L, Dummy) + T where T is used as an expiry time for all wireless systems and devices to get a new K
- K will then be stored in the database system and will be updated just before time T is up.

The communication module will then

- Authenticate and set-up communication link as per normal communication channel establishment, send a one block of K to all wireless system and devices. Tear down and set up communication link again once the wireless system and devices have re-configured themselves using those security parameters. Move to data transfer stage. Continue sending the blocks of K.
- Activate all wireless system and devices when T is up for new K to be sent.

Example:
Seq = (A1, A2, N4, S2), C = 5, L = 10sec, T = 5 minutes

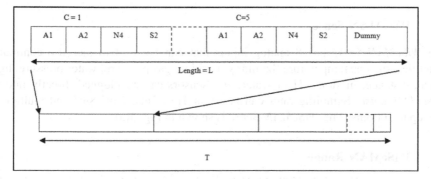

Fig. 2. Data Structure

A block of K will be received from Key Management System (KS). WS will be based on some or all relevant security parameters information from the received K to re-configure itself. Once done, WS will have to be in synchronisation with the KS before communication link is set up and data transfer takes place at the first instance. In normal data transfer stage, synchronisation of new key will be taking place between the KS and WS at the background.

During data transfer stage, if there is any intruder being detected, recording of intruder packets and activities will be activated to allow the option of playback at a later stage. Important files or documents can be deleted if required/necessary. These actions will further enhance the security level of the wireless systems and devices.

Peer to peer communications between different WS can also take place based on the same concepts explained earlier between WS and KS. In another way, WS can also make use of the K obtained from KS to establish connections with other WS subsequently.

4 IoT "Things" Created

Two "*things*" for the secured IoT network, namely, the *WiSeMAN-Sense* and *WiSeMAN-Range*, were created that embed the CSND routing protocol described in Section 3. The alphabets "*Wi*" "*Se*" and "*MAN*" corresponds to the words, "*Wi*reless", "*Se*cured" and "*M*obile *A*d-hoc *N*etwork", respectively. The *WiSeMAN-Sense* is created for the IoT network relating to sensors. On the other hand, the main purpose of *WiSeMAN-Range* is to extend the IoT network range. The hardware specifications of both the *WiSeMAN-Sense* and *WiSeMAN-Range* are shown in Table 1.

Table 1. Hardware Specifications of *WiSeMAN-Sense* and *WiSeMAN-Range*

CPU	Samsung S3C2440
Core	ARM920T
CPU speed	400 MHz
SDRAM	64 MB
Operating System	Debian Linux

4.1 WiSeMAN-Sense

The *WiSeMAN-Sense* has 8 analog inputs that can be used to measure continuous quantities, such as temperature, humidity, position, gas pressure, water pressure, light, sound, distance, motion, pH of water, *etc.* Sensors can be plugged directly into the *WiSeMAN-Sense*. Sampling rates can be set at 1ms, 2ms, 4ms, 8ms and multiple of 8ms up to 1000ms. The *WiSeMAN-Sense* is shown in Fig. 3(a).

4.2 WiSeMAN-Range

The main objective of the *WiSeMAN-Range* is to extend the reachability of an IoT network. In addition, the *WiSeMAN-Range* is able to provide an entry point for an end-user device such as an iPad or iPhone to access a secured IoT network without any software installation. Figure 3(b) shows the *WiSeMAN-Range*.

Fig. 3. (a) WiSeMAN-Sense (b) WiSeMAN-Range

5   Field Tests

5.1   Field Test 1 (Indoor) – Education and Training

Most networked e-Learning solutions require physical network infrastructure such as routers, switches, firewalls and/or wireless access points. In addition, the effort of laying the network cables for establishing communication channels between the teacher's computer and students' laptops is required. These e-Learning solutions are not mobile and do not allow the flexibility of having education outside of the classroom. On the contrary, the concept of Internet-of-Things technology allows a collaborative network to be formed spontaneously by mobile wireless devices without any fixed network infrastructure. These mobile devices can communicate with each other directly and at the same time relay information amongst them. The students' laptops and mobile devices such as tablets and smart phones can establish communication to the teacher's laptop or tablet without any infrastructure cost.

This field test was done in Lab S.441, Level 4, Block S at the Nanyang Polytechnic, Singapore. One interactive plasma display (installed with slot PC running on Windows 7 embedded OS), two *WiSeMAN-Range* (running on Debian), two Windows 7 tablets and two iPads were used for this field test. The interactive plasma display and the two *WiSeMAN-Range* were installed with the CSND routing protocol to form a secured IoT network among the devices. The other mobile devices functioned as end-user nodes that did not require any software installations. The interactive plasma display was used by the teacher for classroom content delivery. As the teacher wrote on the display, the screen of the display was shown in real-time on all students' mobile devices (*e.g.*, iPads and tablets). Tests on communication establishment, network range, bandwidth and security were conducted. The results are as follows:

Communication Establishment
A *VNC Sever* [16] was installed in the interactive plasma display. On the other hand, the tablets and iPads were each installed with the *VNC Viewer* [16]. It was verified that communications were established instantaneously among the devices when the tablets and iPads attempted to join the IoT network. The screen of the interactive plasma display could be shared among the tablets and iPads as shown in Fig. 4.

Fig. 4. Screen of Interactive Plasma Display Shared Amongst the Mobile Devices

Network Range

An interactive plasma display, two *WiSeMAN-Range*, one iPad and one Windows 7 tablet were used in this section. This test covered an indoor floor area of estimated 100m by 50m. The locations of the devices are shown in Fig. 5.

The CSND routing protocol was running in the slot PC (interactive plasma display) and the two *WiSeMAN-Range* so that a secured IoT network was formed automatically among the devices. All the walls partitions were of half-glass and half-timber. The distances between each of the devices were about 30 meters. The communication paths of the devices are presented in Fig. 6.

Figure 7 shows the devices in various locations used in this field test. Using the *VNC Viewer* software installed on the iPad (192.168.0.31) in Fig. 7(d), the iPad was able to view the directories and files on the slot PC that was located in lab S.441 via the two *WiSeMAN-Range* with IP addresses 192.168.0.54 (Fig. 7(b)) and 192.168.0.55 (Fig. 7(c)). Figure 8 shows the *"tracert"* command on the slot PC (interactive plasma display). The iPad was replaced by a Windows 7 tablet that exhibited the same network range capability.

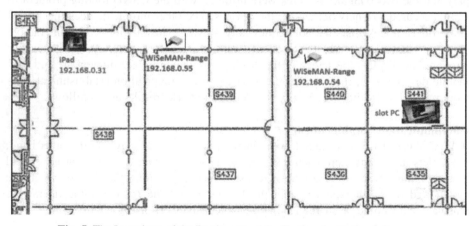

Fig. 5. The Locations of the Devices Used for Testing the Network Range

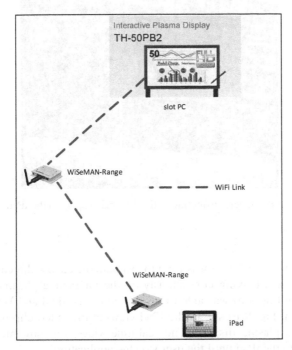

Fig. 6. Communication Paths of the Devices

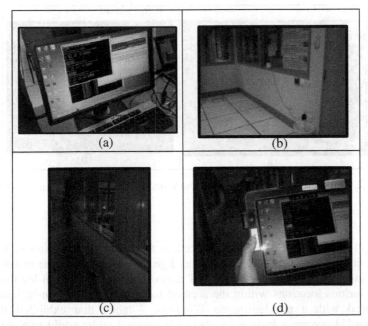

Fig. 7. (a) Location of Slot-PC in S.441 (b) Location of *WiSeMAN-Range* (192.168.0.54) (c) Location of *WiSeMAN-Range* (192.168.0.55) (d) Location of iPad (192.168.0.31)

Fig. 8. "tracert" from Interactive Plasma Display to the iPad

Bandwidth

The same setup shown in Fig. 5 was used to evaluate the bandwidth capacity. A video clip (30MB) that was playing continuously on the interactive plasma display in lab S.441 as shown in Fig. 9(a) was able to be viewed on the iPad and Windows 7 tablet without any glitch. Fig. 9(b) shows the iPad viewing the video clip that was playing on the interactive plasma display. The real-time video was streamed continuously from the display to the iPad until the user stop the application.

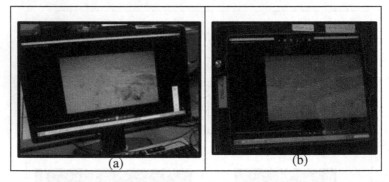

Fig. 9. (a) Interactive Plasma Display Playing the Video Clip (b) iPad Viewing the Video Clip

Security

A non-legitimate iPad was then introduced to the secured IoT setup in Fig. 5. A "…connection attempt timed out" error message was obtained on the non-legitimate iPad. All attempts to join the secured IoT network failed as the non-legitimate iPad moved to various locations within the secured IoT network. The non-legitimate iPad was replaced with a non-legitimate Windows 7 tablet that exhibited the same phenomenon. However, a legitimate iPad or Window 7 tablet could join the secured IoT network instantaneously.

The four areas of field tests, namely in the area of communication establishment, network range, bandwidth and security were conducted successfully at Level 4, Block S in the Nanyang Polytechnic campus. The results showed that the information on the interactive plasma display could be shared among the mobile devices using a secured IoT network formed by using CSND routing protocol. Thus, demonstrating that the CSND routing protocol can be integrated with large display unit that for education and training purposes.

5.2   Field Test 2 (Outdoor) – Construction Site Communication

Currently most construction sites use the walkie-talkie and/or 3G/4G as the basic communication tools. However, the walkie-talkie has the limitation of only allowing voice communication. On the other hand, social media applications such as WhatsApps and Facebook though allowing text, photographs and videos transmissions but require establishing a connection to a server in the Internet through 3G/4G connectivity. Sensors' data such as temperature, humidity and air quality levels at certain areas of the construction sites also uses the 3G/4G wireless technology for transmission to a monitoring station. However, 3G/4G signals may not be available in certain parts of the construction sites, *e.g.*, underground. In this field test, we show how multimedia communication and sensors' data transmission can be accomplished by using a secured IoT network.

This field test was done along the semi-open corridor of Level 4, Block S at the Nanyang Polytechnic, Singapore. The equipment used were two Windows 7 laptops, each installed with the Hawking's hi-gain 15dBi outdoor omni-directional antenna [17], two Ubuntu laptops, each installed with Openmeetings[18],   one Ubuntu tablet for sensors' data monitoring, one *WiSeMAN-Range,* one *WiSeMAN-Sense* and one iPad as shown in Fig. 10.

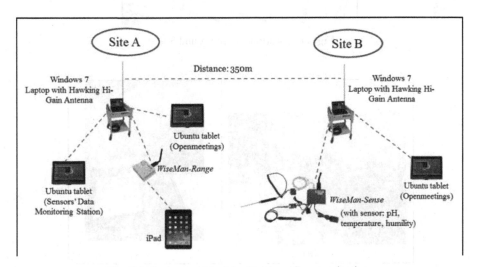

Fig. 10. Equipment Used for Outdoor Communication

The CSND routing protocol was installed in all the devices (except the iPad) to form a secured IoT network among the devices. The iPad was an end-user device that did not require any software installation. The Openmeetings [18] was for text, video and voice communication between the two Ubuntu tablets. The distance between Site A and Site B was about 350m. Figure 11 shows the locations of the both sites within the Nanyang Polytechnic campus which span across three buildings, namely Blocks S, R and P. The locations of the devices are shown in Fig. 12.

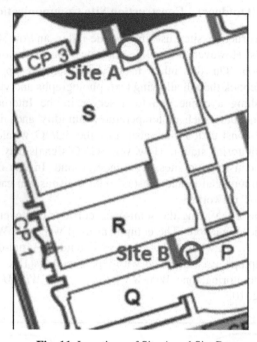

Fig. 11. Locations of Site A and Site B

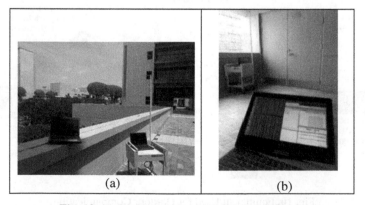

Fig. 12. (a) Devices at Site A (b) Devices at Site B

The test results are as follows: The two Ubuntu installed with Openmeetings [18] at Site A and Site B were able to establish communication to each other using text, whiteboard sharing and video conferencing as shown in Fig. 13.

The *WiSeMAN-Sense* collected the various sensors' data at Site B and transmitted the data to a monitoring station at Site A as shown in Fig. 14. The *WiSeMAN-Range* extended the secured IoT network in Fig. 11, such that the iPad that was not within the transmission range of the hi-gain antenna at Site A was still able to obtain the sensors' information from Site B via the monitoring station.

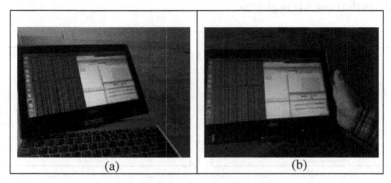

| (a) | (b) |

Fig. 13. (a) Communication Openmeetings at Site A (b) Communication using Openmeetings at Site B

Fig. 14. Sensors' information Obtained by the Monitoring Station at Site A

The round-trip-delay (RTD) and bandwidth of the CSND were obtained using the experimental setup depicted in [15]. The results of the RTD and bandwidth performances of the CSND were compared against the Secure Multi-Hop Routing Protocol (SMRP) [15] as shown in Fig. 15.

From the results of the 10 RTD tests conducted, the RTD obtained by SMRP and CSND were comparable. The RTD obtained by SMRP ranged from 7.3 msec. to 9.5

msec. On the other hand, the RTDs obtained by CSND were within 7.1 msec. to 10.2 msec. On the average, SMRP had a RTD of 8.13 msec. whereas the average RTD of CSND is 8.54 msec. The difference of 0.41 msec. is insignificant.

The bandwidth performances of the two algorithms are depicted in Fig. 15(b). Out of the 10 tests conducted with each protocol, the SMRP obtained a bandwidth performance between 2.7Mbit/sec to 4.9 Mbit/sec. The bandwidth performance obtained was comparable when CSND was used, with a bandwidth ranging from 2.7 Mbit/sec to 4.5 Mbit/sec. The average bandwidth for SMRP and CSND is 3.99Mbit/sec and 3.62 Mbit/sec, respectively. The difference of 0.37Mbit/sec in the average bandwidth performance is negligible.

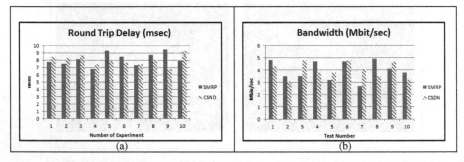

Fig. 15. (a). Round Trip Delay Test Results (b). Bandwidth Test Results

6   Conclusion

In this paper, a scalable cross-layer secured network and device architecture (CSND) that enhances IoT devices' security is presented. The proposed CSND merges the routing and authentication processes for forming a secured IoT network without incurring significant overheads. Two devices created for the IoT network that embeds the CSND routing protocol, namely, the *WiSeMAN-Sense* and *WiSeMAN-Range* are introduced. Our field tests done for both indoor and outdoor environments showed that the CSND produces a secured cross-layer, secured IoT communication network.

Acknowledgement. The work presented in this paper is part of the outcome from the 4th and 11th Singapore National Research Foundation grant. The authors would like to thank the Singapore National Research Foundation for providing the funding available for this research.

This material is based on proof-of-concept research project supported by the Singapore National Research Foundation. Any opinions, findings, and conclusions or recommendations expressed in this material are those of the authors and do not necessary reflect the views of the Singapore National Research Foundation.

References

1. Kortuem, G., Kawsar, F., Fitton, D., Sundramoorthy, V.: Smart objects as building blocks for the internet of things. IEEE Internet Computing 14, 44–51 (2010)
2. Kranz, M., Holleis, P., Schmidt, A.: Embedded interaction: Interacting with the internet of things. IEEE Internet Computing 14, 46–53 (2010)
3. Atzori, L., Iera, A., Morabito, G.: The internet of things: A survey. Computer Networks 54, 2787–2805 (2010)
4. Internet of Things (IoT) & Machine-To-Machine (M2M) Communication Market by Technologies & Platforms (RFID, Sensor Nodes, Gateways, Cloud Management, NFC, CEP, SCADA, ZigBee), M2M Connections, IoT Components - Worldwide Market Forecasts (2014 - 2019) (March 2014), http://marketsandmarkets.com
5. Sullivan, F.A.: Internet of Things–Technology Penetration and Roadmapping (Technical Insights) (December 31, 2012)
6. Miorandi, D., Sicari, S., De Pellegrini, F., Chlamtac, I.: Internet of things: Vision, applications and research challenges. Ad Hoc Networks 10, 1497–1516 (2012)
7. Weber, R.H.: Internet of Things–New security and privacy challenges. Computer Law & Security Review 26, 23–30 (2010)
8. Loh, R.C., Kan, W.Y., Kan, S.L.: A User-Controllable Multi-Layer Secure Algorithm for MANET. In: 2012 8th International Wireless Communications and Mobile Computing Conference (IWCMC), pp. 1080–1084 (2012)
9. Kan, S.L.: Method and system for secured service-oriented nodes discovery and route determination in mobile ad-hoc network (Patent: PCT/SG2009/000032) (2009)
10. Kan, S.L., Ang, K.W.: Method and System for Securing Wireless Systems and Devices (Patent: 154827) (2007)
11. Welbourne, E., Battle, L., Cole, G., Gould, K., Rector, K., Raymer, S., et al.: Building the internet of things using RFID: the RFID ecosystem experience. IEEE Internet Computing 13, 48–55 (2009)
12. Guinard, D., Trifa, V., Karnouskos, S., Spiess, P., Savio, D.: Interacting with the soa-based internet of things: Discovery, query, selection, and on-demand provisioning of web services. IEEE Transactions on Services Computing 3, 223–235 (2010)
13. Broll, G., Rukzio, E., Paolucci, M., Wagner, M., Schmidt, A., Hussmann, H.: Perci: Pervasive service interaction with the internet of things. IEEE Internet Computing 13, 74–81 (2009)
14. Zhou, L., Chao, H.-C.: Multimedia traffic security architecture for the internet of things. IEEE Network 25, 35–40 (2011)
15. Loh, R.C., Kan, S.L.: A Secure Multi-Hop Routing for IoT Communication. Presented at the IEEE World Forum on Internet of Things 2014, Seoul, Korea (2014)
16. Richardson, T., Stafford-Fraser, Q., Wood, K.R., Hopper, A.: Virtual network computing. IEEE Internet Computing 2, 33–38 (1998)
17. Hawking Technology: Hi-Gain 15dBi Outdoor Omni-Directional Antenna, http://hawkingtech.com/products/hawking_products/outdoor_wireless_solutions/hao15sip.html
18. Openmeetings – Open Source Web Conferencing (2006), https://code.google.com/p/openmeetings/

Inference of Opponent's Uncertain States in Ghosts Game Using Machine Learning

Sehar Shahzad Farooq, HyunSoo Park, and Kyung-Joong Kim*

Department of Computer Science and Engineering, Sejong University, South Korea
{sehar146,hspark8312}@gmail.com, kimkj@sejong.ac.kr

Abstract. Among many categories, board games can be classified into two main categories: Games with perfect information and games with imperfect information. The first category can be represented by the example of "Chess" game where the information about the board is open to both players. The second category can be determined with the "Ghosts" game. Players can see the position of the opponent's pieces on the board whereas the identity of the ghost pieces (good or bad) is hidden, which makes this game uncertain to apply search state space based technique. In this work, we have investigated the opponent game state with uncertainty for Ghosts using machine learning algorithms. From last year competition replay data, we extracted several features and apply various machine learning algorithms to infer game state. Also, we compare our experimental results to the previous prototype based approach. As a result, our proposed method shows more accurate results.

Keywords: Ghosts challenge, Uncertainty, Game AI, Machine learning, Feature extraction.

1   Introduction

Games have been considered as one of the main source of digital entertainment now a day. There has been different type of games that are played against the other player or against the game AI (artificial intelligence). The purpose of playing games is not only to exercise the brain like making some strategies and winning/finishing the game but also to express the explicit thinking of the human mind. Therefore with the help of the games, the behavior of the human can be evaluated. There are several games in which the player wishes to play against other human player rather than playing against the AI. This is because of the limitations for computer-controlled opponents to build the strategy based decisions like humans. Computer-controlled opponent is a background program which is capable of automatically playing the game and can give the human players the feeling that they are interacting with other human players. It requires an enormous design effort in terms of strategies and interaction options. However, there have been a lot of game AI developed to predict future game state and can defeat the human in many games (i.e. Chess) [1].

---

* Corresponding author.

© Springer International Publishing Switzerland 2015   335
H. Handa et al. (eds.), *Proc. of the 18th Asia Pacific Symp. on Intell. & Evol. Systems – Vol. 2,*
Proceedings in Adaptation, Learning and Optimization 2, DOI: 10.1007/978-3-319-13356-0_27

There are some board games that have been solved so perfectly that any program or human cannot win against the computer generated program [2]. However, there are still some board games which are under observation where the strategy of the human players cannot be easily evaluated. This is because of the imperfect information type of the board games. Since the information about the board game is missing, several search state space based techniques cannot be applied straightforward for strategy prediction. It can be possible to identify the game state and the opponent's strategy by applying the machine learning techniques using game play data [3, 4].

In this paper we found out the game state for the uncertain game named "Ghosts" using machine learning algorithms by collecting its game play logs. Although the Ghosts is a very simple board game, it is difficult to play because of uncertainty of opponent ghost's identity. We collected game play data over 1,400 games and applied various machine learning algorithms to build ghost identification inference model. Also we compare its results to previous approach used in [5]. As a result, our results show more accurate results.

2   Ghosts Challenge

"Ghosts" is a simple board game invented by Alex Randolph [6]. Its German name is Geister and is played between two players. Each player has a total of eight ghosts which are equally divided into two categories, good ghosts and bad ghosts. The identity of which are good ghosts and which are bad ghosts is hidden from the opponent as it is marked at the back side of the ghost which can only be seen by its own player. These ghosts have to be placed at middle of the least two rows on a 6 × 6 board as can be seen from the fig. 1.

The players can move their ghosts alternatively. The ghosts have limitation of not to take a step diagonally instead they can move one square forward, backward and sideways. The ghosts can capture the opponent ghosts (regardless of any identity) by landing onto the opponent's ghost position. Upon moving a ghost onto the same space, the nature of the latter ghosts is revealed to the capturing player. On the other hand, the player (whose ghost is captured) couldn't realize the identity of the capturing ghost.

Different winning strategies can be adopted as there are diverse conditions for winning: A player can win the game if it is eating/capturing all the good ghosts of the opponent. A player can win the game if the opponent eats/captures all the bad ghosts of the player (it can be possible to adopt a strategy so that the opponent is given a choice to eat our bad ghosts i.e. bluffing). A player can win the game if it reaches to the opponent's corner space (moving off the board) with its good ghost. Each corner of the board is marked with an arrow sign which indicate that the ghost (if it is good) reaching that corner is moving off the board and finishing the game. The length of the game is limited to 100 plies where a "ply" means a single move of a player. The game is considered as tie if it reaches a length of 100 plies.

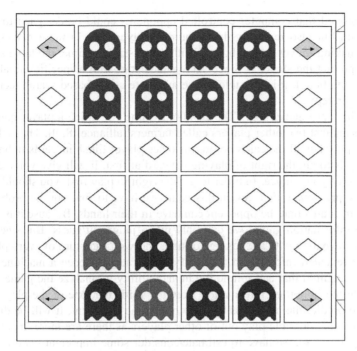

Fig. 1. Initial board setup for Ghosts (Top: Opponent)

There have been a lot of game artificial intelligence competitions now a days organized by many game related international conferences all over the world. These competitions include first-person shooting games, real-time strategy games, board games and many other genres of games. The purpose of these competitions is to create autonomous bots/agents to play the game automatically without human intervention. "Ghosts Challenge[1]" is one of the recent simple board game competition based on "Ghosts" game organized by IEEE CIS Student Games-based Competition Committee in 2013. The competition continues its series and will hold again in 2014. The purpose of the competition is to develop an autonomous agent in order to play the game using computational intelligence techniques.

3   Background and Related Works

Games have different genres like platform games, arcade games, board games, card games, social games, real time strategy based games. On the other hand, there are other categories of games like perfect and imperfect information games. Focusing only to the imperfect information games in our study, there are different types of games where the players don't have the clear information about the state of the game.

Research on game AI character strategy and decision making emerged from the design of AI opponents in two-player games such as checkers and Othello. Othello in

---

[1]   https://ghosts-challenge.math.unipd.it/

particular proved that computer-controlled opponents could be designed to not only compete with but also regularly defeat human players [7]. In these games, players take turns making moves on the board. With the passage of time, the table status can be used to predict the strategy. Since such games start with a specific initial position of pliers on the board, it is possible to use any state space based search technique to analyze the strategy.

Some other card games like "The Landlord Game" in which a player named Landlord fights against two other players called farmers' alliance [8, 9]. The task of each player is to play out all the cards before the other player finish its cards in hands. The best strategy is to get the right of playing the card at first. It will give you a chance to play the card of your choice. In order to get the right to play first, you should suppress others in the previous round. So, in this game we have to find out the probability of the type of the cards that the opponent can have in their hands. Because it is a kind of incomplete information game, we can only make judgment/guess. But since the information is revealing with the passage of time and with each turn of every player, we can use the revealed information to modify our strategy. Hence a machine learning approach is required that can mimic the human ability to analyze the game state and can evaluate the important information from the cards that one of the players has in his hands to predict the next possible play out of the opponent. It will be difficult to estimate the exact possible play out of other players as there are more than two players in this game. So, we have to estimate/consider some important features that can be used to train the system and the system then can predict the next outcome.

In Ghosts game, the information of identity of all the ghosts of the opponent is hidden, therefore, we need to use a heuristic judgment to play the game without the human. It is possible if we use some information from the board (i.e. the position of the ghosts and playing pattern of the opponent); we can plan the next possible move in the game and hence can evaluate the strategy of the opponent.

The first competition of Ghosts challenge held in November 2013 [10]. A total of eight teams participated in this competition. BLISS team took the victory while mutigers were the runner up. The replays of the competition between each of the participants are available at the website of the ghost challenge. BLISS team from China first converted the imperfect information of the Ghosts game to perfect information using the baseline approach and then used Upper Confidence Bounds (UCB) for decision making [11]. Whereas mutigers used hybrid computational intelligence to design their controller. At first they evaluated all the possible actions using goal-based fuzzy inference system, then used neural network to estimate the true nature of the ghosts and finally learned the parameters of the strategy using co-evolutionary system [12, 13]. Aiolli et al in [5] used the simple prototype based approach. He trained the machine learning methodology by considering 17 features and determined the prototype for good and bad ghosts by averaging the features. The badness score for the new feature vector is then calculated using the normalized Euclidean distance between the features of the profile vector and the prototype vector.

4   Proposed Method

To infer the state of game board, we assume that a player usually behaves the same for a particular situation during different games. If we could understand behavior of the player at a particular situation, we could use this information to plan a strategy against that player. Depending on the previous moves, the player has taken; we can analyze the type of the ghost and can use a suitable style to compete the opponent. There can be different playing styles like aggressive playing, attacking the opponent, defending from being killed and bluffing the opponent. It can be assumed that a player could adopt the same playing style. The current availability of in-game data (board position) and behavior style of the opponent can support the researcher to learn and predict the strategy using any machine learning algorithm.

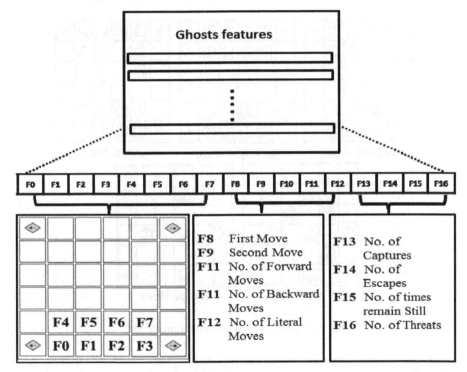

Fig. 2. Feature vector for Ghosts

To investigate the ghosts of the opponent in a particular game, we consider 17 features that can be used to profile the ghost. The impact and importance of these features is explained by Aiolli *et.al.* [5] Who used prototype based approach for the ghosts prediction. These features have been extracted from the replays of the previous year Ghosts challenge competitions. These replays are available at the website of Ghosts Challenge[2] in the form of XML format. These replays contain all the game

---

[2] https://ghosts-challenge.math.unipd.it/2013/matches

logs played between each participant. In total there are 28 logs (with 50 games between two players in each log). With the help of these game logs, we can evaluate the
behavior, analyze the player strategies and train an AI system to learn player strategies.

Based on these features, we created two standard 17-D vectors to describe the good
and bad ghosts. Among these 17 features, first eight features represent the initial position of the board. We believe that the initial setting of the ghosts in the board is the
most important part of the strategy. Since we are not sure about the identities of the
opponent ghosts (even though the identities are revealed in the game logs), we try to
extract the initial position from the initial setup of the game. It is a rule that the ghosts
has to set up initially in the middle of the least two rows in the board, we have fixed
the least row dimensions as their initial configuration for any ghost as can be seen in
fig. 2. The position of the pliers (ghosts) is represented with the binary values (0 or 1).

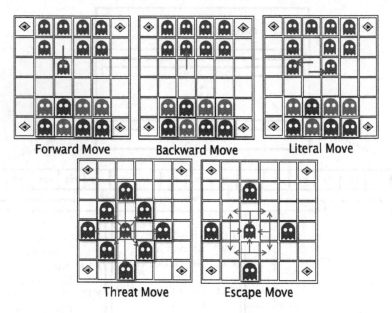

Fig. 3. Ghosts moves and behavior prediction

Next five features represent the movements of the pliers on the board in the game
session: if this piece is moved at first move, if this piece is moved as second move,
how many numbers of moves does the piece moved forward?, how many numbers of
moves does the piece moved backward?, how many numbers of moves does the piece
moved sideways? In order to find out how many number of moves does a ghost taken,
we use the configuration of the table after each ply. The table configuration provides
information about the latest position of ghosts after each turn. By comparing the two
consecutive table configurations, the movement of the ghosts are identified and
marked.

Last four features represent the behavior of the pieces: How many numbers of
pieces are stalked by the piece? (Capturing the opponent's ghost), how many times

does the piece take a move to escape from the opponent's attack?, how many times does the piece remain still? (No move) and how many times does the piece take a move to threat the opponent ghost? The number of captured pieces and the number of still moves are calculated by counting the missing pieces and no moves for each ghost, respectively. The number of threats are counted out by checking the second space of each ghosts in all directions (forward, backward and literal) and first diagonal space. The number of escapes is counted by checking the first space around the ghost. The initial positions, the moves and the behavior of the ghosts can be seen in fig. 3.

The features are extracted based on the XML data provided in the website. In order to extract the data, XML format file is first converted into an excel format for a quick and better understanding of the data. The Game IDs, Initial position and Table columns are then used to design and play the game. While playing the game, the features (movements and behavior) are calculated using the technique explained above. We have created 16 feature vectors (consisting of 17 features each) for every ghost in one game as can be seen in fig.2. A data set of 22,400 × 16 is then used for our experiments.

5   Experimental Results

Instead of setting up new programming environments or designing a prototype based approach, we use built-in open source software named "Weka" which is a well-suited for data mining tasks. Weka[3] contains a collection of machine learning algorithms that are suitable for classification [14]. We have considered the most promising machine learning algorithms in our research. These algorithms are K-Star, Bagging, PART (decision list), J48 (C4.5), RSS (Random Subspace), RC (Random Committee), LMT(Logistic Model Tree), CART (Classification and Regression Tree), IBK(K-Nearest Neighbor classifier) and RF (Random Forest), We run the experiment several times with different size of data sets. To measure the accuracy of the machine learning algorithms, we adopt a ten-fold cross validation. Since we extracted the features from the game replays and these game replays are available up-till the end of the game, we also extract the features for half-length of the game, first 10-turns length of the game and first 5-turns length of the game in order to validate the accuracy of machine learning algorithms. We also run the experiment using our data set for a prototype based algorithm explained in [5]. The results are explained below.

5.1   Evaluation with Full-Length Game Replays

In this experiment, we use the data set of the complete game. Fig. 4 shows the percentage of the correct instances for each machine learning algorithm. The correct instance means the system recognized the good ghost as good and bad ghost as bad. In this experiment, we have considered all the games between each player. Among many

---

[3] http://www.cs.waikato.ac.nz/ml/weka/

machine learning algorithms in Weka, we have considered top ten algorithms based on their performance. K- Star machine learning algorithm showed the highest performance in this experiment. K-Star is an instance based classifier that determines similar instances by using Entropy based distance function. Normally probabilistic approaches (Naïve Bayesian, Bayesian logistic Regression, naïve Bayes Updateable and so on) are promising in uncertainty handling. However, in our experiments, they have shown very low performance than those shown in the figures.

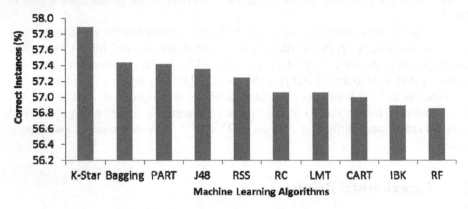

Fig. 4. Performance with complete game replays

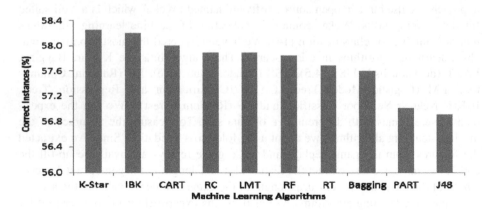

Fig. 5. Performance with half-length game replays

5.2   Evaluation with Half-Length Game Replays

In this experiment, we have extracted the features up-till half of the length of the game. Fig. 5 shows the percentage of the correct instances for each machine learning algorithm. It is seen that the performance of these experiments is not very promising (maximum performance is 58%). This is because we have considered the game replays of all the participants in the previous year competition. However, some bots performed very low in the Ghosts challenge.

5.3   Evaluation with Ten-Turn Length Game Replays

In this experiment, we have extracted the features up-till first ten turns of each game. The purpose of this experiment was to train our system with very little information about the features of the ghosts and to predict the identity of the ghost within the game. In previous experiments (i.e. Full-length and half-length), the length of each game is different. Few games finished very early while few games were draw because none of the team could win against each other. In this experiment, we decided to fix the length of each game and hence we consider first ten turns in each game. Fig. 6 shows the percentage of the correct instances for each machine learning algorithm. The results are somewhat related to the previous experiments. This is because most of the features (Initial positions (binary), first move (binary), second move (binary), threats (very less threats in first few moves), escapes (very less escapes), still moves (since the length of the game is very short, so there are very less still moves), captures (very few captures) are common in almost all the games.

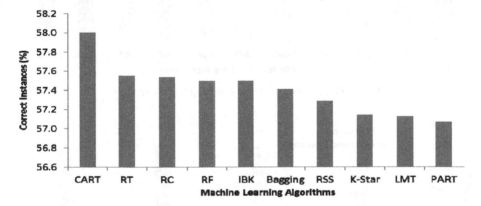

Fig. 6. Performance with ten-turn length game replays

5.4   Evaluation with Five-Turn Length Game Replays

In this experiment, we have extracted the features up-till first five turns of each game. The main focus of this experiment was to predict the identity of the ghosts based on the initial positions in order to understand the importance of the initial ghosts' settings. Since the game is in its initial stages and movement features and the behavior features of the ghost are not identified at this early stage of the game, we can say that the system can predict the identity of the opponent ghost based on the initial set-up of the ghosts. Fig. 7 shows the percentage of the correct instances for each machine learning algorithm.

5.5   Evaluation and Comparison Using Prototype Based Approach

We also use our data set to implement the prototype based approach discussed in [5]. The prototype for good or bad piece is determined by taking the average among the

344   S.S. Farooq, H. Park, and K.-J. Kim

feature vectors and a badness score is calculated using normalized Euclidean distance between the average feature vector and the new profile vector. We used ten-fold cross validation in this prototype based approach in order to compare the performance results with other machine learning algorithms. We also compare the results of prototype approach with our all experiments. In the prototype based approach, the prediction is made based on the normalized Euclidean distance between the profile vector of the unknown ghosts and the average feature vector defined for good and bad ghosts.

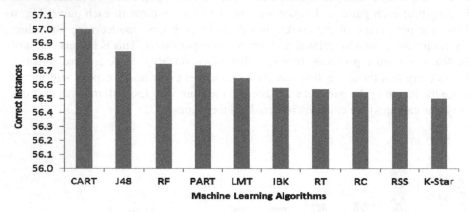

Fig. 7. Performance with five-turn length game replays

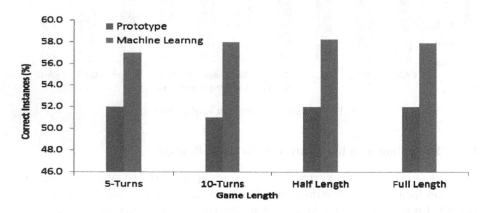

Fig. 8. Comparison of performance of Prototype based approach vs. machine learning algorithm

It can be seen that the performance of the machine learning algorithms is less in five turns and 10 turns experiment because of the less information about the features and the game states. However, the performance also decreased at the full length experiment. This is because the performance of the bots participated in the Ghosts challenge is not similar. Few are very good (like BLISS or MuTigers) while some have shown very poor performance (like Tsengine and WAIYNE1). The comparison of prototype based approach and the machine learning algorithms is shown in fig. 8.

6   Conclusion and Future Works

In this work, we have investigated the uncertain opponent game state for Ghosts game using machine learning algorithms. We use last year Ghosts competition game play data and apply various machine learning algorithms to infer uncertain game state. Also we compare our experimental results to previous prototype based approach. As a result, our proposed method shows more accurate result about six percent than the prototype based approach.

Game designers are creating highly skilled computer-controlled players that can provide challenging opportunities to game players. Instead of encoding classical AI rules, it is possible to design adaptive computer-controlled opponents which are capable of learning by imitating human players. We tried to infer game state in Ghosts game by training our system with the previous played game replays. Since the replays in the Ghosts Challenge are not human players, and the strategies that are adapted by previous year participants based on their individual learning techniques, it is challenging to realize the strategy in these replay games. However, with the help of the replays and using machine learning algorithms, we can at least train our system for a certain level to predict the unknown ghost's identity based on the feature vectors. In this work, we have used different length game replays to find out the identity of the ghosts using built in machine learning algorithms in Weka.

The performance was based on the identification of the correct instances by the algorithms. Different machine learning algorithms showed different performance on the same data. CART performance was the highest in five-turn and ten-turn length game replays while K-Star showed highest performance in half-length and full-length game replays. In this experiment, we have used all the game replays which include the replays of those participants whose bot didn't perform well in the last year competition which cause the reduction of overall performance. Also, in this experiment, we only have used 17 features. We can also find some obscure features that can help to correctly identify the ghosts.

Our long-term goal is to design a computer-controlled opponent that can learn player strategies, styles and employ them in game bot against human players. Since these game replays are not played by humans, instead the bots designed by humans, we are not sure to imitate human strategies exactly. Further experiments can be done on the data sets extracted using only the final match (i.e. BLISS vs. Mutigers) or by collecting the data using human players. It is also possible to implement further state-of-the-art machine learning techniques on the extracted datasets to find out the most important features among the feature vectors.

Acknowledgements. This work was supported by the National Research Foundation of Korea (NRF) grant funded by the Korea government (MSIP) (2013 R1A2A2A01016589, 2010-0018950).

References

1. Campbell, M., Hoane Jr., A.J., Hsu, F.: Deep Blue. Artif. Intell. 134, 57–83 (2002)
2. Schaeffer, J., Burch, N., Björnsson, Y., Kishimoto, A., Müller, M., Lake, R., Lu, P., Sutphen, S.: Checkers is solved. Science 317, 1518–1522 (2007)
3. Weber, B.G., Mateas, M.: A data mining approach to strategy prediction. In: IEEE Symposium on Computational Intelligence and Games (CIG), pp. 140–147 (2009)
4. Cho, H.-C., Kim, K.-J., Cho, S.-B.: Replay-based strategy prediction and build order adaptation for StarCraft AI bots. In: 2013 IEEE Conference on Computational Intelligence in Games (CIG), pp. 1–7 (2013)
5. Aiolli, F., Palazzi, C.E.: Enhancing artificial intelligence in games by learning the opponent's playing style. In: Ciancarini, P., Nakatsu, R., Rauterberg, M., Roccetti, M. (eds.) New Frontiers for Entertainment Computing. IFIP, vol. 279, pp. 1–10. Springer, Boston (2008)
6. Aiolli, F., Palazzi, C.E.: Enhancing artificial intelligence on a real mobile game. International Journal of Computer Games Technology, Article ID 456169 (2009)
7. Hsieh, J.-L., Sun, C.-T.: Building a player strategy model by analyzing replays of real-time strategy games. In: IEEE International Joint Conference on Neural Networks (IJCNN), pp. 3106–3111 (2008)
8. Han, A., Zhuang, Q., Han, F.: A strategy based on probability theory for poker game. In: IET International Conference on Information Science and Control Engineering, pp. 1–5 (2012)
9. Ponsen, M., Gerritsen, G., Chaslot, G.: Integrating opponent models with Monte-Carlo tree search in Poker. In: Workshops at the Twenty-Fourth AAAI Conference on Artificial Intelligence (2010)
10. Ghosts Challenge (2013), `https://ghosts-challenge.math.unipd.it/2013/`
11. Brief Description, Team "BLISS", `https://ghosts-challenge.math.unipd.it/public/docs/2013/bliss.pdf`
12. Geister Implementation Strategy, Team, MU Tigers, `https://ghosts-challenge.math.unipd.it/public/docs/2013/mutigers.pdf`
13. Buck, A., Banerjee, T., Keller, J.: Evolving a fuzzy goal-driven strategy for the game of geister. In: IEEE International Congress on Evolutionary Computation (CEC) (July 2014)
14. Hall, M., Frank, E., Holmes, G., Pfahringer, B., Reutemann, P., Witten, I.H.: The WEKA Data Mining Software: An Update. SIGKDD Explor. Newsl. 11, 10–18 (2009)

A Re-constructed Meta-Heuristic Algorithm for Robust Fleet Size and Mix Vehicle Routing Problem with Time Windows under Uncertain Demands

Kusuma Soonpracha[1,*], Anan Mungwattana[1], and Tharinee Manisri[2]

[1] Department of Industrial Engineering, Kasetsart University, Bangkok, Thailand
soonpracha@gmail.com,
fenganm@ku.ac.th
[2] Department of Industrial Engineering, Sripatum University, Bangkok, Thailand
tharinee.ma@spu.ac.th

Abstract. Recent work of the fleet size and mix vehicle routing problem with time windows mostly assumes that the input variables are deterministic. Practice in the real world, however, faces considerable uncertainty in the data. But recent research studies lack emphasis on this uncertainty. This paper focuses to contribute to a new challenging study by considering the customer demand as uncertain. This characteristic increases the difficulty for solving. The meta-heuristic algorithms are developed consisting a modification of a genetic algorithm and an adaptation of a greedy search hybridized with inter-route neighborhood search methods. Because this paper relates to uncertain customer demands, decision making is performed using the robust approach based on worst case scenarios. The final results are evaluated by using the extra cost and the unmet demand against the deterministic approach to balance the decision making.

Keywords: Robustness, Fleet size and mix vehicle routing problem, Uncertain demand, Meta-heuristic, Modified genetic algorithm, Adapted greedy search.

1 Introduction

The fleet size and mix vehicle routing problem (FSMVRP) is one of the specific problems of the classical vehicle routing problems (VRPs) in which heterogeneous fleets are composed. The heterogeneous fleets deal with real-world problems with more than the single kind of the vehicles. Moreover, the capacity of the vehicles is not the only factor used to consider the route assignment; the other variants such as time windows, split deliveries, etc. are constraints in practice. These extensions make the problems more complex and are much harder to solve than the classical VRP. This paper focuses on the time windows constraint, the FSMVRP is extended to be the fleet size and mix vehicle routing problem with time windows (FSMVRPTW).

---

*Corresponding author.

© Springer International Publishing Switzerland 2015

347

H. Handa et al. (eds.), *Proc. of the 18th Asia Pacific Symp. on Intell. & Evol. Systems – Vol. 2*,
Proceedings in Adaptation, Learning and Optimization 2, DOI: 10.1007/978-3-319-13356-0_28

Real situations include very often uncertainty. The considerable input parameters, for example, customer demands, traveled times, etc., can vary during the period of time. The solutions for future are difficult to describe precisely if the uncertain characteristics of the concerned parameters are involved. The customer demands, for example, can be revised by some reasons such as an emergency shutdown of one of a customer's production lines which might reduce the actual demands. A vehicle that is planned to serve such a customer might have some remaining spaces. Therefore, the previous route assignments should be recalculated to yield the optimal solution. In this study, the customer demands are under the assumption of the uncertainty.

Many researchers put effort into developing heuristic/meta-heuristic algorithms instead of using exact methods to handle complex problems such as VRPs. The recent algorithms published have been constructed based on the iterated local search, greedy search, tabu search, ant colony optimization, genetic algorithm, scatter search, for instance. In this research work, the algorithm is proposed a re-constructed meta-heuristic algorithm for robust vehicle routing problem with time windows under uncertain demands.

Due to the demand characters are non-deterministic, the authors propose the robustness approach in robust decision making. The output can be represented as the average, the best, or the worst solutions. It depends upon the judgment of decision makers to investigate an appropriate way for individual situation. But this work emphasizes the fact that any unexpected situations which might occur in the real world future, that the decisions must be realized even in the worst case. Therefore, the worst case scenario approach is applied.

The further details of the problems, the literature review, the proposed methodologies, the computation results, and the conclusions are discussed in next sections.

2   Literature Review

This paper is the continuation of our previous work [1] that surveys the heterogeneous vehicle routing problems and has been constructed as an overview structure which puts a special emphasis on robustness approach. The heterogeneous fleet is classified into two major classes: the heterogeneous fleet vehicle routing problem (HFVRP) and the fleet size and mix vehicle routing problem (FSMVRP). The limitation on the vehicle numbers, i.e. limited and unlimited, is the borderline to divide the problems into both types [2,3]. The FSMVRP with the unlimited transportation devices and its additional variants are focused in this study.

In 1996, Osman and Salhi have presented the vehicle fleet size and mix problem under the assumption of all concerned input data such as customer demands, number of customers, traveled times, geographical locations, service times, etc. are known with certainty [4]. It is because the characteristics of the considerable input parameters are fixed values, the obtained solutions are unique. Years later, the FSMVRP and its extensions attract the academicians to attempt the difficult talks of developing the methodologies for achieving the optimal solutions, i.e. the total cost (fixed and variable costs) minimization.

The results of the literature review show that the FSMVRPs and the extensions lack attention in the research of non-deterministic variables. The summarization matrix is shown in Table 1. The second column describes the additional variants of the FSMVRPs, the abbreviations of TW, SD, MD, and dash stand for time windows, split deliveries, multiple depots, and classical problems, respectively. The next part of the matrix indicates the sets of the model parameters consist of customer demands (CD), number of customers (NC), traveled times (TT), geographical locations (GL), service times (ST), vehicle productivities (VP), and vehicle availability (VA). The characteristics of these input variables can be classified as certainty (C) and uncertainty (U). In this study, the decision approach as shown in the last column is categorized into 3 groups: deterministic (D), stochastic (S), and robustness (R). The publication works of the same group of the considerable input parameters studies are summarized in the shading rows. The names are sorted by the year of publication.

Table 1. Fleet size and mix vehicle routing problems (FSMVRPs) and the variants literature review summarization matrix

#	Authors	Additional Variants	Considerable input parameters							Decision approach
			CD	NC	TT	GL	ST	VP	VA	
1	[4,5]	-	C	C	C	C	C	-	-	D
2	[6,7,8,9,10,11,12,13]	-	C	C	-	C	-	-	-	D
3	[14]	-	C	C	C	C	-	-	-	D
4	[15,16,17,18,19,20]	TW	C	C	C	C	C	-	-	D
5	[21,22]	TWSD	C	C	C	C	C	-	-	D
6	[23]	MD	C	C	C	C	-	-	-	D
7	[24]	MD	C	C	-	C	-	-	-	D

Table 2. Robust and stochastic vehicle routing problems (RVRPs and SVRPs) and the related research studies literature review summary matrix

#	Authors	Problems	Considerable input parameters							Decision approach
			CD	NC	TT	GL	ST	VP	VA	
1	[25]	Robust Fleet Sizing - Transport Freight	U	C	C	-	-	-	-	R
2	[26]	RVRP	U	C	C	-	-	-	-	R
3	[27]	VRPTWST	C	C	U	-	C	-	-	S
4	[28]	Road Network	C	C	U	U	C	-	-	R
5	[29,30]	RVRP	U	C	C	C	-	-	-	R
6	[31]	RVRPTWSD	U	C	U	C	-	-	-	R
7	[32]	SVRP	U	C	-	C	-	-	-	S
8	[33]	RVRPTW	C	C	U	-	C	-	-	R
9	[34]	RVRP	U	C	-	C	-	-	-	R
10	[35]	SVRP	U	C	-	C	-	-	-	S

The stochastic or uncertainties of the input variables are employed in the other specific problems of VRPs but not for the FSMVRPs and its extensions. Table 2 demonstrates the survey summary in the same way as mentioned above in Table 1. But, the second column shows the names of the specific VPRs instead of the FSMVRP variants. The matrix of the robust and stochastic vehicle routing problems and the related research studies indicate that when the problems investigate insight in the real world by modeling one of the parameters as non-deterministic, the robust or stochastic decision making approaches are referred to handle the unknown futures.

It is well known that the VRPs have a complexity of NP-hard problems, so the FSMVRPs have as well. Therefore, most researchers put emphasis on developing heuristic/metaheuristic algorithms to produce (near)-optimal solutions. As this research is an extension of previous research that has concentrated on the literature review, the chronological order of the proposed methodologies can be concluded as shown in Table 3.

Table 3. Methodology approaches for the fleet size and mix vehicle routing problems (FSMVRPs) and the variants

Year	Authors	Problem	Algorithm
1996	[4]	FSMVRP	Route perturbation procedure and tabu search
1997	[23]	FSMVRPMD	A multi-level (p-level) composite heuristic
1999	[6]	FSMVRP	Generalized insertion and unstringing/ stringing, tabu search using sweep procedure, adaptive memory procedure
2002	[5]	FSMVRP	Sweep-based algorithm approach and suborders of petals selection
2002	[7]	FSMVRP	Tabu search (TS) mixed with reactive TS concepts, variable neighborhoods, data-memory structures, and hashing functions
2002	[15]	FSMVRPTW	Adapted combine savings, adapted optimistic opportunity savings, adapted realistic opportunity savings
2007	[16]	FSMVRPTW	Insertion-based parallel approach and a meta-heuristic procedure that adopts the ruin and recreate paradigm for current solution improvement
2007	[17]	FSMVRPTW	Scatter search approach
2008	[18]	FSMVRPTW	Multi-restart deterministic annealing with 3 phases algorithm
2009	[8]	FSMVRP	Tabu search and the generalized insertion and neighborhood reductions
2009	[9]	FSMVRP	Genetic algorithm (GA) applied local search mutation
2009	[10]	FSMVRP	GA hybridized with a local search and distance measure in solution space
2009	[19]	FSMVRPTW	Three-phase hybridized meta-heuristic
2009	[21]	FSMVRPTWSD	Scatter search approach
2009	[24]	FSMVRP and FSMVRPMD	Exact algorithm based on the set partitioning formulation using 3 types of bounding procedures
2010	[20]	FSMVRPTW	Adaptive memory programming solution approach, semi-parallel construction heuristic, and tabu search
2011	[12]	FSMVRP	Iterated local and a set partitioning formulation
2011	[14]	FSMVRP	Hybridized heuristic based on iterated local search useing a variable neighborhood descent procedure, with a random neighborhood ordering
2012	[13]	Fleet composition	9-step meta-heuristic based on evolutionary algorithms and local search
2012	[36]	Fleet composition	A ring radial topology continuous model to define vehicle zones and types
2013	[22]	FSMVRPTWSD	Scatter search approach

3   Problem Description

The problem description is divided into two major parts: the fleet size and mix vehicle routing problem with time windows (FSMVRPTW), and the robust fleet size and mix vehicle routing problem with time windows (RFSMVRPTW).

3.1   Fleet Size and Mix Vehicle Routing Problem with Time Windows

The fleet size and mix vehicle routing problem with time windows is an extension of the fleet size and mix vehicle routing problem (FSMVRP) that is a specific problem of the classic vehicle routing problems (VRPs). The FSMVRPTW can be formed on the directed graph $G = (N, A)$. N represents the node set consisting of the customers and a depot. $N = \{0, 1, ..., n+1\}$. The depot is denoted by two nodes of $\{0\}$ and $\{n+1\}$, the remaining set of nodes $C = \{1, 2, ..., n\}$ is a given customer set. A is the arc set that design the routes.

In this paper, the FSMVRPTW is classified into two points of view: customers and fleet size and mix vehicles. The problem formulations are based on the models as proposed by several researchers [16],[17],[18],[20]. The customers are represented by the node set, where $C = \{1, 2, ..., n\}$, a location of each individual customer scatters around a depot represented by the graph $G = \{N, A\}$. Each customer must be visited by exactly one vehicle. In this FSMVRPTW, the time windows are determined by the customer i denoted as the earliest and latest arrival time, $[e_i, l_i]$ and e_i is less than or equal to l_i. The time windows are supposed to be of the hard type in this paper. It means that the customers do not allow any services to violate the time windows constraint. The truck that arrives at the destination too early has to wait until the earliest time permission is opened [15]. In this research, the demand of the customers, d_i is stochastic or uncertain and is modeled as shown in the next topic.

The fleet size and mix vehicle routing problem is approached at the strategic level, the problem is initiated from the assumption that there are unlimited number of available vehicles [3]. The fleet is heterogeneous with K different types of vehicles, such type $K = \{1, 2, ..., k\}$, are composed in order to serve all customers' uncertain demand d_i, that is particular for this paper. Each vehicle type k is an element of K, and has a capacity q^k and $q^1 < q^2 < ... < q^K$. Alike other general capacitated VRPs, each truck can carry a maximum of its capacity. Repoussis and Tarantilis [20] present two addition formulations that eliminate all possible infeasible sub-tour. The accumulated carried weight of each vehicle k, a_i^k is computed and to make sure that the truck k will not service the customer j if its total weight moment at the customer j, in the route from i to j, is exceeded the maximum capacity of the truck.

Recall the graph $G = (N, A)$, there is only one depot, it is the centralized node of the graph. The depot is represented by two nodes of $N = \{0\}$ and $\{n+1\}$. Every arc, called route of the vehicle fleet, must be started at the depot, linked to the other assigned customer(s), and ended a loop by returning to the depot. Once the first customer in a path is visited, a vehicle has to leave from that place, continues to the next one until the tasks are completed. The depot is also determined the time interval restriction, denoted by $[e_0, l_0] = [e_{n+1}, l_{n+1}]$. As described in the above part of customer definition, when a truck k arrives at customer i, it is allowed to begin the unloading services, denoted by service beginning time y_i^k, within the time windows of such customer. Two possible events of the truck arrival time are considered, 1) too early arrival and 2) in due time arrival, it is noted that time windows are hard, too late arrival is not permitted. In the case of a truck arriving early, the useless activity is considered as waiting time. Anyhow, both cases can be modeled the moment at which service begins at customer j, by setting $y_j^k = \max\{0,\ e_j - (y_i^k + st_i + t_{ij})\}$. The feasible schedule for each vehicle route can be guaranteed by forces y_i^k to zero whenever customer i is not visited by vehicle k [20] and be imposed a minimum time for beginning the customer service j in a determined route with no subtours guarantee [17]. The further trick linearizes the formulation by using the big M-method, that may be replaced by $\max\{l_i + st_i + t_{ij} - e_j,\ 0\}$ for all (i, j) are the elements in A and for all k are the members of K [18].

In each route, two types of cost are considered for the total transportation cost, a fixed acquisition cost and a variable cost. The fixed acquisition cost, f^k is for a vehicle of type k where $f^1 < f^2 < \ldots < f^K$. The variable cost, c_{ij} is the cost of traveling from customer i to j, (i, j) is a set of arc A. The travel distance dt_{ij} and the travel time t_{ij} are given and can be obtained by joining a pair of node. The symmetry and deterministic properties are imposed for both parameters. Further, a unit of distance is assumed to be equaled to one and has the same unit of the travel time, t_{ij} [15]. Because the time windows are stated of being constrained in this case, assume the variable cost equals to the total time spending along the determined route. The total time spending is computed by considering three types of usage time consisting of 1) traveling time between a pair of nodes (t_{ij}), 2) service times (st_i) that the truck spends for performing a loading or unloading activity at each customer site and 3) waiting time (w_i) that can occur only if the truck arrives the customer i before the permitted earliest time. In the general problem, not yet the robustness case, the objective function of the FSMVRPTW may consider three components, 1) fixed acquisition cost, 2) variable traveling cost and 3) waiting time and/or service time consideration as the ,en route' cost. In this paper, the FSMVRPTW is the total summation of the fixed cost obtained from vehicle fleet composition acquisition and the sum of total times spending including waiting times, both components are demonstrated in the first and second term of the FSMVRPTW objective function (equation 1), respectively.

$$Z_{FSMVRPTW} = \min \sum_{k \in K} f_k \sum_{j \in N} x_{0j}^k + \sum_{k \in K} (y_{n+1,k} - y_{0,k}) \tag{1}$$

3.2   Robust Fleet Size and Mix Vehicle Routing Problem with Time Windows (RFSMVRPTW)

In this paper, the term "*robustness*" refers to the solution robustness in which the obtained solution remains *close* to optimal for all scenarios. The specific definition of the *robustness* is applied the definitions as definited by Kouvelis and Yu [37], Manisri *et al.* [33], and Moghaddam *et al.* [30] as following:

Definition 1: A scenario s is a set of customer demands realizations, Ud. A whole system S is a combination of individual scenario in which $s_1 \cup s_2 \cup \ldots \cup s_n \in S,\ \forall s \in S$.

Definition 2: A scenario s_i is a representation of a system in which the customers' demands are uncertain by the impact of individual customer's behavior based on risk aversion $(\beta^s \alpha^s)$. The permutation percentage, β^s, of each customer is randomly generated, assuming as uncertainty represented by the uniform distribution of $[\beta_l^s, \beta_u^s]$, and the normal distribution of $N(\mu, \sigma)$. The β_l^s and β_u^s are the lower and upper bounds of an avoidance of the risk. The symbols μ and σ are the mean and the variance of the risk. The customer demands of each scenario are either lower or higher than the expected values. The independent random value, α^s, of -1 and 1 are assumed as generated randomly to indicate the direction of the uncertain demand that makes the value of the demand becomes lower or upper the expected value.

Definition 3: By the definitions 1 and 2, a scenario is a set of uncertain customers' demands Ud, modeled as $Ud_i = (1+\beta^s\alpha^s)d_0$ where $i = 1, 2, ..., n$; d_0 is an expected demand of customer i, and $s \in S$.

A mathematical formulation for the RFSMVRPTW belongs to the FSMVRPTW but the customer demand (d_i) is replaced by the set of uncertain demand (Ud) model as modified [30]. The original model assigns one single fixed value of the percentage deviation of the risk averse for all customers and for each scenario, but in this paper the risk averse depends upon each customer's behavior which is represented by a random value with the uniform and normal distribution. The uncertain demand is possible to be lower or higher than the expected demand (d_0) that depends on the independent random variable and has a value between [-1,1].

As presented in the robust handbook of Kouvelis and Yu [37], this concept is applied in some research such as [31] and [33]. In this research, the robust decision making framework is adapted the concept of Kouvelis and Yu [37] but the final result is evaluated against the the deterministic approach [26]. Thus it can balance between the expensive cost when a robust approach is applied and the unmet need when the deterministic approach has to suffer if the worst case happens.

Even this research assumption considers the uncertain input variables, the robust discrete optimization is suggested by using the minimax criterion to reduce the complexity of the problems. The minimax criterion is one of the worst case approaches. The criterion aims to evaluate the highest level of cost taken across all possible future input data scenarios to be as low as possible, as a result that the outcome can protect the worst that might happen [37].

Referring to Kouvelis and Yu [37], let X be the set of the decision variables and D^s denotes the instance of the input data that corresponds to scenario s. The notation F^s stands for the set of all feasible decisions when the scenario s is realized. The function $f(X, D^s)$ is used for evaluating the robustness quality of the decision $X \in F^s$. Then, the optimal single scenario decision X^{s^*} for the input data instance D^s is the solution to a deterministic optimization problem and it satisfies Equation 2.

$$z^s = f(X^{s^*}, D^s) = \min_{X \in Fs} f(X, D^s) \qquad (2)$$

The proactive robustness approach is focused to benefit in long run planning by hedging against all scenarios. The absolute robustness is one of the proactive robustness approaches that is applied for RFSMVRPTW. The absolute robust decision X_A is defined as the one that minimizes the maximum total cost, among all feasible decisions over all realizable input data scenarios. The absolute robust decisions are of a conservative nature, as they are based on the anticipation that the worst might happen. One way to motivate such a criterion is for competitive situations where the parameters of the decision model are affected by competitors' actions. The main uncertainty to this RFSMVRPTW concentrates on the customer demands. The demands vary over a pre-specified planning horizon. Thus, the fleet and routing designs are decided over a long period of time for reducing the impact on the system effectiveness. It means that the solutions obtained from the decision making are good enough for a variety of future operating scenarios and this is referred to the term *robustness* [27].

Recall $s \in S$ be the input data scenario index and S be the set of all possible scenarios. The objective function of RFSMVRPTW is to minimize the maximum total cost of FSMVRPTW (Equation 1) is in placed by the absolute robustness as shown in Equation 3. The robust objective function is subject to the constraints as common used in the fleet size and mix vehicle routing problems. The comprehensive meaning of the robust optimization solution i.e. solution of the total transportation cost is good for all possible data uncertainty and hedge against the worst case.

$$Z_A(RFSMVRPTW) = \min_{X,Y}\max_{s \in S}(\sum_{k \in K}f_k\sum_{j \in N}x_{0j}^k + \sum_{k \in K}(y_{n+1,k} - y_{0,k})) \tag{3}$$

4   Solution Approach

Resulting from the reviews, the heuristic and metaheuristic algorithms can be either renewed or innovated to solve the RFSMVRPTW under uncertainty of an input parameter such as travel time, demand, etc. In this paper, the uncertain characteristics of the customer demands are focused, so the modification of randomized search heuristics based on genetic algorithms is suggested. The heuristics and me-tageuristic is reconstructed by performing three major phases. The first phase is to build an initial solution, and pass to the next phase for improvement. The robustness is generated in the last phase based on worst case scenarios.

Even the customer demands in this paper are considered as uncertainty, this complexity is reduced by converting the uncertainty to deterministic scenario-based approach. A single scenario is a representation of a set of input data uncertainty to the decision model, and all total assumed cases represent a whole system. The number of realizable scenarios over a pre-specified planning horizon normally depends upon a person who takes charge of the strategic planning task or a person who gets involved in managing the customers' demands information. The scope of this research does not involve finding the potential number of scenarios, so several numbers of scenarios will be assumed to be the representations of realistic situations. The first and second phases are processes until all scenarios are completely solved. All results are passed through the last phase for finding the robustness solution.

4.1   Phase I: Initial Solution Construction

The re-constructed meta-huristic based on modified genetic algorithm (mGA) which adapts the algorithm [38] is used to construct an initial solution of a giant tour. Two kinds of operations in genetic algorithms are induced: crossover and mutation. Kirk [38] developed the GA process using a one-point crossover with an order-based operator. The one-point crossover operator randomly selects one crossover point and then copies everything before this point from the first parent and then everything after the crossover point copy from the second parent. The mutation process uses a frame-shift based on flip operation. The pair of genes substitution is applied by a swap operation. By believing that the best solution can heal the weak cells, the original program has been modified by memorizing the best solution. After the solutions have been generated for several generations, the best solution will take place either a randomized solution or the worse solution. According to this assumption, the next generation result will be improved by then.

4.2   Phase II: Solution Improvement

Step 1: Route Insertion Based TSP Ordered Hybridized Shift Operation

In this phase, the single route of the TSP is converted into multiple routes of the VRP. A single customer node inserts to each route by TSP ordered but the algorithm is designed to check the lower and upper bounds of the time intervals. If the ordered node that is inserted into the route violates the time windows constraint, the next city will be considered. The algorithm is programmed by applying the shift procedure. Meanwhile, the algorithm considers the result of a single iteration and uses the best result obtained from each shift operation to form the next arc of the current route. The tour is constructing continuously until time windows constraint is violated. The other routes are created until all customers are completely assigned by a single vehicle.

Step 2: Route Merging

It is according to one component of the objective function is to achieve the minimization of the 'en route' time travelled. The waiting time is the critical key that impacts the objective. It means that the customer who has the least open time window with least waiting time should be prioritized to serve first. Thus, the procedure is designed to perform ascending the order of the waiting times once the random tours are merged. After having combined the routes and sorted the waiting times, a greedy search is activated. The greedy algorithms build up a solution piece by piece, always choosing the next piece that offers the most obvious and immediate benefit [39].

Step 3: Fleet Sizing and Mixing

In this step, the number of scenarios is determined. The input variable data sets are generated according to the uncertain demand model as described by Definition 3. The process is executed in a one by one scenario. The total uncertain customer demand of each assigned tour is calculated. Vehicle matching is performed by selecting the best fit between the total demands and the vehicle type in which the remaining spaces after loaded is as less as possible. The total transportation cost composes the fixed vehicle cost and the total en route time travelled is performed the calculation after the fleet size selection is done. The process is continued and is terminated after the predetermined number of iterations is reached.

Step 4: Inter-route Neighborhood Search Methods

The neighborhood search procedures using the concept of the inter-route moves based on relocation, exchange, and cross is applied. The major scheme is to randomize the sets of a total number of routes, the route orders, and the customer nodes for performing the inter-route moves. The procedure is designed either for a pair-route and a set-route of the randomized sets as mentioned previously. If an infeasible solution is obtained, then steps 1-3 are recalled for regenerating a feasible result. The best solution is memorized and will be replaced by the current solution if its outcome is better. The program is terminated when the determined iteration number is reached.

4.3   Phase III – Robustness Decision Making

The proactive robustness is concentrated in this paper by assuming the planning is decided over a long period of time for reducing the impact by the demand uncertainty on the system. The decision making in this research is supposed to perform before the fact and using the expected demands of the original fleet size and mix vehicle routing problem represent the actual realized data. From phase I and II, the scenarios of the input uncertain demands of the customers are created. The previous process seeks for a solution of each run for each scenario in which the total transportation cost is minimized. The robust solution is evaluated using three criteria: absolute difference robustness criteria, relative difference robustness criteria, and variable or deviation robustness criteria. That is to select the maximum solution among all decisions of each scenario and to perform the calculation among Equation 3. The worst-case implementation is proposed to find the solution which is hedges against the worst of all possible scenarios.

In this research, the evaluation of the robust solution results uses the extra cost comparison and the unmet demand indicators [26]. The extra cost performance measurement indicator (ratio x) quantifies the relative extra cost of the robust with respect to the cost of the deterministic. It means that once the robust approach is selected, there are the additional costs caused by the worst case based consideration. If the deterministic approach is purposed, this extra cost is not suffered. The extra cost ratio is calculated using the Equation 4.

$$\text{Extra cost ratio, } x = (Z_{RFSMVRPTW} - Z_{FSMVRPTW}) / Z_{RFSMVRPTW} \qquad (4)$$

$$\text{Unmet demand ratio, } u = \max Ud / \sum_{i \in C} d_i^0 \qquad (5)$$

The unmet demand performance measurement indicator is used to show the effect if the deterministic optimization is applied to the uncertain data problems. The unmet demand denoted by the ratio u is a performance indicator used to measure the demand when facing with the worst case. The deterministic approach is chosen to solve the problem under the expected demand, but the demands are vary in the real situations. The unmet demand indicator quantifies the relative maximum unsatisfied demand (max Ud) of the uncertainty based scenarios with respect to the total expected demands of the deterministic problem. The unmet demand ratio is calculated using the Equation 5. It is because the robust approach based worst case scenarios is designed to protect against the worst of all possible scenarios, thus the unmet demand of the robust decision making is equaled to zero.

Two proposed robust performance indicators: extra cost and unmet demand, are used for balancing between the expensive cost when a robust approach is applied and the unsatisfied demands when the deterministic approach has to suffer once the worst case happens. The additional solution performance measurement is compared with some benchmark problem sets. Due to the non-existence of recent published papers on the RFSMVRPTW, the solutions resulted from some FSMVRPTW research works, for example, [18] and [20] are used to examine the competitive performance obtained from the proposed methodology.

5   Computational Results

The proposed metchologogy of the modification of a genetic algorithm and the adaptation of a greedy search hybridized with inter-route neighborhood search methods has been programmed in MATLAB and has ran on an Intel(R) Core(TM) i5-3337U CPU@1.80GHz 8.00GB-RAM.

In this paper, the first trial of the re-constuctued meta-heuristic algorithm for RSFMVRPTW is tested on the data set of the first problem, i.e. R101 with 100-customer, of the well-known benchmark problem sets generated by Solomon. The geographical data are randomly generated in R101, a short scheduling horizon and only a few customers per route allowing are the characteristics of this problem as stated by Solomon [40]. The cost structure of Liu and Shen [42], as referred in [18],[20] is the benchmark problem using for the experiment. The performance of the algorithm is compared with three previous outputs of the other authors, i.e. (A) the best known solution that have collected from the survey [20], (B) the result obtained by the adaptive memory programming [20], and (C) the multirestart deterministic annealing meta-heuristic [18].

On the real business, the scenarios are determined by the decision makers who are authorized in the decision making of the indiviual problem. In this paper, the algorithm has been operated to handle three scenarios. The first scenario (Scen-1) uses the expected demands using the original given demands of the benchmark problem set. The second scenario (Scen-2) assumes that the demands are patterned as the uniform distribution of the risk averse $(\beta^s \alpha^s)$ within [-1,1] interval are used for generating β^s, further about the α^s, only the upper interval of 100 is determined. The last scenario (Scen-3) has the same characteristic of β^s, but α^s is represented as the normal distribution using the the concept of the weekly demand as proposed in [41] where the expected demand is converted based on weekly i.e. the monthly expected and the variance of the demand is divided by 4.33 (52 weeks per year/12 months) and the square root of 4.33, respectively. The results are determined as the mean and variance, respectively. It is because 1) the algorithm has been programmed based on random permutation selection in Phase I and Phase II, and 2) Scenario 2 and Scenario 3 assume that the customer demands are not certain, the output is different for each run. Thus, each scenario is repeated the execution for ten runs.

The computation results demonstrate in a cost form (*1,000) of {DC;FC;TC} where the abbrevations in the bracket denote the distance cost (DC), fleet cost (FC), and total cost (TC) of all three scenarios, respectively. The average outcomes of each scenratios are as following: Scen-1: {4.01;2.74;6.75}, Scen-2: {4.01;2.77;6.78}, and Scen-3: {4.09;2.71;6.80}. The total cost deviation of three scenarios are eqaul to 0.08, 0.27, and 0.10 for Scen-1, Scen-2, and Scen-3, respectively. The Scen-1, Scen-2, and Scen-3 have the average CPU run time (RT) as eqaul to 78.125, 78.996, and 90.507, respectively. Recall Formulation 3, the summarization of the maximum total cost of FSMVRPTW among all three scenatios for all 15 runs can be listed as following: {6.71;6.93;6.79;6.87;6.85;6.86;7.28;7.10;6.85;6.77}, it results to obtain the minimize the maximum total cost, i.e. $Z_{RFSMVRPTW}$, equals to 6,706.35. This result is called the robustness solution. Such solution of the total transportation cost is good for all possible data uncertainty and hedge against the worst case.

The second report shows the evaluation of the robust solutions used the extra cost comparison and the unmet demand indicators [26]. The robust solution is obtained from the whole scenarios, i.e. $Z_{RFSMVRPTW} = 6,706.35$ and the deterministic solution is based on Scen-1, i.e. $Z_{FSMVRPTW} = 6,621.83$. The extra cost ratio (x) equals to 0.0126, it means that when the robustness solution is selected, this extra cost percentage is on top of the normal case. The maximum uncertain demand (maxUd) as equals to 1,628 is selected from the whole runs that the program has generated based on the uncertain demand model. The total value of the expected demands (totalD_o), i.e. 1,458, is the base case based on the data set as determined in the benchmark problem. The result of unmet demand ratio (u) is equal to 1.1166. It indicates that if the deterministic solution is chosen, it has to plan for suffering the unknown demands with such extra ratio of the expected demand that might be occurred in some periods of time.

The last performance indicators show the evaluation of the percent improvement among the previous outputs of the other authors. The letters, A,B, and C represent the best known solutions that have been collected from the survey [20], the result obtained from the adaptive memory programming [20], the multirestart deterministic annealing mera-heuristic [18], respectively. The comparison between the deterministic and deterministic solution of ten experiments for each scenario shows the improvement of 4.41%, 4.54%, and 1.95% in average when compare with A, B, and C, respectively. Further, the result comparisons of the robust solution against the deterministic solution indicate that the outcomes of 4.96%, 5.10%, and 2.52% are improved in average when compare with A, B, and C, respectively.

6   Conclusions and Discussions

In this paper, the re-constructed meta-heuristic aglorithm based on the modification of the genetic algorithm and the adaptation of the greedy search hybridized with inter-route neighborhood search methods are proposed to solve the robust fleet size and mix vehicle routing problem with time windows under uncertain demands. The results indicate that by applying this technique, the average total cost of each scenario is not quite different. The deviation of the solutions of Scenario 1 (base case) and Scenario 3 (demands with normal distribution) are less than Scenario 2 (demands with uniform distribution). The robust decision making is performed based on the worst case scenario, the minimum of the maximum over the whole scenario is evaluation as the robustness solution in this experiment. The solution is preferred to select for hedging against the worst of all possible scenarios that might occur in the unknown future if the uncertain situtation(s) is involved in the problem. In this paper the robustness performance is evaluated by using the extra cost and the unmet demand, both criterions are for the decision maker to consider when the robust approach is implemented. The final results are compared to three benchmark solutions. In the base case (scenario-1), the total cost is reduced compared among all three previous known. When the robustness approch is implemented, the solutions are still significant in the competitive performance.

As mentioned in the previous section, this is the first experiment of this proposed algorithm. Thus, in order to illustrate that the proposed technique is more efficient and competitive, the main recommendation is that the addition future researches should be conducted as following: 1) the other problem sets of the well-known benchmark problem have to be tested, 2) the experiment of this developed methodologies should be performed and implemented in the real business cases 3) the number of trial runs have to be re-considered, and 4) the other meta-heuristic algorithms such as fuzzy logic, differential evolution, ant corony optimization, particle swarm optimization, articificial neural network, hybrid evolution swarm with local search methods, etc. should be cosidered for the comparative studies.

References

1. Soonpracha, K., Mungwattana, A., Janssens, G.K., Manisri, T.: Heterogeneous VRP Review and Conceptual Framework. In: The International MultiConference of Engineers and Computer Scientists, APIEMS 2014, Hong Kong, pp. 1052–1059 (2014)
2. Hoff, A., Andersson, H., Christiansen, M., Hasle, G., Lokketangen, A.: Industrial aspects and literature survey: fleet composition and routing. Computers & Operations Research 37(12), 2041–2061 (2010)
3. Baldacci, R., Battarra, M., Daniele, V.: Routing a heterogeneous fleet of vehicles. Technical Report DEIS OR.INGCE. 1 (2007)
4. Osman, I.H., Salhi, S.: Local search strategies for the vehicle fleet mix problem. In: Rayward-Smith, V., Osman, I., Reeves, C.R., Smith, G. (eds.) Modern Heuristic Search Methods, pp. 131–154. John Wiley & Sons, Chichester (1996)
5. Renaud, J., Boctor, F.F.: A sweep-based algorithm for the fleet size and mix vehicle routing problem. European Journal of Operational Research 140, 618–628 (2002)
6. Gendreau, M., Laporte, G., Musaraganyi, C., Taillard, É.D.: A tabu search heuristic for the heterogeneous fleet vehicle routing problem. Computers & Operations Research 26, 1153–1173 (1999)
7. Wassan, N.A., Osman, I.H.: Tabu search variants for the mix fleet vehicle routing problem. Journal of the Operational Research Society 53, 768–782 (2002)
8. Brandão, J.: A deteministic tabu search algorithm for the fleet size and mix vehicle routing problem. European Journal of Operational Research 195, 716–728 (2009)
9. Liu, S., Huang, W., Ma, H.: An effective genetic algorithm for the fleet size and mix vehicle routing problems. Transportation Research Part E 45, 434–445 (2009)
10. Prins, C.: Two memetic algorithms for heterogeneous fleet vehicle routing problems. Engineering Applications of Artificial Intelligence 22, 916–928 (2009)
11. Baldacci, R., Mingozzi, A.: A unified exact method for solving different classes of vehicle routing problems. Mathematical Programming 120, 347–380 (2009)

12. Subramanian, A., Penna, P.H.V., Uchoa, E., Ochi, L.S.: A Hybrid Algorithm for the Fleet Size and Mix Vehicle Routing Problem. In: International Conference on Industrial Engineering and Systems Management (2011)

13. Redmer, A., Żak, J., Sawicki, P., Maciejewski, M.: Heuristic approach to fleet composition problem. Social and Behavioral Sciences 54, 414–427 (2012)

14. Penna, P.H.V., Subramanian, A., Ochi, L.S.: An iterated local search heuristic for the heterogeneous fleet vehicle routing problem. Journal of Heuristics 19(2), 201–232 (2011)

15. Dullaert, W., Janssens, G.K., Sörensen, K., Vernimmen, B.: New heuristics for the fleet size and mix vehicle routing problem with time windows. Journal of the Operational Research Society (2002)

16. Amico, M.D., Monaci, M., Pagani, C., Vigo, D.: Heuristic approaches for the fleet size and mix vehicle routing problem with time windows. Transportation Science 41(4), 516–526 (2007)

17. Belfiore, P.P., Fávero, L.P.L.: Scatter search for the fleet size and mix vehicle routing problem with time windows. Central European Journal of Operations Research 15, 351–368 (2007)

18. Bräysy, O., Dullaert, W., Hasle, G., Mest, D.: An effective multirestart deterministic annealing metaheuristic for the fleet size and mix vehicle routing problem with time windows. Transportation Science 42(3), 371–386 (2008)

19. Bräysy, O., Porkka, P.P., Dullaert, W., Repoussis, P.P., Tarantilis, C.D.: A well-scalable metaheuristic for the fleet size and mix vehicle routing problem with time windows. Expert Systems with Applications 36(4), 8460–8475 (2009)

20. Repoussis, P., Tarantilis, C.: An effective multirestart deterministic annealing metaheuristic for the fleet size and mix vehicle routing problem with time windows. Transportation Research Part C 18, 695–712 (2010)

21. Belfiore, P., Yoshizaki, H.T.: Scatter search for a real-life heterogeneous fleet vehicle routing problem with time windows and split deliveries in Brazil. European Journal of Operational Research 199, 750–758 (2009)

22. Belfiore, P., Yoshizaki, H.T.: Heuristic methods for the fleet size and mix vehicle routing problem with time windows and split deliveries. Computers & Industrial Engineering 64, 589–601 (2013)

23. Salhi, S., Sari, M.: A multi-level composite heuristic for the multi-depot vehicle fleet mix problem. European Journal of Operational Research 103, 95–112 (1997)

24. Baldacci, R., Mingozzi, A.: A unified exact method for solving different classes of vehicle routing problems. Mathematical Programming 120, 347–380 (2009)

25. List, G.F., Wood, B., Nozick, L.K., Turnquist, M.A., Jones, D.A., Kjeldgaard, E.A., Lawton, C.R.: Robust optimization for fleet planning under uncertainty. Transportation Research Part E 39, 209–227 (2002)

26. Sungur, I., Ordónez, F., Dessouky, M.: A robust optimization approach for the capacitated vehicle routing problem with demand uncertainty. Inst. Ind. Eng. Trans. 40(5), 509–523 (2008)

27. Janssens, G.K., Caris, A., Ramaekers, K.: Time Petri nets as an evaluation tool for handling travel time uncertainty in vehicle routing solutions. Expert Systems with Applications 36, 5987–5991 (2009)

28. Yin, Y., Madanat, S.M., Lu, X.-Y.: Robust improvement schemes for road networks under demand uncertainty. European Journal of Operational Research 198, 470–479 (2009)

29. Sörensen, K., Sevaux, M.: A Practical Approach for Robust and Flexible Vehicle Routing Using Metaheuristics and Monte Carlo Sampling. Journal of Mathematical Modelling and Algorithms 8, 387–407 (2009)

30. Moghaddam, B.F., Sadjadi, S.J., Seyedhosseini, S.M.: Comparing mathematical and heuristic methods. International Journal of Research and Reviews in Applied Sciences 2(2), 108–116 (2010)
31. Zhu, J., Gao, M., Huang, J.: A robust approach to vehicle routing for medical supplies in large-scale emergencies. In: International Symposium on Emergency Management (ISEM 2009) (December 2009)
32. Aguirre, A., Coccola, M., Zamarripa, M., Méndez, C., Espuña, A.: A robust MILP-based approach to vehicle routing problems with uncertain demands. In: 21st European Symposium on Computer Aided Process Engineering - ESCAPE 21, pp. 633–637 (2011)
33. Manisri, T., Mungwattana, A., Janssens, G.K.: Minimax optimisation approach for the robust vehicle routing problem with time windows and uncertain travel times. International Journal of Logistics Systems and Management 10(4), 461–477 (2011)
34. Moghaddam, B.F., Ruiz, R., Sadjadi, S.J.: Vehicle routing problem with uncertain demands: An advanced particle swarm algorithm. Computers & Industrial Engineering 62, 306–317 (2012)
35. Goodson, J.C., Ohlmann, J.W., Thomas, B.W.: Cyclic-order neighborhoods with application to the vehicle routing problem with stochastic demand. European Journal of Operational Research 217, 312–323 (2012)
36. Jabali, O., Gendreau, M., Laporte, G.: A continuous approximation model for the fleet compositon problem. Transportation Research Part B 46, 1591–1606 (2012)
37. Kouvelis, P., Yu, G.: Robust discrete optimization and its applications. Kluwer Academic Publishers, Dordrecht (1996)
38. Kirk, J.: Mathlab Central (2007), http://www.mathworks.com
39. Dasgupta, S., Papadimitriou, C., Vazirani, U.V.: Algorithms, 1st edn., McGraw-Hill Science/Engineering/Math. (2006)
40. Solomon, M.M.: VRPTW Benchmark Problems (2005), http://w.cba.neu.edu/~msolomon/problems.html
41. Cachon, G., Terwiesch, C.: Matching Supply with Demand: An Introduction to Operations Management, 3rd edn. McGraw-Hill (2011)
42. Liu, F.H., Shen, S.Y.: The fleet size and mix vehicle routing problem with time windows. Journal of the Operational Research Society 50, 721–732 (1999)

The Influence of Elitism Strategy on Migration Intervals of a Distributed Genetic Algorithm

Takeshi Uchida[1], Teruo Matsuzawa[2], and Yasushi Inoguchi[2]

[1] Salesian Polytechnic, Tokyo, Japan
uchida@salesio-sp.ac.jp
[2] Japan Advanced Institute of Science and Technology, Ishikawa, Japan
{matuzawa,inoguchi}@jaist.ac.jp

Abstract. A distributed genetic algorithm is an important technique on practical use. To parallelize distributed genetic algorithms, researchers have been discussing various advanced algorithms with reduced migrations. An interesting previous study shows that both too long migration interval and too short migration interval cause a degraded performance in finding solutions. This paper assumes that a cause degrading the performance is a behavior of elites and also discusses a modified elitist model. The experiments show that a modified elitist model improves the performance even if the migration intervals are not set appropriately. These results seem to be design guides for discussing distributed genetic algorithms with reduced migrations.

Keywords: evolutionary computation, genetic algorithm, island model, multiple populations, migration, elitist model, convergence.

1 Introduction

Genetic algorithms are stochastic search methods that have been successfully applied in many search, optimization and machine learning problems [1]. Genetic algorithms have also been powerful tools for finding solutions in a reasonable time and of an acceptable quality. Recently, a need for solving complicated large-scale problems [2] drives the evolutionary computation community towards advanced models of evolutionary algorithms. An advanced model of genetic algorithms is a distributed genetic algorithm [3].

A distributed genetic algorithm has multiple populations (islands) which evolve simultaneously and interact with each other by means of migrations. These migrations are very important factors to find better solutions. There exist a lot of theoretical studies and empirical studies on the island model, interconnections between islands and so on [3,4,5]. In studies on parallelization of distributed genetic algorithms, researchers have been discussing migration schemes to reduce the amount of migration [7,8,9].

An interesting study by Skolicki [5] shows inspiring results about an island model. The experimental results indicate that (a) the migration interval seems to be a dominating factor and the migration size is playing a minor role with

© Springer International Publishing Switzerland 2015
H. Handa et al. (eds.), *Proc. of the 18th Asia Pacific Symp. on Intell. & Evol. Systems – Vol. 2*,
Proceedings in Adaptation, Learning and Optimization 2, DOI: 10.1007/978-3-319-13356-0_29

regard to finding the best solution, and (b) too frequent migrations and rare migrations cause a degraded performance due to losing the global diversity and the slow convergence respectively.

This paper assumes that one of the causes degrading the performance in finding solutions is a behavior of elites existing in the populations. To eliminating the cause, this paper discusses a modified elitist model which improves a conventional elitist model [10]. In the experiments, this paper also confirms the modified elitist model improves the performance in finding solutions.

2   A Distributed Genetic Algorithm

This paper uses a typical distributed genetic algorithm. The distributed genetic algorithm consists of well-known genetic operators. Fig. 1 shows a procedure of the distributed genetic algorithm used in the following experiments. In the remainder of this paper, the following notations are used.

t	t-th generation
N_{pop}	total number of individuals
N_{is}	number of islands
N_{elite}	number of elites on each island
L	chromosome length
p_c	crossover rate
p_m	mutation rate
$P_i(t)$	set of individuals on the i-th island at t-th generation
M_r	migration rate
M_i	migration interval

The distributed genetic algorithm divides N_{pop} individuals into multiple populations (N_{is} islands) which evolve simultaneously and interact with each other by means of migrations. Each island maintains and evolves candidate solutions through selection and variation. New populations are generated by reproduction, recombination (crossover) and mutation.

The individuals are evaluated to determine how well they solve the problem with a fitness function. The individuals with better fitness values are selected as ones of the next generation in reproduction. The distributed genetic algorithm creates new individuals using simple randomized operators that resemble sexual recombination and mutation in natural world. The new individuals are evaluated with a fitness function, and the cycle of reproduction, crossover and mutation is repeated until a user-defined termination criterion is satisfied.

The typical implementation of migrations assumes sending $M_r \times N_{pop}/N_{is}$ copies of individuals (migrants) every M_i generations. These two parameters, M_r (migration rate) and M_i (migration interval), are important ones to control the quantitative aspect of migrations. It is also necessary to determine how to select migrants and how to exchange migrants for existing individuals. In this paper, a well-known policy is used in order to choose random individuals

```
begin
    t := 0
    for i := 1 to N_is
        initialize population P_i(0) randomly
        Evaluate(P_i(0))
    end for
    repeat
        for i := 1 to N_is
            P'_i(t) := Reproduction(P_i(t))
            if the migration conditions are met then
                // sending migrants to j-th island (j ≠ i)
                SendMigrant(P'_i(t), j)
                // receiving migrants from k-th island (k ≠ i)
                ReceiveMigrant(P'_i(t), k)
            end if
            Crossover(P'_i(t))
            Mutation(P'_i(t))
            Evaluate(P'_i(t))
            P_i(t + 1) := P'_i(t)
            t := t + 1
        end for
    until the termination conditions are met
end
```

Fig. 1. A pseudo-code of a distributed genetic algorithm used in the following experiments

from a source island and replace random individuals in a target island (*random-random policy*). This policy should not increase the selection intensity and thus seems appropriate for studying other migration parameters. Another migration parameter is needed for deciding a migration topology. The migration topology has pairs of a source island and a target island, and is defined for establishing destinations of migrating individuals. In Fig. 1 , a target island j and a source island k are decided by such migration topology.

The migration parameters strongly influence the performance of finding better solutions, thus there are many previous studies concerning migrations. Especially, Skolicki's experimental study [5] showed that the migration interval was playing much bigger role than the migration rate. The study also said that a degradation in the solution search performance was observed if the migration interval is too small or too large.

The goal of this paper is to control these phenomena using another elitist model in the reproduction operator, and to get better performance in finding solutions.

3   A Modified Elitist Model in Reproduction

An elitist model is usually used in a reproduction operator of a typical distributed genetic algorithm. This section shows two elitist models which are used in the following experiments. The first model was proposed in De Jong's Dissertation [10] and has been used in various implementation of genetic algorithms. The second model has improved timing in which the elite is returned to the population on each island in order to get better performance in finding solutions.

3.1   Reproduction with De Jong's Elitist Model

An elitist model keeps an individual (an elite) which has the best fitness value so far. Figure 2 shows a procedure of a typical reproduction operator with De Jong's elitist model. The procedure of this figure assumes that the number of elites is one, and it's easy to apply the procedure in other cases.

```
begin
    e(t) := GetEliteFromPopulation(P_i(t), e(t − 1))
    P'_i(t) := RouletteSelection(P_i(t))
    if e(t) ∉ P'_i(t) then
        w := GetIndividualRandomly(P'_i(t))
        P'_i(t) := P'_i(t) − {w} ∪ {e(t)}
    end if
end
```

Fig. 2. A pseudo-code of a reproduction operator with De Jong's elitist model

De Jong's elitist model returns the elite $e(t)$, which has been preserved in advance, to the population $P'_i(t)$ only if the population does not have the elite.

When the elite $e(t)$ is returned to the population $P_i'(t)$, the elite should be actually exchanged for an individual w, which is randomly selected from the population $P_i'(t)$, in order to keep the number of individuals on the population constant. Note that De Jong's elitist model in this paper keeps a total number of individuals whereas original De Jong's elitist model will increase a total number of individuals by 1 because of adding an elite to the population.

3.2 Reproduction with a Proposed Elitist Model

This paper tries an easy modification of De Jong's elitist model to improve the performance in finding solutions. De Jong's elitist model always maintains the elite in the population after the reproduction operator. Therefore, there should be the elite in subsequent operators such as migration, crossover and mutation. The elite in subsequent operators seems to be part of the reason for a degradation in the solution search performance.

Figure 3 shows a procedure of a reproduction operator with a simply modified elitist model. In this procedure, the proposed elitist model returns the elite $e(t)$, which has been preserved in advance, to the population $P_i(t)$ before the population $P_i'(t)$ is generated. Therefore the population $P_i'(t)$ including candidate of the next generation does not always have the elite $e(t)$.

```
begin
    e(t) := GetEliteFromPopulation(P_i(t), e(t − 1))
    P_i(t) := P_i(t) ∪ {e(t)}
    P_i'(t) := RouletteSelection(P_i(t))
end
```

Fig. 3. A pseudo-code of a reproduction operator with a proposed elitist model

4 Experiments and Results

In this section, we show the results to confirm good effects on improving the solution search performance when the amount of migration is too small or too large.

4.1 Test Functions

To experiment the modified elitist model, we use 3 function optimization problems defined by (1), (2) and (3).

$$f_{\text{rastrigin}}(\boldsymbol{x}) = 10n + \sum_{i=1}^{n} \left(x_i^2 - 10\cos(2\pi x_i) \right) \tag{1}$$

$$(-5.12 \le x_i < 5.12)$$

$$f_{\text{griewank}}(\boldsymbol{x}) = 1 + \sum_{i=1}^{n} \frac{x_i^2}{4000} - \prod_{i=1}^{n} \cos\left(\frac{x_i}{\sqrt{i}}\right) \tag{2}$$

$$(-512 \leq x_i < 512)$$

$$f_{\text{ridge}}(\boldsymbol{x}) = \sum_{j=1}^{n} \left(\sum_{i=1}^{j} x_i\right)^2 \tag{3}$$

$$(-64 \leq x_i < 64)$$

Rastrigin function and Griewank function are n-dimensional multi-modal functions. On the other hand, Ridge function is a n-dimensional uni-modal function. All functions are well-known for benchmark examples which examine the quality of convergence. In the experiments the dimension is set to 10 for each function.

4.2   Experimental Conditions

The experiments are preceded by choosing parameters in such a way that the problems are semi-difficult in order to observe improvements and degradations of the modified elitist model. Table. 1 shows parameters of a distributed genetic algorithm used in the experiments.

Table 1. Parameters used in experiments

Parameter	Value
Chromosome length: L	100
The total number of individuals: N_{pop}	512
One-point crossover rate: p_c	1.0
Mutation rate: p_m	$1 / L$
The number of elites on each island: N_{elite}	1
The number of islands: N_{is}	4, 16, 64
Migration interval: M_i	5, 25, 50
Migration rate: M_r	0.25
Migration topology	Random-ring topology[1]

Each individual is typically represented by a single chromosome which is a string of zeroes and ones. The chromosome encodes a solution candidate to the problem as a gray code. In the experiments each variable in a test function is represented by a 10 bits gray code, thus a chromosome is 100 bits long in total. The experiments use one-point crossover at a rate of 1.0 and mutation at a rate of $1/L$. The mutation acts by altering some bits according to the probability of mutation.

The experiments varied the migration interval using the values of 5, 25 and 50. A run with $M_i = 5$ considers the environment with too small amount of migration. A run with $M_i = 25$ considers the environment with a moderate

---

[1] The pairs of islands are chosen randomly at each generation.

amount of migration. A run with $M_i = 50$ considers the environment with too large amount of migration.

The algorithm can be run for a maximum 10^6 generations, though runs will be stopped earlier if it found the optimum solution. Because of the stochasticity of the process, the algorithm should be run 10 times for each combination of two parameters, the number of islands and the migration interval. To make it easier to compare, for different combinations the runs were starting from the same set of random seeds. Each of the runs observes the generation at which algorithm finds the optimum solution.

4.3   Comparison of Convergence

We show results in Fig. 4, Fig. 5 and Fig. 6 respectively in case of Rastrigin function, Griewank function and Ridge function. Each chart shows values of convergence generation at which an algorithm finds the optimum solution. Each value in the charts is mean value of 10 runs.

For all setups for Rastrigin function, the results in Fig. 4 look quite similar, which suggests that

- even though the amount of migration is too small ($M_i = 5$), a degradation in the solution search performance can not be observed both with De Jong's elitist model and with the proposed elitist model. In the case of infrequent migration, the proposed elitist model can find solutions earlier than De Jong's one.
- comparing the results with $M_i = 25$ and $M_i = 50$, in both cases the proposed elitist model can find solutions earlier than De Jong's one. The proposed elitist model can also make the performance hold when the number of islands is small ($N_{is} = 4$ or $N_{is} = 16$).

For all setups for Griewank function, the results with both setups for $N_{is} = 4$ and for $N_{is} = 16$ present a clear distinction from ones with setups for $N_{is} = 64$. The results with setups for $N_{is} = 64$ look quite similar to ones with same setups for Rastrigin function. The results with setups for $N_{is} = 4$ and $N_{is} = 16$ suggest that

- De Jong's elitist model worsens the performance when the amount of migration is too small ($M_i = 5$ and $N_{is} = 4$, or $M_i = 5$ and $N_{is} = 16$). In these cases, the proposed elitist model can also find solutions earlier than De Jong's one.
- comparing the results with $M_i = 25$ and $M_i = 50$, De Jong's elitist model worsens the performance, but the proposed one improves the performance except for the results with $N_{is} = 64$.

For all setups for Ridge function, the results with setups for $N_{is} = 4$ present a clear distinction from ones with both setups for $N_{is} = 16$ and for $N_{is} = 64$. The results with setups for $N_{is} = 64$ look quite similar to ones with same setups for Rastrigin function. The results with setups for $N_{is} = 4$ and $N_{is} = 16$ suggest that

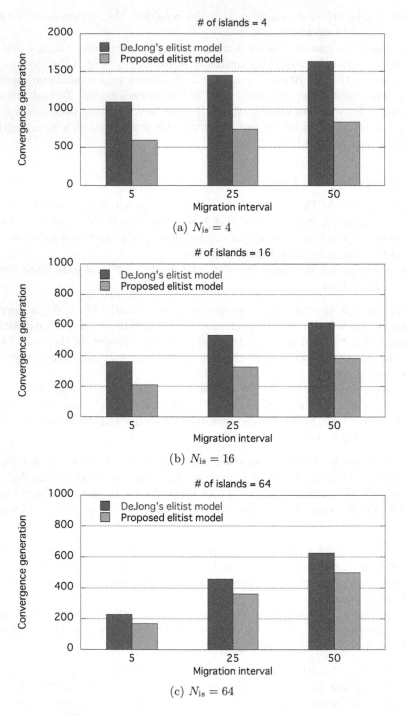

(a) $N_{is} = 4$

(b) $N_{is} = 16$

(c) $N_{is} = 64$

Fig. 4. Average results from 10 runs on Rastrigin

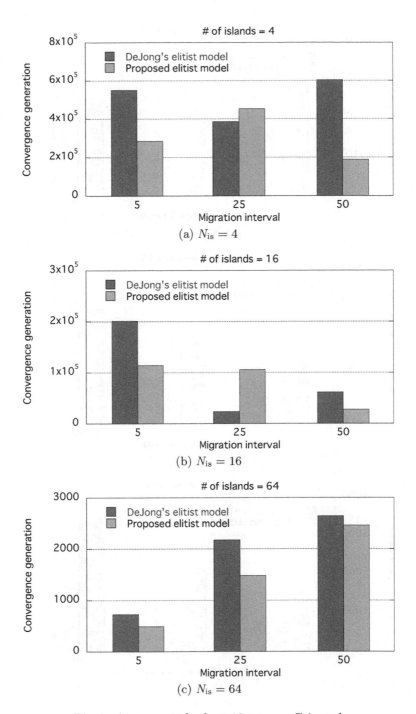

(a) $N_{is} = 4$

(b) $N_{is} = 16$

(c) $N_{is} = 64$

Fig. 5. Average results from 10 runs on Griewank

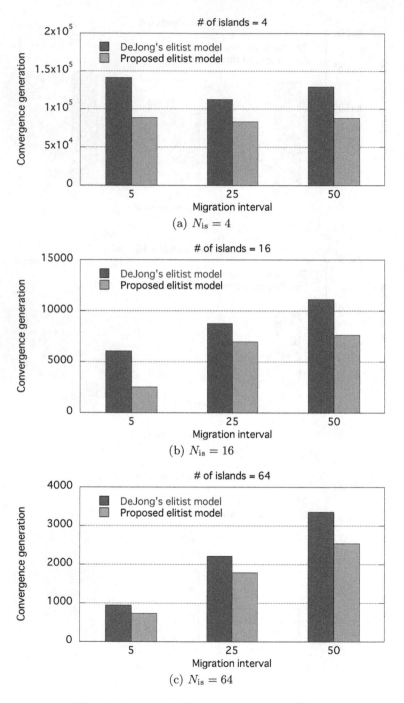

Fig. 6. Average results from 10 runs on Ridge

- De Jong's elitist model worsens the performance when the amount of migration is too small ($M_i = 5$ and $N_{is} = 4$). In this case, the proposed elitist model can also find solutions earlier than De Jong's one.
- comparing the results with $M_i = 25$ and $M_i = 50$, De Jong's elitist model worsens the performance, but the proposed one improves the performance except for the results with $N_{is} = 64$.

5 Related Works

We can find many studies about migration schemes in distributed genetic algorithms. In early studies, Tanese showed capabilities to find solutions in milt-population genetic algorithms with migration for the first time [3]. Cantú-Paz discussed migration rates, migration topologies and relationships between migrations and selection pressure in his work [4].

As distributed genetic algorithms came to be known as one of the most powerful tools to heuristically solve optimization problems, many empirical studies about parameters to control the amount of migration were reported. Hiroyasu et. al [11] proposed a migration scheme with randomized migration rate to find satisfactory solutions in a smaller amount of time. Munetomo et. al [12] and Kojima et. al [9] studied an asynchronous migration scheme to reduce unnecessary migrations. Nakamura et. al [7,8] discussed the necessity to perform periodical migrations in their empirical studies. Furthermore, experimental results by Skolicki et. al [5,6] showed that the migration interval was playing much bigger role than the migration rate and there is a degradation in search performance for large or too small migration intervals.

There also exist interesting theoretical studies to lead to new insights into the usefulness of periodical migrations. Lässig et. al [13] performed a rigorous runtime analysis for island models using the panmictic $(\mu+1)$ EA (Evolutionary Algorithm) and the parallel $(1+1)$ EA with migrations. They also performed experiments to complement the theoretical results. Both the theoretical results and the experimental results came to the same conclusions in terms of the migration interval as ones by Skolicki et. al [5].

6 Conclusions

This paper examined a distributed genetic algorithm with a modified elitist model different from the conventional model. The experiments confirmed the modified elitist model could find solutions earlier than the conventional model even if the amount of migration has not been set properly.

From these results, we empirically found that

- the elitist model had a major role in finding solutions under the inadequate migration intervals, and
- deciding how to return an elite to populations was a design guide in order to improve the performance of finding solutions with long migration intervals.

We should address the following future issues and problems.

- We need to clarify the elite role in finding solutions from the viewpoint of a population diversity.
- We did not have a heavy discussion about the migration topology. Therefore we need to discuss the migration topology according to a particular parallel system.

References

1. Goldberg, D.E.: Genetic Algorithm in Search, Optimization and Machine Learning. Addison Wesley Longman Publishing, Boston (1989)
2. Gen, M., Cheng, R.: Genetic Algorithms and Engineering Optimization. Wiley-Interscience Publication, New York (1999)
3. Tanese, R.: Distributed Genetic Algorithms. In: Schaffer, J.D. (ed.) Proceedings of the 3rd International Conference on Genetic Algorithms, pp. 434–439. Morgan Kaufmann Publishers, Virginia (1989)
4. Cantú-Paz, E.: Efficient and Accurate Parallel Genetic Algorithms. Springer, New York (2000)
5. Skolicki, Z., De Jong, K.A.: The influence of migration sizes and intervals on island models. In: Beyer, H., O'Reilly, U. (eds.) Proceedings of the 2005 Conference on Genetic and Evolutionary Computation, GECCO 2005, pp. 1295–1302. ACM, Washington DC (2005)
6. Skolicki, Z.: An analysis of island models in evolutionary computation. In: Beyer, H., O'Reilly, U. (eds.) Proceedings of the 2005 Conference on Genetic and Evolutionary Computation, GECCO 2005, pp. 386–389. ACM, Washington DC (2005)
7. Gong, Y., Guan, S., Nakamura, M.: Migration Effects of Parallel Genetic Algorithms on Line Topologies of Heterogeneous Computing Resources. IEICE Transactions 91-A(4), 1121–1128 (2008)
8. Miyagi, H., Tengan, T., Mohamed, S., Nakamura, M.: Migration Effects on Tree Topology of Parallel Evolutionary Computation. In: Proceedings of TENCON 2010 - 2010 IEEE Region 10 Conference, pp. 1601–1606. IEEE, Fukuoka (2010)
9. Kojima, K., Ishigame, M., Chakraborty, G., Hatsuo, H., Makino, S.: Asynchronous Parallel Distributed Genetic Algorithm with Elite Migration. International Journal of Computational Intelligence 4(2), 105–111 (2008)
10. De Jong, K.A.: An Analysis of the Behavior of a Class of Genetic Adaptive Systems. Ph.D Thesis, University of Michigan (1975)
11. Hiroyasu, T., Miki, M., Negami, M.: Distributed Genetic Algorithms with Randomized Migration Rate. In: Proceedings of 1999 IEEE International Conference on Systems, Man, and Cybernetics Conference, vol. 1, pp. 689–694. IEEE, Tokyo (1999)
12. Munetomo, M., Takai, Y., Sato, Y.: An Efficient String Exchange Algorithm for a Subpopulation-Based Asynchronously Parallel Genetic Algorithm and Its Evaluation. Transactions of IPSJ 35(9), 1815–1827 (1994) (in Japanese)
13. Lässig, J., Sudholt, D.: Design and Analysis of Migration in Parallel Evolutionary Algorithms. Soft Computing 17(7), 1121–1144 (2013)

cuSaDE: A CUDA-Based Parallel Self-adaptive Differential Evolution Algorithm

Tsz Ho Wong[1], A.K. Qin[1], Shengchun Wang[2], and Yuhui Shi[3]

[1] School of Computer Science and Information Technology, RMIT University,
Melbourne, Victoria 3001, Australia
[2] Department of Computer Education, Hunan Normal University, Changsha, China
[3] Department of Electrical and Electronic Engineering, Xi'an Jiaotong-Liverpool
University, Suzhou, China

Abstract. Differential evolution (DE) is a powerful population-based stochastic optimization algorithm, which has demonstrated high efficacy in various scientific and engineering applications. Among numerous variants of DE, self-adaptive differential evolution (SaDE) features the automatic adaption of the employed search strategy and its accompanying parameters via online learning the preceding behavior of the already applied strategies and their associated parameter settings. As such, SaDE facilitates the practical use of DE by avoiding the considerable efforts of identifying the most effective search strategy and its associated parameters. The original SaDE is a CPU-based sequential algorithm. However, the major algorithmic modules of SaDE are very suitable for parallelization. Given the fact that modern GPUs have become widely affordable while enabling personal computers to carry out massively parallel computing tasks, this work investigates a GPU-based implementation of parallel SaDE using NVIDIA's CUDA technology. We aim to accelerate SaDE's computation speed while maintaining its optimization accuracy. Experimental results on several numerical optimization problems demonstrate the remarkable speedups of the proposed parallel SaDE over the original sequential SaDE across varying problem dimensions and algorithmic population sizes.

1 Introduction

Evolutionary algorithms (EAs) [1] are a broad class of nature-inspired population-based metaheuristic optimization algorithms, which have shown high efficacy for solving diverse real-world optimization problems. However, they typically require considerable computation time to achieve satisfactory solutions. Consequently, EAs are less favored by those applications with stringent computational budgets.

EAs comprise a population of candidate solutions and explore a given solution space using various nature-inspired computational operations, such as selection, reproduction and replacement, to gradually evolve the population to search for global optima. This type of algorithms is highly suitable for parallelization because each population member is typically imposed with the same set of operations. However, in the past, hardware and software platforms that facilitate

© Springer International Publishing Switzerland 2015
H. Handa et al. (eds.), *Proc. of the 18th Asia Pacific Symp. on Intell. & Evol. Systems – Vol. 2*,
Proceedings in Adaptation, Learning and Optimization 2, DOI: 10.1007/978-3-319-13356-0_30

parallel computing were not widely available and affordable. Accordingly, most of existing EAs had been designed and implemented in a sequential way.

In recent years, the graphics processing units (GPUs) have become a powerful computing device that can support general-purpose massively data-parallel computation. Nowadays, modern GPUs are characterized by inexpensive prices, general-purpose parallel computing infrastructures, and easy-to-use programming models, which have transformed common personal computers (PCs), i.e., desktops and laptops, into powerful parallel computing platforms.

Among existing parallel computing platforms based on GPUs, NVIDIA's compute unified device architecture (CUDA) [2,3] provides an intuitive and scalable programming model based on an extended C programming language: CUDA-C. Developers can write a C-style routine to process one data element, which then gets distributed across hundreds of streaming processors (SPs) for thousands of threads to concurrently process different data elements. It requires less effort for developers already familiar with the C language to learn CUDA-C. Nowadays, many state-of-the-art algorithms in different scientific and engineering fields have been redesigned based on CUDA to speed up their computation. However, computational efficiency of CUDA-C applications highly depends on comprehensive consideration of various technical properties of GPUs during development and implementation. Without delicate consideration, parallel programs written in CUDA-C might even run slower than their sequential counterparts.

Differential evolution (DE) [4] is one of the most promising EAs, which has consistently demonstrated superiority for handling challenging optimization tasks. Several CUDA-based DE algorithms had already been proposed and shown superior computational efficiency [5,6,7,8,9,10,11]. Self-adaptive differential evolution (SaDE) [15,16] is a well-known DE variant, capable of gradually adapting the employed search strategy and its associated parameters by online learning the preceding behavior of already applied strategies and their accompanying parameter settings. Since its invention, SaDE has received consideration attention as evidenced by its high citation counts and the winner of the 2012 *IEEE Transactions on Evolutionary Computation* outstanding paper award. The original SaDE is a CPU-based sequential algorithm, though SaDE is highly parallelizable owing to its data-parallel algorithmic structure. This work investigates a CUDA-based implementation of parallel SaDE, called cuSaDE, to speed up SaDE while retaining its optimization accuracy by refactorizing its major algorithmic modules for the efficient parallel execution on GPUs. We evaluate the performance of cuSaDE on several numerical optimization problems, and compare cuSaDE with the original sequential SaDE implemented in the C language in terms of both the optimization accuracy and computation time. Experiment results show that cuSaDE has the similar optimization accuracy but much faster computation speed in comparison to the original sequential SaDE across varying problem dimensions and algorithmic population sizes.

The remaining paper is organized as follows. Section 2 introduces CUDA-based GPU computing, followed by the description of the original SaDE and the proposed CUDA-based implementation of SaDE in Section 3. Experimental results are reported and analyzed in Section 4. Finally, Section 5 draws conclusions and mentions some future work.

2   CUDA-Based GPU Computing

Besides the dedicated support for graphics and gaming applications, modern GPUs can provide a powerful and budget general-purpose parallel computing environment in favor of massively data-parallel applications, i.e. different parts of data are subjected to the same set of operations. Nowadays, common PCs have been widely equipped with GPUs having hundreds of SPs, and thus become very suitable for parallel application development.

Compared to the central processing unit (CPU) that contains one or several sophisticated processors working at the high clock speed, GPU consists of hundreds of SPs having simplified structures and working at the lower clock speed. Although CPU can rapidly tackle many general-purpose tasks owing to its high clock speed, operation re-scheduling ability and large cache memory, it is less efficient in massively data-parallel applications. In contrast, GPU that operates based on the single instruction multiple threads (SIMT) model [2] can simultaneously invoke many threads to concurrently execute the same set of operations on different data elements, and thus can lead to high computation speed in data-parallel applications.

As the graphics hardware advances, its application programming interfaces (APIs) keep being improved to facilitate the development of general-purpose parallel computing applications. Since GPUs were originally designed for graphics computing, graphics shader language based APIs were used in the early time for general-purpose parallel computing development. Accordingly, developers had to transform scientific calculations into problems that can be represented by geometry primitives so as to solve them on GPUs using shader language based APIs for programming [12]. Nowadays, user-friendly APIs specifically designed for general-purpose parallel computing had been developed to reduce the programming difficulty and thus allow developers without expertise in graphics to leverage GPU computing.

NVIDIA's CUDA technology [2,3] provides a parallel computing architecture on modern NVIDIA's GPUs in which hundreds of SPs are grouped into several streaming multiprocessors (SMs). Each SM contains a number of SPs that share the on-chip control logic units, shared memory with low latency, registers, and so on. All SMs share the global memory with high latency. The number of SMs and the size of the global memory vary as per GPU models and brands. However, in each SM, the number of SPs, the size of the shared memory and the number of registers depend on GPU's compute capability [2]. The hardware architecture of SMs has been gradually changed across three major generations so far, i.e., Tesla-class, Fermi-class and Kepler-class. The latest Kepler-class hardware is very different from previous generations, which was named as SMX (next-generation SM) by NVIDIA. Compared to SMs of previous generations, SMX featuring the much larger number of SPs is more computation-powerful, energy-efficient and programmable. The newest Kepler-class GK110 GPU also has the increased numbers of special function units and double-precision units in each SMX to enhance the computing horsepower. Our work uses NVIDIA's GK110-based Geforce GTX Titan GPU, which contains 15 SMXs with 192 SPs, 64 double-precision units and 32 special function units per SMX.

CUDA also refers to an intuitive and scalable programming model based on an extended C programming language, called CUDA-C [2,3]. This model unifies CPU and GPU, so-called host and device, into a heterogeneous computing system to make the best use of both of them. Specifically, CUDA-C contains three types of functions: (1) host functions, called and executed only by the host, which are exactly the same functions available in the C language; (2) kernel functions, only called by the host and executed by the device; (3) device functions, called and executed only by the device. In general, the sequential operations should be programmed as host functions that are executed on CPU. The parallelizable operations should be programmed as kernel or device functions that are executed on GPU. Both host and kernel functions will be encapsulated and called in one main host function.

In fact, each kernel function will be executed on GPUs by a large number of threads at the same time in the single instruction multiple data (SIMD) [2] fashion. In the CUDA programming model, these threads are organized into a grid of thread blocks with each block containing a certain number of threads. The grid can have up to three dimensions of blocks. Its size is denoted by a predefined struct variable gridDim with three fields x, y and z storing the block numbers in three dimensions respectively. Each block in a grid can be indexed by a predefined struct variable blockIdx with three fields x, y and z storing the position of the corresponding block in the grid. The block can have up to three dimensions of threads with its size denoted by a predefined struct variable blockDim having three fields x, y and z storing the thread numbers in three dimensions respectively. Each thread in a block can be indexed by a predefined struct variable threadIdx with three fields x, y and z storing the position of the corresponding thread in the block. All threads in one block will be grouped into warps. Each warp contains 32 threads with consecutive thread IDs. These 32 threads will always be executed together in the SIMD fashion.

The dimension and size of the grid and block usually depend on the characteristics of problems and algorithms. Once a grid of blocks is determined, each block of threads will be executed on one SM (SMX). Since each SM (SMX) has the maximally allowed number of resident thread blocks which is determined by the device's compute capability, once all SMs (SMX) are fully occupied the remaining blocks have to wait for any available slots released by the completed blocks on any SMs (SMXs). This scheme ensures the transparent scalability of CUDA-C programs executed on the future generations of GPUs that contain more SMs (SMXs).

When launching a kernel function, its associated kernel execution configuration parameters, such as gridDim and blockDim, must be specified within "$<<< ... >>>$". When determining configuration parameters, we typically have to consider: each thread block has the maximally allowed number of threads (or warps) which is determined by the device's compute capability; each SM (SMX) has the limited shared memory size and register number, which will influence the allowed number of threads per block and the allowed number of blocks per SM (SMX); all threads in one block can access the same data stored on the shared memory of low latency while threads in different blocks can only communicate via the global memory of high latency.

3   Parallel SaDE Using CUDA-Based GPU Computing

3.1   Overview of SaDE

DE emerged as a simple and powerful EA more than a decade ago and has now developed into a very promising research area in the field of evolutionary computation. However, DE's performance highly depends on the employed trail vector generation strategy and its associated parameter setting, which usually invokes the tedious trial-and-error efforts of selecting the most suitable strategy and its accompanying parameters.

SaDE [15,16], as one of the most well-known DE variants, avoids the time-consuming strategy and parameter setting selection task. It features a pool of potentially effective yet complementary trial vector generation strategies. During the population's evolution, with respect to each target vector in the population at the current generation, one strategy will be selected from this pool according to strategy selection probabilities, which are computed according to the success rate of each strategy for generating promising trial vectors (those that can enter the population for the next generation) within the number of LP (learning period) preceding generations. This selected strategy is then applied to the corresponding target vector to generate the trial vector. For the parameter setting associated with this selected strategy, SaDE adapts CR while randomizing F. Specifically, it archives the CR values associated with each strategy which had generated promising trial vectors within the preceding LP generations. The mean of those recorded CR values with respect to each strategy is computed at the end of the current generation, and used as the mean value of the normal distribution with standard deviation 0.1 to generate the CR values to be used by the corresponding strategy in the next generation. Note that different from [16], we employ the mean value for CR here as in [15] to facilitate the GPU-based implementation with less influence on the optimization accuracy. The value of F is randomly sampled from the normal distribution with mean value 0.5 and standard deviation 0.3 to maintain both exploration (large F values) and exploitation (small F values) in the entire course of the search. SaDE leaves NP as a manually specified parameter to be determined based on the available problem knowledge and computational budget.

In SaDE, the initial LP generations accumulate the search behavior to be learnt. During this period, all strategy selection probabilities are set to be equal and the mean value of the normal distribution for generating CR values is set to 0.5. To avoid invalid selection probabilities when the success rates of all strategies are zero or when a strategy is never chosen within the preceding LP generations, a small constant value (0.01) is introduced as illustrated in Algorithm 1. More details about SaDE's implementation can be found in [15,16].

3.2 CUDA-Based Implementation of Parallel SaDE

SaDE is highly parallelizable due to its data-parallel algorithmic structure, which motivates us to develop a CUDA-based parallel SaDE algorithm, called cuSaDE. Fig. 1 illustrates the flowchart of the proposed cuSaDE, which encapsulates the major algorithmic modules of SaDE into separate kernel functions that are communicated via the global memory. Specifically, there are six kernel functions detailed as follows:

kernel(I): initializes the population and writes it into the global memory. This kernel utilizes uniform random numbers generated by the GPU-based random number generator contained in the NVIDIA CUDA random number generation library (cuRAND) [13]. The number of threads per thread block is set to 1024 with each thread handling one element of one population member.

kernel(E): evaluates the objective function values of population members and writes them into the global memory. The objective function is defined according to the problem being solved and thus has the varying complexity. The SMX of the GPU card used in our work contains 4 warp schedulers and thus can concurrently invoke 128 threads. Therefore, we set the number of threads per thread block to 128. We do not choose a larger thread block so as to enable each thread in a thread block to have sufficient resources in terms of the share

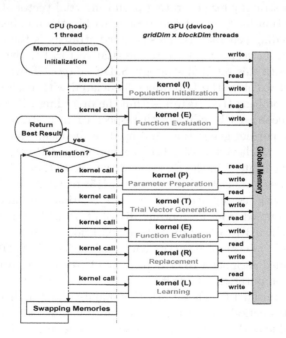

Fig. 1. Flow chart of cuSaDE

Algorithm 1. The Original SaDE Algorithm (Input: NP, LP)

1: Initialize generation counter $g = 0$, strategy selection probability $stPb_{k,g} = 1/4, k = 1, \ldots, 4$, and $CRm_{k,g} = 0.5, k = 1, 2, 3$; Empty success and failure archives.

2: Initialize population $\mathbf{P}_g$ of NP D-dimensional individuals: $\mathbf{P}_g = \{\mathbf{x}_{1,g}, \ldots, \mathbf{x}_{NP,g}\}$ with $\mathbf{x}_{i,g} = \{x_{i,g}^1, \ldots, x_{i,g}^D\}$.

3: Objective function evaluation of each individual in $\mathbf{P}_g$, i.e., $f(\mathbf{x}_{i,g}), i = 1, \ldots, NP$.

4: **while** the predefined termination criteria are not met **do**

5:     **for** $i = 1 \to NP$ **do**

6:        Select a strategy index k_i in $\{1, 2, 3, 4\}$ based on $stPb_{k,g}, k = 1, \ldots, 4$ using the stochastic universal sampling.

7:        Randomly generate a F value using normal distribution $rand_n(0.5, 0.3)$.

8:        Randomly select in $\{1, \ldots, NP\}$ five mutually exclusive indices $r_m, m = 1, \ldots, 5$ that are distinct from i.

9:        Generate a mutant vector $\mathbf{v}_{i,g} = \{v_{i,g}^1, \ldots, v_{i,g}^D\}$:

         if $(k_i == 1)$ then $\mathbf{v}_{i,g} = \mathbf{x}_{r1,g} + F \cdot (\mathbf{x}_{r2,g} - \mathbf{x}_{r3,g})$

         if $(k_i == 2)$ then $\mathbf{v}_{i,g} = \mathbf{x}_{i,g} + F \cdot (\mathbf{x}_{gbest,g} - \mathbf{x}_{i,g}) + F \cdot (\mathbf{x}_{r2,g} - \mathbf{x}_{r3,g})$

         if $(k_i == 3)$ then $\mathbf{v}_{i,g} = \mathbf{x}_{r1,g} + F \cdot (\mathbf{x}_{r2,g} - \mathbf{x}_{r3,g}) + F \cdot (\mathbf{x}_{r4,g} - \mathbf{x}_{r5,g})$

         if $(k_i == 4)$ then $\mathbf{v}_{i,g} = \mathbf{x}_{i,g} + rand_u(0, 1) \cdot (\mathbf{x}_{r1,g} - \mathbf{x}_{i,g}) + F \cdot (\mathbf{x}_{r2,g} - \mathbf{x}_{r3,g})$

10:       Generate a trial vector $\mathbf{u}_{i,g} = \{u_{i,g}^1, \ldots, u_{i,g}^D\}$:

         if $(k_i == 1)$ or $(k_i == 2)$ or $(k_i == 3)$ **then**

         Randomly generate a CR value using $rand_n(CRm_{k_i}, 0.1)$ with $CR \in [0, 1]$

         $j_{rand} = ceil(rand_u(1, D))$

         for $j = 1 \to D$ **do**

$$u_{i,g}^j = \begin{cases} v_{i,g}^j & \text{if } rand_u(0,1) \le CR \text{ or } j = j_{rand} \\ x_{i,g}^j & \text{otherwise} \end{cases}$$

         end for

         else

            $\mathbf{u}_{i,g} = \mathbf{v}_{i,g}$

         end if

11:       Evaluate the objective function value of the generated trial vector $\mathbf{u}_{i,g}$.

12:       if $(f(\mathbf{u}_{i,g}) \le f(\mathbf{x}_{i,g}))$ **then**

13:         $\mathbf{x}_{i,g+1} = \mathbf{u}_{i,g}$, and store the tuple (g, k_i, CR) (if $k_i == 1, 2$ or 3) or (g, k_i) (if $k_i ==$ 4) into the success archive.

14:       **else**

15:         $\mathbf{x}_{i,g+1} = \mathbf{x}_{i,g}$, and store the tuple (g, k_i) into the failure archive.

16:       **end if**

17:     **end for**

18:     if $(g \ge LP)$ **then**

19:       if $(g > LP)$ **then**

20:         Remove those tuples with the first elements smaller or equal to $g - LP$ from the success and the failure archives.

21:       **end if**

22:       Calculate $S_{k,g}$ and $F_{k,g}, k = 1, \ldots, 4$ as the number of tuples having the second elements equal to k in the success and failure archives, respectively.

23:       if $(S_{k,g} + F_{k,g} > 0)$ **then**

24:         $stPb_{k,g} = S_{k,g}/(S_{k,g} + F_{k,g}) + 0.01$.

25:       **else**

26:         $stPb_{k,g} = 0.01$.

27:       **end if**

28:       $stPb_{k,g} = stPb_{k,g}/\sum_{k=1,\ldots,4} stPb_{k,g}$.

29:       Calculate $CRm_{k,g}, k = 1, 2, 3$ as the mean value of the third elements in those tuples in the success archive having the second elements equal to $k, k = 1, 2, 3$.

30:     **end if**

31:     Increase the generation counter: $g = g + 1$.

32: **end while**

NOTE: (1) $rand_u(a, b)$ is a uniform random number generator sampling in $[a, b]$; (2) $rand_n(a, b)$ is a Gaussian random number generator with mean a and standard deviation b; (3) $ceil(c)$ takes on the smallest integer larger than or equal to c.

---

memory and registers. Meanwhile, this can avoid the time-consuming thread synchronization that may happen within a thread block. When loading population members from the global memory to the shared memory, we set each thread to simultaneously load two double-precision elements of a population member via one global memory access. The parallel reduction procedure is frequently used in the calculation of objective function values for adding up multiple data. If this procedure is not carefully designed (especially when a thread block will handle parallel reduction for multiple population members), serious shared memory bank conflicts may happen, which will dramatically degrade computational efficiency. To avoid such an issue, we divide the threads in a thread block into multiple segments with each segment dealing with parallel reduction for one population member. Additionally, we set the number of threads in each segment to the minimal integer power of 2 that is not smaller than the problem dimension, and pad zeros into the shared memory for extra threads. For example, 16 threads will be used for a 10D population member. The first 10 threads correspond to each element of this population member stored in the shared memory, and the last 6 threads correspond to the shared memory spaces following this population member, which are set to zeros. In addition to the avoidance of bank conflicts, this approach can also avoid the branching and divergence procedure for determining whether the number of data involved in parallel reduction is even or odd, which can lead to the further computational speedup. It is worth of noting that the objective function evaluation could be the most time-consuming module in EAs, especially when the population size and the problem dimension go large. Its effective parallelization on GPUs can lead to the remarkable computational speedup.

kernel(P): generates the strategy ID, the values of two control parameters CR and F, the random index *jrand*, the mutually exclusive indices of randomly sampled population members for each population member to generate its corresponding trial vector. These generated auxiliary parameters will be written into the global memory. The number threads per thread block is set to the population size with each thread handling the generation of all auxiliary parameters with respect to one population member. The strategy IDs are generated as per each strategy's selection probability at the current generation using a GPU-based stochastic universal sampling algorithm, which features the time-efficient atomic addition operation [2]. The values of CR and F for all population members are generated concurrently. The mutually exclusive indices are generated using multiple threads via several iterations. This kernel utilizes both uniform and gaussian random numbers generated by the GPU-based random number generator from cuRAND [13].

kernel(T): generates trail vectors for each population member using its respective strategy ID, CR and F values, *jrand* and mutually exclusive indices generated by kernel(P). These auxiliary parameters used for the trial vector generation are loaded from the global memory to the shared memory for fast access. The generated population of trial vectors will be written into the global memory. This kernel utilizes uniform random numbers generated by the

GPU-based random number generator from cuRAND [13]. The number threads per thread block is set to the integer multiplier of the problem dimension (1000 in our work) with each thread handling one element in one population member. Such a configuration ensures any trial vector to be generated by the same thread block.

kernel(R): compares each member in the current population with its corresponding trial vector generated by kernel(T) in terms of their objective function values, and writes the fitter one and its corresponding objective function value into the global memory. Meanwhile, it updates the success and failure archives in the global memory with the success and failure counts as well as the sum of the successful CR values using the atomic add operation [2]. The number threads per thread block is set to the population size with each thread handling one population member.

kernel(L): learns from the success and failure experience recorded in the success and failure archives to update each strategy's selection probability and each strategy's CR mean value (if applies). This kernel only uses eight threads contained in one thread block to handle success and failure archives with respect to four strategies. We use a simplified pointer-based success (and failure) accumulation scheme to avoid the time-consuming operation of summation over rows in the archive.

At the end of each generation, two memory addresses for storing new and old populations (as well as two memory addresses for storing their corresponding objective function values) will be swapped by the host. Moreover, to accelerate memory access speed, the constant memory is used to store those values that can be accessed by multiple threads, such as shifting vectors, constant matrices and vectors defined in some objective functions (Section 4.2).

After necessary memory allocation (host and device memories) and initialization (e.g, cuRAND random number generators, algorithmic variables, etc.), kernel(I) and kernel (E) are invoked in sequence by the host. Then, the host repeatedly invokes the sequence of kernel(P), kernel(T), kernel(E), kernel(R), kernel(L) and the swapping of memory addresses until the main loop terminates.

4 Experiments

We compare the original sequential SaDE and the proposed parallel cuSaDE in terms of both the optimization accuracy and computation time on four numerical optimization problems of 10D, 50D and 100D respectively. Different algorithmic population sizes (P30, P50, P100, P300 and P500) are examined.

4.1 Experimental Setup

Experiments are conducted on a PC equipped with an Intel I5-3470 CPU at 3.2 Ghz and a NVIDIA Geforce GTX Titan GPU with 6GB of GDDR5 global memory. GTX Titan supports compute capability 3.5, which has 2880 SPs evenly deployed in 15 SMXs, i.e. each SMX consists of 192 SPs. Our development environment is made of Windows 7 operating system, CUDA toolkit 5.5 and Microsoft Visual Studio 2008.

In SaDE, the learning period parameter LP is set to 50 as suggested in [16]. Population sizes are set to 30, 50, 100, 300 and 500 respectively for each test case.

For each test problem of a specific dimension, each SaDE implementation under a specific parameter setting is executed 15 times starting from different random number generator seeds while all of four implementations share the same initial random number generator seed for any individual run. The algorithm terminates once the maximal number of function evaluations is reached, which is set to 10^4 times the problem dimension, e.g. for a 10D problem, the maximal number of function evaluations is 10^5.

4.2   Test Problems

We use four numerical optimization problems that are adapted from the test functions F1, F6, F10 and F12 defined in [14]:

- F1: Sphere function

$$F1(x) = \sum_{i=1}^{D} z_i{}^2 + f_1{}^*, \quad \mathbf{z} = \mathbf{x} - \mathbf{o} \tag{1}$$

- F2: Rotated Rosenbrock Function

$$F2(x) = \sum_{i=1}^{D-1}(100(z_i{}^2 - z_{i+1})^2 + (z_i - 1)^2) + f_2{}^*,$$

$$\mathbf{z} = \mathbf{M}_1(\frac{2.048(\mathbf{x} - \mathbf{o})}{100}) + 1 \tag{2}$$

- F3: Rotated Griewank's Function

$$F3(x) = \sum_{i=1}^{D} \frac{z_i{}^2}{4000} - \prod_{i=1}^{D} \cos(\frac{z_i}{\sqrt{i}}) + 1 + f_3{}^*,$$

$$\mathbf{z} = \Lambda^{100}\mathbf{M}_1(\frac{600(\mathbf{x} - \mathbf{o})}{100}) \tag{3}$$

- F4: Rotated Rastrigin's Function

$$F4(x) = \sum_{i=1}^{D}(z_i{}^2 - 10\cos(2\pi z_i) + 10) + f_4{}^*,$$

$$\mathbf{z} = \mathbf{M}_1\Lambda^{10}\mathbf{M}_2 T_{asy}{}^{0.2}(T_{osz}(\mathbf{M}_1\frac{5.12(\mathbf{x} - \mathbf{o})}{100})) \tag{4}$$

Table 1. Optimization accuracy comparison between the original sequential SaDE (denoted as "C") and the proposed parallel cuSaDE (denoted as "G") in terms of the mean value (standard deviation in bracket) of the best EFVs achieved when the algorithm terminates over 15 runs with respect to five population sizes (P30, P50, P100, P300, P500) on four test problems at three dimensions (10D, 50D and 100D).

		F1			F2			F3			F4		
		10D	50D	100D	10D	50D	100D	10D	50D	100D	10D	50D	100D
P30	C	0.000 (0.000)	0.000 (0.000)	0.000 (0.000)	0.402 (0.147)	45.451 (13.491)	144.903 (34.918)	1.000 (0.000)	1.000 (0.000)	1.000 (0.000)	25.462 (3.142)	372.332 (10.524)	893.437 (18.327)
	G	0.000 (0.000)	0.000 (0.000)	0.000 (0.000)	0.169 (0.079)	43.952 (13.085)	144.856 (28.452)	1.000 (0.000)	1.000 (0.000)	1.000 (0.000)	28.092 (3.551)	388.443 (18.170)	948.944 (22.228)
P50	C	0.000 (0.000)	0.000 (0.000)	0.000 (0.000)	0.483 (1.039)	42.622 (1.890)	163.264 (58.593)	1.000 (0.000)	1.000 (0.000)	1.000 (0.000)	28.554 (3.873)	378.171 (18.952)	879.183 (28.994)
	G	0.000 (0.000)	0.000 (0.000)	0.000 (0.000)	0.167 (0.074)	71.674 (2.233)	138.115 (31.631)	1.000 (0.000)	1.000 (0.000)	1.000 (0.000)	28.874 (4.647)	400.427 (7.486)	921.488 (142.033)
P100	C	0.000 (0.000)	0.000 (0.000)	0.000 (0.000)	0.377 (0.107)	44.914 (1.686)	118.224 (45.603)	1.000 (0.000)	1.000 (0.000)	1.001 (0.001)	30.837 (5.221)	388.528 (13.785)	891.054 (17.666)
	G	0.000 (0.000)	0.000 (0.000)	0.000 (0.000)	0.429 (0.172)	44.140 (1.379)	109.034 (24.435)	1.000 (0.000)	1.000 (0.000)	1.000 (0.000)	30.132 (2.325)	406.200 (9.221)	970.332 (22.902)
P300	C	0.000 (0.000)	0.000 (0.000)	0.000 (0.000)	3.455 (0.490)	46.889 (0.888)	113.559 (33.608)	3.042 (1.116)	3.486 (2.790)	4.916 (2.550)	31.656 (3.389)	396.577 (18.952)	924.825 (24.703)
	G	0.000 (0.000)	0.000 (0.000)	0.000 (0.000)	3.306 (0.330)	46.477 (0.799)	97.759 (6.5730)	2.012 (0.628)	1.000 (0.000)	1.000 (0.000)	32.204 (2.916)	404.842 (10.282)	956.798 (13.156)
P500	C	0.000 (0.000)	0.000 (0.000)	0.000 (0.000)	5.963 (0.512)	47.155 (0.984)	104.445 (19.974)	102.788 (35.756)	43.087 (31.756)	70.423 (17.588)	30.487 (5.442)	402.733 (11.350)	919.219 (33.770)
	G	0.047 (0.128)	0.581 (1.780)	0.000 (0.000)	5.737 (0.541)	56.987 (16.766)	104.055 (17.490)	64.415 (17.953)	37.165 (17.993)	1.000 (0.000)	30.505 (6.015)	405.820 (10.477)	995.496 (19.891)

Here, the search ranges for F1, F2, F3 and F4 are defined as [-100, 100], [-100, 100], [-600, 600] and [-5, 5] respectively. The global optimal value (f^*) for F1, F2, F3 and F4 are defined as -450, 390, -180 and -330, respectively. The shifted global optimum o, rotation matrices M_1 and M_2, diagonal matrices Λ^α, asymmetric transformation function $T_{asy}{}^\beta$ and oscillation transformation function T_{osz} are detailed in [14].

4.3   Results

We evaluate and compare the optimization accuracy and computation time of the original sequential SaDE and the proposed parallel cuSaDE on four numerical optimization problems under different problem dimensions and algorithmic population sizes.

Table 1 compares the optimization accuracy of the CPU-based sequential SaDE and the GPU-based parallel cuSaDE, measured by the mean value and standard deviation of the best error function values (EFVs), i.e. the difference of objective function values between the best solution found so far and the global optimum, achieved when the algorithm terminates over 15 runs. It can be observed that the discrepancy of the optimization accuracy between the sequential SaDE and the parallel cuSaDE is very small and not biased towards one certain algorithm, which verifies the functional correctness of cuSaDE. In fact, such discrepancy is unavoidable due to the use of distinct random number generators in the sequential and parallel SaDE. Moreover, we can observe that larger population sizes may not always result in higher accuracies, which might be explained by the fact that the current maximal number of function evaluations is insufficient to ensure the convergence of a large population towards a promising solution.

Table 2. Computation speed comparison between the original sequential SaDE (denoted as "C") and the proposed parallel cuSaDE (denoted as "G") in terms of the mean value (standard deviation in bracket) of computation time (in seconds) when the algorithm reaches the maximal number of function evaluations (i.e., $10^4 \cdot D$) over 15 runs and the speedup ratio (denoted as "Rt") calculated using the average computation time with respect to five population sizes (P30, P50, P100, P300, P500) on four test problems at three dimensions (10D, 50D and 100D).

		F1			F2			F3			F4		
		10D	50D	100D	10D	50D	100D	10D	50D	100D	10D	50D	100D
P30	C	1.312 (0.011)	7.496 (0.063)	17.666 (0.072)	1.286 (0.014)	8.841 (0.058)	28.43 (0.137)	1.374 (0.008)	10.60 (0.054)	35.295 (0.084)	1.413 (0.009)	13.342 (0.041)	54.675 (0.163)
	G	0.364 (0.017)	1.989 (0.103)	4.171 (0.100)	0.390 (0.001)	2.389 (0.006)	5.707 (0.014)	0.403 (0.014)	2.437 (0.098)	6.117 (0.038)	0.485 (0.001)	3.220 (0.035)	11.559 (0.039)
	Rt	3.61	3.77	4.24	3.29	3.70	4.98	3.41	4.35	5.77	2.91	4.14	4.73
P50	C	1.039 (0.009)	6.158 (0.036)	14.886 (0.091)	0.998 (0.007)	7.451 (0.043)	25.601 (0.119)	1.088 (0.004)	9.185 (0.037)	32.432 (0.108)	1.132 (0.006)	11.943 (0.063)	51.768 (0.187)
	G	0.235 (0.008)	1.280 (0.045)	2.538 (0.148)	0.235 (0.014)	1.585 (0.013)	4.319 (0.019)	0.237 (0.015)	1.595 (0.099)	4.630 (0.012)	0.304 (0.012)	2.298 (0.010)	9.875 (0.018)
	Rt	4.42	4.81	5.87	4.25	4.70	5.93	4.59	5.76	7.00	3.72	5.20	5.24
P100	C	0.941 (0.008)	5.784 (0.046)	14.218 (0.101)	0.897 (0.005)	7.003 (0.041)	24.722 (0.085)	0.969 (0.006)	8.737 (0.039)	31.606 (0.078)	1.031 (0.004)	11.462 (0.055)	50.902 (0.226)
	G	0.146 (0.006)	0.750 (0.001)	1.508 (0.068)	0.156 (0.000)	0.923 (0.012)	3.357 (0.011)	0.153 (0.011)	0.961 (0.024)	3.644 (0.013)	0.181 (0.008)	1.669 (0.012)	8.765 (0.020)
	Rt	6.45	7.71	9.43	5.76	7.59	7.36	6.34	9.09	8.67	5.70	6.87	5.81
P300	C	1.267 (0.006)	8.198 (0.030)	20.128 (0.111)	1.273 (0.007)	9.481 (0.041)	30.479 (0.102)	1.343 (0.007)	11.196 (0.054)	37.336 (0.081)	1.398 (0.007)	13.84 (0.054)	56.493 (0.173)
	G	0.063 (0.002)	0.384 (0.012)	0.899 (0.051)	0.066 (0.004)	0.611 (0.014)	2.849 (0.013)	0.064 (0.005)	0.636 (0.012)	3.126 (0.011)	0.082 (0.005)	1.283 (0.014)	8.161 (0.012)
	Rt	20.11	21.35	22.39	19.37	15.52	10.70	21.12	17.60	11.94	17.05	10.79	6.92
P500	C	1.634 (0.007)	11.121 (0.031)	26.048 (0.082)	1.637 (0.006)	12.624 (0.051)	36.858 (0.158)	1.702 (0.008)	14.337 (0.049)	43.663 (0.108)	1.7588 (0.008)	16.969 (0.056)	62.784 (0.190)
	G	0.023 (0.002)	0.157 (0.006)	0.425 (0.014)	0.028 (0.000)	0.381 (0.008)	2.393 (0.012)	0.027 (0.002)	0.410 (0.013)	2.651 (0.007)	0.038 (0.002)	1.055 (0.015)	7.699 (0.013)
	Rt	72.39	70.83	61.29	58.42	33.13	15.40	63.45	34.97	16.47	46.61	16.08	8.15

Table 2 reports, for each of four test problems, the average computation time over 15 runs of the CPU-based sequential SaDE and the GPU-based parallel cuSaDE and the corresponding speedup ratios across varying algorithmic population sizes (P30, P50, P100, P300 and P500) and under different problem dimensions (10D, 50D and 100D). Major observations are as follows:

1. Different test problems demonstrate similar patterns in terms of the result comparison with the magnitude of computation time depending on problem complexity.
2. For all test problems, the parallel cuSaDE consistently demonstrates the superior computation speed over the sequential SaDE with respect to any population size and any problem dimension under test.
3. Given a specific problem dimension, as the population size increases, the speedup ratio of the parallel cuSaDE over the sequential SaDE keeps increasing remarkably.
4. Given a specific population size, as the problem dimension increases, the speedup ratio of the parallel cuSaDE over the sequential SaDE is not consistently increasing or decreasing. When the population size is small (e.g., 30 and 50), the ratio is increasing. When the population size is large (e.g., 300 and 500), the ratio is decreasing.
5. For test problems of higher complexity (e.g., F4), the speedup ratio is less significant in comparison of test problems of lower complexity (e.g., F1).

Given a specific problem dimension that corresponds to a specific maximal number of function evaluations, the population size determines the total number generations (main loops) and accordingly the execution times of those kernels involved in the main loop. As the population size increases, the number of main loops decreases, which potentially reduces the total computation time. However, the augmented population size increases the amount of computation carried out by each kernel (via the increased number of threads) involved in the main loop, which potentially increases the total computation time. When a GPU card contains many computing units (e.g., SMs (SMXs), SPs, etc.) like the GTX Titan GPU used in our experiment, the increased amount of computation with respect to a kernel in the main loop can be somewhat hidden by the thread-level and instruction-level parallelism [2]. Consequently, the time-saving due to the reduced number of main loops dominates, which leads to the observation (3).

Given a specific population size, the total number generations (main loops) is fixed. In such a case, as the problem dimension increases, the amount of computation carried out by each kernel (via the increased number of threads) involved in the main loop increases. However, when the population size is small, the GPU is able to reduce such increased computation amount via the thread-level and instruction-level parallelism [2], which slows down the increase of computation time. On the other hand, when the population size is large, the GPU may become more overloading as the problem dimension further increases, which may lead to the remarkable decrease of computation speed. In contrast, computation time of the CPU-based sequential SaDE gradually increases as the problem dimension increases. These facts explains the observation (4).

For test problems of high computational complexity, e.g., F4 involving several rotation operations, the execution of kernel(E) is very time-consuming especially when the problem dimension becomes large. In such a case, even when the population size is increased to reduce the number of main loops, the time-saving due to the reduced number of main loops may be compromised by the much increased kernel execution time. Therefore, the speedup ratio is less significant in comparison of test problems of lower complexity, e.g., F1 involving no rotation operations.

5   Conclusions and Future Work

We proposed a CUDA-based parallel SaDE algorithm to speed up the original sequential SaDE while maintaining its optimization accuracy. The similar optimization accuracy and consistent speedup of the proposed parallel cuSaDE in comparison to the original sequential SaDE is verified using several numerical optimization problems across varying problem dimensions and algorithmic population sizes. Our future work includes but is not limited to: analyzing and improving the performance of each individual kernel; adopting streams to concurrently execute multiple kernels; investigating the influence of the CUDA random number generators in an in-depth manner.

Acknowledgments. This work was partially supported by NSFC under Grant No. 61005051 and 61271264, SRFDP under Grant No. 20100092120027.

References

1. De Jong, K.A.: Evolutionary Computation: A Unified Approach. The MIT Press (2006)
2. NVIDIA CUDA C Programming Guide Version 5.5
3. Kirk, D., Hwu, W.-M.: Programming Massively Parallel Processors: A Hands-on Approach. Morgan Kaufmann (2010)
4. Price, K., Storn, R., Lampinen, J.: Differential Evolution: A Practical Approach to Global Optimization. Springer, Berlin (2005)
5. Veronese, L.D.P., Krohling, R.A.: Differential evolution algorithm on the GPU with C-CUDA. In: Proc. of the 2010 IEEE Congress on Evolutionary Computation (CEC 2010), Barcelona, Spain, July 18-23 (2010)
6. Zhu, W., Li, Y.: GPU-accelerated differential evolutionary Markov Chain Monte Carlo method for multi-objective optimization over continuous space. In: Proc. of the 2nd Workshop on Bio-inspired Algorithms for Distributed Systems, New York, NY, USA, June 7-11 (2010)
7. Kromer, P., Platos, J., Snasel, V., Abraham, A.: A comparison of many-threaded differential evolution and genetic algorithms on CUDA. In: Proc. of the 2011 World Congress on Nature and Biologically Inspired Computing (NaBIC 2011), Salamanca, October 19-21 (2011)
8. Kromer, P., Platos, J., Snasel, V.: Differential evolution for the linear ordering problem implemented on CUDA. In: Proc. of the 2011 IEEE Congress on Evolutionary Computation (CEC 2011), New Orleans, LA, USA, June 05-08 (2011)
9. Kromer, P., Snasel, V., Platos, J., Abraham, A.: Many-threaded implementation of differential evolution for the CUDA platform. In: Proc. of the 2011 Genetic and Evolutionary Computation Conference (GECCO 2011), Dublin, Ireland, July 12-16 (2011)
10. Qin, A.K., Raimondo, F., Forbes, F., Ong, Y.S.: An Improved CUDA-Based Implementation of Differential Evolution on GPU. In: Proc. of the 2012 Genetic and Evolutionary Computation Conference (GECCO 2012), Philadelphia, USA, July 7-11 (2012)
11. Wang, H., Rahnamayan, S., Wu, Z.J.: Parallel differential evolution with self-adapting control parameters and generalized opposition-based learning for solving high-dimensional optimization problems. Journal of Parallel and Distributed Computing 73(1), 62–73 (2013)
12. Fok, K.-L., Wong, T.T., Wong, M.-L.: Evolutionary computing on consumer-level graphics hardware. IEEE Intelligent Systems 22(2), 69–78 (2007)
13. CURAND Library Programming Guide Version 5.5
14. Liang, J.J., Qu, B.-Y., Suganthan, P.N., Hernandez-Diaz, A.G.: Problem Definitions and Evaluation Criteria for the CEC 2013 Special Session and Competition on Real-Parameter Optimization, Technical Report 201212, Computational Intelligence Laboratory, Zhengzhou University, Zhengzhou China and Technical Report, Nanyang Technological University, Singapore (January 2013)
15. Qin, A.K., Suganthan, P.N.: Self-adaptive differential evolution algorithm for numerical optimization. In: Proc. of the 2005 IEEE Congress on Evolutionary Computation (CEC 2005), Edinburgh, UK, September 2-5 (2005)
16. Qin, A.K., Huang, V.L., Suganthan, P.N.: Differential evolution algorithm with strategy adaptation for global numerical optimization. IEEE Transactions on Evolutionary Computation 13(2), 398–417 (2009)

Mathematical Theory for Social Phenomena to Analyze Popularity of Social Incidents Quantitatively Using Social Networks

Akira Ishii[1], Takuma Koyabu[1], Koki Uchiyama[2], and Tsukasa Usui[3]

[1] Department of Applied Mathematics and Physics, Tottori University,
Koyama, Tottori 680-8552, Japan
[2] Hottolink, Kanda-nishikicho, Chiyoda-ku, Tokyo 101-0054, Japan
[3] M Data Co.Ltd., Toranomon, Minato-ku, Tokyo 105-0001 Japan

Abstract. A mathematical theory for social events is presented based on a former mathematical model for the hit phenomenon in entertainment as a stochastic process of interactions of human dynamics. The model uses only the time distribution of advertisement budget as an input, and word-of-mouth (WOM) represented by posts on social network systems is used as data to compare with the calculated results. The unit of time is a day. The calculations of intention of people in Japanese society for the scandal of cell of stimulus-triggered acquisition of pluripotency (also known as STAP) agree very well with the twitter posting distribution in time.

Keywords: social phenomena, hit phenomena, stochastic process, STAP.

1 Introduction

Human interaction in real society can be considered in the sense of ＂many body＂ theory where each person can be treated as atoms or molecules in the ordinary many body theory of theoretical physics. Thus, even if we cannot use Hamiltonian because of lacking of energy conservation rules in social phenomena, we can use many body theory to the modeling of social phenomena in some sense. On the other hand, with the popularization of social network systems (SNS) like blogs, Twitter, Facebook, Google+, and other similar services around the world, interactions between accounts can be stocked as digital data. Though the SNS society is not the same as real society, we can assume that communication in the SNS society is very similar to that in real society. Thus, we can use the huge stock of digital data of human communication as observation data of real society [1,2,3,4]. Using this observed huge data (so -called ＂Big Data＂), we can apply the method of statistical physics to social sciences. Since word-of-mouth (WOM) is very significant in marketing science [5,6,7,8], such analysis and prediction of the digital WOM in the sense of statistical physics become very important today. Recently, we present a mathematical theory for hit phenomena where effect of advertisement and propagation of reputation and rumors due to human communications are included as the statistics physics of human dynamics [9]. This theory has also been applied to the analysis of the local entertainment events in Japan successfully [10]. Our model has been also applied to "general election" of a Japanese pop girls gourd AKB48 [11], music concert [12,13] and even to a Kabuki player of 19th century[14].

Our model was originally designed to predict how word-of-mouth communication spread over social networks or in the real society, applying it to conversations about movies in particular, which was a success. Moreover, we also found that when they overlapped their predictions with the actual revenue of the films, they were very similar.

In the model [9], the key factors to affect the mind of the persons in the society are three: advertisement or public announcement effects, the word-of-mouth (WOM) effects and the rumor effects. Recognizing that WOM communication, as well as advertising, has a profound effect on whether a person pays attention to the something or not, whether this is talking about it to friends (direct communication or WOM) or overhearing a conversation about it in a caf (indirect communication or the rumor), we accounted for this in our calculations. The difference between our theory and the previously presented researches [15,16,17,18,19,20,21,22,23,24,25,26,27,28,29,30] are discussed in ref.[9].

We found that the effects of advertisements and WOM are included incompletely and the rumor effect is not included in the previous works [15,16,17,18,19,20,21,22,23,24,25,26,27,28,29,30]. Therefore, from the point of view of statistical physics, we present in our previous paper a model to include these three effects: the advertisement or public announcement effect, the WOM effect, and the rumor effect. The previous model called "mathematical model for hit phenomena" has been applied to the motion picture business in the Japanese market, and we have compared our calculation with the reported revenue and observed number of blog postings for each film.

However, in the recent our several works, we found that our theory can be applicable not only for box office but also other social entertainment like local events[10], animation drama on TV[31],"general election" of a Japanese pop girl group AKB48[11], online music[32], play[33], music concerts[12,13], Japanese stage actors[34], Kabuki players of 19th century[14] and TV drama[35]. In these works, we have used an extended mathematical theory for hit phenomena for applying to general entertainments in societies. In this paper, we show the extension of our theory as *mathematical theory for social phenomena* and apply it also to non-entertainment social phenomena; the scandal of the stimulus-triggered acquisition of pluripotency cell (known as STAP cell). In this paper, the responses in social media are observed using the social media listening platform presented by Hottolink. Using the data set presented by M Data Co.Ltd monitors the exposure of each films.

2   Theory

2.1   Intention for Individual Person

Based on the observation of posting on blog or twitter, we present a theory to explain and predict hit phenomena. First, instead of the number of potential persons who feel attention to the certain fact $N(t)$, we introduce here the integrated intention of individual persons, $J_i(t)$ defined as follows,

$$N(t) = \sum_i J_i(t) \tag{1}$$

here the suffix i corresponds to individual person who has attention to the event he/she concern.

The daily intention is defined from $J_i(t)$ as follows,

$$\frac{dJ_i(t)}{dt} = I_i(t) \tag{2}$$

The number of integrated customers or incoming people to the event can be calculated using the intention as follows,

$$N(t) = \int_0^t \sum_i I_i(\tau)d\tau \tag{3}$$

Since the purchase intention of the individual customer increase due to both the advertisement and the communication with other persons, we construct a mathematical model for social phenomena as the following equation.

$$\frac{dI_i(t)}{dt} = advertisement(t) + communication(t) \tag{4}$$

2.2   Advertisement and Communication

Advertisement is the very important factor to increase intention of the customer in the market. Usually, the advertisement campaign is done at TV, newspaper and other media. We consider the advertisement effect as an external force to the equation of intention as follows,

$$\frac{dI_i(t)}{dt} = \sum_\xi c_\xi A_\xi(t) + \sum_j d_{ij} I_j(t) * \sum_j \sum_k p_{ijk} I_j(t) I_k(t) \tag{5}$$

where $A_\xi(t)$ is advertisement of each media like TV, newspaper, internet news sites, etc for each day in unit of counts or the exposure time, c_ξ is the the factor of the effect of each advertisement and the suffix ξ means the type of the media like TV, newspaper, internet news sites, etc. The factor c_ξ are determined by using the Monte Carlo like method to adjust the observed data as we introduced in ref.[9]. d_{ij} is the factor for the direct communication and p_{ijk} is the factor for the indirect communication[9]. The suffix "j, k" are summed up for all friends for

Fig. 1. Advertisement via TV to consumers

each person. Because of the term of the indirect communication, this equation is a nonlinear equation.

The image of the mass media effect, the direct communication and the indirect communication can be illustrated in fig.1, fig.2 and fig.3.

Fig. 2. Direct communication

2.3   Mean Field Approximation

To solve the equation (5), we introduce here the mean field approximation for simplicity. Namely, we assume that the intentions of every persons are similar so that we can introduce the averaged value of the individual intention.

$$I = \frac{1}{N_p} \sum_j I_j(t) \tag{6}$$

Fig. 3. Indirect communication

where introducing the number of potential customers N_p. We obtain the direct communication term from the person who do not watch the movie as follows,

$$\sum_j d_{ij} I_j(t) = d \sum_j I_j(t) = N_p d I(t) = D I(t) \tag{7}$$

where $D = N_p d$ is the averaged factor of the direct communication. Similarly, we obtain the indirect communication,

$$\sum_j \sum_k p_{ijk} I_j(t) I_k(t) = p \sum_j \sum_k I_j(t) I_k(t) = N_p^2 p I^2(t)$$
$$= P I^2(t) \tag{8}$$

where $P = N_p p$ is the averaged factor of the indirect communication.

Substituting the above, we obtain the following equation as the equation of intention,

$$\frac{dI(t)}{dt} = \sum_\xi c_\xi A_\xi(t) + D I(t) + P I^2(t). \tag{9}$$

The above equation can be also drived as the following way. If we consider the unknown function for the effect of human communication as $F(I(t))$ to write

$$\frac{dI(t)}{dt} = \sum_\xi c_\xi A_\xi(t) + F(I_0 + I(t)), \tag{10}$$

we can expand the function F by $I(t)$ as follows,

$$\frac{dI(t)}{dt} = \sum_\xi c_\xi A_\xi(t) + F(I_0) + \frac{dF}{dI} I(t) + \frac{1}{2} \frac{d^2 F}{dI^2} I^2(t) + \frac{1}{3!} \frac{d^3 F}{dI^3} I^3(t) + \cdots \tag{11}$$

Thus, we can recognize that the equation (9) corresponds to the equation (11) of the second-order.

The equations (9) is based on the equation we presented in the previous work for the motion picture entertainment market [9] where the equation is derived using the stochastic processes. Thus, the equation is the nonlinear differential equation. However, since the handling data is daily, the time difference is one day, we can solve the equation numerically as a difference equation.

When the indirect communication term is dominant in (9), we obtain that

$$\frac{dI(t)}{dt} = PI^2(t).$$ (12)

The solution of this equation is

$$I(t) = -\frac{1}{Pt} + const$$ (13)

Thus, we obtain the power law like divergence of intention if the coefficient P is positive. On the other hand, if the case after the event, we can obtain the power-law like tail if the P is negative.

Therefore, the equation (9) can be considered to be very general equation of intention of people in societies. If necessary, we can extend the model adding extra terms to (9).

2.4   Determination of Parameters

For the purpose of the reliability, we introduce here the so-called gR-factorh (reliable factor) well-known in the field of the low energy electron diffraction (LEED) experiment [36]. In the LEED experiment, the experimentally observed curve of current vs. voltage is compared with the corresponding theoretical curve using the R-factor. For our purpose, we define the R-factor for our purpose as follows,

$$R = \frac{\sum_i (f(i) - g(i))^2}{\sum_i (f(i)^2 - g(i)^2)}$$ (14)

where the functions $f(i)$ and $g(i)$ correspond to the calculated $I(t)$ and the observed number of posting to blog or Twitter. The smaller R, the function f and g show that better matches. Thus, we use the random number to search the best parameter set which gives us the minimum R. This random number technique is similar to the Metropolis method.[37] This method has been introduced in ref.[9] We use this R-factor as guide to get best adjustment of our parameters for each calculation in this paper.

Actually, the parameters $c_\xi AD$ and P in eq.(9) can be considered as a function of time, because the attention of people can be changed as a function of time. However, if we introduce the functions $c_\xi(t)AD(t)$ and $P(t)$, we can adjust any phenomena by adjusting these functions. Thus, we try to keep the parameters $c_\xi AD$ and P to be constant value in order to examine that the equation (9) can be apply to any social phenomena or not.

3   Results

We apply the mathematical theory for social phenomena to the scandal of STAP cell research project. Here, the numbers of blog posting and Twitter posting are collected using the Kuchikomi Kakaricyo service of the Hottolink co. Ltd. and the exposure data on television in unit of second is obtained by the M Data Co. Ltd. The Twitter data used in this paper is 1/10 sampling data.

3.1   Scandal of STAP Cell

In 2014, the scandal of the stimulus-triggered acquisition of pluripotency cell (known as STAP cell) happen in Japanese famous research institute RIKEN. In January of 2014, the observation of STAP cells was reported on Nature [38,39]. The research was mainly done by Dr.H Obokata and a part of the experiment was done by Prof. T Wakayama of Yamanashi University Japan. The STAP project was strongly supported and promoted by Prof. Y Sasai who is very famous for the researches on embryonic stem cells. However, soon after the publication of the paper, it was pointed out by many researchers in the world that the conclusion of the papers are not confirmed by their experiments. In March, one of the key authors, Prof. T Wakayama proposed to withdrawn the papers of Nature because of questionable points on the papers. On 1 April 2014, RIKEN concluded that Obokata had falsified data to obtain her results. Finally, their papers have been withdrawn.

At 29 January, RIKEN was announced the research results on STAP cell at a press conference with a bang. Thus, even after failed investigations to confirm the STAP cell existence by many scientists, a lot of press reports are issued on TV or on newspapers or on internet in Japan because of very high popularity of the STAP cell problem. Therefore, we use the STAP cell incident as an application of the mathematical model for social phenomena.

3.2   Detail of Calculation

We perform the calculation on the popularity of STAP cell using eq.(9) . D and P in eq.(9) are the coefficient of the direct and the indirect communication, respectively. C_ξ in eq.(9) are the strength of advertisement effect for advertisement on TV C_{adv-TV} and advertisement on internet news $C_{adv-news}$. These 4 coefficients are determined by using the random number technique similar to Metropolis method[37] in the molecular dynamics commented in the above section.

First, we perform the calculation to adjust the four coefficients D, P, C_{acv-TV} and $C_{adv-news}$ to reproduce the observed number of posting on blog or Twitter including the word "STAP" from 24 January 2014 to 10 May 2014 in figs.4 and 5. The notation A, B, C, D, E and F correspond to the events listed in the following section. The four coefficients D, P, C_{acv-TV} and $C_{adv-news}$ are assumed to be constant throughout the whole period.

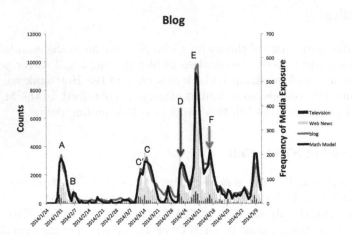

Fig. 4. Observation of the comment to blog for STAP cell and corresponding calculation with the assumption that the four coefficients D, P, C_{acv-TV} and $C_{adv-news}$ are constant throughout the whole period. A, B, C', C, D, E, F are the events explained in the text.

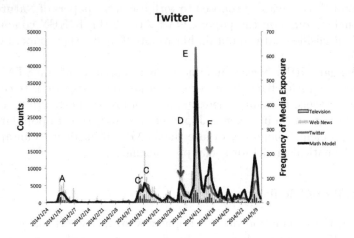

Fig. 5. Observation of the comment to Twitter for STAP cell and corresponding calculation with the assumption that the four coefficients D, P, C_{acv-TV} and $C_{adv-news}$ are constant throughout the whole period.

From the figures, we found that the calculated result can adjust the observed blog and Twitter data very well even under the assumption that the four coefficients D, P, C_{acv-TV} and $C_{adv-news}$ are constant throughout the whole period.

It should be commented that we consider only the strength of popularity for the STAP cell problem for all Japanese people who commented on that problem using blogs and Twitter. The posted comments are not separated into the positive and negative comments. It should be significant that we can adjust the observed data using the mathematical model for social phenomena even for neglecting the difference between positive and negative. The approach to distinguish the positive and negative comments are also possible using the mathematical model for social phenomena and its approach is now going on.

3.3   Calculation for Popularity of STAP Cell

On the above section, we calculate under the assumption that the four coefficients D, P, C_{acv-TV} and $C_{adv-news}$ are constant throughout the whole period. However, the popularity of the STAP cell scandal can be divided into several stages, so that it would be better to calculate for each stage separetely.

The time series of significant events on the STAP cell problem are listed up as follows.

A press conference of RIKEN for STAP cell (28 Jan)
B revelation of the suspicion (5 Feb)
C' proposal of withdrawn of Nature paper by Prof.Wakayama(10 March)
C RIKEN released an interim report of the investigation on the suspicion (14 March)
D RIKEN released the final report for STAP cell suspicion (1 April)
E press conference by Dr. H Obokata (9 April)
F press conference by Prof. Y Sasai (16 April)
G Publication of the research notebooks of Obokata
H Decease of Prof. Y Sasai

Thus, we divide the time series of events on the STAP cell scandal into the following 5 terms. In fig.6, we show the 5 stage. The graph is the popularity of the word "Obokata" on Twitter. For each press conferences above were fully relayed by television in Japan in that time.

① before A
② A to C'
③ C' to E
④ E to F
⑤ F to 10 August

The calculation for 5 stages separately are shown bellow. In fig.7 and fig.8, we show the observation of the comments including the word "Obokata" on blogs and Twitter. In fig.9 and fig.10, we show the observation of the comments including the word "STAP cell" on blogs and Twitter. We found a very sharp peak at the press conference by Obokata. It would be the effect that Dr.H Obokata is very beautiful young female scientist.

The results show us that the calculation using the mathematical model for social phenomena reproduce the observed number of daily postings on blog and Twitter both very well. For separating into 5 stages, the agreement is improved. We also do the calculation to reproduce the daily postings including the word "Sasai" in the same way.

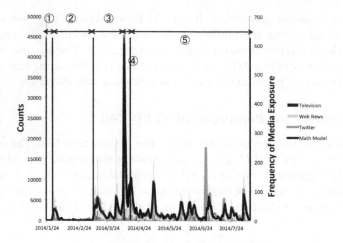

Fig. 6. The 5 stages ①, ②, ③, ④ and ⑤. The graph is the observation of the comment to Twitter for "Obokata" and corresponding calculation

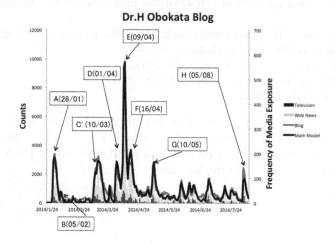

Fig. 7. Observation of daily posting to blog for "Obokata" and corresponding calculation. The blue curve is the daily posting number and red curve is the calculation. The histograms are exposure on TV and internet news. The significant events are also shown in the figure.

4   Discussion

In the above calculations, we can obtain the adjusted coefficients D, P, C_{acv-TV} and $C_{adv-news}$ which are adjusted by the Monte Carlo like method. We found that the indirect communication coefficient P is very large at the stage before the press conference. P for "Obokata" and "STAP cell" are large before the press conference by Dr. H Obokata. P for "Sasai" is large just before his press

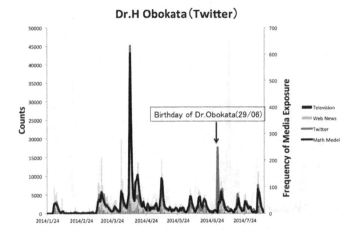

Fig. 8. Observation of daily posting to Twitter for "Obokata" and corresponding calculation. The blue curve is the daily posting number and red curve is the calculation. The histograms are exposure on TV and internet news.

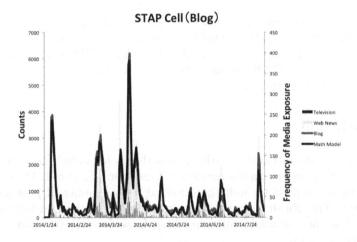

Fig. 9. Observation of daily posting to blog for "STAP cell" and corresponding calculation. The blue curve is the daily posting number and red curve is the calculation. The histograms are exposure on TV and internet news.

conference. Since the indirect communication is very significant as rumor in societies, we found that the rumor in the Japanese society is extremely high just before the press conference. After the press conference, the corresponding P value goes down. It means that the society were satisfied in some sense with the corresponding press conference so that the popularity settled down after the corresponding press conference.

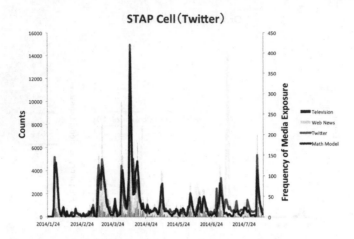

Fig. 10. Observation of daily posting to Twitter for "STAP cell" and corresponding calculation. The blue curve is the daily posting number and red curve is the calculation. The histograms are exposure on TV and internet news.

The direct communication is considered as communication between well-known friends. In this case of STAP cell problem, the direct communication coefficient D are very large for the stage ④ and ⑤ for the word "STAP cell". It means that the discussion on the STAP cell were very popular in these terms after the press conference by Dr. H Obokata.

5   Conclusion

In this paper, we show that the mathematical theory for social phenomena can be applied to social phenomena like academic scandal on STAP cell. The calculation using the theory reproduce well the observed daily number of posting on blog and Twitter. The obtained indirect communication strength using the theory shows us that the popularity on something in societies settle down after the press conference on it.

Thus the mathematical theory for social phenomena can be applied to analyze many social popular phenomena.

Aknowledgement. The three illustrations are drawn by Ms. Nobue Inomoto, the former secretary of one of the author, A.I.

References

1. Allsop, D.T., Bassett, B.R., Hoskins, J.A.: J. Advertising Research 47, 398 (December 2007)
2. Kostka, J., Oswald, Y.A., Wattenhofer, R.: Word of Mouth: Rumor Dissemination in Social Networks. In: Shvartsman, A.A., Felber, P. (eds.) SIROCCO 2008. LNCS, vol. 5058, pp. 185–196. Springer, Heidelberg (2008)
3. Bakshy, E., Hofman, J.M., Mason, W.A., Watts, D.: Proceedings of the Fourth ACM International Conference on Web Search and Data Mining
4. Jansen, B.J., Zhang, M., Sobel, K., Chowdury, A.: J. Am. Soc. Inform. Sci. Tech. 60, 2169 (2009)
5. Brown, J.J., Reingen, P.H.: Journal of Consumer Research 14, 350 (1987)
6. Murray, K.: Journal of Marketing 55, 10 (January 1991)
7. Banerjee, A.: Quarterly Journal of Economics 107, 797 (1992)
8. Taylor, J.: Brandweek, 26 (June 2, 2003)
9. Ishii, A., Arakaki, H., Matsuda, N., Umemura, S., Urushidani, T., Yamagata, N., Yoshida, N.: New Journal of Physics 14, 063018 (22pp.) (2012)
10. Ishii, A., Matsumoto, T., Miki, S.: Prog.Theor.Phys.: Suppliment No.194, pp.64–72 (2012)
11. Ishii, A., Ota, S., Koguchi, H., Uchiyama, K.: The Proceedings of the, International Conference on Biometrics and Kansei Engineering (ICBAKE 2013), 143–147 (2013), doi 10.1109/978-0-7695-5019-0/13
12. Kawahata, Y., Genda, E., Ishii, A.: The proceedings of the, International Conference on Biometrics and Kansei Engineering (ICBAKE 2013) 208–213 (2013), doi 10.1109/978-0-7695-5019-0/13
13. Kawahata, Y., Genda, E., Ishii, A.: ASIAGRAPH2013 in Kagoshima Proceedings, "Analysis Reputation Prediction of Music Concerts Adopting the Mathematical Model of Hit Phenomena" (in press)
14. Kawahata, Y., Genda, E., Ishii, A.: Possibility of analysis of "Big data" of kabuki play in 19th century using the mathematical model of hit phenomena. In: Reidsma, D., Katayose, H., Nijholt, A. (eds.) ACE 2013. LNCS, vol. 8253, pp. 656–659. Springer, Heidelberg (2013)
15. Elberse, A., Eliashberg, J.: Marketing Science 22, 329 (2003)
16. Liu, Y.: Journal of Marketing 70, 7 (2006)
17. Duan, W., Gu, B., Whinston, A.B.: Decision Support Systems 45, 1007 (2008)
18. Duan, W., Gu, B., Whinston, A.B.: J. Retailing 84, 233 (2008)
19. Zhu, M., Lai, S.: Proceeding of the 2009 International Conference on Electronic Commerce and Business Intelligence (2009)
20. Goel, S., Hofman, J.M., Lahaie, S., Pennock, D.M., Watts, D.J.: PNAS 107, 1786 (2010)
21. Karniouchina, E.V.: International Journal of Research in Marketing 28, 62 (2011)
22. Sinha, S., Raghavendra, S.: Eur. Phys. J. B42, 293 (2004)
23. Pan, R.K., Sinha, S.: New J. Phys. 12, 115004 (2010)
24. Asur, S., Huberman, R.A.: aiXiv:1003.5699v1
25. Ratkiewicz, J., Fortunato, S., Flammini, A., Menczer, F., Vespignani, A.: Phys. Rev. Lett. 105, 158701 (2010)
26. Eliashberg, J., Jonker, J.-J., Sawhney, M.S., Wierenga, B.: Marketing Science 19, 226 (2000)
27. Bass, F.M.: Management Science 15, 215–227 (1969)

28. Bass, F.: The Adoption of a Marketing Model: Comments and Observations. In: Mahajan, V., Wind, Y. (eds.) Innovation Diffusion Models of New Product Acceptance, Ballinger (1986)
29. Dellarocas, C., Awad, N.F., Zhang, X.: 2004 working paper. MIT Sloan School of Management (2004)
30. Dellarocas, C., Zhang, X., Awad, N.F.: J. Interactive Marketing 21(4), 23–45 (2007)
31. Ishii, A., Furuta, K., Oka, T., Koguchi, H., Uchiyama, K.: Intelligent Decision Technologies. In: Neves-Silva, R., Watada, J., Phillips-Wren, G., Jain, L.C., Howlett, R.J. (eds.) Frontiers in Aritificial Intelligence and Applications, vol. 255, pp. 267–276. IOS press (2013)
32. Ishii, A., Fujimoto, H., Fukumoto, W., Koguchi, H., Uchiyama, K.: Presentation in WEHIA 2012 in Paris (2012)
33. Kawahata, Y., Genda, E., Ishii, A.: ASIAGRAPH2013 in Kagoshima Proceedings, "Analysis of Stage Actors of Japan Using Mathematical Model of Hit Phenomena" (in press)
34. Kawahata, Y., Genda, E., Koguchi, H., Uchiyama, K., Ishii, A.: International Journal of Affective Engineering 13, 89–94 (2014)
35. Ishii, A., Ota, S., Tanimura, T., Kitao, A., Arakaki, H., Uchiyama, K., Usui, T.: In Proceedings of WEIN2014 in AAMAS 2014 (2014)
36. Pendry, J.B.: J. Phys. C3 937 (1980)
37. Metropolis, N., Rosenbluth, A.W., Rosenbluth, M.N., Teller, A.H., Teller, E.: Equations of State Calculations by Fast Computing Machines. Journal of Chemical Physics 21, 1087–1092 (1953)
38. Obokata, H., Wakayama, T., Sasai, Y., Kojima, K., Vacanti, M.P., Niwa, H., Yamato, M., Vacanti, C.A.: Nature 505, 641–647 (2014)
39. Obokata, H., Sasai, Y., Niwa, H., Kadota, M., Andrabi, M., Takata, N., Tokoro, M., Terashita, Y., Yonemura, S., Vacanti, C.A., Wakayama, T.: Nature 505, 676–680 (2014)

In-Game Action Sequence Analysis for Game BOT Detection on the Big Data Analysis Platform

Jina Lee[1], Jiyoun Lim[2], Wonjun Cho[3], and Huy Kang Kim[4,*]

[1] Target Ads Research Team, Daum Communications, Seoul, Republic of Korea
jina.lee320@gmail.com
[2] SW Future Research Team, ETRI, Daejon, Republic of Korea
kusses@gmail.com
[3] Distributed Data Processing Team, NC Soft, Inc., Seongnam, Republic of Korea
wonjun.cho@gmail.com
[4] Graduate School of Information Security, Korea University, Seoul, Republic of Korea
cenda@korea.ac.kr

Abstract. Nowadays a game BOT becomes a major threat in the online game industry. Game BOTs make the users get more experience points and become a higher-level easily. Therefore, normal users feel unfair and finally leave the game. Accordingly, there have been many efforts to distinguish game BOTs from normal users. Among them, action sequence analysis is effective way compared to other BOT detection methods. However, previous works could not use large datasets because of the limitation of computing power and accessibility to handle large dataset so far. In this paper, we analyzed the full action sequence of users on the big data analysis platform. We evaluated the BOT detection accuracy with real game service data, Blade and Soul. As a result, we showed that full sequence analysis gives the power to detect BOTs precisely. The values for precision and recall were 100%.

Keywords: BOT detection, sequence data mining, Naïve Bayesian classifier, online game security, Big data analysis.

1 Introduction

A game BOT is a well-crafted AI program that can play a game without any human interaction or control. Game BOTs annoy normal users because BOTs continuously consume in-game resources (e.g., sweeping out all monsters that provide hunting experience to other users, collecting rocks and harvesting crops before other normal users have a chance to harvest them). These BOTs can cause complaints from normal players, and damage the online game service provider's reputation. Additionally, game BOTs cause inflation in a game economy because a user can acquire experiences, items, and game money easily by simply installing a game BOT on his/her desktop computer. Consequently, game BOTs shorten the life cycle of the game [1].

---

* Corresponding author.

© Springer International Publishing Switzerland 2015
H. Handa et al. (eds.), *Proc. of the 18th Asia Pacific Symp. on Intell. & Evol. Systems – Vol. 2*,
Proceedings in Adaptation, Learning and Optimization 2, DOI: 10.1007/978-3-319-13356-0_32

Therefore, a game BOT is one of the main threats because it can harm a legitimate users' gaming experience.

There are numerous studies to effectively detect game BOTs. As well as academia, commercial products are deployed in the field. In addition, many game companies also have developed their own private methodologies still not publicly open.

In general, for the purpose of debugging or user behavior analysis, most MMORPG (Massively Multiplayer Online Role Playing Game) servers generate event logs whenever a user performs an action (e.g., hunting, harvesting, farming, and chatting). Therefore, these logging features can serve as the best data analysis arena for researching user behaviors.

As initial countermeasures against game BOTs, many game companies adopted client-side detection methods that analyzed BOT signatures (e.g., the BOT program's specific name, process information, and PC's memory status). This is similar to the manner in which most antivirus programs detect computer viruses. However, client-side detection can easily be evaded by BOT-makers. In addition, client-side detection and prevention programs can degrade the PC's performance. Hence, many counter-measure based on this approach, such as commercial anti-BOT products, are not preferred today.

Nowadays, many online game service providers employ server-side detection me-thodologies that use data mining techniques. Server-side detection has a low-key profiling ability to distinguish BOT users from normal users. After analysis, game service providers can selectively block BOT users whom they want to ban. This is why most online game service providers prefer server-side detection methods. While many practices have been applied in the industry, there are few researches pertaining to server-side detection in academia because of the inaccessibility of log-data.

We collaborated with NC Soft, Inc., one of the largest MMORPG service compa-nies, so that we could fully analyze long-term user activity logs with the aid of a big data analysis platform. In this paper, we propose a methodology based on users' full activity sequence analysis, to improve the accuracy of game BOT detection beyond that of the results achieved in previous research.

2   Related Work

The studies for detecting game BOTs have been classified into three categories: serv-er-side, network-side, and client-side [1, 2]. Network-side BOT detection assumes that traffic information, such as traffic explosiveness, network response, and traffic interval time, represents disparate characteristics of human players and BOTs [3]. Client-side detection methods acquire action data directly through user involvement. Thus, client-side methodology has been adopted for BOT detection even though it requires significant network resources that interfere with the user's game [4]. Server-side BOT detection methodology is based on data mining techniques that analyze log data from game servers. In general, it is hard to modify server-side logs with delibe-rate intent, unless hackers penetrate and compromise the game servers. Thus, server-side detection methods are more secure than client-side methods. Previous studies on server-side BOT detection methods used various action logs as a data source. These studies selected features from a vast array of user actions, and applied data mining techniques as shown in Table 1.

Table 1. Server-side BOT detection research

Research	Data source	Employed Methodology
M.A. Ahmad et al., 2009 [5]	Sequence data of game user's demographic, Character Stats, Economic Features, Anonymized social interaction	Naïve Bayesian, KNN, Bayesian network, Decision tree algorithm
K. Chen et al., 2009 [6]	Traffic pattern	Ensemble schemes
Z. Chu et al., 2009 [7]	Web-page embedded loggers, server-side detector components	Binary classifier
S. Mitterhofer et al., 2009 [8]	Waypoints	K-means clustering
C. Platzer, 2011 [11]	Combat sequence	Levenshtein distance
A. Kang et al., 2012 [9]	Chat log	Random forest, Lazy learning, Logistic regression
J. Woo et al., 2012 [10]	Network log	Decision tree algorithm
A. Kang et al., 2013 [1]	Party play log	Neural network

Most of these studies used one-dimensional log data due to computational complexity in spite of the fact that game BOT traces are found in multi-dimensional log data.

A game server writes a user's every single action with timestamp information into the server-side log as shown in Fig. 1. Every action can be mapped with a discrete event in the log file. For example, when a user hunts a monster in the game field, all sub-actions (moves to approach the monster, hits the monster with a sword, defends against the monster's counterattack with a shield, heals the player by drinking a healing-potion, etc.) are recorded with a timestamp for each sub-action. Therefore, all of the user's continuous play sequences can be expressed by a series of discrete actions. This kind of sequence data has rarely been used for BOT detection analysis, although sequence data has been proven important for BOT detection.

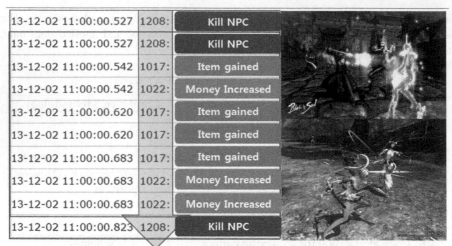

Ex) where k=9
1st seq: **1208-1208**-1017-1022-1017-1017-1017-1022-1022
2nd seq: **1208**-1017-1022-1017-1017-1017-1022-1022-**1208**

Fig. 1. Illustration of the user's action sequence data, when k=9

Platzer (2011) used combat sequences extracted from WOW and compared their Levenshtein distances to detect game BOT [11]. Ahmad et al. (2009) also exercise user behaviors sequence data to detect gold farmers using data mining methods [5]. These studies show that sequence data sheds the light on the study about detect BOT detection. However, they extracted the sequence data from emulated game servers and by simulating game BOTs, not in real game service situations.

Previous studies about BOT detection have examined sequential data on their analysis and found its significancy for game related studies [5, 11], nevertheless it had been barely used prior research. In addition, these previous studies have limitations because of the small amount and limited variety of sequence data used in the analysis. Time-series sequence data is defined as a dataset recorded at successive points, at uniform time intervals, and within a given period. They have been used for classification of stock prices [12], web users [13], DNA, and proteins [14]. Therefore, we examined sequential log-data based on big data platform that have not been easily accessible to academic researchers.

In this study, we collected real data from a live game server and iterated various sequence datasets for analysis. The sequence data used in this paper is actual data obtained from the "Blade and Soul" game server, the third largest MMORPG service in Korea. This log contains discrete and successive actions of characters, although the data is not collected at uniform time intervals. The sequence data with time-stamp information is higher-dimensional information. Although previous studies utilized these sequence datasets that include time information, they have not considered time-related information such as the event-occurrence frequency [5, 15]. In this paper, we employ an analysis methodology for minimizing the loss of information from big data on a Hadoop system, in order to improve the performance of extracting sequence data.

As shown in Table 1 various data mining methodologies are applied for BOT detection. Thurau et al. (2003) analyze the BOT user's behavior with self-organizing map derived from Kohonen Network [22]. Mitterhofer et al. (2009) employs K-mean algorithm with CAPTCHA [8]. Woo et al. (2012) decision tree algorithm for analyzing network feature to identify theft detection [10]. Especially analyzing sequence data, Ahmad et al. (2009) apply various data mining techniques such as Naïve Bayesian, KNN, Bayesian Network and Decision tree classifier [5]. Among them, the Bayesian classifier minimizes the probability of misclassification. These previous studies claim that data mining is a proper tool to analyze data extracted from a game database. Additionally the results of previous studies show that the Bayesian classifier is one of the empirically verified methods for this study. The Naïve Bayesian classifier is a simple Bayesian classifier based on the assumption that variables are independent of each other; thus, it is used for approximated classification [16, 17]. It is adopted in this study because the log data used is relatively large in terms of both its count and noise levels compared to that in earlier studies.

3 Experiment with Log Data of Blade & Soul Game

3.1 System Design

In this study, we design and implement big data platform to perform complex calculations faster than that in previous studies, by using the MapReduce program, as shown in Fig. 2. Raw data is stored in 159 clusters of an HDFS (Hadoop Distributed File System), and we use "Pig", "Python", and "R" for collecting and analyzing data.

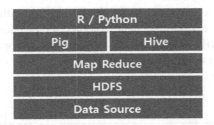

Fig. 2. Big data analysis platform architecture

Fig. 3 shows the overall design of full sequence analysis system for Blade & Soul. About 600 TB of logs generated by games servers (It includes all kinds of game logs. E.g. users' action log, chatting log, battle log) are transferred and analyzed by the analysis module on the big data analysis platform. We can trace all action of an arbitrary user from login to logout. As a result, the detected result is delivered to game operators (i.e. Game Masters).

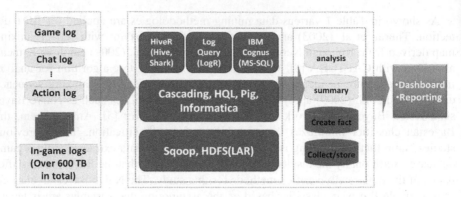

Fig. 3. Blade & Soul log analysis system's overall design for this experiment

Our system is working at the server-side. That is, this system does not make any side-effects as client-side detection programs usually do. It silently analyzes the BOT users' activities and game companies can control the number of accounts to be banned. Also, the detection rules are not exposed to malicious BOT users and makers.

3.2   Data Description

The BOT detection model proposed in this paper was applied to log data from the "Blade and Soul" server. (See the official web pages: http://bns.plaync.com, http://bns.qq.com/, http://us.bladeandsoul.com/en/). The game log data used in the analysis represents the behavior of characters in the game. 110 logs of characters were used for creating sequence patterns.

The data from a single game server for the period from 8th August, 2012 to 15th August, 2012 was used for training data. Another server dataset from 15th August, 2012 to 22nd August, 2012 was used to test algorithms and optimize k, s values. The size of the training and test data is shown in Table 2.

Table 2. dataset used for this study

	# of characters	# of records	# of BOTs
Training data	41,467	750 million	1,187
Test data	40,364	570 million	829

Before applying the sequence data for training, data preprocessing is needed to eliminate the sequence data generated by some users that are not significant for BOT detection. We conduct the previous research to determine the optimal sequence length and the number of sequence patterns as follows.

The optimal sequence length "k" and the significant data frequency "s" are determined by the Apriori algorithm that makes explicit the relationship between data, based on the data frequency [18].

The data are separated into "all sequence patterns" and "limited number of sequence patterns" to compare the performance of prediction rate based on the amount of information. The "all sequence patterns" are literally all the sequence patterns generated in the dataset which have a higher value than the minimum support rate. The number of sequence patterns is decided by the algorithm suggested by Ahmad et al. (2009) [5]. We evaluated classification performance according to the number of sequence patterns obtained by comparing the results of the variety of all sequence patterns and that of the limited number of sequence patterns. Finally, twelve kinds of sequence patterns were selected on the basis of the number of characters in a single server.

3.3   Analysis Based on a Simple Scoring Algorithm

Because BOTs main purpose is to gain items more efficiently, BOTs' actions are usually repetitive and focused on doing some specific tasks and quests such as farming, hunting. That means, game BOTs have significantly different action sequence pattern than other normal users.

In this paper, the simple scoring algorithm used for classification [20] was applied to BOT detection by analyzing sequence data. The scoring algorithm for BOT detection is composed of three stages as shown is Fig. 4.

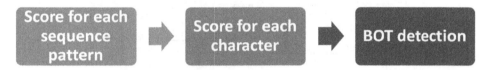

Fig. 4. Scoring Process

The BOT score for each sequence pattern is calculated from the sequence patterns of the BOTs in training data. The score is calculated by comparing a normal user's count of sequence patterns and the BOT's count as shown in (1). This value is represented as a Boolean variable, i.e., either "0" or "1" as shown in (2).

For example, [1100-1200-3100] is a 3-length sequence pattern. 11 BOTs have this pattern, and they generate it 21 times. In case of normal accounts, they generate 130 times and there are 23 normal accounts. BOT probability of this pattern can be calculated as $11/(23+11) = 0.324$ and normal probability as 0.676. Normal probability is larger than BOT one, so final BOT score of this pattern is 0.

$$P(A) = \frac{Actor \;\; count}{Total \;\; count}$$

(1)

Where, BOT score for sequence pattern i is as in (2)

$$BOT\ Score\ i\ =\ \begin{cases} 0, & P(Normal) > P(BOT) \\ 1, & P(Normal) < P(BOT) \end{cases} \tag{2}$$

A character's BOT score is estimated by taking the average value of each character's sequence patterns as shown in (3). We decided on a cutoff value based on the distribution of a normal character's BOT score and the BOT character's score. Fig. 5 shows the BOT score distribution in the case of a 9-length sequence. Normal users' scores were widely distributed between 0 and 1, but most of the BOT characters had a BOT score greater than 0.8. The minimum score of the BOT characters excepting outliers is the value that is used to determine if the character is a BOT. In this case, characters having a BOT score greater than 0.76 are identified as BOT characters.

$$P(actor) = \frac{\sum The\ BOT\ score\ of\ sequence\ pattern}{\sum The\ frequency\ of\ sequence\ pattern} \tag{3}$$

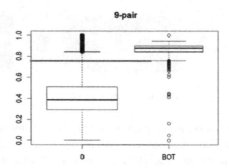

Fig. 4. Score Distribution of Normal Users and BOTs

3.4   Analysis Based on the Naïve Bayesian Algorithm

The Naïve Bayesian algorithm was applied to email spam filters in the study by Androutsopoulos, et al. [21]. In their research, the authors assumed that if an email contains many keywords that are usually found in spam, then that email has a high probability of being spam mail. This assumption can be adapted for a BOT's behavior patterns, and hence, we applied the Naïve Bayesian algorithm in our analysis. This algorithm compares the pattern occurrence probability of both a normal user and a BOT. Each probability of BOTs and normal users are calculated with (4) and (5). P(x) can be skipped due to the same values of both BOTs and normal users.

$$Pi \; = \; \frac{frequency \; of \; pattern \; i \; of \; BOT \; actors}{\sum i \; (frequency \; of \; pattern \; i \; of \; BOT \; actors)} \tag{4}$$

$$P(B|x) \; = \; \frac{P(B)P(x|B)}{P(x)}, \; P(N|x) \; = \; \frac{P(N)P(x|N)}{P(x)} \tag{5}$$

Where, x: actor, B: BOT class, N: Normal class

4   Experiment Result

We analyzed the sequence data with the simple scoring algorithm and Naïve Bayesian algorithm. The optimal value of the length of the sequence pattern (k) and the minimum support rate (s) were decided, as shown in Table 3. When the scoring algorithm was used, the number of sequence patterns that satisfied the minimum support rate 400 and the length 9 was 71,362. When the Naïve Bayesian algorithm was used, 17,252 sequence patterns were used for analysis. The sequence data of 139 characters, decided by three GM (Game Master) experts, were used for the evaluation of the algorithms. 79 characters were proved to be BOTs with no false positives (i.e., all 3 experts pointed out that these characters were BOTs by conducting a thorough manual total review). Some of the manual inspections are shown in Table 4, 60 out of 139 characters were normal users. The full results are in the following URL. https://sites.google.com/a/hksecurity.net/pds/full-seq-analysis

Table 3. The BOT detection result of each algorithm

Algorithm	k and s	The # of sequence patterns	Precision	Recall	F-measure
Simple scoring	k = 9 s = 400	12	3.8%	87.2%	7.4%
		73,362	100%	78.5%	87.9%
Naïve Bayesian	k = 4 s = 500	12	32.5%	68.6%	44.1%
		17,252	100%	100%	100%

There was no significant difference between precision values whereas the Naïve Bayesian algorithm achieved better performance for recall (100%). When comparing the results of the twelve sequence patterns that are restricted by their length, all sequence patterns showed better performance in BOT precision. The values for re-call were similar to that obtained by Ahmad et al. (2009) [5]. When the length of the sequence pattern is relatively small, i.e., when the pattern's length is one or two-length

Table 4. Sample results of BOT detection by experts' manual review

User (anonymized)	A	B	C	User (anonymized)	A	B	C
5B9790EF	X	X	X	6CC3F125	X	X	X
C8DEECEA	X	X	X	4DBF7AFE	O	O	O
FED144EB	X	X	X	65BF7AFE	O	O	O
CF0872EB	X	X	X	3D98E5AF	X	X	X
96BE40F0	X	X	X	5AAE9351	O	O	O
765063D4	X	X	X	7AC7EEBF	O	O	O
4B90AD8D	X	X	X	F05113FD	O	X	X
87A486FE	X	X	X	78C9CAFB	X	X	X
F981C3F2	X	X	X	23E30F04	O	O	O
2C8013FB	X	X	X	E2568CF4	O	O	O

of the sequence, the proposed algorithm exhibits low accuracy in classifying BOTs and normal users. Therefore, the amount of input data has to be large enough to include a variety of information for the training algorithm.

5   Conclusion

A BOT detection model based on analyzing sequence data by applying the Naïve Bayesian algorithm is proposed in this study. This model was empirically applied to game log data in MMORPGs. There are many studies that have attempted to distinguish between BOTs and normal users, using proposed classifiers. However, the data size used for analysis was not large enough. In addition, the performances of the classifiers were not evaluated properly. To the best of our knowledge, this is the first study that analyzes the full sequence of action with complete log data for BOT detection. Various sequence pattern support rates from a big data analysis environment were considered. Finally, this is the experiment based on the real dataset from famous MMORPG, and also the result of classifiers was fully reviewed and validated by human experts.

We compared the results of all sequence patterns that had a value higher than that of the minimum support rate. The comparison was based on big data analysis and twelve-length sequence patterns. As a result, our analysis, which used all kinds of sequence patterns, was more effective in identifying BOTs. Accordingly, we can conclude that more information can be extracted from a large number of sequence patterns, and this in turn, has a positive effect on training data. In addition, the values for recall and precision were equivalent to 100% of the values that are applicable in the actual service area.

Notwithstanding the contributions of this paper, in future work, the results can be compared to the results from analyses using other classifiers. In addition, an algorithm that can be substituted in place of human judgment for evaluating the performance of the BOT detection model can be studied in the future.

Acknowledgment. A preliminary version of this work was appeared in ACM Net-Games 2013 [23].

We would like to thank NC Soft, Inc. for providing the data for this research. This research was supported by Basic Science Research Program through the National Research Foundation of Korea (NRF) funded by the Ministry of Science, ICT & Future Planning (2014R1A1A1006228). Also, this research was supported by Ministry of Culture, Sports and Tourism (MCST) and Korea Creative Content Agency (KOCCA) in the Culture Technology (CT) and Research Development Program 2014.

References

1. Kang, A., Woo, J., Park, J., Kim, H.: Online game bot detection based on party-play log analysis. Computers and Mathematics with Applications 65, 1384–1395 (2013)
2. Woo, J., Kim, H.: Survey and research direction on online game security. In: Workshop at ACM SIGGRAPH ASIA 2012, pp. 19–25 (2012)
3. Thawonmas, R., Kashifuji, Y., Chen, K.: Detection of MMORPG bots based on behavior analysis. In: ACE 2008 Proceedings of the 2008 International Conf. on Advances in Computer Entertainment Technology, pp. 91–94 (2008)
4. Chen, K., Pao, H.K., Chang, H.: Game bot identification based on manifold learning. In: NetGames 2008 Proceedings of the 7th ACM SIGCOMM Workshop on Network and System Support for Games, pp. 21–26 (2008)
5. Ahmad, M.A., Keegan, B., Srivastava, J., Williams, D., Contractor, N.: Mining for gold farmers: automatic detection of deviant players in MMOGs. In: International Conference on Computational Science and Engineering, CSE 2009, vol. 4, pp. 340–345 (2009)
6. Chen, K., Jiang, J., Huang, P., Chu, H., Lei, C., Chen, W.: Identifying MMORPG Bots: a traffic analysis approach. EURASIP Journal on Advances in Signal Processing (2009)
7. Chu, Z., Gianvecchio, S., Koehl, A., Wang, H., Jajodia, S.: Blog or block: detecting blog bots through behavioral biometrics. Computer Networks 57(3), 634–646 (2013)
8. Mitterhofer, S., Kruegel, C., Kirda, E., Platzer, C.: Server-side bot detection in massively multiplayer online games. IEEE Security & Privacy 7(3), 29–36 (2009)
9. Kang, A., Kim, H., Woo, J.: Chatting pattern based game BOT detection: do they talk like us? 6(11), 2866–2879 (2012)
10. Woo, J., Choi, H., Kim, H.: An automatic and proactive identity theft detection model in MMORPGs. Appl. Math. Inform. Sci. 6(1S), 291S-302S (2012)
11. Platzer, C.: Sequence-based bot detection in massive multiplayer online games. In: Communications and Signal Processing (ICICS) 2011 8th International Conference on. Information. IEEE (2011)
12. Shatkay, H., Zdonik, S.B.: Approximate queries and representations for large data sequences, data engineering. In: Proceedings of the Twelfth International Conf. on, pp. 536–545 (1996)
13. Wang, W., Zaiane, O.R.: Clustering web Sessions by sequence alignment. In: 13th Int. Workshop on Database and Expert Syst. Applications, France (2002)
14. Morzy, T., Wojciechowski, M., Zakrzewicz, M.: Scalable hierarchical clustering method for sequences of categorical values. Proc. 5th Pacific-Asia Conf. KDD, Hong Kong (2001)
15. Kim, H., Hong, S., Kim, J.: Detection of auto programs for mMORPGs. In: Zhang, S., Jarvis, R.A. (eds.) AI 2005. LNCS (LNAI), vol. 3809, pp. 1281–1284. Springer, Heidelberg (2005)

16. Domingos, P., Pazzani, M.: On the optimality of the simple Bayesian classifier under zero-one loss. Mach. Learn. 29(2-3), 103–130 (1997)
17. Famoni, M., Senastiani, P.: Robust Bayes classifiers. Artificial Intelligence 125, 209–226 (2001)
18. Agrawal, R., Srikant, R.: Fast algorithms for mining association rules in large databases. In: Proceedings of the 20th International Conference on Very Large Data Bases, Santiago, vol. 1215, pp. 487–499 (1994)
19. Sahami, M., Dumais, S., Heckerman, D., Horvitz, E.: A Bayesian approach to filtering junk e-Mail. In: Proceedings of the AAAI Workshop, pp. 55–62 (1998)
20. Chen, P.L., Lee, C.C., Li, C.Y., Chang, C.M., Lee, H.C., Lee, N.Y., Wu, C.J., Shih, H.I., Tang, H.J., Ko, W.C.: A simple scoring algorithm predicting vascular infections in adults with nontyphoid Salmonella bacteremia. Clin. Infct. Dis. 55(2), 194–200 (2012)
21. Androutsopoulos, I., Koutsias, J., Chandrinos, K.V., Paliouras, G., Spyropoulos, C.D.: An evaluation of naive bayesian anti-spam filtering. In: 11th European Conference on Machine Learning, Barcelona, pp. 9–17 (2000)
22. Thurau, C., Bauckhage, C., Sagerer, G.: Combining Self Organizing Maps and Multilayer Perceptrons to Learn Bot-Behavior for a Commercial Game. In: Proceedings of GAME-ON 2003, pp. 119–123 (2003)
23. Lee, J., Lim, J., Cho, W., Kim, H.: I know what the BOTs did yesterday: Full action sequence analysis using Naïve Bayesian algorithm. Network and Systems Support for Games, pp.1–2 (2013)

Multi-agent Based Bus Route Optimization for Restricting Passenger Traffic Bottlenecks in Disaster Situations

Sayaka Morimoto[1,*], Takahiro Jinba[1], Hiroto Kitagawa[2], Keiki Takadama[1],
Takahiro Majima[3], Daisuke Watanabe[4], and Mitsujiro Katsuhara[5]

[1] The University of Electro-Communications, Tokyo, Japan
{sayaka,jinba}@cas.hc.uec.ac.jp
keiki@inf.uec.ac.jp
[2] TIS Inc., Tokyo, Japan
kitagawahiroto0829@gmail.com
[3] National Maritime Research Institute, Tokyo, Japan
majy@nmri.go.jp
[4] Tokyo University of Marine Science and Technology, Tokyo, Japan
daisuke@kaiyodai.ac.jp
[5] SocioTechData, Tokyo, Japan
kat-151@mail.bbexcite.jp

Abstract. This paper focuses on the passenger traffic *bottlenecks* occurred in the bus route network in disaster situations and proposes the multi-agent based bus route optimization method to resolve such bottlenecks by generating the networks which can effectively transport many stranded persons including ones who wait around the station as the passenger traffic bottlenecks. For this purpose, the proposed method modifies the bus route networks generated as usual conditions to suitably pass many bus lines to and redistribute the buses among the bus lines according to the number of passengers. The intensive simulations have revealed the following implications: (1) the proposed bus route network optimization method generates the route network which is suitable for passenger traffic bottlenecks; (2) the proposed method decreases a risk of the bottlenecks; and (3) our method transports the passengers faster than those by the conventional one in various virtual disaster situations.

Keywords: route optimization, multi-agent system, traffic bottleneck, disaster, stranded persons.

1 Introduction

When disaster occurs, a large number of persons cannot return to their home due to the suspended most public transportations [1]. In such a situation, a bus service has the one of candidates to transport many stranded persons it instead of the other transportations (*e.g.*, the rail transport). However, the bus service has a problem called *bottleneck*, *i.e.*, the number of passengers exceeds usual demands which accumulates

© Springer International Publishing Switzerland 2015   415
H. Handa et al. (eds.), *Proc. of the 18th Asia Pacific Symp. on Intell. & Evol. Systems – Vol. 2*,
Proceedings in Adaptation, Learning and Optimization 2, DOI: 10.1007/978-3-319-13356-0_33

the number of stranded persons around a bus stop. Although it is necessary not to cause the passenger traffic bottlenecks in advance to prevent secondary disasters such as the traffic accident and congestion, the conventional optimization methods such as a minimization of the costs of passengers (*i.e.*, the total movement time) and bus companies (*i.e.*, the number of buses) [2] or a minimization of the damage of the road destruction [3], cannot tackles the problem of the passenger traffic bottlenecks. Furthermore, some of them have no restrictions of the number of buses meaning that required buses can be provided to transport all passengers, which is impractical in the disaster situations.

To tackle the above problem, *i.e.*, the passenger traffic bottlenecks, this paper proposes a new multi-agent based bus route network optimization method which can generate the networks for effectively transporting many stranded persons including ones who wait around the station as the passenger traffic bottlenecks. For this purpose, the proposed method concentrates the bus lines and the number of buses near the bottleneck station by modifying the bus route network generated as usual conditions by the conventional method. To investigate the effectiveness of the proposed method, this paper compares our methods with the conventional one through the benchmark problem of the bus route network optimization [4].

This paper is organized as follows. Section 2 introduces the conventional route network optimization method, and Section 3 describes the mechanism of our method which is based on the conventional method. The experiments are conducted in Section 4. Their results are discussed in Section 5. Finally, our conclusions are given in Section 6.

2   Route Network Optimization Method Based on Multi-agent System

2.1   Bus Route Optimization Problem

The bus route optimization problem [2] includes the map with positions of the bus stops (referred to as the stations) and the roads, an OD (Origin-Destination) table which indicates the number of passengers between each station, the capacity of a bus (fixed), and the speed of a bus (fixed). We attempt to obtain the operating route (referred to as the line) and the number of used buses as the bus route network. A line is expressed as the set of the stations sorted stopping order and the bus stops in them one by one and runs back and forth.

2.2   Algorithm

The flowchart of the Majima's conventional route optimization method based on the multi-agent system (MAS) [2] is shown in Fig. 1. First, the primary route network which consists of many lines and covers all of stations is read (#1-1) or generated (#1-2). Next, we calculate the route passengers use to travel home for each OD using Dijkstra's algorithm (#2). Some unused lines are deleted if there are any (#3). Next, for each living lines we estimate the evaluation value P based on the number of passengers and buses and sort the lines in descending order based on P (#4) for preparation of lines evolving. Next, order k, the number which points the target of lines evolving, is set to 0 (#5) and the k-th line agent (now it is one which has highest P) attempts to evolve to maximize own evaluation value P (#6-1). Lines evolving means to either add or remove any station (see Section 2.4). When the line agent evolves (#6-1 YES) we return to calculate the route (#2), or when not evolves (#6-1 NO) order k is incremented and if k is below the number of lines (#6-2 NO) k-th line agent also attempts to evolve (#6-1). Finally when order k exceeds the number of lines (*i.e.*, any line cannot increase own evaluation value P through the evolution) the optimization is terminated.

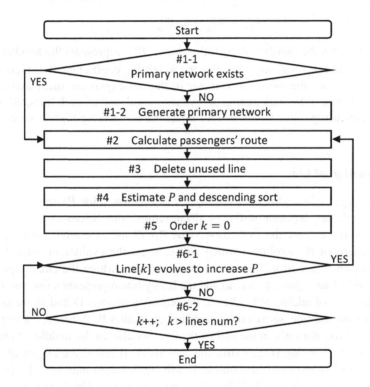

Fig. 1. Flowchart of the route optimization method based on MAS

2.3   Evaluation of Network

The evaluation value Z of the route network is calculated by

$$Z = \sum_{s_i \neq s_j} T_{s_i, s_j} D_{s_i, s_j} + \omega_1 \sum_{L_k} B_{L_k} \tag{1}$$

where $T_{S_i S_j}$ represents the travel time of the passengers from origin station S_i to destination station S_j, $D_{S_i S_j}$ represents the demand of the passengers which occur per unit time from S_i to S_j, B_{L_k} represents the number of buses in the line L_k and ω_1 indicates the weighting coefficient. The lower the evaluation value Z shows, the more superior the route network is; the travel time is short and the number of buses is few. In this paper this estimation value is only used to generate the primary route network and set the number of buses each line has.

2.4   Selection of Line

The evaluation value P used for deciding evolved line is calculated by

$$P_{L_k}^n = R_{L_k}^n - \omega_2 B_{L_k}^n \tag{2}$$

where n represents the number of evolution steps, $R_{L_k}^n$ represents the number of passengers in the line L_k and ω_2 indicates the weighting coefficient. The higher evaluation value P shows, the more superior line is; the passengers are many and the number of buses is few. After calculating all evaluation value P of each lines, all lines are sorted in descending order based on P to select the line having high evaluation value to evolve.

2.5   Evolution of Line

The top line agent (*i.e.*, which has the highest evaluation value P) attempts to evolve either (1) to add any adjacent station or (2) to remove one belonging to the line. The line agent estimates how the evaluation value P will increase after evolving and selects the pattern of the evolution which will increase the evaluation value P higher. Fig. 2 shows an image of adding or removing a station where the circle represents a station, the thin line represents a road and the heavy line represents a line passing over the road. In case of adding (Fig. 2 (a)), the adjacent stations D and E are targets of adding and now the line agent evolves adding the station E. A line agent can evolve adding any station not only in the end of the line but also in the middle. On the other hand in case of removing (Fig. 2 (b)), the stations A, B and E are targets of deleting (if the station C is removed, the station E will be isolated) and now the line agent evolves removing the station B. When the station in the middle is removed, the stations which in both sides are connected.

The line agent selects the pattern of evolution based on estimating how the evaluation value P will increase. Increase of the evaluation value A is calculated by

$$A^n = \left(R_{L_k}^{n+1} - R_{L_k}^n\right) - \omega_2 (B_{L_k}^{n+1} - B_{L_k}^n) \tag{3}$$

where $R_{L_k}^{n+1}$ and $B_{L_k}^{n+1}$ are the estimated values after evolving. The line agent compares all of the patterns with their A and selects the pattern which has the highest A.

After that we return to calculate the route of passengers. When all A is less than 0, the next line agent attempts to evolve likewise. When all of the line agents cannot increase their P, the optimization is terminated.

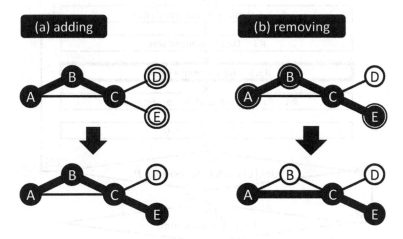

Fig. 2. Patterns of the line evolution

3 Route Network Optimization Method Based on Bus Redistributing and Passengers Weighting to Restrict Bottlenecks

3.1 Outline

Only route network optimization method based on MAS cannot consider a bottleneck. To restrict it we attempt to concentrate the lines and buses on crowded spot. Our proposed method consists of two changes. One is that we weight the passengers who pile up in the station neighbor upon the bottleneck (referred to as the bottleneck adjacent station) (see Section 3.2), and the other is that in evolution the line agent which has extra buses gives them to the other line agent which has deficient buses (see Section 3.3).

3.2 Algorithm

The flowchart of the proposed method is shown in Fig. 3. Important differences are emphasized by bold type and heavy bordered box.

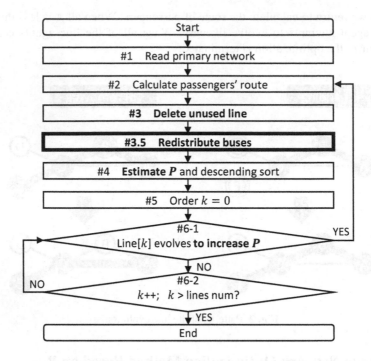

Fig. 3. Flowchart indicating proposed method

3.3   Weighting Passengers in Evolution of Line

To concentrate many lines on crowded station because of bottleneck, we change Eq. 2 in Section 2.4 into Eq. 4.

$$P_{L_k}^n = \alpha R_B^n + \beta R_N^n - \omega_2 B_{L_k}^n \tag{4}$$

where R_B^n represents the number of passengers in the bottleneck adjacent stations of line L_k, R_N^n represents one in the other stations, α ($\alpha > 1$) and β ($\beta < 1$) indicates the weighting coefficient. The higher we set the value α, the higher evaluation value P of the line which contains the bottleneck adjacent stations shows. Due to this change, the line which may restricts the bottleneck become easily to be selected for evolution.

Furthermore, we change Eq. 3 in Section 2.5 into Eq. 5.

$$A^n = \left(R_{L_k}^{n+1} - R_{L_k}^n\right) - \omega_2 R_{L_k}^n (T_{L_k}^{n+1} - T_{L_k}^n) \tag{5}$$

where $T_{L_k}^n$ represents the waiting time of passengers in the line L_k. In this method, because in the case of disaster adding some buses from the outside is difficult, we fix the sum total of buses around the route network. Therefore we delete the fluctuation

$B_{L_k}^{n+1} - B_{L_k}^n$ which is not changed and add the fluctuation $T_{L_k}^{n+1} - T_{L_k}^n$ to make the lines less increasing the waiting time. Then, to consider waiting for the passengers we conduct the term $\left(R_{L_k}^{n+1} - R_{L_k}^n\right)$ of Eq. 5 in two different cases: where the station which will be added or removed is bottleneck adjacent or not such as Eq. 6 and Eq. 7.

$$\left(R_{L_k}^{n+1} - R_{L_k}^n\right) = \begin{cases} \alpha\left(R_{L_k}^{n+1} - R_{L_k}^n\right), if\ bottleneck & (6) \\ \beta\left(R_{L_k}^{n+1} - R_{L_k}^n\right), else & (7) \end{cases}$$

In evolution if the station which the line will add or remove is bottleneck adjacent then the number of passengers is weighted higher ($\alpha > 1$), and if it is not then weighted lower ($\beta < 1$). The images of influence of this change upon adding the station is shown in Fig. 4 and removing is Fig. 5. In these images, the station E means to the bottleneck adjacent stations. When adding the bottleneck adjacent stations (Fig. 4 (a)), due to the weighting coefficient α the fluctuation of the number of passengers (*i.e.*, the evaluation A) is regard as high, therefore adding the bottleneck adjacent station become easily to be selected for evolution (Eq. 6). Meanwhile when adding the bottleneck nonadjacent stations (Fig. 4 (b)), due to β the evaluation A is regard as low, therefore adding the bottleneck nonadjacent station become difficulty to be selected for evolution (Eq. 7). Similarly when removing the bottleneck adjacent stations (Fig. 5 (a)), due to α the evaluation A is regard as low, therefore removing the bottleneck adjacent station become difficulty to be selected for evolution (Eq. 6). Meanwhile when removing the bottleneck nonadjacent station (Fig. 5 (b)), due to β the evaluation A is regard as high, therefore removing the bottleneck nonadjacent station become easily to be selected for evolution (Eq. 7). Consequently, the line which has the bottleneck adjacent station shows high evaluation A.

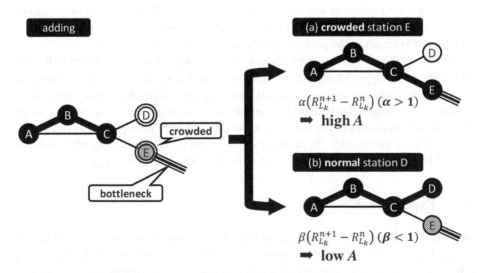

Fig. 4. Image of adding the station of proposed method

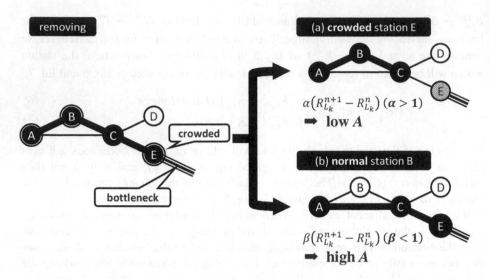

Fig. 5. Image of removing the station of proposed method

3.4   Redistribution of Buses between Lines

To concentrate many buses on crowded line because of bottleneck, we define the number of staying persons for each station calculated by

$$C_{L_k} = R_{L_k} - (Capacity \times B_{L_k}) \tag{8}$$

where *Capacity* represents the capacity of a bus (fixed). When C_{L_k} shows negative, the number of persons who the line L_k can carry exceeds the number of passengers and the line is regard as no problem giving some buses to the other line.

Before evolution starts, the line which has least C_{L_k} gives some buses to the line which has most C_{L_k}. The number of buses is fixed. Hence the frequency of bus coming for each station of the line which is gave becomes high, the waiting time of the passengers becomes shortly and the line can carry them quickly. Nevertheless we fix the sum total of buses around the route network, with only giving phase it enables the line agents to redistribute the number of buses properly. Moreover, when deleting unused line, then the line gives all buses to the line which has most C_{L_k}.

4   Experiment

Our experiments start by investigating the effectiveness of the proposed method by comparing with the conventional one in a typical disaster situation, then investigating it in an extreme disaster situation. For this purpose, our experiments are composed of two cases: (case 1) typical disaster situation; (case 2) extreme disaster situation.

4.1   Case 1: Typical Disaster Situation

4.1.1   Experiment Setup

To investigate the effectiveness of the proposed method, we conducted the simulation using the Mandl's urban transport bench mark problem [4]. This problem consists of 15 stations, 21 roads (*i.e.*, the link between each stations) and OD table in time of peace. The physical infrastructure network is shown in Fig. 6 where the circles indicate the stations, the lines indicate the roads, the circle numbers indicate the ID number, and the number near the lines indicate the time running the bus between the two stations.

We compared the three results of the route network: (A) generated by conventional method as usual conditions, (B) optimized by conventional method based on (A) for the disaster, and (C) optimized by proposed method based on (A) for the disaster. Now we approximated the sum total of buses in the route network (B) to the one of (A) using the weighting coefficient ω_1 to compare on equal terms. Specifically, first we generated the route network (A) using the conventional method with the *normal* OD table. Next, we made the *high* OD table to multiply the number of passengers who depart from one of the stations by 10 for reproducing the disaster. After that we optimized (A) using the conventional method as (B) and using the proposed method as (C). The conventional method is Majima's method [2] on which proposed method is based and the ID number of the stations which is multiplied is 9 because this OD value is the most of all.

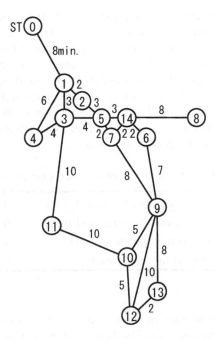

Fig. 6. Infrastructure network of Mandl's urban transport bench mark problem

4.1.2 Evaluation Index and Parameters

As evaluation criteria, all simulation results are evaluated by the overflow of the passengers and the critical movement time (hereafter we call it as the *CriticalTime*). The overflow means the number of passengers who cannot travel home in a fixed time through the route network. *CriticalTime* is defined as follows: first we prepared the total movement time (hereafter we call it as the *TTT*) of all passengers traveling home. *TTT* is calculated by

$$TTT = \sum_{S_i \neq S_j} (TravelTime_{S_i,S_j} + WaitTime_{S_i,S_j} + TransTime_{S_i,S_j}) \qquad (9)$$

where *TravelTime* represents the time of passengers who travel from S_i to S_j riding the bus, *WaitTime* represents the time of them waiting the bus, and *TransTime* represents the time of them getting on and off the bus. After calculating all simulation results are evaluated by the *CriticalTime* calculated by

$$CriticalTime = TTT - MinimumTime \qquad (10)$$

where *MinimumTime* represents the minimum movement time which cannot be shortened physically by any contrivance and calculated by the number of passengers and the infrastructure network (*i.e.*, *CriticalTime* means the movement time which can be shortened by the efficacy of the route network). The shorter *CriticalTime* shows, the more superior the route network is; the total movement time is short and the bottleneck can be restricted.

The experiment parameters are shown in Table 1. When generating the route network by conventional method as usual conditions (A), we set the parameter ω_1 and ω_2 to same value as prior study [2] which make the result most superior. When optimizing by conventional method (B), we set ω_1 and ω_2 so that the number of buses of the route network optimized by this method is closest to one of the route network as usual conditions (*i.e.*, so that the condition of the route networks is closest to one before optimizing). When optimizing by proposed method (C), we set ω_1, ω_2, ω_3 to 2.0, α to 1.5, β to 0.5 and the number of buses which the lines give together in evolution to 3 which make the result most superior.

Table 1. Parameters of the route optimization

The route network generated or optimized by	ω_1	ω_2
(A) Conventional method as usual conditions	0.7	1.0
(B) Conventional method for the disaster	4.5	1.0
(C) Proposed method for the disaster	0.7	1.0

4.1.3 Results

Fig. 7 shows the result of each method. The left graph shows the overflow of the passengers in each route network. The horizontal axis indicates the route network (A) generated by conventional method as usual conditions (left), (B) optimized by conventional method for the disaster(center) and (C) optimized by proposed method for the disaster (right). The vertical axis indicates the number of the overflow.

The right graph shows each time. The vertical axis indicates *TTT* (left) and *CriticalTime* (right). The number on data elements indicates each *CriticalTime* respectively. These optimization method can produce several results, therefore we select each one of them which shows the shortest total movement time (*i.e.*, the most superior one). Then *MinimumTime* was figured out at 903 hours and *CriticalTime* is calculated by them.

From this result, through the route network as usual conditions 3730.5 persons are overflow, however the network optimized both methods can carry all people. The total movement time of the proposed method is shorter than the conventional method. In fact, the proposed method reduces 18% of the critical movement time.

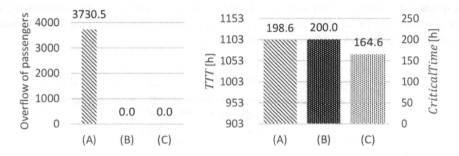

Fig. 7. Comparison of usual and optimized route networks

4.2   Case2: Extreme Disaster Situation

4.2.1   Experiment Setup
To investigate whether proposed method has truly effectiveness for restricting the bottleneck, we conducted the extreme disaster simulation with Mandl's bench mark problem. We compared the 8 results of the route network: optimized by the conventional and proposed method with the 4 patterns of high OD table to multiply the number of passengers in the station which ID is 9 by 10, 11, 12 and 13.

4.2.2   Evaluation Index and Parameters
As evaluation criteria all simulation results are evaluated by *CriticalTime* by the same way as Section 4.1.2. The experiment parameters in the conventional method are shown in Table 2. Then we also set these parameters to equalize the conditions. For all cases, in proposed method ω_1 is set to 0.7, ω_2 is 1.0, ω_3 is 2.0, α is 1.5, β is 0.5 and the number of buses which the lines give together in evolution is 3.

Table 2. Parameters of the conventional route optimization.

The high OD of the conventional method	ω_1	ω_2
Multiplied by 10	4.5	1.0
Multiplied by 11	8.0	1.0
Multiplied by 12	7.0	1.0
Multiplied by 13	13.0	1.0

4.2.3 Results

Fig. 8 shows the result of both methods. The horizontal axis indicates how much the number of passengers in the station 9 of each high OD is multiplied by. The vertical axis indicates *CriticalTime*. The number on data elements indicates each *CriticalTime* respectively. Here we also select each one of them which shows the shortest total movement time. From this result, the critical movement time of the proposed method is shorter than the conventional method in all cases.

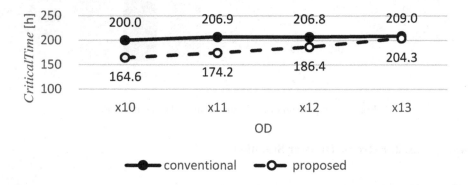

Fig. 8. Comparison in various OD of the conventional method and the proposal method

5   Discussion

First we discuss the results of the experiment in Section 4.1 (typical disaster situation). The route networks of (A), (B) and (C) is shown in Fig. 9. The arrows indicate the lines, the black-filled circle indicates the station which OD is multiplied and the heavy line indicates the road in which the bottleneck occurs. Now the roads and the ID number of each station are omitted unlike Fig.6.

Looking at the route network (B), the green line has evolved adding the bottleneck adjacent station 7. Although due to adding the length of the green line has become longer and due to the number of buses approximation it is considered that the number of buses has set unsuitable one, the waiting time of the passengers is considered to become longer and *CriticalTime* also become longer. Looking at the route network (C), it shows the shortest *CriticalTime* and has evolved removing of two lines. There is no concentrating the lines to near the bottlenecks, however due to removing the stations the length of lines has become shorter and due to giving the buses between the lines it is considered that the number of buses has set suitable one. Therefore the waiting time of the passengers is considered to become shorter and *CriticalTime* also become shorter.

Next we discuss the results of the experiment in Section 4.2 (extreme disaster situation). The proposed method has succeeded in optimizing more proper than the conventional one in the case of disaster. As mentioned the Majima's method can control the number of buses in generated or optimized route network using the weighting coefficient ω_1, however there is limitation. Furthermore in such control the conventional method cannot exhibit its maximum performance. By contrast, nevertheless the proposed method prohibit additional buses, due to only two changes it has fitted for the disaster.

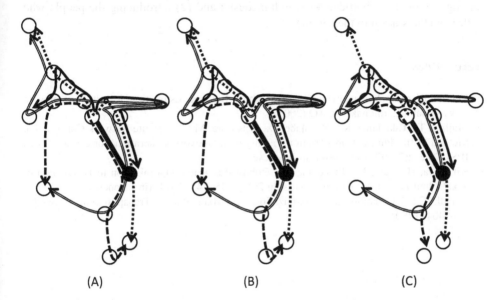

(A)	(B)	(C)

Fig. 9. Route networks of the conventional method and the proposal method

6   Conclusion

This paper focused on the passenger traffic *bottlenecks* occurred in the bus route network in disaster situations and proposed the multi-agent based bus route network optimization method to solve such bottlenecks by generating the networks which can effectively transport many stranded persons including ones who wait around the station as the passenger traffic bottlenecks. For this purpose, the proposed method aims at modifying the bus route networks generated as usual conditions to suitably pass many bus lines to and redistribute the buses among the bus lines according to the number of passengers. Such the bus line and number modification contributes to resolving the passenger traffic bottleneck and transporting stranded passengers quickly. Through the simulations on the Mandl's urban transport benchmark problem, we have revealed the following implications: (1) our proposed method succeeded in optimizing the bus route networks which do not cause the passenger traffic bottlenecks; and (2) our method can reduce 18% of the passengers' movement time without additional buses in comparison with the conventional method. These results suggest that the proposed method gives the great potential of transporting many stranded persons in the disasters situations.

What should be noticed here is that these results have only been obtained from one test-bed problem, *i.e.,* the Mandl's urban transport benchmark problem. Therefore, further careful qualifications and justifications, such as an analysis of results using other but route networks, are needed to generalize our results. Such important directions must be pursued in the near future in addition to the following research: (1) dispersing two or more bottlenecks simultaneously; and (2) introducing the people who walk from the station to their home.

References

1. Central Disaster Management Council: About Damage Estimation Result Pertaining to Tokyo Inland Earthquakes. CAO (2005)
2. Majima, T., Takadama, K., Watanabe, D., Katsuhara, M.: The Route Network Construction Method of the Public Transportation Facility by a Network Generation/Correction Model. IPSJ TOM 2(2), 92–102 (2009) (in Japanese)
3. Kitagawa, H., Sato, K., Takadama, K.: Robust Bus Route Optimization to Destruction of Roads. Journal of Information Processing 2(22), 235–242 (2014) (in Japanese)
4. Mandl, C.: Evaluation and Optimization of Urban Public Transportation Networks, 396–404 (1980)

Simplified Swarm Optimization to Solve the K-Harmonic Means Problem for Mining Data

Wei-Chang Yeh[1] and Chia-Ling Huang[2]

[1] Integration and Collaboration Laboratory
Department of Industrial Engineering and Engineering Management
National Tsing Hua University
P.O. Box 24-60, Hsinchu 30013, Taiwan, ROC
yeh@ieee.org
[2] Integration and Collaboration Laboratory
Department of Logistics and Shipping Management
Kainan University
Taoyuan 33857, Taiwan, R.O.C
cl.hfirst@gmail.com

Abstract. This paper used an efficient hybrid data mining approach, called gSSO proposed by Yeh in 2014 [1], is a modification introduced to simplified swarm optimization and based on K-harmonic means (KHM) algorithm to help the KHM algorithm escape from local optimum. To test its solution quality, the proposed gSSO is compared with other recently introduced KHM-based Algorithms in a larger size dataset named car dataset in the UCI database. The experimental results conclude that the proposed gSSO outperforms other algorithms in the solution quality of all aspects including average, minimum, maximum, and standard deviation for space and stability.

Keywords: K-harmonic means, KHM, Simplified Swarm Optimization, SSO.

1   Introduction

Clustering is one of the most scientific techniques in data mining. Many conventional algorithms have been proposed for classification. Among all classification algorithms, K-means (KM) is an algorithm with long history that was first presented over three decades ago [5] with two serious drawbacks, sensitivity to initial starting points and convergence to the local optimum. Therefore, many researches were done to overcome these problems. Zhang [6-7] proposed K-harmonic means (KHM) algorithm in 1999 to solve the problem of sensitivity to initial starting points. But the problem of convergence to the local optimum also exits.

Many scholars kept proposing methods based on KHM or Particle Swarm Optimization (PSO), which is one of soft computing methods have been utilized to find optimal or good quality solutions to complex optimization problems, to overcome the problem of convergence to the local optimum such as simulated annealing K-harmonic means clustering (SA) [4], Tabu K-Harmonic means (Tabu) [3], hybrid data clustering algorithm based on PSO and KHM (PSOKHM) [8], an improved gravitational search algorithm (IGSA) [12], the Ant clustering algorithm with K-harmonic means clustering (ACA) [10], a basic variable neighborhood search for KHM clustering (VNS) [11], multi-start local search for KHM clustering (MLS) [11], and candidate groups search combining with K-harmonic mean (CGS) [2]. Hung et al. [2] showed the computational results of CGS is better than VNS especially for large data size but CGS still cannot guarantee to escape from local optimum.

Swarm-intelligence is an artificial intelligence, primarily inspired by the social behavior patterns of self-organized systems, that considers the interactions among large groups of individuals [13-15, 21-23, 25-30]. The simplified swarm optimization (SSO) proposed by Yeh is a population-based stochastic optimization technique that belongs to the swarm-intelligence category [13-15] and is also an evolutionary computational method inspired by PSO [13]. Also known as discrete (DPSO), SSO was originally proposed to overcome the drawbacks of PSO for discrete-type optimization problems [13]. Yeh [1] proposed an efficient hybrid data mining approach called gSSO is a modification introduced to simplified swarm optimization and based on KHM algorithm to help the KHM escape from local optimum. The principal goal of this work is to demonstrate the efficiency of the proposed gSSO algorithm by testing on a larger size dataset named car dataset in the UCI database and the proposed gSSO algorithm is superior to all of the VNS, MLS, and CGS algorithms.

The rest of the paper is organized as follows. Section 2 provides a description of the KHM and the existing KHM algorithms. Section 3 introduces to the SSO. The gSSO proposed by Yeh [1] is introduced in Section 4. The experiments of the comparisons and the ANOVA-test on the larger size dataset named car dataset in the UCI datasets performed and demonstrate the effectiveness of the gSSO proposed by Yeh [1] are in Section 5. Finally, Section 6 makes conclusions.

2   The Khm and the Existing Khm Algorithms

The KHM design is to determine the clusters which defined by K centers by a center-based clustering algorithm that was proposed by Zhang [10, 11] in 1999. The first step of the KHM procedure is to calculate the distance from a data point to all the centers for the second step to search the minimum distance. There are two categories of methodology for the KHM to calculate the distance from a data point to all the centers, one is the Euclidean distance and the other is the Manhattan distance, let $C = \{c_k \mid k = 1,...,K\}$ be K centers and $(X,Y) = \{x_i, y_i \mid i, j = 1,...,N\}$ be N data points with two dimension, which is formulated as the following:

The Euclidean distance:

$$d(x_i, y_j; c_k) = \sqrt{(x_i - c_k)^2 + (y_i - c_k)^2} \, , for \, k = 1,...K \qquad (1)$$

The Manhattan distance:

$$d(x_i, y_j; c_k) = \|x_i - c_k\| + \|y_i - c_k\|, for \, k = 1,...K \qquad (2)$$

And the K-Harmonic Means' performance function is

$$f_{KHM}(K,p) = \min \sum_{i=1}^{N} \frac{K}{\sum_{l=1}^{K} \frac{1}{\|x_i - m_l\|^p}} \tag{3}$$

where p is the pth power of the Manhattan distance

The Euclidean distance of the KHM is not a population based soft computing algorithms (SCs). However, the SCs with population base are extremely important and easy to use in order to search randomly and achieve optimal or good quality solutions for complex optimization problems [13-23]. Hence, there are many new researches of data mining using methodologies based on SCs to achieve optimize solutions, such as the genetic algorithm (GA, a biology-inspired SC) [19, 23-24] and particle swarm optimization (PSO, a swarm-intelligence SC) [8, 13-15, 22-23, 25-27].

The PSO is a new population based optimization algorithm, which was first introduced by Kennedy and Eberhart [8]. The objective of this algorithm is to optimize various continuous nonlinear functions. The concept of PSO is based on the interactions among large groups of individuals originated with the social behavior patterns of self-organized systems [8, 13-15, 22-23, 25-27]. The simplified swarm optimization (SSO) proposed by Yeh is a population-based stochastic optimization technique that belongs to the swarm-intelligence category [13-15] and is also an evolutionary computational method inspired by PSO [13]. The SSO was originally proposed to overcome the shortages of PSO for discrete-type optimization problems and the detailed contents of SSO are introduced in next section.

3   Introduce to the SSO

The SSO is an algorithm with the advantages of simplicity, efficiency, and flexibility - that was developed and first introduced by Yeh [13-16]. The algorithm solves efficiently the redundancy allocation problem by way of simplifying the traditional PSO process to defeat the shortages of PSO for discrete variables [13]. Before discussing the proposed SSO, the detailed contents of SSO are introduced in this section first.

The SSO is also initialized with a population of random solutions inside the problem space like most soft computing approaches, and then searches for optimal solutions by updating generations. In the update mechanism of SSO, solution $X_i^{t+1} = (x_{i1}^{t+1}, x_{i2}^{t+1}, \ldots, x_{i,\text{DIM}}^{t+1})$ is a compromise of the current solution $X_i^t = (x_{i1}^t, x_{i2}^t, \ldots, x_{i,\text{DIM}}^t)$, the pBest $P_i = (p_{i1}, p_{i2}, \ldots p_{i,\text{DIM}}) \in \{X_i^1, X_i^2, \ldots, X_i^t\}$, the gBest $G = (g_1, g_2, \ldots, g_{\text{DIM}}) \in \{P_1, P_2, \ldots, P_{\text{POP}}\}$, and a random movement as follows after C_w, C_p, and C_g are given [25-28].

$$x_{ij}^{t+1} = \begin{cases} x_{ij}^t & \text{if } \rho \in [0, C_w) \\ p_{ij} & \text{if } \rho \in [C_w, C_p) \\ g_j & \text{if } \rho \in [C_p, C_g) \\ x & \text{if } \rho \in [C_g, 1) \end{cases} \qquad (4)$$

where $\rho \in Uniform(0, 1)$ and $x \in Uniform(l_j, u_j)$, respectively.

In Eq.(4), both concepts of *pBest* and *gBest* are adopted directly from the traditional PSO. P_i is a local best such that $F(P_i) \geq F(X_i^j)$ for $j=1,2,\ldots,t$, to push X_i^t to climb the local optimum (i.e., local exploitation); G is a global best such that $F(G) \geq F(P_i)$ for $i=1,2,\ldots,POP$ to guide the search towards unexplored regions to find the global optimizer (i.e., global exploration). To maintain population diversity and enhance the capacity of escaping from a local optimum, a random movement is added to the update mechanism of SSO [26-29].

The above update mechanism of SSO considers the interaction between individuals and also maintains the diversity between individuals. It is extremely simpler, more efficient, and more flexible than other major SC techniques, for example, PSO which has to compute both the velocity and position functions, GA which claims genetic operations such as crossover and mutation, EDA of which was difficult and complicated to build an appropriate probability model, and IMA which is lack of the consider the interaction of variables [26]. According to the discussion above, the overall steps of SSO are listed as follows.

PROCEDURE SSO

STEP S0. Generate X_i randomly, let $t=1$, $P_i = X_i$, and $G = P_j$, where $F(P_j) = \underset{i}{Max}$

$\{F(P_i)\}$ for $i=1,2,\ldots,POP$.

STEP S1. Let $i=1$.

STEP S2. Update X_i based on Eq.(1) and calculate $F(X_i)$.

STEP S3. If $F(X_i) > F(P_i)$, let $P_i = X_i$; else go to STEP S5.

STEP S4. If $F(P_i) > F(G)$, then let $G = P_i$.

STEP S5. If $i < POP$, then let $i = i+1$ and go to STEP S2.

STEP S6. If $t = GEN$, then G is the final solution and halt; otherwise let $t = t+1$ and go to STEP S1.

4   The Proposed gSSO

This section outlines the gSSO proposed by Yeh [1] in the following components:

4.1   The Solution Structure

To present each centroid of clusters easily, the matrix denoted by $\Xi_{i,j} = \begin{bmatrix} X_{i,j,1} \\ \vdots \\ X_{i,j,K} \end{bmatrix}$ with

size $K \times (Nvar)$ is used to define the jth solution obtained in the ith generation of the proposed gSSO. The kth row $X_{i,j,k}=(x_{i,j,k,1}, x_{i,j,k,2}, \dots, x_{i,j,k,\text{Nvar}})$ in $\Xi_{i,j}$ denoted the centroid of the kth cluster and $x_{i,j,k,l}$ is the value of the lth coordinate of such centroid. For example,

$$\Xi_{0,2}=\begin{bmatrix} X_{0,2,1} \\ X_{0,2,2} \\ X_{0,2,3} \\ X_{0,2,4} \end{bmatrix}=\begin{bmatrix} 4.6 & 3.6 & 1 & 0.2 \\ 6.9 & 3.1 & 5.1 & 2.3 \\ 5.3 & 3.7 & 1.5 & 0.2 \\ 6.2 & 2.9 & 4.3 & 1,3 \end{bmatrix} \tag{5}$$

is the 2nd solution in the 0th generation (i.e., the initial population) and the centroid of the 1st, 2nd, 3rd, and 4th clusters are $X_{0,2,1}=(4.6, 3.6, 1, 0.2)$, $X_{0,2,2}=(6.9, 3.1, 5.1, 2.3)$, $X_{0,2,3}=(5.3, 3.7, 1.5, 0.2)$, and $X_{0,2,4}=(6.2, 2.9, 4.3, 1.35)$.

4.2   The Record-Based Initial Solutions

Each row of the ith initial solution (note that the first subscript is zero) $\Xi_{0,i}=\begin{bmatrix} R_{k_1} \\ R_{k_2} \\ \vdots \\ R_{k_K} \end{bmatrix}$ is simply selected any record randomly in the dataset such that the centroid of the jth cluster is the k_j-th record to reduce the running time. For example,

$$\Xi_{0,2}=\begin{bmatrix} X_{0,2,1} \\ X_{0,2,2} \\ X_{0,2,3} \\ X_{0,2,4} \end{bmatrix}=\begin{bmatrix} R_{22} \\ R_{141} \\ R_{48} \\ R_{97} \end{bmatrix}=\begin{bmatrix} 4.6 & 3.6 & 1 & 0.2 \\ 6.9 & 3.1 & 5.1 & 2.3 \\ 5.3 & 3.7 & 1.5 & 0.2 \\ 6.2 & 2.9 & 4.3 & 1,3 \end{bmatrix} \tag{6}$$

given in Section 4.1 is actually the 2nd solution in the initial population and the centroid of the 1st, 2nd, 3rd, and 4th clusters are selected randomly from the 22nd, 141st, 48th, and 97th records, i.e., $R_{22}=X_{0,2,1}$, $R_{141}=X_{0,2,2}$, $R_{48}=X_{0,2,3}$, and $R_{97}=X_{0,2,4}$, in the IRIS dataset [2-3, 13, 15]. Note that there are 150 records and each record includes 4 features in IRIS dataset [2-3, 13, 15].

4.3   The Proposed Local Search ε_g-KHM

It is well-known that a good local search is able to improve the solution quality under the cost of the efficiency. Like the existing related KHM algorithms, the proposed gSSO is also adapted the KHM proposed in [13, 15] to be the local search to improve the solution quality after each new solution is obtained. There are two stopping criteria of the traditional KHM:

1. The absolute difference of the centroids in the same cluster between two consecutive solution are all less than a given value ε_p (called the individual tolerance here), i.e., $|X_i - X_i^*| \le \varepsilon_p$ where X_i and X_i^* are the centroids of the ith cluster in the current solution and new updated solution for $i=1,2,\dots,K$.
2. The heuristic number is greater than a specific integer number.

The first stopping criterion may take an unexpected long time to hold especially for a dataset with higher dimensions and the last one may resulted in the heuristic is terminated too early to obtain a good quality solution. Therefore, the global tolerance ε_g is introduced to the proposed KHM named the ε_g-KHM of the proposed gSSO. If the Euclidean distance between two consecution updated solutions is less than or equal to ε_g, the ε_g-KHM is halt and the current solution will be the final solution obtained from updated procedure or in the initial solution procedure.

PROCEDURE ε_g-KHM
STEP 0. Let ε_g be the global tolerance ε_g and Ξ be a given solution.
STEP 1. Calculate $F_{\text{KHM}}(\Xi)$, m, and w based on Eq.(3), Eq.(7), and Eq.(8), respectively.
STEP 2. Update Ξ to Ξ^*
STEP 3.   If $|\Xi^*-\Xi| \leq \varepsilon_g$, then halt; otherwise let $\Xi=\Xi^*$ and go to STEP 1

$$m_{KHM}(c_j / x_i) = \frac{\|x_i - c_j\|^{-p-2}}{\sum_{j=1}^{K}\|x_i - c_j\|^{-p-2}}, \quad \forall \ i=1,...,N, \quad \forall \ j=1,...,K \quad (7)$$

$$w_{KHM}(x_i) = \frac{\sum_{j=1}^{K}\|x_i - c_j\|^{-p-2}}{(\sum_{j=1}^{K}\|x_i - c_j\|^{-p})^2}, \quad \forall \ i=1,...,N \quad (8)$$

$$c_j^{(new)} = \frac{\sum_{i=1}^{N} m_{KHM}(c_j / x_i) \cdot w_{KHM}(x_i) \cdot x_i}{\sum_{i=1}^{N} m_{KHM}(c_j / x_i) \cdot w_{KHM}(x_i)}, \quad \forall \ j=1,...,K \quad (9)$$

4.4   The Proposed Update Mechanism (UM)

The UM is the core of each Soft Computing method. Each Soft Computing method has its own UM, e.g., the crossover and mutation UM operators in GA and the inequality-based UM in SSO. There is always a need to adjust and/or modify the related UM to fit different problems in various situations when the related Soft Computing method is implemented. Due to the specific factor that the K-Harmonic Means is essentially insensitive to the initialization of the centers, a new UM is proposed in this study by simplifying the inequality-based UM in SSO as follows:

1.   Mixed Population-Based Strategy
Only in the initial population, the proposed gSSO is still a population-based algorithm. After finding the *gBest* the initial population, the proposed gSSO is changed to the single-population based method and the *gBest* is the only solution used in the rest

procedure of gSSO to update until it is replaced with a new *gBest* which with better solution.

2. *gBest*-only Strategy

Since the *K*-Harmonic Means is essentially insensitive to the initialization of the centers, the *pBest* proposed in the traditional SSO and each solution should be closed to the *gBest*. Based on the above special characteristic and the proposed mixed population-based, the roles of C_p and C_w are removed totally and the UM in Eq.(7) of is revised according as follows:

$$G_k^* = \begin{cases} G_k & \text{if } \rho_{[0,1]} < C_g \\ R & \text{otherwise} \end{cases} \tag{10}$$

4.5 The Algorithm of the Proposed gSSO

The detail of the proposed algorithm is described in the following steps:

PROCEDURE gSSO

STEP 0. Generate $\Xi_{0,i} = \begin{bmatrix} X_{0,i,1} \\ : \\ X_{0,i,K} \end{bmatrix}$ randomly, let gen=1 and the *gBest*

$\Gamma = \Xi_{0,j} = \begin{bmatrix} G_1 \\ : \\ G_K \end{bmatrix} = \begin{bmatrix} X_{0,j,1} \\ : \\ X_{0,j,K} \end{bmatrix}$, where the centroid of *k*th cluster in the *i*th solution

$X_{0,i,k} \in \{R_1, R_2, \dots, R_{\text{Nrec}}\}$ and $F(\Gamma) \leq F(\Xi_{0,i})$ for $i,j=1,2,\dots,$Nsol and $k=1,2,\dots,K$.

STEP 1. Let $i=j=1$ and $\Gamma^* = \Gamma$.

STEP 2. Update $G_k^* = \begin{cases} G_k & \text{if } \rho_{[0,1]} < C_g \\ R & \text{otherwise} \end{cases}$, where G_k is the centroid of the *k*th cluster

in Γ, and both $k \in \{1, 2, \dots, K\}$ and $R \in \{R_1, R_2, \dots, R_{\text{Nrec}}\}$ are selected randomly.

STEP 3. If $j<i$, let $j=j+1$ and go to STEP 2.

STEP 4. If $F(\Gamma^*)<F(\Gamma)$, let $\Gamma=\Gamma^*$.

STEP 5. If $i<K$, let $i=i+1, j=1$, and go to STEP 2.

STEP 6. If gen<Ngen, then let gen=gen+1 and go to STEP 1; otherwise Γ is the final solution and halt.

5   The Car Dataset and Numerical Examples

To evaluate the quality and performance for data mining, all experimental variants and the gSSO is applied to a larger size   dataset named car dataset adopted from the UCI database. The information of the dataset used in this study is referenced Table 1.

Table 1. Information of dataset used in the study

Datasets	K (Number of group)	N (Size of dataset)
Car	4	1728

5.1   Experimental Design

All experimental variants denoted: VNS, CGS, and MLS, respectively and the gSSO are implemented in C programming language and run on an Intel Core i7 3.07 GHz PC with 6 GB memory. The runtime unit is CPU seconds. Moreover, c_w=.45, c_p=.40, c_g=.15 (in Eq.(4)). The stopping criterion is error term ε=0.1. Meanwhile, the corresponding p (in Eq.(3)) equal 1.5, 2.0, 2.3, 2.5, 3.0, 3.5, 4.0 respectively. From the output of fitness of performance and runtime, we have Nrun=30 (*i.e.* Each value with 30 replicates) and let the notation AVG, MIN, MAX, and STDEV denote the average, minimal (the best), maximal (the worst), and standard deviation of the related values for general observations judgment first.

5.2   General Observations

The fitness of performance and the running time of the experiment results for the four methods are summarized in Table 2 on AVG, MIN, MAX, and STDEV.

Table 2. Summary of the experiment results of four methods

P	Statistics	Fitness				Time			
		gSSO	VNS	CGS	MLS	gSSO	VNS	CGS	MLS
1.5	AVG	**7069.2771**	7070.2097	7133.0678	7097.9826	0.831	0.774	**0.003**	0.217
	MIN	**7068.7388**	7069.2076	7106.9507	7082.7433	0.812	0.765	**0.000**	0.203
	MAX	**7070.5515**	7071.0388	7157.7272	7106.0419	0.844	0.782	**0.016**	0.219
	STDEV	**0.4572**	0.4652	12.2690	5.5070	0.009	0.007	**0.006**	0.004
2.0	AVG	**1474.74444**	11480.8865	11604.1485	11533.9598	0.607	0.547	**0.002**	0.151
	MIN	**1472.2676**	11477.6071	11538.2028	11507.5100	0.593	0.531	**0.000**	0.140
	MAX	**1480.9036**	11485.2059	11648.1954	11549.3547	0.625	0.563	**0.016**	0.157
	STDEV	2.2288	**1.7536**	23.5158	13.2656	0.009	0.007	**0.005**	0.007
2.3	AVG	**5385.3690**	15394.6868	15560.8331	15458.5587	0.901	0.822	**0.003**	0.218
	MIN	**5380.0064**	15388.8592	15456.1720	15426.9756	0.875	0.797	**0.000**	0.203
	MAX	**5392.6243**	15398.8962	15609.7014	15482.2393	0.921	0.844	**0.016**	0.219
	STDEV	3.1411	**2.6840**	33.0301	16.1997	0.009	0.012	**0.006**	0.003
2.5	AVG	**8701.9981**	18716.4013	18927.7192	18797.9813	0.914	0.832	**0.003**	0.217
	MIN	**8692.8598**	18699.7114	18791.0535	18745.2270	0.891	0.812	**0.000**	0.203
	MAX	**8723.7886**	18730.1093	18977.8004	18823.7894	0.938	0.844	**0.016**	0.219
	STDEV	7.7417	**6.1016**	42.2116	18.7797	0.010	0.009	**0.006**	0.004
3.0	AVG	**30465.0912**	30476.9047	30861.4250	30606.5069	0.938	0.844	**0.003**	0.216
	MIN	**30448.5533**	30465.1758	30664.4759	30526.0326	0.921	0.828	**0.000**	0.203
	MAX	**30502.2117**	**30492.4415**	30960.9213	30647.7001	0.954	0.860	**0.016**	0.219
	STDEV	13.8882	**7.2101**	84.2247	33.2922	0.010.	0.011	**0.006**	0.005
3.5	AVG	**49391.3454**	49396.6581	50234.0497	49607.9063	0.950	0.859	**0.003**	0.216
	MIN	**49367.2154**	49374.7364	49731.5439	49518.7848	0.937	0.828	**0.000**	0.203
	MAX	**49436.4573**	**49432.2977**	50659.5575	49659.2411	0.969	0.891	**0.016**	0.219
	STDEV	19.3467	**14.5022**	262.7617	36.3203	0.009	0.012	**0.006**	0.005
4.0	AVG	79617.4443	**79595.5449**	81674.7172	79935.9780	0.675	0.598	**0.002**	0.152
	MIN	79523.3054	**79520.9196**	80097.6407	79787.0760	0.656	0.578	**0.000**	0.140
	MAX	79808.9365	**79654.6640**	83065.5680	80146.5716	0.688	0.610	**0.016**	0.157
	STDEV	61.6804	**31.2025**	792.7755	100.0552	0.007	0.008	**0.005**	0.006

From Table 2, the solution quality of CGS is the worst, then is MLS. The solution quality of the gSSO is the best in all aspects (AVG, MIN, MAX, and STDEV). Followed by, VNS is better than MLS and CGS methods. Hence, we focus on the comparison of gSSO and VNS.

To further compare the average fitness between the gSSO and VNS by the plot in Figure 1. The larger size value of p, the difference between gSSO and VNS is more significant.

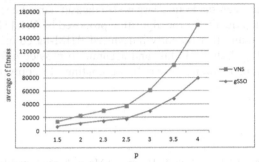

Fig. 1. Plot for average fitness of the gSSO and VNS

5.3 Test the Comparison between gSSO and VNS

Since gSSO and VNS are the first and the second among the four methods in the solution quality no matter the size of p, a detailed comparison of gSSO and VNS is discussed with the ANOVA test to AVG and the test results shown in Table 3.

Table 3. The comparison of gSSO and VNS with the ANOVA test to AVG

Dataset: Car (N=1728, K=4)	
p	p-Value
1.5	0.000^*
2.0	0.000^*
2.3	0.000^*
2.5	0.000^*
3.0	0.000^*
3.5	0.234
4.0	0.088

* gSSO is significant better than VNS at critical value=0.05

In Table 3, the test results indicate gSSO is significant better than VNS in the solution quality for p=1.5, 2.0, 2.3, 2.5 and 3.0. There is no significant difference between gSSO and VNS in the solution quality for p=3.5 and 4.0.

Overall, from the general observations and the ANOVA test, the gSSO proposed by Yeh [1] is all much better than the rest of the methods for most of solution quality of all aspects in accuracy and stability.

6   Conclusions

In this work, a new methodology proposed by Yeh [1] introducing a new KHM based SSO to help KHM algorithm escape from local optimum. A comprehensive comparative study of the algorithm proposed by Yeh [1] and other recently introduced KHM-based algorithms has been made in a larger size dataset named car dataset in the UCI database. The solution quality of the algorithm proposed by Yeh [1] is clearly superior to and more robust than other recently introduced KHM-based algorithms. Thus, the algorithm proposed by Yeh [1] is fast and accurate to provide effective and efficient data mining approach. In future research, we continue to apply the proposed algorithm to more datasets.

References

1. Yeh, W.-C., Huang, C.-L.: A New K-Harmonic Means based Simplified Swarm Optimization for Data Mining. In: IEEE Symposium Series on Computational Intelligence Conference, Orlando, Florida, USA, December 9-12 (2014)
2. Hung, C.H., Chiou, H.M., Yang, W.N.: Candidate groups search for K-harmonic means data clustering. Applied Mathematical Modelling 37, 10123–10128 (2013)
3. Gungor, Z., Unler, A.: K-harmonic means data clustering with tabu-search method. Applied Mathematical Modelling 32, 1115–1125 (2008)
4. Gungor, Z., Unler, A.: K-harmonic means data clustering with simulated annealing heuristic. Applied Mathematics and Computation 184, 199–209 (2007)
5. Forgy, E.W.: Cluster analysis of multivariate data: efficiency versus interpretability of classifications. Biometrics 21(3), 768–769 (1965)
6. Zhang, B., Hsu, M., Dayal, U.: K-harmonic means – a data clustering algorithm, Technical Report HPL-1999-124, Hewlett–Packard Laboratories (1999)
7. Zhang, B.: Generalized k-harmonic means – boosting in unsupervised learning, Technical Report HPL-2000-137, Hewlett–Packard Laboratories (2000)
8. Kennedy, J., Eberhart, R.C.: Particle swarm optimization. In: Proceedings of the 1995 IEEE International Conference on Neural Networks, pp. 1942–1948. IEEE Press, New Jersey (1995)
9. Yang, F., Sun, T., Zhang, C.: An efficient hybrid data clustering method based on K-harmonic means and Particle Swarm Optimization. Expert Systems with Applications 36, 9847–9852 (2009)
10. Jiang, H., Yi, S., Li, J., Yang, F., Hu, X.: Ant clustering algorithm with K-harmonic means clustering. Expert Systems with Applications 37, 8679–8684 (2010)
11. Alguwaizani, A., Hansen, P., Mladenovic, N., Ngai, E.: Variable neighborhood search for harmonic means clustering. Applied Mathematical Modelling 35, 2688–2694 (2011)
12. Yin, M., Hu, Y., Yang, F., Li, X., Gu, W.: A novel hybrid K-harmonic means and gravitational search algorithm approach for clustering. Expert Systems with Applications 38, 9319–9324 (2011)
13. Yeh, W.: Study on quickest path networks with dependent components and apply to RAP, Report,NSC 97-2221-E-007-099-MY3, 2008-2011
14. Yeh, W.: A two-stage discrete particle swarm optimization for the problem of multiple multi-level redundancy allocation in series systems. Expert Systems with Applications 36(5), 9192–9200 (2009)

15. Yeh, W., Chang, W., Chung, Y.: A new hybrid approach for mining breast cancer pattern us-ing discrete particle swarm optimization and statistical method. Expert Systems with Applications 36(4), 8204–8211 (2009)
16. Yeh, W., Lin, H.: A soft computing algorithm for disassembly sequencing. In: International Conference on Engineering and Computational Mathematics (2009)
17. Yeh, W.C.: Novel swarm optimization for mining classification rules on thyroid gland data. Information Sciences 197, 65–76 (2012)
18. Chang, W.-W., Yeh, W.-C., Huang, P.-C.: A hybrid immune-estimation distribution of al-go-rithm for mining thyroid gland data. Expert Systems with Applications 37, 2066–2071 (2010)
19. Chen, T.-C., Hsu, T.-C.: A GAs based approach for mining breast cancer pattern. Expert Systems with Applications 30, 674–681 (2006)
20. Dasgupta, D.: Advances in artificial immune systems. IEEE Computational Intelligence Magazine 1, 40–49 (2006)
21. Gandhi, K.R., Karnan, M., Kannan, S.: Classification rule construction using particle swarm optimization algorithm for breast cancer data sets. In: Proceedings of the 2010 International Conference on Signal Acquisition and Processing, pp. 233–237 (2010)
22. Das, S., Abraham, A., Konar, A.: Particle swarm optimization and differential evolution algorithms: technical analysis, applications and hybridizationperspectives. In: Liu, Y., et al. (eds.) Advances of Computational Intelligence in Industrial Systems. SCI, vol. 116, pp. 1–38. Springer, Berlin (2008)
23. Das, S., Abraham, A., Konar, A.: Swarm intelligence algorithms in bioinformatics. Computational Intelligence in Bioinformatics 147, 113–147 (2008)
24. Goldberg, D.E.: Genetic Algorithms in Search, Optimization, and Machine Learning. Addison-Wesley (1989)
25. Abraham, A., Das, S., Roy, S.: Swarm intelligence algorithms for data clustering. Soft Computing for Knowledge Discovery and Data Mining, 279–313 (2008)
26. Alatas, B., Akin, E.: Multi-objective rule mining using a chaotic particle swarm optimization algorithm. Knowledge-Based Systems 22, 455–460 (2009)
27. Zhao, X., Zeng, J., Gao, Y., Yang, Y.: Particle swarm algorithm for classification rules generation. In: Proceedings of the 6th International Conference on Intelligent Systems Design and Applications, vol. 2, pp. 957–962 (2006)
28. Holden, N., Freitas, A.A.: A hybrid PSO/ACO algorithm for discovering classification rules in data mining, Journal of Artificial Evolution and Applications (2008), doi:10.1155/2008/316145
29. Liu, Y., Qin, Z., Shi, Z., Chen, J.: Rule Discovery with Particle Swarm Optimization. In: Chi, C.-H., Lam, K.-Y. (eds.) AWCC 2004. LNCS, vol. 3309, pp. 291–296. Springer, Heidelberg (2004)
30. Shi, Y., Eberhart, R.C.: Evolutionary computation: empirical study of particle swarm optimization. In: Proceedings of the Congress on Evolutionary Computation, pp. 1945–1950 (1999)

A Novel Evolutionary Multi-objective Algorithm Based on S Metric Selection and M2M Population Decomposition

Lei Chen[1], Hai-Lin Liu[1,*], Chuan Lu[1,2], Yiu-ming Cheung[3,4], and Jun Zhang[5]

[1] Guangdong University of Technology, Guangzhou, China
[2] University of Electronic Science and Technology of China, Chengdu, China
[3] Department of Computer Science,
Hong Kong Baptist University, Hong Kong, China
[4] United International College,
Beijing Normal University – Hong Kong Baptist University, Zhuhai, China
[5] Sun Yat-Sen University, Guangzhou, China
{hlliu@gdut.edu.cn}

Abstract. The excellent performance of evolutionary multi-objective algorithms based on S metric selection (SMS) has been identified by many researchers. However, huge computational effort of S metric calculation has limited the full application of those algorithms. This paper proposes a novel S metric selection evolutionary algorithm (SMS-M2M) based on the population decomposition strategy MOEA/D-M2M. In SMS-M2M, SMS is conducted in each subpopulation instead of the whole population, which can avoid the S metric calculation of the total population. The purpose of population decomposition is to directly reduce the huge computational effort of calculating S metric and thus to give a simple but effective method to improve the effectiveness of SMS based algorithm. SMS-M2M utilizes the same SMS with a popular SMS based evolutionary algorithm SMS-EMOA. Numerical studies of SMS-M2M and SMS-EMOA have shown that the M2M population decomposition can effectively reduce the computational effort of SMS, meanwhile the theoretic analysis identifies the efficiency and effectiveness of SMS-M2M.

Keywords: Evolutionary Algorithm, Multi-objective, S Metric Selection (SMS), Population Decomposition.

1 Introduction

S metric [1] is also called Hypervolume measure [2], or Lebesgue measure [3]. S metric is defined as the size of dominated space when firstly proposed by Zitzler and Thiele, because it measures the size of the region dominated by the populations in the objective space. At first, S metric is only used as a metric for comparing the performance of different evolutionary multi-objective algorithm.

---

* Corresponding author.

© Springer International Publishing Switzerland 2015
H. Handa et al. (eds.), *Proc. of the 18th Asia Pacific Symp. on Intell. & Evol. Systems – Vol. 2*,
Proceedings in Adaptation, Learning and Optimization 2, DOI: 10.1007/978-3-319-13356-0_35

Because of its ability to measure both the closeness of the population to the Pareto Front (PF) and the diversity of population on the PF, S metric has been becoming popular for selection [6]. Compared with the other selection metrics, SMS has the following advantages:

☐ S metric can measure convergence and diversity of a set of solutions at the same time, while the other selection metrics normally need additional strategy, such as crowding distance in NSGA-II [4], to measure the diversity of solutions.

☐ S metric is calculated based on population other than individuals. It can detect even the smallest difference between two subsets, and thus make the selection more precise.

Evolutionary multi-objective algorithms based on SMS can benefit these advantages: only by optimizing the S metric, we can get a set of individuals with both good approximation to PF and good distribution on PF. Both experiments and theoretical analysis have shown its excellent performance even in the optimization of problems with complex PF [6]. However, the computation of SMS is quite time-consuming [11]. This weakness of SMS limits the practical application of the algorithms based on SMS. A lot of researches focusing on how to reduce the computational effort [5]-[12] have emerged in the past few years.

Recently, Nicola Beume et al. have proposed a new algorithm based on SMS (SMS-EMOA) [5] which attracts a lot of attentions because of its high efficiency and excellent performance. The main idea of SMS-EMOA is that the total computational effort can be reduced by reducing the number of individuals involved in the computation of S metric. The $(\mu + 1, \mu)$ selection strategy is used in SMS-EMOA, i.e. only one offspring is generated by crossover and mutation, and only the worst individual is discarded. In the process of selection, SMS-EMOA does not calculate every individual exclusive S metric. On the contrary, it makes best use of the advantages of nondominated sorting in NSGA-II [4]. Firstly, all the nondominated fronts are identified by nondominated sorting. Then, the SMS is executed only in the final front. Unfortunately, as the evolving of population or the increasing of objective space dimension, the percent of nondominated solutions increases drastically. This situation leads to the useless of the nondominated sorting, and the computational effort cannot be reduced. So SMS-EMOA cannot be said to really avoid the expensive calculation of SMS.

The idea of population decomposition has been widely studied in the literature [12]-[15]. MOEA/D-M2M [16] is a framework for population decomposition. It has strong ability to maintain the diversity of the population [16]. In MOEA/D-M2M, the multi-objective optimization problem is decomposed into a number of uncomplicated multi-objective optimization subproblems, which are then solved in a collaborative manner. A lot of selection strategies can be adopted to evolve the population in MOEA/D-M2M framework.

The motivation of the proposed algorithm SMS-M2M is the same as SMS-EMOA, but we use a different way in this paper. In the process of selection, when an individual is discarded, the exclusive S metric of every remained individual needs to be recalculated to find the next individual to be discarded. A straightforward idea is that, since we cannot avoid the calculation of S metric, we can reduce the number of individuals involved in the S metric calculation. The calculation of S metric is limited to the relatively small subpopulation by means of population decomposition. Consequently, the computational effort of discarding an individual is reduced, and then the total computational effort of selection is decreased. The main characteristic of the SMS-M2M can be summarized as:

☐ SMS-M2M inherits the advantages of SMS.

☐ SMS-M2M can significantly reduce the computational effort of SMS by population decomposition.

☐ SMS-M2M has strong ability to maintain the diversity of the population.

The remainder of this paper is organized as follows: Section II describes the main idea of the proposed algorithm and gives the design of SMS-M2M in detail. In Section III, we study the performance of SMS-M2M by comparing its simulation result of seven widely used ZDT and DTLZ series test instances with that of the SMS-MOEA. The experimental results are shown and the corresponding theoretical analysis is made. Finally, we conclude the paper in Section IV.

2   The Main Idea and Algorithm

2.1   Multi-objective Optimization Problem

In this paper, we consider the following continuous multi-objective optimization problem(MOP):

$$\text{minimize } F(x) = (f_1(x), \ldots, f_m(x)) \tag{1}$$

$$\text{subject to} \qquad x \in \prod_{i=1}^{n} [a_i, b_i]$$

where $\prod_{i=1}^{n} [a_i, b_i]$ is the decision space and R^m is the objective space. F is mapping from n-dimension decision space to m-dimension objective space consisting of m real-value continuous objective functions $f_1, f_2, ..., f_m$. A solution x is a vector of n decision variable. It is assumed that these objectives conflict with one another, so no solution can optimize them at the same time. Let $u = (u_1, u_2, ..., u_m)$ and $v = (v_1, v_2, ..., v_m)$, u dominates v if and only if $u_i \leq v_i$ for every i and there exists one index j such that $u_j < v_j$, denoted as $u \prec v$. If there is no $x \in [a, b]^n$ such that $F(x)$ dominated $F(x^*)$, then x^* is called Pareto-optimal point and $F(x^*)$ is the Pareto-optimal objective vector in objective space. The set of all Parato-optimal points is called Parato set (PS) and the set of all Parato-optimal objective vector is called PF.

2.2   Population Decomposition in MOEA/D-M2M

For simplicity, we assume that all the individuals objective function $f_1, f_2, ..., f_m$ are nonnegative, otherwise we can make a linear transportation replace f_i by $f_i + M$, where M is a large enough positive number so that $f_i + M > 0$.

Population decomposition-based algorithm has been proved to function well in preserving population diversity. This paper studies its another potential ability to reduce the computational effort.

To decompose the population, firstly K unit vectors $v^1, ..., v^K$ in R_+^m are generated from the unit circle in the first quadrant. Then divide R_+^m into K subregions $\Omega_1, ..., \Omega_K$, where Ω_k $(k = 1, ..., K)$ is

$$\Omega_k = \{u \in R_+^m | \langle u, v^k \rangle \leq \langle u, v^j \rangle \text{ for any } j = 1, ..., K\}, \tag{2}$$

where $\langle u, v^j \rangle$ is the acute angle between u and v^j. Therefore, the population is decomposed into a series of subpopulation and each subregion has its own subpopulation.

2.3   The Selection Strategy of SMS-EMOA

During the process of selection in SMS-EMOA, the nondominated sorting is firstly conducted. And then the selection operator of SMS-EMOA is implemented on the final nondominated front. The individual with the minimum exclusive S metric in the final front is discarded. Afterwards, a modified selection method is presented by the authors [5]. The main difference between the original method and the modified method is that the modified method includes the number of dominating points $d(s, P(t))$ in selection, where $d(s, P(t))$ is defined as follows:

$$d(s, P(t)) := |\{y \in P(t) | y \prec s\}| \tag{3}$$

The bigger the $d(s, P(t))$ is, the worse the individual is. The modified method also aims at avoiding the calculation of S metric: whenever the worst individual is found by using $d(s, P(t))$, the calculation of S metric can be avoided. However, the modified method also has the same limitation as nondominated sorting: when the population evolves a lot of generations or the dimension of objective space increases, almost all individuals are nondominated. In this case, individuals are all located in the first front so that the $d(s, P(t))$ sorting will never be used.

2.4   The Framework of SMS-M2M

The main framework of SMS-M2M is given by algorithm 1 and algorithm 2 as follows:

2.5   Performance Metrics

IGD-metric. In our experiments, the IGD-metric is used to measure the quality of a solution set P. Suppose that P^* is a set of points which are uniformly

---

Algorithm 1. SMS-M2M

Input:

1. max_gen: maximum number of generations;
2. K: he number of the subproblems;
3. S: the size of subpopulation;
4. K unit direction vectors: $v^1, \ldots, v^K$;

Output: a set of nondominated solutions in $\bigcup_{k=1}^{K} P_k$.
Initialization: Generate SK initial individuals Q, calculate their objective value and then use them to set $P_1, \ldots, P_K$. Set the current generation: $gen = 1$.
While $gen < max\_gen$
    Generation of New Solutions:
    Set $R = \emptyset$;
    For$k \leftarrow 1$ **To** K
        ForEach $x \in P_k$ Randomly generate a number $r \in [0, 1]$
        If $r < P_c$
        Randomly choose y from Q/P_k;
        Else
        Randomly choose y from P_k.
        End
        Apply genetic operators on x and y to generate a new solution z;
        Compute $F(z)$;
        $R := R \cup \{z\}$.
    End
    $Q := R \cup (\cup_{k=1}^{K} P_k)$;
    According to algorithm 2, use Q to set $P_1, \ldots, P_K$;
    Update $Q = \cup_{k=1}^{K} P_k$.
End
Find all the nondominated solutions in $\cup_{k=1}^{K} P_k$ and output them.
End

Algorithm 2. Allocation of Individuals to Subpopulations

Input: Q: a set of individual solutions in $[a, b]^n$ and their objective values.
Output: $P_1, \ldots, P_K$.
For$k \leftarrow 1$ **To**K
    Initialize P_k as the solutions in Q whose F-values are in Ω_k;
    If$|P_k| < S$
    Randomly select $S - |P_k|$ solutions from Q and add them to P_k;
    Else
    Use SMS to remove from P_k the $|P_k| - S$ solutions one by one.
    End
End

---

distributed along the PF in objective space, and P is an approximation to the PF. The distance between the P^* and P can be defined as:

$$IGD(P^*, P) = \frac{\sum\limits_{v \in P^*} d(v, P)}{|P^*|},$$

where $d(v, P)$ is the minimum Euclidean distance from the point v to P. It is obvious that the smaller value of IGD is, the better the algorithm performs. 500 points for bi-objective test instances and 1000 points for tri-objective test instances are uniformly sampled on the PF to form P^*.

H-metric. H-metric (S metric) is also used to measure the quality of a solution set P. Let $y^* = (y_1^*, \ldots, y_m^*)$ be a refer point in the objective space which is dominated by any Pareto-optimal objective vectors. Let S be the obtained approximation to the PF in the objective space. Then the I_H value of S (with regard to y^*) is the volume of the region which is dominated by S and dominates y^*. We make $y^* = (1, \ldots, 1)$ in our experiments. Obviously, the larger the H-metric is, the better the approximation is.

2.6   Test Instance

The following test instances of ZDT and DTLZ instances are all for minimization. Their search space is $[0, 1]^n$, where n is the dimension of decision space. $n = 30$ for ZDT1, ZDT2, ZDT3 instances and $n = 10$ for others.

2.7   Experimental Setting

☐ The crossover and mutation operators with the same control parameters in [5] are used in the two algorithms for generating new solutions.

☐ The population size is *popsize* $= 100$ for the biobjective instances and *popsize* $= 300$ for triobjective instances.

☐ For the biobjective instances, $K = S = 10$ in SMS-M2M. For the triobjective instances, $K = S = 17$ in SMS-M2M.

☐ Stopping condition: all the three algorithms stop after 3000 generations.

☐ The K direction vectors are uniformly selected from the unit sphere in the first octant in SMS-M2M.

☐ The probability that an individual crossovers with the individual from other subpopulations $P_c = 0.8$ in SMS-M2M;

☐ The mutation probability $P_m = 1/popsize$: .

2.8   Experimental Results and Analysis

The simulation program has been developed within the Matlab programming environment. Both SMS-M2M and SMS-EMOA have been independently run for 30 times for each test instance on identical computers (Intel(R) Core(TM) i3-2100 CPU 3.10 GHZ, 1.82GB memory).

Table 1 shows the mean time and minimum elapsed CPU time consumed by every generation of SMS-EMOA and SMS-M2M in one run for each test instance. The mean time and minimum time reduced by M2M decomposition compared with SMS-EMOA for each test instance are shown in Table 2. From the two tables, we can see that the M2M population decomposition strategy can effectively reduce the computational effort brought by calculating S metric.

Figs. 1-7 plot the distribution of final solutions obtained in the run with the best H-metric value of SMS-MOEA and SMS-M2M. The smallest and the mean of the IGD-metric value of the algorithms for each test instance of 30 independently runs are shown in Table 3. The largest and the mean of the H-metric value of the algorithms for each test instance of 30 independent runs are shown in table 4. From these figures and tables, we can see that although a little worse than SMS-EMOA, the performance of SMS-M2M is still acceptable and it is almost the same with SMS-EMOA. The performance degeneration of SMS selection is almost unavoidable because of population decomposition, but it still under control.

Therefore, we can safely conclude that the M2M population decomposition strategy is helpful in reducing the computational effort of SMS-based algorithm with only a little sacrifice of performance.

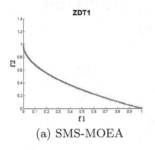
(a) SMS-MOEA

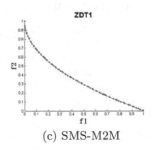
(c) SMS-M2M

Fig. 1. Results of ZDT1

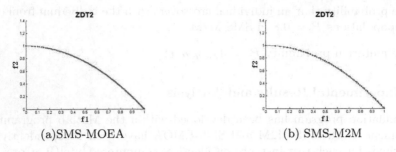

(a)SMS-MOEA

(b) SMS-M2M

Fig. 2. Results of ZDT2

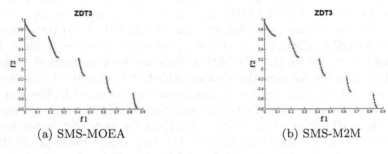

(a) SMS-MOEA

(b) SMS-M2M

Fig. 3. Results of ZDT3

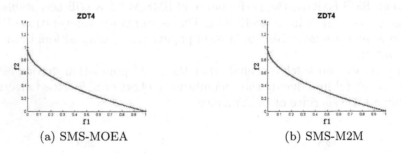

(a) SMS-MOEA

(b) SMS-M2M

Fig. 4. Results of ZDT4

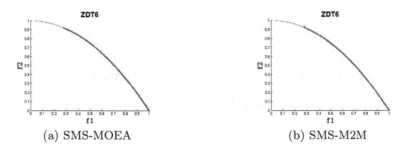

(a) SMS-MOEA

(b) SMS-M2M

Fig. 5. Results of ZDT6

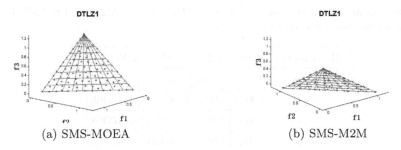

(a) SMS-MOEA   (b) SMS-M2M

Fig. 6. Results of DTLZ1

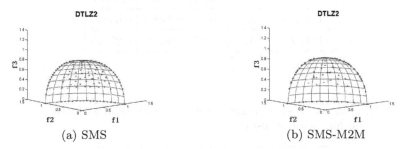

(a) SMS   (b) SMS-M2M

Fig. 7. Results of DTLZ2

Table 1. The mean time and minimum time consumed by every generation of SMS-EMOA and SMS-M2M in one run

Time consumed	SMS-EMOA		SMS-M2M	
Instance	min	mean	min	mean
ZDT1	0.2978	0.3078	0.0297	0.0317
ZDT2	0.3058	0.3258	0.0302	0.0342
ZDT3	0.3167	0.3367	0.0298	0.0318
ZDT4	0.2822	0.3212	0.0279	0.0297
ZDT6	0.2483	0.2638	0.0224	0.0261
DTLZ1	0.7297	0.7501	0.0602	0.0674
DTLZ2	0.7547	0.7607	0.0481	0.0487

Table 2. The mean time and minimum time reduced by M2M decomposition compared with SMS-EMOA

Time reduced	ZDT1	ZDT2	ZDT3	ZDT4	ZDT6	DTLZ1	DTLZ2
min	90.0269%	90.1243%	90.5905%	90.1134%	90.9787%	91.7500%	93.6266%
mean	89.7011%	89.5028%	90.5554%	90.7534%	90.1061%	91.0145%	93.5980%

Table 3. The mean and best of IGD-metric values of SMS-M2M and SMS-EMOA in 30 independent runs for each test instance

IGD-metric	SMS-M2M		SMS-MOEA	
Instance	best	mean	best	mean
ZDT1	0.0041	0.0041	0.003598	0.003469
ZDT2	0.0045	0.0046	0.004357	0.004398
ZDT3	0.0058	0.0059	0.004506	0.004508
ZDT4	0.0041	0.0041	0.003627	0.003659
ZDT6	0.0450	0.0450	0.044573	0.044987
DTLZ1	0.031	0.0324	0.03903	0.03859
DTLZ2	0.0468	0.0493	0.0729	0.07135

Table 4. The mean and best of H-metric values of SMS-M2M and SMS-EMOA in 30 independent runs for each test instance

H-metric	SMS-M2M		SMS-MOEA	
Instance	best	mean	best	mean
ZDT1	0.6615	0.6614	0.662095	0.662235
ZDT2	0.3275	0.3274	0.328817	0.327154
ZDT3	0.7785	0.7785	0.779739	0.780721
ZDT4	0.6614	0.6614	0.662074	0.663021
ZDT6	0.3220	0.3219	0.322639	0.323621
DTLZ1	0.7895	0.7879	0.780918	0.781911
DTLZ2	0.4389	0.4378	0.426442	0.426412

2.9   Theoretic Analysis

The conclusion can also be identified by theory analysis. For simplicity, we only consider the process of S metric calculation. One widely used S metric calculation methods is Hypervolume by Slicing Objectives (HSO) which has the worst-case complexity $O(Popsize^{m-1})$ [10]. It grows exponentially with the number of objectives m. However, if the population is divided into K subpopulations, the worst-case complexity becomes $O(Popsize/K)^{m-1}$. Although the worst-case complexity still grows exponentially, it can be reduced as low as only $1/K^{m-1}$ of the original computational effort. In other words, the computational complexity of SMS can be drastically reduced by population decomposition.

3 Conclusion

In this paper, we present a novel evolutionary algorithm SMS-M2M based on SMS-EMOA and MOEA/D-M2M. M2M-based SMS algorithms can significantly reduce the computation time in the process of selection. It can greatly increase the computational efficiency of SMS-based algorithm without degrading its performance evidently. This study gives a new direction to design faster SMS-based evolutionary multi-objective algorithm.

Acknowledgment. This work was supported by the Natural Science Foundation of Guangdong Province (S2011030002886, S2012010008813), the National Natural Science Foundation of China (NSFC) under grant: 61272366, and the projects of Science and Technology of Guangdong Province (2012B091100033).

References

1. Zitzler, E.: Evolutionary algorithms for multiobjective optimization: Methods and applications. Ph.D. dissertation, Comput. Eng. Netw. Lab., Swiss Federal Instit. Technol (ETH), Zurich, Switzerland (1999)
2. Purshouse, R.: On the evolutionary optimization of many objectives. Ph.D. dissertation, Dept. Automatic Control Syst. Eng., Univ. Sheffield, Sheffield, U.K (2003)
3. Laumanns, M., Zitzler, E., Thiele, L.: A unified model for multiobjective evolutionary algorithms with elitism. In: Proc. Congr. Evol. Comput., pp. 46–53 (2000)
4. Deb, K.: Multi-Objective Optimization using Evolutionary Algorithms. John Wiley & Sons, Chichester (2001)
5. Beume, N., Naujoks, B., Emmerich, M.: SMS-EMOA: Multiobjective Selection Based on Dominated Hypervolume. European Journal of Operational Research, 1653–1669 (2007)
6. Fleischer, M.: The Measure of Pareto Optima. In: Fonseca, C.M., Fleming, P.J., Zitzler, E., Deb, K., Thiele, L. (eds.) EMO 2003. LNCS, vol. 2632, pp. 519–533. Springer, Heidelberg (2003)
7. Emmerich, M., Beume, N., Naujoks, B.: An EMO Algorithm Using the Hypervolume Measure as Selection Criterion. In: Coello Coello, C.A., Hernández Aguirre, A., Zitzler, E. (eds.) EMO 2005. LNCS, vol. 3410, pp. 62–76. Springer, Heidelberg (2005)
8. Fonseca, C.M., Paquete, L., Lopez-Ibanez, M.: An Improved Dimension-Sweep Algorithm for the Hypervolume Indicator. In: IEEE Congress on Evolutionary Computation (CEC 2006), pp. 1157–1163 (2006)
9. Knowles, J.D., Corne, D.W., Fleischer, M.: Bounded Archiving using the Lebesgue Measure. In: Congress on Evolutionary Computation (CEC 2003)l, vol. 4, pp. 2490–2497. IEEE Press (2003)
10. While, L., Bradstreet, L., Barone, L., Hingston, P.: Heuristics for Optimising the Calculation of Hypervolume for Multi-objective Optimisation Problems. In: IEEE Congress on Evolutionary Computation (CEC 2005), pp. 2225–2232 (2005)
11. While, L., Hingston, P., Barone, L., Huband, S.: A Faster Algorithm for Calculating Hypervolume. IEEE Transactions on Evolutionary Computation, 29–38 (2006)

452   L. Chen et al.

12. Liu, H.L., Li, X.: The multiobjective evolutionary algorithm based on determined weight and sub-regional search. In: 2009 IEEE Congress on Evolutionary Computation, Norway, May 18-21, pp. 1928–1934 (2009)
13. Mei, Y., Tang, K., Yao, X.: Decomposition-Based memetic algorithm for multiobjective capacitated arc routing problem. IEEE Trans. Evol. Comput. 15(2), 151–165 (2011)
14. Liu, H.L., Gu, F.: A improved NSGA-II algorithm based on subregional search. In: Proc. Congr. Evol. Comput., pp. 1906–1911 (2011)
15. Huband, S., Hingston, P., Barone, L., While, L.: A review of multiobjective test problems and a scalable test problem toolkit. IEEE Trans. Evol. Comput. 10(5), 477–506 (2006)
16. Liu, H.L., Gu, F., Zhang, Q.: Decomposition of a Multiobjective Optimization Problem into a Number of Simple Multiobjective Subproblems. IEEE Trans. Evol. Comput. (in press, 2013)

A New Equality Constraint-Handling Technique: Unconstrained Search Space Proposal for Equality Constrained Optimization Problem

Yukiko Orito[1], Yoshiko Hanada[2], Shunsuke Shibata[3], and Hisashi Yamamoto[3]

[1] Department of Economics, Hiroshima University,
Higashi-Hiroshima, Hiroshima 739–8525, Japan,
orito@hiroshima-u.ac.jp,
[2] Faculty of Engineering Science, Kansai University,
Suita, Osaka 564–8680, Japan,
[3] Department of System Design, Tokyo Metropolitan University,
Hino, Tokyo 191–0065, Japan

Abstract. For solving an equality constrained optimization problem, it is difficult to search an optimal solution by using any evolutionary algorithms. We propose a new technique which removes an equality constraint in an optimization problem in this paper. The technique transforms an equality constrained search space to an unconstrained search space for a portfolio replication problem. In numerical experiments, we show that evolutionary algorithms can generate good solutions in an unconstrained search space obtained by the technique.

Keywords: Search Spaces Transformation, Unconstrained Equality Constraint, Portfolio Replication.

1 Introduction

In a constrained optimization problem, we have to find an optimal solution in a search space consisting of many feasible solutions and many infeasible solutions. It is difficult to find an optimal solution efficiently by using evolutionary algorithms because the algorithms need to search for only feasible solutions. For optimizing such a problem, many researchers have proposed their techniques in evolutionary algorithms that search efficiently for feasible solutions. Coello[1] provided a survey of popular constraint-handling techniques used with evolutionary algorithms. However, all constraint-handling techniques are applied to a given search space which consisted of many feasible solutions and many infeasible solutions.

In this paper, we propose a new constraint-handling technique which removes a constraint in an equality constrained optimization problem. The technique partially transforms a given constrained search space to an unconstrained search space and thereby has a contribution that evolutional algorithms work well in a search space without constraints.

© Springer International Publishing Switzerland 2015
H. Handa et al. (eds.), *Proc. of the 18th Asia Pacific Symp. on Intell. & Evol. Systems – Vol. 2*,
Proceedings in Adaptation, Learning and Optimization 2, DOI: 10.1007/978-3-319-13356-0_36

Although we show that the technique works very well for a portfolio replication problem; a well known application of an equality constrained optimization problem, our basic idea can be adapted to other constrained optimization problems.

This paper is organized as follows: We propose a new constraint-handling technique in Section 2. Section 3 describes the EDA (Estimation of Distribution Algorithm) with fixed width histogram as our optimization method. Section 4 describes a portfolio replication problem as a real world application of equality constrained problem. Section 5 presents the results of numerical experiments, and we conclude our discussion in Section 6.

2   The Equality Constraint-Handling Technique

We consider the following equality constrained minimization problem.

$$\min f(x_1, \cdots, x_N), \tag{1}$$

$$\text{s.t. } g(x_1, \cdots, x_N) = \sum_{i=1}^{N} x_i$$
$$= 1,$$

where x_i is the ith variable.

The function $f(x_1, \cdots, x_N)$ is an objective function and $g(x_1, \cdots, x_N)$ is an equality constraint. This is one of general allocation problems that a constraint is denoted by the sum total of variables.

While an evolutionary algorithm is applied to the optimization problem given by the equation (1), infeasible solutions generated by the algorithm are removed, evolved into or repaired to feasible solutions satisfying the equality constraint. Therefore, a construction of feasible solution brings about evolutionary stagnation.

In order to overcome such a problem, we propose a new constraint-handling technique which transforms a part of a given constrained search space to an unconstrained search space in this paper. It is expected that an evolutionary algorithm finds good feasible solutions without evolutionary stagnation effectively because an unconstrained search space consists only of feasible solutions.

Let M be the length of binary digits. The number of variables is defined as $N = 2^M$ ($M = 1, 2, \cdots$). When the j-th value of the binary number which converted from the decimal number $i - 1$ ($i = 1, \cdots, N$) is expressed as $a_j \in \{0, 1\}$ ($j = 1, \cdots, M$), the solution $\mathbf{x}$ is defined as follows.

$$\mathbf{x} = (x_1, \cdots, x_N), \tag{2}$$

$$x_i \equiv \prod_{j=1}^{M} \left(\cos^2 \theta_j\right)^{a_j} \left(\sin^2 \theta_j\right)^{1-a_j}, \qquad (i = 1, \cdots, N).$$

From the Pythagorean theorem, that is,

$$\sin^2 \theta_j + \cos^2 \theta_j = 1 \qquad (j = 1, \cdots, M), \tag{3}$$

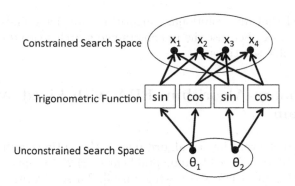

Constrained Search Space

Trigonometric Function

Unconstrained Search Space

Fig. 1. Search Space Transformation

the following equation is satisfied for any value of $\theta = (\theta_1, \cdots, \theta_M)$.

$$\sum_{i=1}^{N} x_i = \prod_{j=1}^{M} \left(\sin^2 \theta_j + \cos^2 \theta_j \right)$$

$$= 1. \tag{4}$$

Therefore, our technique can transform a part of the search space $(x_1, \cdots, x_N)$ with the equality constraint to the search space $(\theta_1, \cdots, \theta_M)$ without constraint.

For example, for $M = 2$ and $N = 2^2 = 4$, the decimal number 0, 1, 2, or 3 is expressed as binary number $00_{(2)}$, $01_{(2)}$, $10_{(2)}$, or $11_{(2)}$, respectively. The search space (x_1, x_2, x_3, x_4) is expressed as follows.

$$\mathbf{x} = (x_1, x_2, x_3, x_4), \tag{5}$$
$$x_1 = \sin^2 \theta_1 \sin^2 \theta_2, \; x_2 = \sin^2 \theta_1 \cos^2 \theta_2,$$
$$x_3 = \cos^2 \theta_1 \sin^2 \theta_2, \; x_4 = \cos^2 \theta_1 \cos^2 \theta_2.$$

Based on the Pythagorean theorem, the following equation is satisfied for any value of θ_1 and θ_2.

$$\sum_{i=1}^{4} x_i = \left(\sin^2 \theta_1 + \cos^2 \theta_1 \right) \left(\sin^2 \theta_2 + \cos^2 \theta_2 \right)$$

$$= 1. \tag{6}$$

The search space transformation from (x_1, x_2, x_3, x_4) to (θ_1, θ_2) is shown in Fig. 1.

Therefore, we re-define the equality constrained minimization problem given by the equation (1) as the following unconstrained minimization problem.

$$\min f(x_1, \cdots, x_N), \tag{7}$$

$$x_i \equiv \prod_{j=1}^{M} \left(\cos^2 \theta_j \right)^{a_j} \left(\sin^2 \theta_j \right)^{1-a_j}, \quad (\boldsymbol{\theta} \in \mathbf{x}, i = 1, \cdots, N).$$

It is expected that an evolutionary algorithm effectively finds good feasible solutions $(x_1, \cdots, x_N)$ because the unconstrained search space $(\theta_1, \cdots, \theta_M)$ consists only of feasible solutions.

3 Evolutionary Algorithm: EDA with Fixed Width Histogram

Among the current evolutionary algorithms to the optimization problems, we find: PBIL (Population-Based Incremental Learning)[2], Compact GA (Compact Genetic Algorithm)[3], EDA with histograms[4]. These algorithms are effective to a problem without dependency between variables included in a solution. In this paper, we employ the EDA with fixed width histogram which proposed by Tsutsui et al.[4].

For our unconstrained search space, as a genetic representation, the k-th individual on the l-th generation's population is represented as

$$\boldsymbol{\theta}^{(l,k)} = \left(\theta_1^{(l,k)}, \cdots, \theta_M^{(l,k)} \right). \tag{8}$$

The function $f(x_1^{(l,k)}, \cdots, x_N^{(l,k)})$ given by the equation (7) is employed as an evaluation function in the EDA. The details of the function are described in Section 4 as a real world application.

The procedure of the EDA consists of processes from Step 1 to Step 5.

1. Initial State
   At an initial generation of $l = 0$, N_p individuals are randomly generated in an initial parents' population. Here, it is defined as

   $$\left(\boldsymbol{\theta}^{(0,1)}, \cdots, \boldsymbol{\theta}^{(0,N_p)} \right). \tag{9}$$

   Each variable is determined by a value on the range of $[0, \pi/2]$.

2. Histograms from Parents' Population
   For the parents' population $\boldsymbol{\theta}^{(l,k)}$ on the l-th generation, the interval of a variable on the range of $[0, \pi/2]$ is divided into H discrete intervals. A histogram consists of N_p individuals in the parents' population. Thereby, Bin h $(h = 1, \cdots, H)$ is a bin of the discrete interval on the range of $[\pi(h-1)/2H, \pi h/2H]$. The histogram with frequencies to the bin h is defined as

   $$v_j^{(l)}[h] = \# \left\{ k \,\middle|\, \frac{\pi(h-1)}{2H} \leq \theta_j^{(l,k)} < \frac{\pi h}{2H} \right\}, \tag{10}$$
   $$(k \in \{1, \cdots, N_p\}, j = 1, \cdots, M, h = 1, \cdots, H).$$

3. Probability Distribution for Offspring Population
   An offspring population consisting of N_o individuals are generated from the current parents' population. We assume that the histogram of j-th variable

defined by the equation (10) is the probability distribution of j-th variable for producing an offspring population.

Let $p_j^{(l)}[h]$ be the probability of j-variable to h on the l-th generation. The probability distribution of j-th variable is defined as

$$p_j^{(l)}[h] = \frac{1}{N_p} v_j^{(l)}[h], \qquad (j = 1, \cdots, M, h = 1, \cdots, H). \qquad (11)$$

For producing offspring population, the h^* of each variable is randomly selected according to the distributions given by the equation (11) and then a new variable is randomly determined from the range of $[\pi(h^* - 1)/2H, \pi h^*/2H]$.

4. Selection

   When the individual $(\theta_1^{(l,k)}, \cdots, \theta_M^{(l,k)})$ is given, the solution $(x_1^{(l,k)}, \cdots, x_N^{(l,k)})$ is obtained through the equation (2) and then the value of the evaluation function is obtained by the equation (7).

   Based on the values of the evaluation function, N_p individuals are selected by the elitism selection and the roulette wheel selection from the current parents' and offspring populations for the next generation.

5. Terminate Criterion

   The operations of producing the parents' population, making the probability distribution, producing the offspring population, and performing the selection are repeated until the maximal number of the repetitions is satisfied.

From the last population we select one solution that has the lowest evaluation value of all. This solution is the optimum or quasi-optimum solution obtained by the EDA.

4   Application: Portfolio Replication Problem

As mentioned in Section 2, our equality constrained-handling technique was proposed in order to optimize a problem like the equation (1). The function $f(x_1, \cdots, x_N)$ is also employed as the evaluation function of the EDA.

In this paper, we deal with a long-short portfolio replication problem, one of allocation problems, as a real world application. Many researchers have used various evolutionary algorithms to optimize their portfolios (for this, see. e.g. [5,6,7,8,9,10]).

A long-short portfolio consists of assets with long positions in which they have bought and been held and with short positions in which they have been borrowed and sold.

The long-short portfolio, represented as the solution $\mathbf{x}$, is defined as follows.

$$\mathbf{x} = (x_1, \cdots, x_N) \qquad (12)$$
$$= \left(x_1^L - x_1^S, \cdots, x_N^L - x_N^S\right),$$
$$\mathbf{x}^L = \left(x_1^L, \cdots, x_N^L\right), \ \mathbf{x}^S = \left(x_1^S, \cdots, x_N^S\right),$$
$$\text{s.t.} \ \sum_{i=1}^{N} x_i^L = 1, \ \sum_{i=1}^{N} x_i^S = 1, \quad (x_i^L \geq 0, x_i^S \geq 0),$$

where $\mathbf{x}^L$ and $\mathbf{x}^S$ are the weights of assets on the long position and the weights of assets on the short position, respectively.

From a practical viewpoint, an asset management firm (Company A) desires to make a replication portfolio to a portfolio of another firm (Company B) when the portfolio of Company B has delivered better performances than the portfolio of Company A. However, Company B opens only the total return of portfolio to the public but other information such as the proportion-weighted combination is closed to the public. Hence, it is difficult for Company A to replicate the portfolio of Company B.

In this paper, we optimize a replication portfolio to a given benchmark portfolio. Let $r_i(t)$ be the return of Asset i $(i = 1, \cdots, N)$ at t. The returns of replication portfolio consisting of N assets over a period between $t = 1$ and $t = T$ are represented as

$$\begin{pmatrix} r_\mathbf{x}(1) \\ \vdots \\ r_\mathbf{x}(T) \end{pmatrix} = \begin{pmatrix} r_1(1) \cdots r_N(1) \\ \vdots \\ r_1(T) \cdots r_N(T) \end{pmatrix} \begin{pmatrix} x_1 \\ \vdots \\ x_N \end{pmatrix}. \tag{13}$$

In the equation (13), the returns of the replication portfolio $(r_\mathbf{x}(1), \cdots, r_\mathbf{x}(T))$ are already given because the replication portfolio has the same returns as a benchmark portfolio. The returns of individual assets $(r_i(1), \cdots, r_i(T))$ $(i = 1, \cdots, N)$ are also given. Therefore, if the number of days data T is equal to the number of assets N, we can obtain an optimal solution $\mathbf{x}$ by solving the simultaneous equations. However, a proportion-weighted combination of benchmark portfolio changes on its rebalancing to keep track of the performances in the future periods. A benchmark portfolio is fixed only in a short period, so T might be very less than N. In case studies of $T < N$, it is difficult to find the real optimal solution. In order to avoid this problem, we try to optimize a replication portfolio such that its return mimics a return of benchmark portfolio. Let $r_{\mathbf{x}^B}(t)$ be the return of benchmark portfolio $\mathbf{x}^B$ at t. For evaluating replication portfolios, we employ the following function consisting of the errors of returns and the rates of changes of returns between replication and benchmark portfolios as an objective function.

$$f(x_1, \cdots, x_N) = \sum_{t=1}^{T} (r_\mathbf{x}(t) - r_{\mathbf{x}^B}(t))^2 + \rho \sum_{t=1}^{T-1} \left(1 - \frac{r_\mathbf{x}(t+1) - r_\mathbf{x}(t)}{r_{\mathbf{x}^B}(t+1) - r_{\mathbf{x}^B}(t)} \right)^2,$$

$$\tag{14}$$

where ρ is a parameter given in the experiments.

Based on the equations (7) and (14), the proposed equality constrained-handling technique is applied to optimize the replication portfolio. The equation (14) is employed as the evaluation function of EDA. From the equation (12), however, the solution $\mathbf{x}$ is obtained by the weights of assets on the long position $\mathbf{x}^L$ and the weights of assets on the short position $\mathbf{x}^S$. Thereby there are two individuals, $\mathbf{x}^L$ and $\mathbf{x}^S$, and then the algorithm of EDA is applied to each of both these individuals in this paper.

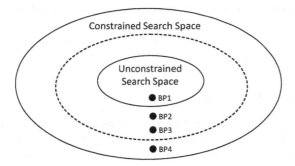

Fig. 2. The locations of BP1, BP2, BP3, and BP4 in the search space

5 Numerical Experiments

Our technique can transform a part of the search space $(x_1, \cdots, x_N)$ with the equality constraint to the unconstrained search space $(\theta_1, \cdots, \theta_M)$. To investigate the ability of the proposed equality constrained-handling technique, in the experiments of this section, we compare the case studies that an optimal solution is located in the unconstrained search space and is not located in the unconstrained search space. As mentioned in Section 4, the optimal solution is the benchmark portfolio for the portfolio replication problem. The benchmark portfolio (optimal solution) consists of N weights of assets (variables). From this context, we employ the following four kinds of benchmark portfolios, from BP1 to BP4, as the different optimal solutions. The locations of BP1, BP2, BP3, and BP4 in the search space are shown in Fig. 2.

1. BP1: Benchmark portfolio consisting of N variables in the unconstrained search space
   The values of M variables of $(\theta_1, \cdots, \theta_M)$ are randomly given in the unconstrained search space and then change to the values of N variables of $(x_1, \cdots, x_N)$ by using the equation (2).
2. BP2: Benchmark portfolio consisting of $N-2$ variables in the unconstrained search space and 2 variables outside the space
   The values of M variables of $(\theta_1, \cdots, \theta_M)$ are randomly given in the unconstrained search space and then change to the values of N variables of $(x_1, \cdots, x_N)$ by using the equation (2). In addition, we select 2 different variables at random and then exchange the values of the selected variables.
3. BP3: Benchmark portfolio consisting of $N-10$ variables in the unconstrained search space and 10 variables outside the space
   The values of M variables of $(\theta_1, \cdots, \theta_M)$ are randomly given in the unconstrained search space and then change to the values of N variables of $(x_1, \cdots, x_N)$ by using the equation (2). In addition, we select 2 different variables at random and then exchange the values of the selected variables. We repeat this exchange operation five times.

4. BP4: Benchmark portfolio consisting of N variables outside the unconstrained search space
   The values of N variables of $(x_1, \cdots, x_N)$ are randomly given in the constrained search space.

5.1 Experimental Setting

In the numerical experiments, we employ N assets with high turnover on the Tokyo Stock Exchange. Each training phase consists of $T = 20$ days data from 2005 to 2010. We call them Phase 1 through Phase 13 respectively.

The setting of the optimization problem was as follows.

- The number of variables for individual $\boldsymbol{\theta}$: $M = 7$
- The number of variables for solution $\mathbf{x}$: $N = 2^M = 128$
- Length of Phase: $T = 20$
- Parameter of the equation (14): $\rho = 1.0E - 08$

The parameters of the EDA were set as follows.

- Parents' population size: $N_p = 100$
- Offspring population size: $N_o = 200$
- Elitist rate: 0.1
- The number of bins: $H = 100$ (The width of bin is set to $\pi/2H$.)
- Generation size: 100
- Algorithm run: 10

5.2 Main Results

We optimized the replication portfolio by performing the EDA to each of the four benchmark portfolios from BP1 to BP4. In the experiments, we compare the replication portfolios obtained by the EDA using our technique with the replication portfolios obtained by the EDA of not using our technique.

For BP1, the value of evaluation function (EF) and the mean absolute error (MAE) of values of variables between replication and benchmark portfolios are shown in Figs 3 and 4, respectively. For BP4, EF and MAE of values of variables between replication and benchmark portfolios are shown in Figs 5 and 6, respectively.

Figs 3 and 4 say that, for BP1, the value of evaluation function and the mean absolute error of solution obtained by the EDA with our technique is smaller than those of the EDA without our technique through almost all phases. Of course, the benchmark portfolio BP1 is located in the unconstrained search space, thereby the EDA with our technique can find the good solution with small value of evaluation function.

On the other hand, Figs 5 and 6 say that, for BP4, the mean absolute error of solution obtained by the EDA with our technique is smaller than that of the EDA without our technique through almost all phases though the value of

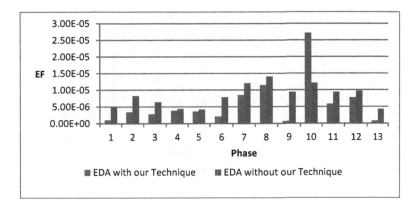

Fig. 3. Replication portfolio of BP1: Evaluation function (EF)

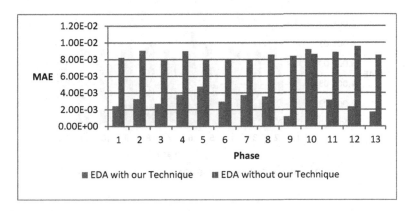

Fig. 4. Replication portfolio of BP1: Mean absolute error (MAE) of values of variables between replication and benchmark portfolios

evaluation function obtained by the EDA with our technique is larger than that of the EDA without our technique. The benchmark portfolio BP4 is not located in the unconstrained search space, however the EDA with our technique can find the good solution. We can conclude that, therefore, an evolutionary algorithm with our technique finds good feasible solutions without evolutionary stagnation effectively.

Now, we focus on a location of optimal solution. The benchmark portfolio BP3 is located near the unconstrained search space and BP2 is located very near the unconstrained search space. We optimized the 127 replication portfolios by performing the EDA to the randomly determined 127 benchmark portfolios of BP2 and BP3, respectively. For BP2 and BP3, the value of evaluation function (EF) and the mean squared error (MSE) of values of variables between replication and benchmark portfolios are shown in Figs 7 and 8, respectively.

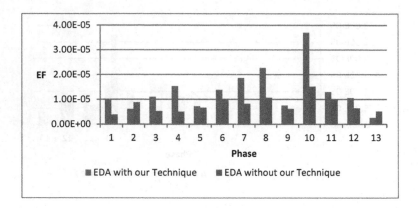

Fig. 5. Replication portfolio of BP4: Evaluation function (EF)

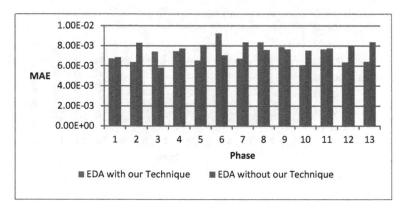

Fig. 6. Replication portfolio of BP4: Mean absolute error (MAE) of values of variables between replication and benchmark portfolios

From Fig 7, the variances of values of evaluation function and the mean squared errors of solution obtained by the EDA with our technique is smaller than those of the EDA without our technique, respectively.

On the other hand, Fig 8 says that the mean squared error of solution for BP3 is larger than that for BP2 because BP3 is located more in the distance from the unconstrained search space than BP2 is. However, the variances of values of evaluation function and the mean squared errors of solution obtained by the EDA with our technique are smaller than those of the EDA without our technique. Therefore, we can conclude that the EDA with our technique works better than the EDA without our technique does even if an optimal solution is located outside the unconstrained search space.

Of course, the EDA with our technique cannot find an optimal solution if the optimal solution is not located in the unconstrained search space. We improve the

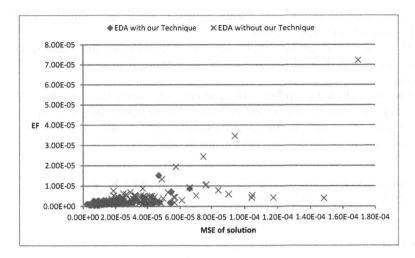

Fig. 7. Replication portfolio of BP2: the value of evaluation function (EF) and the mean squared error (MSE) of solution

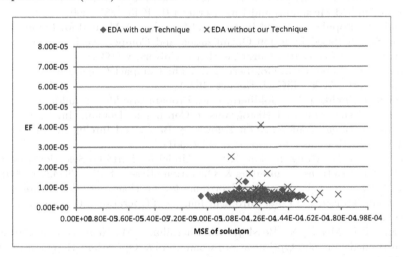

Fig. 8. Replication portfolio of BP3: the value of evaluation function (EF) and the mean squared error (MSE) of solution

technique and then make a new technique to estimate a location of an optimal solution. This is our future work.

6   Conclusion

In this paper, we proposed a new constraint-handling technique which removes a constraint in an equality constrained optimization problem. The technique partially transforms a given constrained search space to an unconstrained search space and thereby has a contribution that evolutional algorithms work well in a search space without constraints.

In the numerical experiments, we showed that an evolutionary algorithm with our technique found good feasible solutions without evolutionary stagnation effectively.

We need to show the diffence of landscapes between the constrained search space and our unconstrained search space. In addition, our technique cannot find an optimal solution if the optimal solution is not located in a search space without constraints. These are our future works.

Acknowledgment. This work was supported by JSPS KAKENHI Grant Numbers #25730148 and #26330290.

References

1. Coello, C.A.C.: Theoretical and Numerical Constraint-handling Techniques Used with Evolutionary Algorithms: A Survey of the State of the Art. Computer Methods in Applied Mechanics and Engineering 191, 1245–1287 (2002)
2. Baluja, S.: Population-based Incremental Learning: A Method for Integrating Genetic Search Based Function Optimization and Competitive Learning. Technical Report. CMU-CS-94-163, Carnegie Mellon University (1994)
3. Harik, G.R., Lobo, F.G., Goldberg, D.E.: The Compact Genetic Algorithm. Technical Report. 97006, IlliGAL Report (1997)
4. Tsutsui, S., Pelikan, M., Goldberg, D.E.: Probabilistic Model-building Genetic Algorithms Using Marginal Histograms in Continuous Domain. In: Proceedings of the International Conference on Knowledge-Based and Intelligent Information & Engineering Systems 2001, pp. 112–121 (2001)
5. Xia, Y., Liu, B., Wang, S., Lai, K.K.: A Model for Portfolio Selection with Order of Expected Returns. Computers & Operations Research 27, 409–422 (2000)
6. Lin, C.C., Liu, Y.T.: Genetic Algorithms for Portfolio Selection Problems with Minimum Transaction Lots. European Journal of Operational Research 85(1), 393–404 (2008)
7. Chang, T.J., Meade, N., Beasley, J.E., Sharaiha, Y.M.: Heuristics for Cardinality Constrained Portfolio Optimization. Computers & Operations Research 27, 1271–1302 (2000)
8. Crama, Y., Schyns, M.: Simulated Annealing for Complex Portfolio Selection Problems. European Journal of Operational Research 150, 546–571 (2003)
9. Oh, K.J., Kim, T.Y., Min, S.: Using Genetic Algorithm to Support Portfolio Optimization for Index Fund Management. Expert Systems with Applications 28, 371–379 (2005)
10. Orito. Y., Yamamoto, H., Tsujimura, Y.: Equality Constrained Long-Short Portfolio Replication by Using Probabilistic Model-building GA. In: Proceedings of WCCI 2012 IEEE World Congress on Computational Intelligence, IEEE Congress on Evolutionary Computation, pp. 513–520 (2011)

Extended Local Clustering Organization with Rule-Based Neighborhood Search for Job-shop Scheduling Problem

Yasumasa Tamura, Hiroyuki Iizuka, and Masahito Yamamoto

Graduate School of Information Science and Technology, Hokkaido University, Japan
{tamura,iizuka,masahito}@complex.ist.hokudai.ac.jp

Abstract. Job-shop scheduling problem (JSP) is one of the hardest combinatorial optimization problems. Local clustering organization (LCO) is proposed by Furukawa et al. to solve such a combinatorial optimization problems as the metaheuristic algorithm. Its effectiveness for the JSP is verified by the comparison with genetic algorithm. However, since LCO is based on the greedy search, the solution is often trapped in local minima. To improve the problem, this study proposes a novel neighborhood search method using priority rules. This paper also shows the extended LCO integrated with the search method.

Keywords: job-shop scheduling problem, local clustering organization, dispatching rules, kicking techniques.

1 Introduction

Job-shop scheduling problem (JSP) is a combinatorial optimization problem to determine a feasible and efficient schedule to process multiple jobs on multiple machines [1]. The JSP is commonly described as follows. A set of jobs $J = \{J_1, J_2, \cdots, J_n\}$ and a set of machines $M = \{M_1, M_2, \cdots, M_m\}$ are given, where the notation n and m correspond to the number of jobs and the number of machines, respectively. Each job J_i consists of the sequence of consecutive operations $(o_{i1}, o_{i2}, \cdots, o_{im})$ and an operation corresponds to each process of the job on each machine. The sequential order of the processes is generally called *technological sequence* and independently given to each job. The feasible solutions in the JSP are restricted by the following constraints.

- All of operations are processed in accordance with the technological sequence.
- A machine cannot process multiple operations simultaneously.
- Multiple operations belong to the same job cannot be processed on different machines simultaneously.
- Each operation cannot be interrupted and resumed its process while being processed.

© Springer International Publishing Switzerland 2015
H. Handa et al. (eds.), *Proc. of the 18th Asia Pacific Symp. on Intell. & Evol. Systems – Vol. 2,*
Proceedings in Adaptation, Learning and Optimization 2, DOI: 10.1007/978-3-319-13356-0_37

The efficiency of a schedule is generally evaluated by the *makespan*, which means the maximum completion time of all jobs.

The JSP has been studied as an important subject in the field of the operations research, the computer science and so on [2,3,4]. Garey et al. showed that the computational complexity theory classifies JSP into *NP-hard* class as same as many other combinatorial optimization problems [5,6]. In addition, it is shown that the JSP is more difficult problem than the *travelling salesman problem* (TSP), a typical combinatorial optimization problem, by the comparison with the *flow-shop scheduling problem* (FSP), a kind of the restricted JSP. Because of these characteristics, many recent studies to solve the JSP focus on the approximate algorithms such as *heuristic* or *metaheuristic methods* instead of the optimization algorithms such as *branch and bound*.

There are two typical heuristic methods for the JSP, a *shifting bottleneck procedure* proposed by Adams et al. [7] and the *dispatching priority rules* [1]. They are known as the practical solution methods since they can find near-optimal solutions in the reasonable computational time. Metaheuristics can also find high quality solutions in the practical computation time. In particular, *neighborhood search* (also called *local search*) with the *critical paths* or the *critical blocks* are known as the effective solution methods for the JSP. The critical paths (blocks) are defined as sets of consecutive operations whose order directly affects the makespan. Van Laarhoven et al. proposed a solution method based on *simulated annealing* (SA) [8]. They also showed that the neighborhood structure using the critical paths can be used to search for the near-optimal solutions efficiently. Furukawa et al. proposed *local clustering organization* as a metaheuristic method without using critical paths [9]. The effectiveness of LCO for the JSP is verified in comparison to genetic algorithm (GA). The authors also proposed the hybrid algorithm based on LCO and SA without critical paths [10]. The study showed that LCO is difficult to escape from local optima because of its searching mechanism based on the greedy search. The hybrid algorithm improves the problem and can search for better solutions than the original LCO. On the other hand, the hybrid algorithm also requires relatively longer computational time than the original LCO because of the annealing processes.

This paper shows another approach to improve the problem in LCO. This paper proposes an extended LCO integrated with a novel neighborhood search mechanism using priority rules. The proposed neighborhood search mechanism searches effective solutions by applying a large-scale changes to the current solution without the annealing processes. In addition, to search good solutions efficiently, this paper also proposes a novel mechanism to apply the priority rules to the neighborhood search, in which the rules are adaptively optimized along with the search of the solution. By the collaboration between the large-scale neighborhood search performed by the proposed method and the small-scale neighborhood search performed by LCO, it is expected that the extended LCO will search for good solutions efficiently. The effectiveness of the extended LCO and the proposed neighborhood search method is verified by some numerical experiments.

Table 1. An instance of the JSP

—	1	2	3
J_1	$(M_1, 11)$	$(M_2, 9)$	$(M_3, 16)$
J_2	$(M_2, 25)$	$(M_1, 10)$	$(M_3, 13)$
J_3	$(M_2, 9)$	$(M_3, 11)$	$(M_1, 15)$
J_4	$(M_3, 12)$	$(M_2, 14)$	$(M_1, 11)$

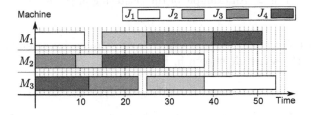

Fig. 1. A schedule for the instance shown in Table 1

2 Local Clustering Organization for the JSP

Local Clustering Organization is a probabilistic metaheuristic algorithm inspired by the mechanism of the Self-Organizing Map (SOM), a kind of neural networks proposed by Kohonen [11], and it is originally proposed to solve the TSP [12]. The searching mechanism of LCO is fundamentally based on the selection and the modification of the *local* which means a part of the solution. In LCO, the modification processe is also named *clustering*.

This section briefly explains the overview of LCO for the JSP. The original application of LCO to the JSP is studied in [9]. The work uses the *permutation with repetition* as a solution representation. In addition, it provides some clustering methods to solve the JSP efficiently.

2.1 Solution Representation : Permutation with Repetition

The solution representation for the JSP based on the permutation with repetition was originally proposed by Bierwirth[13]. The permutation consists of the sequence of jobs, in which each job occurs m times and x^{th} occurrence of the job J_i corresponds to the operation j_i^x. Decoding the permutation to a schedule is performed by scanning it from left end to right sequentially. For example, the permutation with repetition **S** in eq. (1) represents one of the solutions for the example problem given by Table 1, where $(J_1, 1) = (M_1, 11)$ means the first operation of J_1 has to be processed on M_1 and it takes 11 units to process.

$$\mathbf{S} = (J_1, J_3, J_2, J_4, J_2, J_3, J_2, J_4, J_1, J_3, J_4, J_1) \tag{1}$$

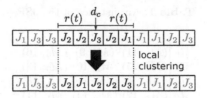

Fig. 2. An example of the clustering in the local

In this case, **S** is decoded into the sequence of consecutive operations $\mathbf{L}_o$ and it also represents the schedule shown in Fig. 1.

$$\mathbf{L}_o = (o_{11}, o_{31}, o_{21}, o_{41}, o_{22}, o_{32}, o_{23}, o_{42}, o_{12}, o_{33}, o_{43}, o_{13}) \qquad (2)$$

The sequence $\mathbf{L}_o$ can also be used for solution representations. However, when some entities in $\mathbf{L}_o$ are swapped recklessly, it can represent an unfeasible schedule. This problem makes the application of the solution algorithm difficult and complex. On the other hand, the solution representation using the permutation with repetition is always decoded to a feasible schedule.

2.2   Overview of the Algorithm

In LCO for the JSP, a subsequence of the consecutive operations on the solution representation is randomly selected as the local in each searching step. The following notations are used to determine the subsequence.

d_c : the index of the center entity in selected local
$r(t)$: the clustering radius in t^{th} step

In the above notations, d_c is determined at random in each step. The local corresponds to a subsequence of the consecutive operations within a radius $r(t)$ from the selected entity d_c.

The clustering method rearranges the sequence of the operations in the local with the hill climbing search. Figure 2 partly shows the rearrangement performed by the clustering method. In general, LCO uses multiple clustering methods in combination and one of them is stochastically selected in each searching step (*mixed clustering*). According to some preliminary experiments, this paper uses three clustering methods, *SEM*, *SIM* and *IEM*. SEM tries to rearrange the sequence of operations in the local by swapping two entities iteratively and SIM inserts an entity into another place iteratively in the local. In addition, IEM tries to rearrange the sequence of operations in the local by inverting some subsequence iteratively in the local. It is important that these clustering methods rearrange the sequence of operations in the local along with the greedy search mechanism. Therefore, the current solution cannot become worse by each clustering method. The detail of them is described in our previous work [10] or the original paper of LCO for the JSP [9].

The following procedure briefly shows the algorithm of LCO, where S_c means the current solution. The notation S_c is basically used to describe the current solution in this paper.

1. Set step $t \leftarrow 1$.
2. Initialize S_c at random.
3. Select a local on S_c.
4. Select a clustering method stochastically.
5. Apply the selected clustering method to the selected local on S_c.
6. Replace t with $t + 1$.
7. If termination conditions of the algorithm are satisfied, stop the procedure. Otherwise, go back to step 3.

In LCO, the clustering radius $r(t)$ has a great effect on accuracy and computational time of optimization. The smaller radius sometimes generates inaccurate solutions because the widely covered optimization in the whole solution is not performed. On the other hand, the bigger radius causes the delay of computational time. In addition, the probabilities to select the clustering methods are also important parameters of LCO to acquire high accurate solutions.

3   Extended LCO with Rule-Based Neighborhood Search

This paper integrates a novel large-scale neighborhood search method into LCO. In this section, firstly, the neighborhood structure and the searching mechanism using priority rules are described. By combining the proposed neighborhood search method with LCO, this section finally proposes the extended LCO.

3.1   Neighborhood Structure on the Permutations with Repetition

To define the large-scale neighborhood structure, the *rearrangeable blocks* are introduced in this study. The rearrangeable blocks mean the subsequences of the solution representation which consist of the consecutive jobs processed on the same machines. By the evaluation of the solution representation, each job in the permutation with repetition is associated with the machine on which the job is processed. The association between jobs and machines defines the sequence of consecutive jobs which are processed on the same machine as the rearrangeable blocks. For example, Fig. 3.1 shows the rearrangeable blocks on the solution **S** shown in eq. (1). There are four rearrangeable blocks marked with the lines.

The neighborhood in our proposed method is defined as a set of the solutions generated from a solution by only applying rearrangement to the sequences of consecutive jobs in the rearrangeable blocks on the permutation with repetition. In addition, as shown in Fig. 3.1, the rearrangement of the jobs in the rearrangeable blocks causes the change of the schedule (note that the rearrangements in the solution representation does not necessarily changes the schedule due to the redundancy of the representations).

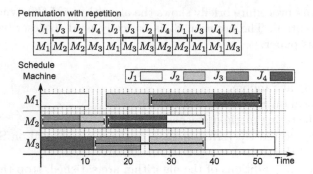

Fig. 3. The rearrangeable blocks on the solution representation and the schedule

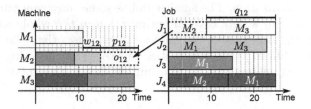

Fig. 4. An example of rule-based schedule construction

3.2   Rearranging the Sequence of Operations Using Priority Rules

Since there are generally multiple rearrangeable blocks in a solution, the number of neighborhood solutions generated from a solution sometimes increases exponentially. The proposed method uses the priority rules to limit the number of neighborhood solutions and to acquire the effective neighborhood solutions efficiently.

We proposed the *composite priority rule* in the previous work [14]. The composite priority rule consists of some simple priority rules, such as the *shortest processing time* (SPT) and *least work remains* (LWKR) [1], and it gives the degree of relative priority $P(o_{ix})$ to each operation o_{ix}. This study uses following three notations to describe the composite priority rules.

p_{ix} : the processing time of the operation o_{ix}
q_{ix} : the total processing time of the following operations of the operation o_{ix}
w_{ix} : the waiting time of the operation o_{ix} for the machines

The example of those notations for the operation o_{12} is shown in Fig. 4. It is commonly known that p_{ix} is used to describe SPT, q_{ix} is used to describe LWKR and w_{ix} is usd to describe a simple rule called *first come first serve* (FCFS).

In this study, the proposed method uses the linear combination rules as the priority rules. The linear combination rule gives a degree of relative priority

$P(o_{ix})$ to a operation o_{ix} using the linear combination shown in eq. (3), where the notations α_1, α_2 and α_3 correspond to the weights for the linear combination.

$$P(o_{ix}) = \alpha_1 \frac{p_{ix}}{\max p_{ix}} + \alpha_2 \frac{q_{ix}}{\max q_{ix}} + \alpha_3 \frac{w_{ix}}{\max w_{ix}} \tag{3}$$

Before applying the rules to generate the neighborhood solutions, the priorities of operations are originally defined by the current solution. To consider the effects of the original priorities, this paper introduces a threshold model into rearranging mechanism. When two operations o_{ix} and $o_{i'x'}$ in the same rearrangeable block has the original precedence relation $o_{ix} \prec o_{i'x'}$ which means o_{ix} is processed preferentially over $o_{i'x'}$, the precedence relation is inverted if eq. (4) is satisfied. Otherwise, the original precedence relation is kept as it is.

$$P(o_{ix}) - P(o_{i'x'}) \le \theta \tag{4}$$

The notation θ in eq. (4) means the threshold value and this study regards θ as a part of the priority rule. In other words, the priority rule used in this study is dynamically composed of the real coded vector $(\alpha_1, \alpha_2, \alpha_3, \theta)$.

A set of priority rules is created by generating individual vectors $(\alpha_1, \alpha_2, \alpha_3, \theta)$. The number of priority rules is a parameter of the proposed method and it corresponds to the number of neighborhood solutions searched per step. In addition, because the effective rules can be different in each instance and the best rule is generally unpredictable, the rules are optimized along with the searches of the solutions (schedules). In this study, the optimization of the rules is performed by genetic algorithm (GA). The fitness of each rule is evaluated by the makespan of the solution generated by the rule. From this mechanism, some effective rules which generate effective neighborhood solutions are adaptively obtained in searches, and the optimization of the schedule is progressed using the effective rules.

3.3 Integration into LCO

The extended LCO alternately applies the local clustering and the large-scale neighborhood search using rules to the current solution. The following procedure briefly shows the algorithm.

1. Generate an initial solution $\mathbf{S}_c$ at random.
2. Generate initial rules at random and evaluate each rule.
3. Select a local on $\mathbf{S}_c$ and a clustering method stochastically.
4. Apply the selected clustering method to the selected local on $\mathbf{S}_c$.
5. Apply genetic operations, selection, crossover and mutation, to the rules.
6. Generate neighborhood solutions from $\mathbf{S}_c$ using each rule and replace the best neighborhood solution with $\mathbf{S}_c$.
7. Evaluate each rule.
8. If termination conditions of the algorithm are satisfied, stop the procedure. Otherwise, go back to step 3.

In the procedure, the steps 3-4 correspond to the searches performed by the conventional LCO and the step 6 corresponds to the proposed neighborhood search. The step 5 and step 7 correspond to the optimization processes for the rules in the procedure. In addition, because the current solution can become worse in the step 6, the best solution found in searches should be memorized.

4   Numerical Experiments

To examine the effectiveness of the proposed method, this section provides the experimental results and discussion. In the experiments, this paper uses 25 well-known benchmark problems introduced by Lawrence [15]. To evaluate the effectiveness of the extended LCO in different problem sizes, the benchmarks problems are classified into 5 groups in terms of the problem size. The effectiveness is evaluated using the *relative error rate RE* shown in eq. (5), where C means the obtained makespan by the searching algorithm and C_o means the optimal makespan show in [16].

$$RE = \frac{C - C_o}{C_o} \times 100 \quad [\%] \tag{5}$$

In this paper, the effectiveness of the extended LCO is compared with the conventional LCO and SA proposed in [8]. All of methods are tested 100 times in each benchmark problems. The conventional LCO and the extended LCO are terminated when 10 seconds have passed from the algorithm is started.

The parameters for LCO, the clustering radius $r(t)$ and the selection probabilities for each clustering method, are determined by some preliminary experiments. The clustering radius $r(t)$ is determined by eq. (6) in each search step, where URAND(a, b) generates a uniform random number in the range $[a, b)$, n means the number of jobs and m means the number of machines.

$$r(t) = \text{URAND}\left(\frac{1}{4}, \frac{1}{3}\right) \times n \times m \tag{6}$$

The effectiveness of each individual clustering method, SEM, SIM and IEM is also verified in the preliminary experiments. Considering the effectiveness of LCO, the selection probabilities should generally be set to (SIM) $\geq$ (SEM) $\gg$ (IEM), where (SIM), (SEM) and (IEM) respectively mean the selection probabilities of SIM, SEM and IEM. In this study, the selection probabilities are set as eq. (7).

$$(\text{SIM}) : (\text{SEM}) : (\text{IEM}) = 50\% : 40\% : 10\% \tag{7}$$

These parameters are also used in the extended LCO as well as the conventional LCO. In addition, the extended LCO generates 10 neighborhood solutions using 10 kinds of the priority rules in the searching processes to apply the proposed neighborhood search method.

Table 2. Comparison with the Conventional LCO

Instances	Jobs × Machines	Extended LCO		Conventional LCO	
		Best [%]	Avg. [%]	Best [%]	Avg. [%]
la16 - la20	10 × 10	0.160	0.410	0.730	2.23
la21 - la25	15 × 10	0.760	1.15	0.920	2.36
la26 - la30	20 × 10	1.25	1.71	1.52	2.18
la31 - la35	30 × 10	0.000	0.000	0.000	0.000
la36 - la40	15 × 15	2.09	3.49	2.78	4.43
Avg.		0.850	1.35	1.19	2.24

Table 3. Comparison with SA shown in [8]

Instances	Jobs × Machines	Extended LCO		SA		
		Best [%]	Avg. [%]	Best [%]	Avg. [%]	T [sec]
la16 - la20	10 × 10	0.160	0.410	0.707	1.27	715.20
la21 - la25	15 × 10	0.760	1.15	1.23	2.02	2095.6
la26 - la30	20 × 10	1.25	1.71	1.83	2.41	4319.0
la31 - la35	30 × 10	0.000	0.000	0.000	0.744	1740.6
la36 - la40	15 × 15	2.09	3.49	1.67	2.50	5450.4
Avg.		0.850	1.35	1.09	1.79	2864.2

4.1   Comparison with the Conventional LCO and SA

Table 2 compares the results of the extended LCO and the conventional LCO and Table 3 compares the results of the extended LCO and SA. To compare the mean performance of each method, this paper shows the mean of the best RE (the column *Best*) and the average RE (the column *Avg.*) for each instance over 100 trials. In addition, the column T [sec] in the results of SA shows the average computational time of SA. Because the architecture of computers used in [8] is far different from the one used in this study, the column T [sec] cannot be used to the statistical comparison. However, the values can be used to understand trends of the search performed by SA.

From Table 2, the extended LCO can obviously search for better solutions than the conventional LCO, which means the extended LCO is averagely more effective than the conventional method in all problem sizes. The results suggest that the proposed neighborhood search method can improve the searching mechanism of LCO. On the other hand, in Table 3, while most results show the similar trends as Table 2, the effectiveness of the extended LCO is less than that of SA in the group la36-la40. Because SA requires much long computational time for the instances la36-la40, it is considered that the extended LCO has not much computational time for those instances in the experiments. The performance of the proposed method and the extended LCO is investigated in following discussion, including the cause of the ineffectiveness of the extended LCO shown in the comparison with SA.

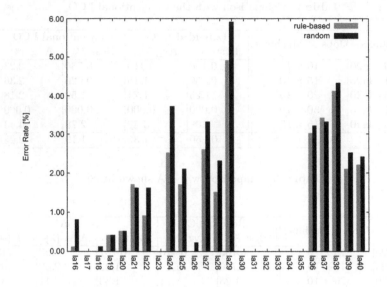

Fig. 5. Comparison with the random selection

4.2   The Effectiveness of the Rule-Based Neighborhood Search

To evaluate the effectiveness of the rule-based neighborhood search method, this paper compares it with the neighborhood search not using the priority rules. The comparison search method generates neighborhood solutions based on the same neighborhood structure as the rule-based one. On the other hand, the comparison method selects the limited number of neighborhood solutions (in this case 10 solutions) at random and the current solution is replaced with the best one in the selected solutions. Figure 5 shows the results of the comparison for each instance, where the red boxes correspond to the average RE of rule-based selection and the blue ones correspond to the average RE of random selection for each instance. The average RE is calculated from the results obtained over 100 trials as well as the previous experiments. Figure 5 explains that the rule-based neighborhood search leads better results than the random neighborhood search in most cases. The results suggest that the proposed method can search for effective neighborhood solutions efficiently when the number of generated neighborhood solutions is limited.

On the other hand, Fig. 5 also shows that there is hardly any difference between the effectiveness of the rule-based neighborhood search and that of the random neighborhood search in the group la36-la40. It is inferred that this result is related with the fact that the extended LCO is less effective than SA for the instances in the group la36-la40. The rest of this section discusses the behavior of the proposed method for the instances la36-la40.

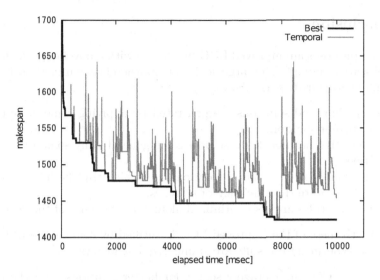

Fig. 6. Changes of makespan in a trial for la37

Figure 6 shows the changes of makespan which the best solution and the temporal solution have, where the temporal solution corresponds to the current solution in searches. From Fig. 6, while the makespan of the temporal solution decreases in the early stage of the search (approximately from 0 to 1000 [msecs]) as well as the best solution, it often increases in the middle and late part of search. This result means the proposed method can assist LCO in obtaining effective solution in the early part of search, which is considered as an effective advantage of the proposed method. In addition, while the temporal solution become worse, the best solution is improved in the middle stage of the search. It means the proposed method successfully help the current solution to escape from local minima as a kind of *kicking methods*. Although these results show the advantages of the proposed method, the behavior of the method in the latter stage of the search (after 8000 [msecs]) shows its disadvantage. The makespan of the temporal solution is much worse than the best solution the latter stage of the search (after 8000 [msecs]), which makes the search ineffective. From Fig. 6, such a problem is caused by the difficulties of obtaining effective rules or the performance limit of the rules. If the problem is caused by the difficulties of obtaining the effective rules, the problem can be improved by extending the computational time or devising the optimization algorithm for the rules. It is also considered that if the problem caused by the performance limit of the rules, we should revise the searching mechanism. Since the consecutive application of the rule-based neighborhood search seems to make the current solution worse, the alternate application of LCO and the proposed method should particularly be revised. Our future work is to verify those consideration and to improve the proposed method.

5   Conclusion

This paper proposes an improved LCO integrated with a novel large-scale neighborhood search method. The large-scale neighborhood search method has some significant characteristics as follows:

- The neighborhood solutions are generated by applying some large changes to the current solution.
- Some effective neighborhood solutions are generated by some dynamic priority rules.
- The number of neighborhood solutions are limited by the number of rules not to deteriorate the efficiency of searches.
- The priority rules are also optimized along with searches of the schedule.

The effectiveness of the improved LCO is verified by some numerical experiments. The experimental results are summarized as follows:

- Improved LCO can averagely search for better solutions than the conventional LCO.
- Rule-based selection of the neighborhood solutions is more effective than random selection, when the number of neighborhood solutions is limited.
- The results suggest the proposed neighborhood search method improves the current solution in the early part of search, and it also performs the kicking mechanism in the middle part of search.

Acknowledgement. This study was supported by JSPS KAKENHI, Grant-in-Aid for JSPS Fellows, 26·1342. This study was also supported by the members of the Laboratory of Autonomous Systems Engineering, Hokkaido University. We would like thank them for their numerous suggestions.

References

1. French, S.: Sequencing and scheduling: an introduction to the mathematics of the job-shop. Ellis Horwood, Chichester (1982)
2. Conway, R.W., Maxwell, W.L., Miller, L.W.: Theory of scheduling. Courier Dover Publications (2003)
3. Coffman, E.G., Bruno, J.L.: Computer and job-shop scheduling theory. John Wiley & Sons (1976)
4. Zweben, M., Fox, M.: Intelligent Scheduling. Morgan Kaufmann Publishers Inc. (1994)
5. Garey, M.R., Johnson, D.S., Sethi, R.: The complexity of flowshop and jobshop scheduling. Mathematics of Operations Research 1(2), 117–129 (1976)
6. Garey, M.R., Johonson, D.S.: Computers and Intractability - A Guide to the Theory of NP-Completeness. Freeman and Company (1979)
7. Adams, J., Balas, E., Zawack, D.: The Shifting Bottleneck Procedure for Job Shop Scheduling. Management Science 34(3), 391–401 (1988)
8. Van Laarhoven, P.J.M., Aarts, E.H.L., Lenstra, J.K.: Job Shop Scheduling by Simmurated Annering. Operations Research 40(1), 113–125 (1992)

9. Furukawa, M., Matsumura, Y., Watanabe, M.: Development of Local Clustering Organization Applied to Job-shop Scheduling Problem. Journal of the Japan Society for Precision Engineering (CD-ROM) 72(7), 867–872 (2006)
10. Tamura, Y., Suzuki, I., Yamamoto, M., Furukawa, M.: The Hybrid Approach of LCO and SA to Solve Job-shop Scheduling Problem. Transactions of ISCIE 26(4), 121–128 (2013)
11. Kohonen, T.: The self-organizaing map. Neurocomputing 21(1-3), 1–6 (1998)
12. Furukawa, M., Watanabe, M., Matsumura, Y.: Lcoal Clustering Organization (LCO) Solving a Large-Scale TSP. Journal of Robotics and Mechatronics 17(5), 560 (2005)
13. Bierwirth, C.: A generalized permutation approach to job shop scheduling with genetic algorithms. Operations-Research-Spektrum 17(2-3), 87–92 (1995)
14. Tamura, Y., Yamamoto, M., Suzuki, I., Furukawa, M.: Acquisition of Dispatching Rules for Job-shop Scheduling Problem by Artificial Neural Networks Using PSO. Journal of Advanced Computational Intelligence and Intelligent Informatics 17(5), 731–738 (2013)
15. Lawrence, S.: Resource constrained project scheduling: An experimental investigation of heuristics scheduling techniques. Technical report, GSIA, Carnegie Mellon University (1984)
16. Jain, A.S., Meeran, S.: Deterministic job shop scheduling: past, present and future. European Journal of Operational Research 113(2), 390–44 (1999)

Online Game Bot Detection in FPS Game

Mee Lan Han, Jung Kyu Park, and Huy Kang Kim[*]

Graduate School of Information, Korea University, Seoul, South Korea
{blosst,pjk0825,cenda}@korea.ac.kr

Abstract. There are many different game genres such as MMORPG, FPS, racing games, sports games and mobile games in the online game market. Various forms of security threats exist on the game market. Among the security threats, the use of game bot can cause great damage to the game service. The security threats make that game user lose interest in playing. Also due to the security threats, the cycles of the game could be reduced. The security threats cause baleful influences such as a weakened confidence, pecuniary damage to the game operator. In this paper, we analyzed a special feature of FPS game bot through Point Blank. On the basis of the outcome of the analysis, we composed an analysis method for FPS game bot detection. Because behavioral patterns of game bot user are different with that of normal users, we set the threshold value by analyzing the behavior of game bot user for game bot detection. By applying FPS game bot detection framework proposed in this paper, we could extract the game bot users five times more than game bot user extracted via the commercial bot detection tool and publisher policy.

Keywords: Online Game Security, FPS, Game Bot Detection, Data Mining.

1 Introduction

Online game market in both of PC and mobile computing environment has been increased dramatically. As an unwanted result, many illegal activities such as using cheating tools or game bots are also increased. Game bot is a highly well-crafted AI program that can help users do in-game activities automatically. Usually, game bots are used for illegal activities, such as automatic gold farming or cheating by modifying critical values in PC's memory [1,2,3]. Now game bots become the most-annoying thing to make innocent users feel unfair and decide to leave the game finally.

Innocent users feel unfair and complain if online game companies fail to detect and prevent game bots proactively. However, it is hard to detect and prevent game bots. To increase the detection ratio of game bots become a great burden on the game operators and game publishers.

There were many efforts to combat this problem. For the purpose of banning game bot, the game operator can impose sanctions on the account of game bot users. Many

---

[*] Corresponding author.

© Springer International Publishing Switzerland 2015
H. Handa et al. (eds.), *Proc. of the 18th Asia Pacific Symp. on Intell. & Evol. Systems – Vol. 2*,
Proceedings in Adaptation, Learning and Optimization 2, DOI: 10.1007/978-3-319-13356-0_38

companies use the commercial bot detection tool like GameGuard [4] and Hackshield [5] in other to detect game bot to evolve diversely. The commercial bot detection tools detect and block the game bot's process through the memory and process monitoring methods (similar with Antivirus program's monitoring mechanism) such as signature-based detection. However, these methods can cause other problems. One of the problems was that it created a new account via the account theft. A lot of game developers are trying to find countermeasures but the game bot makers come up with more cunning measures in order to make game bot users not to be detected. That is; spear and shield is continuing the fight.

A game bot gives the benefit to game bot users and provides game bot users with the convenience. In FPS game, game bots such as Aimbot, Wallhack, Speed hack, weapon hack, EXP hack are mainly used. The Aimbot is game bot that it takes aim at the opponent user automatically when user faces the opponent user. It is the key to the victory of the game to make an attack against opponent users quickly. If user uses Aimbot, irrespective of his skill, he can quickly and easily achieve a victory. The Wallhack is game bot that lets user know opponent user's position. Because it can know in advance the position of the opponent user by using the Wallhack, Wallhack can shoot at the opponent user immediately without exposing user position. The speed hack improves the speed of moving much faster than usual. The weapon hack makes user use a weapon that did not buy before and makes user change properties of the weapon. The evolution of a variety of game bot deals a severe blow to the game operators.

FPS game mode is that users fight against opponent users unlike MMORPG. So, it is impossible to find a one-way relationship between user and mobs. It is hard to identify any problems with game bot until finding a two-way relationship that occurred between them. The behavior in FPS game varies and is fast considerably and it is not easy to find suitable features for game bot detection. The game data can be detected by a game bot via a variety of algorithms, such as data mining or logistic regression analysis. However, if we can find the behavioral differences between normal user and game bot user, it will be able to be effective countermeasures to create a rule of game bot detection algorithm.

In this study, we propose a game bot detection method based on user behavior analysis for FPS. In Section 2, we review previous studies related to game bot detection. In Section 3, we present significant features and method for game bot detection. We address the details of our proposed game bot detection. In Section 4, we evaluate our proposed method from a popular FPS game. Finally, in Section 5, we discuss our findings and future work.

2   Related Work

2.1   A Game Bot Detection Based on Data Source

The game bot detection based on user behavior is the method using different behavioral characteristics of normal user and game bot user. Significant features are found via statistical analysis and data mining [6,7,8].

In MMORPG, the difference between the game bot and the normal user is clearly revealed through the frequency and the distribution of data. Fig. 1. shows the number of action of normal user and game bot user in the same time in the MMORPG. It shows a notable difference at the hunting frequency and action of them. It can be seen that game bot does a lot of action and hunts a lot of mobs more than the normal user [9].

Depending on the property of online game, the type and the frequency of the actions performed are different. In FPS game, the accuracy of the aiming to the target, the impact of the weapon, the target time and shoot ratio, aiming hour toward target / target check time can be a feature for game bot detection [10]. The processing speed slows and it is impossible to deal with in real time at the methods based on user behavior, but the methods based on user behavior can control the scale and time of sanction. Also, because it has a high accuracy and it is not exposed on detection rule, the method is used in many.

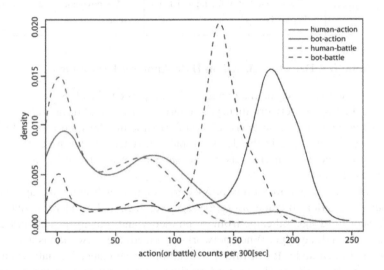

Fig. 1. The distribution of behavior between game bot and normal user in MMORPG

The game bot detection based on the movement routes is a method that can control the scale and time of sanction. It is a method that extracts their routes under the premise that route of the game bot user and the normal user is different [11]. The game bot detection based on the traffic is a method that does not need to modify data of the server or the client. It does not have a lot of data in order to continue the analysis and it does not need many operations in the analysis relatively [12,13]. The game bot detection based on the user event has the two methods. One is a method to observe the input of the user and the other is a method to analyze the window event sequence [14,15,16]. CAPTCHA is more effective to use as a supplementary means rather than using an absolute means [17,18].

Table 1. The pros and cons of the method of   game bot detection in FPS game

Method	Pros	Cons
Based on User behavior[10][11]	- Control of scale and time for sanction - The game bot detection rule is easily unexposed. - Higher accuracy	- Necessity of a large amount of data - Relatively slow speed - The impossibility of real time response
Based on statistical analysis[16][25]	- An unnecessary server patch - The possibility of detecting the other game bot by the same method - Higher accuracy	- The difficulty of choosing the scale and the data
CAPTCHA[18][22]	- The counterfeit of the input data is Impossible.	- A annoying job that has to use a token

2.2   A Game Bot Detection Based on Data Analysis Technique

A game bot detection based on data analysis techniques can be classified by data mining, statistical analysis, network theory, similarity analysis and turing test. Data mining is a technique to find a useful correlation for extracting the significant information in the data base and to use in the decision making. After finding the correlation and behavior pattern between normal user and game bot user, it can be applied to the game bot detection [10].

Statistical analysis is a method expressed by graph after finding average, variance and standard deviation. It can analyze average, variance and standard deviation [16]. Network theory is a method came from graph theory and it analyzes network consisting of the node and the edge. When network is formed, the network has a close interrelationship of each node. It can find a correlation of the respective data via analysis of centrality of the network. It can find an Eigenvector centrality of the network via that mentoring network was formed by mentoring-interaction among a novice and a veteran [19]. Similarity analysis is a method that analyzes the similarity among a game bot and normal users. It can find a certain pattern of game bot by measuring the similarity from user behavior and its frequency. If there is the log about user's skill information and movement information, game bot is able to detect due to the high similarity [20,21]. Turing test is a method that distinguishes the intelligence of a machine based on how machines can react like human. CAPTCHA is used at the online game. Human can easily recognize When saw something but machine cannot easily recognize that [17], [18], [22].

3   Methods for Game Bot Detection

In this paper, we built a game bot detection rule using the domain knowledge of Point Blank, a popular FPS game served by Zepetto. We analyzed the difference between normal user and game bot user by applying statistical method to the user behavior. From joining in on the game until leaving the game, the huge amount of game data is accumulated by events of the game user occurring. There is a complex and hidden relationship between the game data due to a lot of game user simultaneously playing a game. The statistical method is valuable in extracting the efficient data and in observing the significant relationship at the huge amount of game data. Based on the findings of analysis, we found the FPS game bot detection method. The Point Blank is a first-person shooting game served by Zepetto. The Point Blanks is offering service into many countries via ten different game publishers until Jan. 2013. In excess of concurrent users of two hundred thousand people, Point Blank had erected the highest record in the whole country in Indonesia. Three consecutive years since 2010, Point Blank received best-casual game award and FPS game award in TGS (Thailand Game Show) and BIG (Bangkok International Game Festival), a very famous game show in Thailand. Point Blank has set a record that the concurrent user reached approximately fifty thousand people in Brazil. It is leading a Korean wave of FPS game [23,24]. We received the dataset of Point Blank from Zepetto. The dataset includes varied information on the behavior of game users; Login, Logout, Ranking, game money, EXP, battle and so forth. It is that the behavior of game user is recorded in the dataset from 1 February to 30 April day after day. Unique users in the game for 89 days are 1,230,197. In the case of game bot user, we can presume that win and headshot ratio may be higher than the general user. Also, we can presume that the game bot user's game play time may be shorter than the general user's. The exhibitionism is why the game users use a game bot in FPS. The game user using a game bot would like to play a game better than other users. To win games in clan (The clan in FPS is a kind of alliance formed from the game users.) or battle regardless of rank is the biggest purpose of using a game bot. It is efficient that the game user uses a game bot having the function of headshot to win, and they get in a fight with the opponent users by the few shooting. However, the FPS patch processing unlike the MMORPG patch processing is completed in a short time. Therefore, the game user using a game bot play the game for a short time to dodge the spotlight of GM or opponent users. They put themselves in a better position by raising win rate in clan or battle. Based on the variety of presumption, we set threshold value by analyzing a game bot user log through the commercial bot detection tool and game publisher policy. In order to identify the game bot user, the threshold values applied to the normal user log using the average value. The classification of the similar user behavior in FPS game is shown in Table 2. There are only exploration, combat and social activity in FPS game. User behavior as collecting game items or producing game items was not found.

Table 2. Classification of user behavior in FPS game

Exploration	Combat	Socializing
Complete Mission Get EXP Get Point Get Item	PVP Clan Matching	Chatting Clan Item Gift

3.1   Feature Selection

- Feature 1 : Win rate

The result of the online game is divided into win, defeat and draw. User's accumulated results are described by win rate, they can directly or indirectly affect the acquisition of game title and game items or EXP. Because the win rate can affect the clan matching or team matching, user's exhibitionism was incited by the win rate. Game users choose cheating at game rather than having fun. Therefore, we can presume that game bot user's win rate is higher than normal user's. In this paper, we set the threshold value by assigning the actual log of game bot user to the formula of win rate.

$$\text{Win rate} = (\text{Win count} / \text{Total play count}) \times 100 \tag{1}$$

- Feature 2 : Headshot Ratio

Headshot at the FPS game means when shooting at kill the specific part of opponent. Because user has to take aim at the smallest part of opponent user body, headshot technic is different with the other attack and it is difficult rather than them. So, In case of shooting the headshot, attack user can cause great damage and they can get a greater reward. Normal user improves their headshot skill by equipping with a weapon or by practicing the skill, but another users use Aimbot for increasing their skills easily. In this paper, we set the threshold value by measuring the average value using a headshot ratio of total kill count of game bot user.

- Feature 3 : Play Counts and Play Time

Game play time is set to zero when access failure problem occurs and game play time is set to zero when user gives up the game. Except in the case of these problems, all data are included in data for analysis. FPS game is able to patch at any time and is able to change game data. Also, FPS operators are able to react quickly for game bot users. For this reason, game bot user possesses the next characteristics; game bot user ends the game rapidly; game bot user frequently plays; game bot user plays the game without being seen. In this paper, we set the threshold value by measuring the average value of play-count and play-time per day.

- Feature 4 : EXP Variation and Game Money Variation

Game user can get EXP and game money depending on various play action such as mission, rank up, medal, PVP and so on. Because game bot user has a much higher skill for playing game than the normal user, mission result of game bot user is different from mission result of normal user and the win rate of game bot user is higher than the win rate of normal user. Also owing to a variety of reasons, the difference between the game bot user and the normal user as to user behavior affects the generation of EXP and game money. In this paper, we set the threshold value by measuring the variation of EXP and game money during the day.

- Feature 5 : A series of victories

FPS is a game that fights against opponent users. It is not a game that the user can hunt mobs generated by constant pattern like MMORPG. Because there are many obstacles in the game map, it is difficult for users to predict where the opponent users are. Because users frequently have to move through little doors or narrow passages, it is not easy for normal users to win. Therefore, users using a Wallhack or Aimbot are more likely to win regardless of map's hurdle. Wallhack allows user to shoot through solid or opaque objects. Aimbot allows user to lock on the opponent users and kill them. In this paper, we set the threshold value by measuring the straight win count during the day.

3.2   Bot Detection Method

We schematize our proposed the FPS bot detection procedure in Fig. 2. First, we gather game log data of game bot user from data detected by FPS game publisher policy or the commercial bot detection tool. We extract action log centered on user behavior from playing log remained after ending game and then select significant classifiers for analysis. Second, as user behavior analysis stage, we combine various features selected from the previous step. We choose features for analysis of game bot user behavior and establish the threshold for making up a game bot detection rule. Although we can establish higher confidence interval for reducing False-Positive ratio, we applied average value of final features. As the evaluation stage, we extract potential game bot users by applying the game bot detection rule to normal user data. We compare the banned account list with the game bot user list detected by game bot detection rule. Applying the threshold value of features to game bot detection rule is shown in Table 3. It is just a feature that the average Play Time per day is lower than the threshold value. The each value of all the other features is higher than the threshold value. We can extract game bot users only after the threshold values are satisfied.

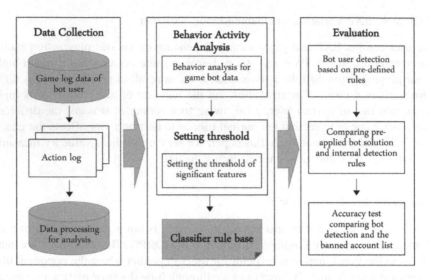

Fig. 2. FPS Bot Detection Procedure

Table 3. Game Bot Detection Rule

Rule-base
Win rate > 0.434 and
Headshot Ratio > 0.354 and
The average daily Play Counts > 15.71 and
The average daily Play Time < 7.47 and
EXP Variation > 204.90 and
Game Money Variation > 160.83
A series of victories > 9.709

4   Result

We extracted potential game bot users from normal user log via the proposed game bot detection rule. The extracted users are a data set that satisfies all threshold values of the rule. If the threshold values are applied conservatively, False-Positive ratio can be diminished by applying higher than average values or by increasing the number of feature for game bot detection rule. But it is difficult to diminish True-Negative ratio. On the other hand, if the threshold values are applied by lower than average values or are applied by decreasing the number of feature for game bot detection rule, True-Negative ratio can be diminished but False-Positive ratio can be increased. Game bot user count detected by the commercial bot detection tool and game publisher policy and game bot user count detected by game bot detection rule proposed in this paper is shown in Table 4. We were able to check the result that game bot detection ratio detected by the proposed rule is 5.45 times higher than game bot detection ratio detected by FPS game publisher policy and the commercial bot detection tool. Detection ratio detected by the commercial bot detection tool and game publisher policy is 1.63 percent. Detection ratio detected by game bot detection rule proposed in this paper is 8.92 percent.

The qualities of a game bot user were not conspicuous when data of game bot user was contained in normal data. We did not know the qualities of normal user and game bot user until we extract the game bot data. Although detection ratio is 8.92 percent of total game user, we could find out the definite difference by measuring the average of play count, playtime and session time. Game play time and game play count about the game bot user and the normal user are shown in Table 5.

Table 4. Game bot detection ratio

	Detected user / Total user	(%)
Game bot user count detected by the commercial bot detection tool and game publisher policy	18,777 / 1,149,166	1.63%
Game bot user count detected by game bot detection rule proposed in this paper	102,482 / 1,149,116	8.92%

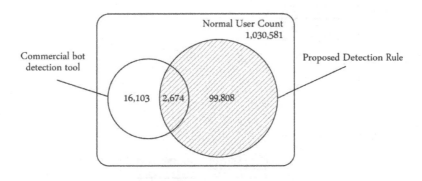

Fig. 3. Game Bot Classification results of the proposed method and the commercial bot detection tool

The average amount of session time spent on game play per day (5.48 minute) was small. The average of game play count per day that detected by game bot detection rule was 3 times more than the normal user. The average amount of time spent on game play per a day (3.77 hours) was longer than the other method.

Table 5. Game play time and game play count about the game bot user and the normal user

	Data analysis of user detected by the game bot detection rule	Normal user	Data analysis of user detected by the commercial bot detection tool	Data analysis of user detected by the game publisher policy
The average of game play count per day	49.55 (times)	15.79 (times)	24.6 (times)	23.6 (times)
The average amount of time spent on game play per day	3.77 (hour)	1.53 (hour)	2.65 (hour)	2.49 (hour)
The average amount of session time spent on game play per day	5.48 (minute)	6.56 (minute)	7.38 (minute)	7.43 (minute)

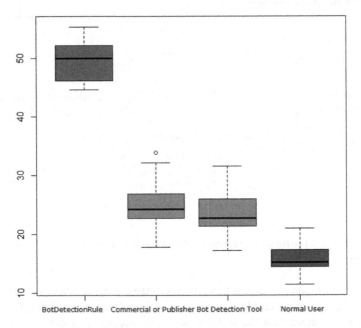

Fig. 4. Game play count about game bot user and normal user

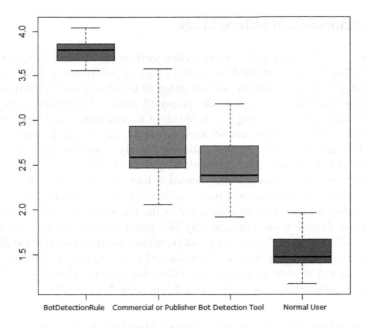

Fig. 5. Game play time about game bot user and normal user

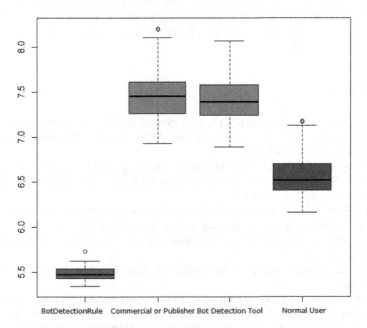

Fig. 6. Game session time about game bot user and normal user

5   Conclusion and Future Work

In this study, we propose a game bot detection method based on user behavior analysis for FPS. The proposed method will help FPS game publishers and the commercial bot detection tool to overcome the shortcomings of client side and network side game bot detection methods. We applied the proposed method to in-game logs of Point Blank in Zepetto. After analyzing the behavior of normal user and game bot user, we selected significant classifiers and set their threshold values and detected game bots based on the rule base lastly. By using statistical analysis and box plot, we conveyed the meanings of the result. We evaluated the game bot detection ratio of users who were detected by the commercial bot detection tool and game publisher policy and detected by game bot detection rule proposed in this paper. We were able to check the result that game bot detection ratio detected by the proposed rule is 5.45 times higher than game bot detection ratio detected by FPS game publisher policy and the commercial bot detection tool. In our future study, we intend to incorporate the distinctive features such as the rank-up log, Get-Item log and Clan log into the existing features. Further, we expect to improve the game bot detection ratio by adding game bot IP and user accounts gathered from web site that can buy game bot program.

Acknowledgement. This research was supported by Basic Science Research Program through the National Research Foundation of Korea (NRF) funded by the Ministry of Science (NRF-20100004910 and NRF-2010-330-B00028), ICT & Future Planning (2014R1A1A1006228), and also supported by Ministry of Culture, Sports and Tourism (MCST) and Korea Creative Content Agency (KOCCA) in the Culture Technology (CT) and Research Development Program 2014, and IT R&D program of MSIP/KEIT (10045459).

References

1. Woo, J., Kang, A.R., Kim, H.K.: The contagion of malicious behaviors in online games. In: SIGCOMM 2013 Proceedings of the ACM SIGCOMM 2013 Conference on SIGCOMM, pp. 543–544 (2013)
2. Woo, J., Kang, A.R., Kim, H.K.: Modeling of bot usage diffusion across social networks in MMORPGs. In: WASA 2012 Proceedings of the Workshop at SIGGRAPH Asia, ACM Digital Library, pp. 13-18 (2012)
3. Seo, D., Kim, H.K.: Detecting Gold-farmers' Groups in MMORPG by connection information. In: Proceedings of the 3th International Conference (2011)
4. nProtect GameGuard, INCA, http://www.nprotect.com/v7/b2b/sub.html?mode=game#ui
5. HackShield, AhnLab, http://hackshield.ahnlab.com/hs/site/en/main/main.do
6. Kang, A.R., Woo, J., Park, J., Kim, H.K.: Online game bot detection based on party-play log analysis. Computers & Mathematics with Applications, 1384–1395 (2013)
7. Ruck, T., Yoshitaka, K., Kuan, T.C.: Detection of MMORPG Bots Based on Behavior Analy-sis. In: Proceedings of the 2008 International Conference on Advances in Computer Entertainment Technology, pp. 91–94 (2008)

8. Yeounoh, C., Chang-yong, P., Noo-ri, K., Hana, C., Taebok, Y., Hunjoo, L., Jee-Hyong, L.: Game Bot Detection Approach Based on Behavior Analysis and Consideration of Various Play Style 35 (2013)

9. Yutaro, M., Kensuke, F., Hiroshi, E.: An analysis of players and bots behaviors in MMORPG. In: IEEE 27th International Conference on Advanced Information Networking and Applications, pp. 870–876 (2013)

10. Hashem, A., Fotos, F., Clifford, N.: Behavioral-Based Cheating Detection in Online First Person Shooters using Machine Learning Techniques. In: Computational Intelligence in Games (CIG), pp. 1–8 (2013)

11. Drachen, A., Sifa, R., Bauckhage, C., Thurau, C.: Guns, Swords and Data: Clustering of Player Behavior in Computer Games in the Wild. In: Computational Intelligence and Games (CIG), pp. 163–170 (2012)

12. Kuan-Ta, C., Jhih-Wei, J., Polly, H., Hao-Hua, C., Chin-Laung, L., Wen-Chin, C.: Identifying MMORPG Bots: A Traffic Analysis Approach. EURASIP Journal on Advances in Signal Processing Archive (2009)

13. Sylvain, H., Hyun-chul, K., Chong-kwon, K.: How to deal with bot scum in MMORPGs. In: IEEE Communications Quality and Reliability (CQR) Workshop, pp. 8–10 (2010)

14. Kim, H., Hong, S., Kim, J.: Detection of Auto Programs for MMORPGs. In: Zhang, S., Jarvis, R.A. (eds.) AI 2005. LNCS (LNAI), vol. 3809, pp. 1281–1284. Springer, Heidelberg (2005)

15. Steven, G., Zhenyu, W., Mengjun, X., Haining, W.: Battle of Botcraft: Fighting Bots in Online Games with Human Observational Proofs. In: Proceedings of the 16th ACM Conference on Computer and Communications Security, pp. 256–268 (2009)

16. Su-Yang, Y., Nils, H., Jeff, Y., Peter, A.: A statistical Aimbot detection method for online FPS games. In: The 2012 International Joint Conference on Neural Networks (IJCNN), pp. 1–8 (2012)

17. Roman, V.Y., Venu, G.: Embedded Non-interactive Continuous Bot Detection. ACM Computers in Entertainment (2008)

18. Philippe, G., Nicolas, D.: Preventing Bots from Playing Online Games. In: Computers in Entertainment (CIE) (2005)

19. Jehwan, O., Borbora, Z.H., Sharma, D., Srivastava, J.: Bot Detection based on Social Interactions in MMORPGs. In: 2013 International Conference on Social Computing (Social Com), pp. 536–543 (2013)

20. Platzer, C.: Sequence-Based Bot Detection in Massive Multiplayer Online Games. In: 2011 8th International Conference on Information, Communications and Signal Processing (ICICS), pp. 1–5 (2011)

21. Muhammad, A.A., Brian, K., Atanu, R., Dmitri, W., Jaideep, S., Noshir, C.: Guilt by Asso-ciation Network Based Propagation Approaches for Gold Farmer Detection. In: 2013 IEEE/ACM International Conference on Advances in Social Networks Analysis and Mining (2013)

22. McDaniel, R.C., Roman, V.Y.: Development of Embedded CAPTCHA Elements for Bot Prevention in Fischer Random Chess. International Journal of Computer Games Technology (2012)

23. Point Blank FPS Game Information, http://www.fps-pb.com/

24. Point Blank FPS Game Information, http://en.wikipedia.org/wiki/Point_Blank_(2008_video_game)

25. Yan, J.: Bot, cyborg and automated turing test. In: Christianson, B., Crispo, B., Malcolm, J.A., Roe, M. (eds.) Security Protocols. LNCS, vol. 5087, pp. 198–201. Springer, Heidelberg (2009)

An Extended SHADE and Its Evaluations

Kazuhiro Ohkura[1], Toshiyuki Yasuda[1],
Masaki Kadota[1], and Yoshiyuki Matsumura[2]

[1] Hiroshima University, Higashi-Hiroshima, Hiroshima, 739-8527, Japan
`kohkura@hiroshima-u.ac.jp`,
[2] Shinshu University, Ueda, Nagano, 386–8567, Japan

Abstract. Differential Evolution (DE) is a population-based stochastic search method for real-valued function optimization. Similar to other meta-heuristic algorithms, DE searches for optimal or near-optimal solutions without a priori knowledge of the function being optimized. However, DE generally shows largely different performances on the basis of the DE parameters adopted. Therefore, many DE variants, including one called SHADE—one of the most highly efficient DEs so far, have been developed to obtain a more stable and better performance. SHADE introduces parameter archives for robust parameter adaptations. In this paper, an extended form of SHADE is proposed, in which parameter archives are managed by a novel strategy with three new operators. Computer simulations are conducted to examine the performance of the proposed DE on the basis of 28 benchmarks.

Keywords: Evolutionary Computation, Differential Evolution, Numerical Optimization.

1 Introduction

Differential evolution (DE) is a simple yet evolutionary algorithm for real-valued function optimization problems [6], [4], [1]. DE has fundamental control parameters, which are the population size (NP), crossover rate (CR), and scaling factor (F). Although DE variants are fixed in classical DE, many DE variants, such as [8], [5], [9], [3], and [7]—in which the control parameters are variables during calculations, have been proposed. The motivation behind this is that control parameter tuning is recognized as one of the key issues for obtaining better optimization performance.

In this paper, a new DE, inspired by JADE [9] and SHADE [7], is proposed, and its performance is examined on the basis of 28 test functions adopted by CEC2013. The remainder of this paper is organized as follows. In the next section, the basic DE algorithm is reviewed. In Section 3, JADE, which is our foundation, is briefly introduced. Section 4 introduces SHADE, which is a DE variant based on JADE. Section 5 presents the details of our proposed DE, RSHADE, which is inspired by SHADE. In Section 6, computer simulations are conducted using a set of 28 benchmarks to examine the performance of the proposed method. Finally, Section 7 shows the conclusions.

© Springer International Publishing Switzerland 2015   493
H. Handa et al. (eds.), *Proc. of the 18th Asia Pacific Symp. on Intell. & Evol. Systems – Vol. 2*,
Proceedings in Adaptation, Learning and Optimization 2, DOI: 10.1007/978-3-319-13356-0_39

2   Differential Evolution

Similarly to conventional evolutionary computation techniques, DE is composed of three procedures: mutation, crossover, and selection. An individual component in DE is represented by a real-valued vector $x_i = \{x_0, x_1, \ldots, x_{D-1}\}$, $(i = 0, 1, 2, \ldots, NP - 1)$, where D and NP are the dimensions defined by a given problem and the population size, respectively. Each individual component is initialized before the evolutionary search.

2.1   Mutation

At each generation G, a mutation vector, $v_{i,G}$, is generated for each individual $x_{i,G}$. For example, three well-known mutation strategies with a mutation vector are described as follows:

"DE/rand/1"

$$v_{i,G} = x_{r1,G} + F \cdot (x_{r2,G} - x_{r3,G}),$$

"DE/best/1"

$$v_{i,G} = x_{\text{best},G} + F \cdot (x_{r1,G} - x_{r2,G}), \text{ and}$$

"DE/current to best/1"

$$v_{i,G} = x_{i,G} + F \cdot (x_{\text{best},G} - x_{i,G}) + F \cdot (x_{r1,G} - x_{r2,G}),$$

where $r1, r2, r3$, and $r4$ are uniformly and randomly chosen integers from the set $\{0, 1, 2, \cdots, NP - 1\} \setminus \{i\}$ and assumed to be different from each other. In addition, $x_{\text{best},G}$ represents the best individual component for the current generation G.

2.2   Crossover

After executing a mutation operation to generate a mutation vector $v_{i,G}$, a crossover operator is applied to generate a trial vector $u_{i,G}$. A common crossover method is a binomial crossover described by the following equation:

$$u_{j,i,G} = \begin{cases} v_{j,i,G} & \text{if } \text{rand}\,[0,1) \leq CR \text{ or } j = j_{\text{rand}} \\ x_{j,i,G} & \text{otherwise} \end{cases}, \tag{1}$$

where $\text{rand}\,[a, b)$ is a uniform random number in the interval $[a, b)$, CR is the crossover rate whose value is $\in [0, 1]$, j is the jth dimension, and j_{rand} is a random integer $\in [0, D - 1]$ generated for each i.

2.3   Selection

After applying crossover, the selection operation is executed. This operation selects the best solution from the parent vector $x_{i,G}$ and the trial vector $u_{i,G}$, from the calculated values $f(\cdot)$. For instance, in the case of the minimization problems,

$$x_{i,G+1} = \begin{cases} u_{i,G} & \text{if } f(u_{i,G}) < f(x_{i,G}) \\ x_{i,G} & \text{otherwise} \end{cases}. \tag{2}$$

After applying the selection operation for all the individual components, the computation for the generation is finished and then reiterated to compute the same procedure again until the termination criterion is satisfied.

3   JADE

JADE [9] is a DE variant that has a mutation strategy "DE/current-to-$p$best" with optional archive, and that independently controls CR and F for each individual component in an adaptive manner. An individual component is represented in the same way as that in the classical DE. In the following, the population is represented by P. When the archive option is used, archive A is defined as the archive that stores the individuals that are discarded at the selection process. Therefore, $A = \emptyset$ at the beginning. The upper limit of the archive is called the archive size $|A|$, which is commonly set to NP. In cases where the archive size reaches the upper limit, an element randomly chosen from A is discarded to store a new candidate.

3.1   Parameter Adaptation

Two variables, μ_{CR} and μ_F, are provided at the initialization process and each is set at 0.5. At each generation G and for each individual i, two parameters, CR_i and F_i, are calculated by the following equations:

$$CR_i = \text{randn}_i(\mu_{CR}, 0.1) \tag{3}$$

$$F_i = \text{randc}_i(\mu_F, 0.1) \tag{4}$$

This means that CR_i is a random number independently generated on the basis of a normal distribution of mean μ_{CR} and standard deviation 0.1. CR_i is then truncated $[0, 1]$. Similarly, F_i is a random number independently generated on the basis of a Cauchy distribution with location parameter μ_F and scale parameter 0.1. F_i is then truncated as 1 when $F_i > 1$, or regenerated when $F_i \leq 0$.

3.2   Mutation

The mutation vector is generated by the strategy "DE/current-to-$p$best/1."

$$v_{i,G} = x_{i,G} + F_i \cdot (x_{p\text{best},G} - x_{i,G}) + F_i \cdot (x_{r_1,G} - x_{r_2,G}), \tag{5}$$

where $x_{r_1,G}$ is an individual component randomly selected from P and $x_{r_2,G}$ is an individual selected randomly from the union set $P \cup A$. When the archive is not used, $A = \emptyset$. $x_{pbest,G}$ is an individual component randomly chosen from the top $100p\%$ individuals in P with $p \in (0, 1]$. p is a constant.

3.3   Crossover

Crossover is applied in the same way as Equation (1) except that CR_i has to be used. Therefore,

$$u_{j,i,G} = \begin{cases} v_{j,i,G} & \text{if } \text{rand}\,[0,1) \le CR_i \quad \text{or} \quad j = j_{\text{rand}} \\ x_{j,i,G} & \text{otherwise} \end{cases}. \tag{6}$$

3.4   Selection

Equation (2) is used for selection. CR_i and F_i for generating $u_{i,G}$ are stored in $S_{CR,G}$ and $S_{F,G}$, respectively, only in cases where $x_{i,G+1} = u_{i,G}$. When the archive option is used, $x_{i,G}$ is saved in A.

When $S_{CR,G}$ and $S_{F,G}$ are not empty, the following parameter adaptations are calculated:

$$\mu_{CR,G+1} = (1 - c) \cdot \mu_{CR,G} + c \cdot \text{mean}_A(S_{CR,G}) \tag{7}$$
$$\mu_{F,G+1} = (1 - c) \cdot \mu_{F,G} + c \cdot \text{mean}_L(S_{F,G}), \tag{8}$$

where $c \in (0, 1)$ is a constant. $\text{mean}_A(S_{CR,G})$ is the arithmetic mean $S_{CR,G}$ and $\text{mean}_L(S_{F,G})$ is the Lehmer mean calculated as follows:

$$\text{mean}_L(S_{F,G}) = \frac{\sum_{F \in S_{F,G}} F^2}{\sum_{F \in S_{F,G}} F} \tag{9}$$

Then, generation is renewed to calculate the next generation.

4   SHADE

SHADE [7] is proposed as an extension of JADE, but it adopts a history of parameter adaptation of JADE. Another new aspect is that a constant p controlling the greediness at the mutation strategy is replaced with a variable pi that is randomly generated for the purpose of reducing the number of parameters that a user has to define.

4.1   Parameter Adaptation

In SHADE, two memory areas, M_{CR} and M_F, are provided for storing the parameters used. Each memory has a maximum memory size H.

$$M_{CR} = \{M_{CR,0}, M_{CR,1}, \ldots, M_{CR,H-1}\} \tag{10}$$
$$M_F = \{M_{F,0}, M_{F,1}, \ldots, M_{F,H-1}\} \tag{11}$$

At the initialization stage, all values are set at 0.5. They are updated in the search process as follows:

$$M_{CR,k,G+1} = \begin{cases} \text{mean}_{WA}\,(S_{CR,G}) \text{ if } S_{CR,G} \neq \emptyset \\ M_{CR,k,G} \qquad\qquad \text{otherwise} \end{cases} \tag{12}$$

$$M_{F,k,G+1} = \begin{cases} \text{mean}_{WL}\,(S_{F,G}) \text{ if } S_{F,G} \neq \emptyset \\ M_{F,k,G} \qquad\qquad \text{otherwise} \end{cases}, \tag{13}$$

where $k \in [0, H-1]$ is the index showing the current place in memory. Therefore, $M_{CR,k,G}$ indicates the kth element in memory area M_{CR} for storing the value of CR at generation G. k is set to 0 at the initialization and incremented as the update occurs. k is set to 0 again when it is incremented when $k = H - 1$. $\text{mean}_{WA}(S_{CR,G})$ and $\text{mean}_{WL}(S_{F,G})$ are weighted means calculated by Equations (14) and (15), respectively.

$$\text{mean}_{WA}\,(S_{CR,G}) = \sum_{q=1}^{|S_{CR,G}|} w_q \cdot S_{CR,G,q}$$
$$w_q = \frac{\Delta f_q}{\sum_{q=1}^{|S_{CR,G}|} \Delta f_q} \tag{14}$$

$$\Delta f_q = |f\,(u_{q,G}) - f\,(x_{q,G})|$$
$$\text{mean}_{WL}\,(S_{F,G}) = \frac{\sum_{q=1}^{|S_{F,G}|} w_q \cdot S_{F,G,q}^2}{\sum_{q=1}^{|S_{F,G}|} w_q \cdot S_{F,G,q}} \tag{15}$$

Therefore, in the same manner as that in Equations (3) and (4) in JADE, M_{CR} and M_F are calculated as follows:

$$CR_i = \text{randn}_i\,(M_{CR,r_i}, 0.1) \tag{16}$$
$$F_i = \text{randc}_i\,(M_{F,r_i}, 0.1), \tag{17}$$

where $r_i \in [0, H-1]$ is a random number.

4.2   Mutation

SHADE has almost the same mutation strategy "DE/current-to-$p$best/1," which is adopted in JADE, except that a constant p in Equation (5) is replaced with:

$$p_i = \text{rand}\,[p_{\min}, 0.2], \tag{18}$$

where $p_{\min}$ is selected such that the mutation is practically calculated (refer to [7] for further details). To confirm the execution, we need the assumption that $NP \geq 10$.

5   Proposed Method: RSHADE

We did several computer simulations with the benchmarks [2] using JADE and SHADE, and confirmed that SHADE gives better results than JADE. However, we found unwanted behavior in SHADE: the parameter adaptation by Equations (12) and (13) sometimes fails to occur and the evolutionary search gets stuck. Therefore, in this paper, the method of parameter adaptation is extended using RSHADE (Robust SHADE).

5.1 Parameter Adaptation

An array, $M_r = \{r_0, r_1, \ldots, r_{NP-1}\}$ is prepared for managing the read from M_{CR} and M_F. $M_{r,i}$ is an integer between 0 and $H-1$. The ith element is the index that defines the read from M_{CR} and M_F. $M_{r,i}$, randomly initialized at the beginning, is managed by the operators described below. Thus, instead of Equations (16) and (17) for SHADE, CR_i and F_i are respectively calculated as follows:

$$CR_i = \mathrm{randn}_i\left(M_{CR,M_{r,i}}, 0.1\right) \tag{19}$$

$$F_i = \mathrm{randc}_i\left(M_{F,M_{r,i}}, 0.1\right) \tag{20}$$

The steps of mutation and crossover are the same as those in SHADE. Selection is also the same as that using DE with Equation (2). For parameter adaptation, instead of using Equations (7) and (8) for JADE, the following three operators are introduced: Op_1, Op_2 and Op_3. In the selection, when a newly generated trial vector survives the selection process, one of these operators is applied with a predefined probability. Otherwise, Op_2 is applied.

5.2 Three New Operators

Suppose that k_G is the index of the writing position to M_{CR} and M_F at generation G.

Op_1: Overwrite the elements of M_{CR} and M_F indexed by k_G with the current values. Increment k_G. Update $M_{r,i}$ with k_G.

$$\begin{aligned}
M_{CR,k_G} &= \mathrm{mean}_{WA}\left(S_{CR}\right)\\
M_{F,k_G} &= \mathrm{mean}_{WL}\left(S_F\right)\\
k_{G+1} &= \begin{cases} 0 & \text{if } k_G = H-1\\ k_G+1 & \text{otherwise}\end{cases}\\
M_{r,i} &= k_G
\end{aligned} \tag{21}$$

Op_2: Overwrite the elements of M_{CR} and M_F indexed by k_G with uniformly random values. Decrement k_G. Update $M_{r,i}$ with a randomly chosen value.

$$\begin{aligned}
M_{CR,k_G} &= \mathrm{rand}[0,1]\\
M_{F,k_G} &= \mathrm{rand}[0,1]\\
k_{G+1} &= \begin{cases} H-1 & \text{if } k_G = 0\\ k_G-1 & \text{otherwise}\end{cases}\\
M_{r,i} &= \mathrm{rand}_i[0, H-1]
\end{aligned} \tag{22}$$

Op_3: Overwrite the elements of M_{CR} and M_F indexed by k_G with the current values. Increment k_G. Update $M_{r,i}$ with a randomly chosen value.

$$\begin{aligned}
M_{CR,k_G} &= \mathrm{mean}_{WA}\left(S_{CR}\right)\\
M_{F,k_G} &= \mathrm{mean}_{WL}\left(S_F\right)\\
k_{G+1} &= \begin{cases} 0 & \text{if } k_G = H-1\\ k_G+1 & \text{otherwise}\end{cases}\\
M_{r,i} &= \mathrm{rand}_i[0, H-1]
\end{aligned} \tag{23}$$

Table 1. Comparetive Results of Benchmarks against JADE when $D = 10$

Func.	$D \times 10000$ SHADE	RSHADE	$D \times 30000$ SHADE	RSHADE	$D \times 50000$ SHADE	RSHADE
1	≈	≈	≈	≈	≈	≈
2	≈	≈	≈	≈	≈	≈
3	+	+	+	+	+	+
4	≈	≈	≈	≈	≈	≈
5	≈	≈	≈	≈	≈	≈
6	≈	≈	≈	≈	≈	≈
7	+	+	+	+	+	+
8	≈	≈	+	≈	+	≈
9	≈	+	≈	+	≈	+
10	+	≈	≈	−	≈	−
11	≈	≈	≈	≈	≈	≈
12	+	+	+	+	+	+
13	+	+	+	+	+	+
14	≈	≈	≈	≈	≈	≈
15	+	≈	≈	≈	≈	≈
16	+	+	+	+	+	+
17	≈	≈	≈	≈	≈	≈
18	+	≈	≈	+	≈	+
19	≈	≈	≈	≈	≈	≈
20	≈	+	≈	+	≈	+
21	≈	≈	≈	≈	≈	≈
22	≈	≈	≈	≈	≈	−
23	≈	≈	≈	≈	≈	≈
24	+	+	+	+	+	+
25	+	+	−	+	−	+
26	+	≈	≈	≈	≈	≈
27	+	+	+	+	+	+
28	≈	≈	≈	≈	≈	≈
+	12	10	8	11	8	11
−	0	0	1	1	1	2
≈	16	18	19	16	19	15

6   Computer Simulations

6.1   Settings

The benchmark set [2] has been employed for our computer simulations. This is composed of 5 unimodal functions, F_1–F_5, 15 multimodal functions, F_6–F_{20}, and 8 composition functions, F_{21}–F_{28}. The search area is $[-100, 100]^D$, where D is the number of dimensions. In this paper, we employ $D = \{10, 30, 50\}$.

The difference between the current best value and the function optimal value is called the error value. An error value smaller than $1.0E - 8$ is recognized as zero. The evolutionary search is terminated when the error value becomes zero or the maximal number of evaluations is calculated. The number of evaluations is set at $D \times 10000$, $D \times 30000$ and $D \times 50000$, while [2] adopts only $D \times 10000$. The number of trials is 51. SHADE and RSHADE are compared with JADE by the Wilcoxon's rank sum test with significance level $p = 0.05$ for the obtained 51 samples.

Table 2. Comparetive Results of Benchmarks against JADE when $D = 30$

Func.	$D \times 10000$ SHADE	RSHADE	$D \times 30000$ SHADE	RSHADE	$D \times 50000$ SHADE	RSHADE
1	≈	≈	≈	≈	≈	≈
2	≈	≈	−	≈	−	≈
3	+	+	≈	≈	+	≈
4	+	+	+	+	+	+
5	≈	≈	≈	≈	≈	≈
6	≈	≈	≈	≈	≈	≈
7	≈	≈	≈	≈	≈	≈
8	+	≈	+	≈	+	+
9	−	−	−	+	−	+
10	−	≈	−	≈	−	≈
11	≈	≈	≈	≈	≈	≈
12	≈	≈	≈	−	≈	−
13	≈	≈	−	≈	−	≈
14	+	≈	+	≈	+	≈
15	≈	+	≈	≈	≈	≈
16	+	+	+	+	≈	+
17	≈	≈	≈	≈	≈	≈
18	+	+	≈	≈	≈	≈
19	+	+	−	−	≈	−
20	≈	≈	≈	≈	≈	≈
21	≈	≈	≈	≈	≈	≈
22	≈	≈	≈	−	≈	−
23	≈	≈	−	≈	−	≈
24	+	+	+	+	+	+
25	+	+	+	+	+	+
26	≈	≈	≈	≈	≈	≈
27	+	+	+	+	+	+
28	≈	≈	≈	≈	≈	≈
+	10	9	7	6	7	7
−	2	1	6	3	5	3
≈	16	18	15	19	16	18

JADE source was implemented originally using C++, according to [9]. The SHADE source was downloaded from the authors' website [1] which was labelled SHADE1.0_CEC2013.zip on May 9, 2014.

The population size NP was set at 100. The archive size $|\boldsymbol{A}|$ was also set at $NP = 100$. The other parameters, $p = 0.05$ and $c = 0.1$ were used to obtain the same condition as that in [9]. We also used $H = NP$, which was the maximum amount of history memory used in SHADE and RSHADE. The ratio of Op_1, Op_2, and Op_3 in RSHADE was set at $2 : 1 : 7$ on the basis of the results of preliminary experiments. In addition to this, we adopted the traditional treatment of boundary condition violation for the trial vector [4]:

$$v_{j,i,G} = \left(x_j^{\min} + x_{j,i,G}\right)/2 \ \text{if} \ \ v_{j,i,G} < x_j^{\min}$$
$$v_{j,i,G} = \left(x_j^{\max} + x_{j,i,G}\right)/2 \ \text{if} \ \ v_{j,i,G} > x_j^{\max} \tag{24}$$

6.2 Results

The results comparison of SHADE and RSHADE with JADE when $D = \{10, 30, 50\}$ are shown in Table 1, 2, and 3, respectively. The symbol "+" or "−" implies that

---

[1] "Ryoji Tanabe's Home Page," https://sites.google.com/site/tanaberyoji/

Table 3. Comparetive Results of Benchmarks against JADE when $D = 50$

Func.	$D \times 10000$ SHADE	$D \times 10000$ RSHADE	$D \times 30000$ SHADE	$D \times 30000$ RSHADE	$D \times 50000$ SHADE	$D \times 50000$ RSHADE
1	≈	≈	≈	≈	≈	≈
2	≈	≈	−	≈	−	≈
3	≈	≈	≈	≈	≈	≈
4	+	+	+	+	+	+
5	≈	≈	≈	≈	≈	≈
6	≈	≈	≈	≈	≈	≈
7	≈	≈	≈	≈	≈	≈
8	+	≈	+	+	+	+
9	−	≈	−	+	−	+
10	−	≈	−	≈	−	≈
11	≈	≈	≈	≈	≈	≈
12	≈	+	−	≈	−	≈
13	−	≈	−	−	−	−
14	≈	≈	≈	≈	≈	≈
15	≈	≈	≈	≈	≈	≈
16	+	−	≈	+	≈	+
17	≈	≈	≈	≈	≈	≈
18	+	+	≈	≈	≈	≈
19	≈	≈	−	−	−	−
20	+	≈	+	+	+	+
21	≈	≈	≈	≈	≈	≈
22	+	≈	≈	+	≈	−
23	≈	≈	−	≈	−	≈
24	+	≈	+	≈	+	≈
25	+	+	+	+	+	+
26	+	+	+	+	+	+
27	+	+	+	+	+	+
28	≈	≈	≈	≈	≈	≈
+	10	6	7	9	7	8
−	3	1	7	2	7	3
≈	15	21	14	17	14	17

the method is superior or inferior to JADE, respectively. The symbol implies that clear superiority is not observed by the comparison. In the lower part of each table, the numbers of these symbols are counted. When $D = 10$, SHADE and RSHADE show clear advantage over JADE for all the cases where the evaluations are $D \times 10000$, $D \times 30000$, and $D \times 50000$. More specifically, RSHADE shows better results than SHADE, except in the case where $D \times 10000$.

This tendency is also observed in Table 2, which summarizes the results for $D = 30$. However, when $D = 50$, the advantage of SHADE against JADE diminishes according to the number of evaluations. We cannot clearly see the superiority of SHADE when $D \times 30000$ and $D \times 50000$, whereas RSHADE consistently shows better results against JADE. To summarize, we can conclude that RSHADE shows better results as the number of evaluations increases. This tendency is much clearer in large dimensional benchmarks. As part of the proof, Tables 4, 5, and 6 show the means and the standard deviations for all the benchmarks for the cases where the number of evaluations is $D \times 50000$, and $D = 10$, 30, and 50, respectively.

Table 4. Test Results of Benchmarks when $D = 10$ and $D \times 50000$ Evaluations

Func.	JADE Mean	Std. dev.	SHADE Mean	Std. dev.		RSHADE Mean	Std. dev.	
1	0.00e+00	0.00e+00	0.00e+00	0.00e+00	≈	0.00e+00	0.00e+00	≈
2	0.00e+00	0.00e+00	0.00e+00	0.00e+00	≈	0.00e+00	0.00e+00	≈
3	3.76e-01	6.69e-01	7.00e-03	2.14e-02	+	4.20e-03	1.70e-02	+
4	0.00e+00	0.00e+00	0.00e+00	0.00e+00	≈	0.00e+00	0.00e+00	≈
5	0.00e+00	0.00e+00	0.00e+00	0.00e+00	≈	0.00e+00	0.00e+00	≈
6	6.73e+00	4.60e+00	7.70e+00	4.08e+00	≈	8.27e+00	3.60e+00	≈
7	8.37e-02	1.31e-01	8.59e-04	2.24e-03	+	1.04e-03	2.49e-03	+
8	2.02e+01	1.33e-01	2.00e+01	7.25e-02	+	2.02e+01	8.70e-02	≈
9	3.04e+00	8.02e-01	2.88e+00	8.99e-01	≈	7.73e-01	8.16e-01	+
10	9.77e-04	2.56e-03	1.94e-03	3.75e-03	≈	1.21e-03	3.09e-03	−
11	0.00e+00	0.00e+00	0.00e+00	0.00e+00	≈	0.00e+00	0.00e+00	≈
12	3.50e+00	8.98e-01	2.87e+00	8.59e-01	+	2.45e+00	1.18e+00	+
13	4.00e+00	1.59e+00	3.21e+00	1.71e+00	+	3.01e+00	1.58e+00	+
14	1.22e-02	3.06e-02	6.12e-03	2.25e-02	≈	1.84e-02	3.60e-02	≈
15	3.16e+02	1.28e+02	2.93e+02	1.13e+02	≈	2.84e+02	1.14e+02	≈
16	3.97e-01	2.64e-01	2.32e-01	7.53e-02	+	2.69e-02	3.79e-02	+
17	1.01e+01	0.00e+00	1.01e+01	0.00e+00	≈	1.01e+01	0.00e+00	≈
18	1.37e+01	1.16e+00	1.34e+01	7.40e-01	≈	1.28e+01	8.91e-01	+
19	1.88e-01	3.10e-02	1.85e-01	2.97e-02	≈	1.87e-01	2.86e-02	≈
20	2.14e+00	4.53e-01	2.01e+00	3.26e-01	≈	1.83e+00	4.26e-01	+
21	4.00e+02	0.00e+00	3.96e+02	2.80e+01	≈	3.96e+02	2.80e+01	≈
22	5.04e-01	2.64e+00	6.88e-01	1.74e+00	≈	1.43e+00	3.05e+00	−
23	3.26e+02	1.55e+02	3.79e+02	1.63e+02	≈	3.14e+02	1.44e+02	≈
24	2.02e+02	3.92e+00	1.99e+02	9.23e+00	+	2.00e+02	1.34e+01	+
25	1.99e+02	1.42e+01	1.99e+02	6.47e+00	−	1.98e+02	1.30e+01	+
26	1.24e+02	3.82e+01	1.20e+02	3.53e+01	≈	1.43e+02	4.82e+01	≈
27	3.09e+02	3.69e+01	3.00e+02	0.00e+00	+	3.00e+02	0.00e+00	+
28	2.92e+02	3.92e+01	3.00e+02	0.00e+00	≈	3.00e+02	0.00e+00	≈

Table 5. Test Results of Benchmarks when D=30 and $D \times 50000$ Evaluations

Func.	JADE Mean	Std. dev.	SHADE Mean	Std. dev.		RSHADE Mean	Std. dev.	
1	0.00e+00	0.00e+00	0.00e+00	0.00e+00	≈	0.00e+00	0.00e+00	≈
2	6.91e+00	1.76e+01	3.59e+01	7.37e+01	−	8.96e+00	4.19e+01	≈
3	7.92e+04	5.45e+05	3.70e+04	2.64e+05	+	8.59e+04	4.81e+05	≈
4	1.33e+03	3.22e+03	0.00e+00	0.00e+00	+	0.00e+00	0.00e+00	+
5	0.00e+00	0.00e+00	0.00e+00	0.00e+00	≈	0.00e+00	0.00e+00	≈
6	5.18e-01	3.70e+00	1.04e+00	5.18e+00	≈	5.18e-01	3.70e+00	≈
7	3.30e+01	3.46e+00	4.49e+00	5.65e+00	≈	2.86e+00	2.88e+00	≈
8	2.08e+01	1.29e-01	2.06e+01	1.46e-01	+	2.06e+01	2.14e-01	+
9	2.58e+01	1.88e+00	2.66e+01	1.78e+00	−	1.84e+01	4.82e+00	+
10	3.85e-02	2.36e-02	8.47e-02	5.65e-02	−	4.66e-02	3.00e-02	≈
11	0.00e+00	0.00e+00	0.00e+00	0.00e+00	≈	0.00e+00	0.00e+00	≈
12	1.80e+01	3.44e+00	1.93e+01	4.94e+00	≈	2.03e+01	4.40e+00	−
13	3.96e+01	1.04e+01	4.41e+01	1.19e+01	−	4.29e+01	1.05e+01	≈
14	2.94e-02	2.43e-02	2.00e-02	2.20e-02	+	2.45e-02	1.94e-02	≈
15	2.81e+03	3.23e+02	2.91e+03	2.10e+02	≈	2.74e+03	3.09e+02	≈
16	5.71e-01	1.82e-01	5.98e-01	1.50e-01	≈	1.73e-01	1.24e-01	+
17	3.04e+01	0.00e+00	3.04e+01	0.00e+00	≈	3.04e+01	0.00e+00	≈
18	5.14e+01	4.53e+00	5.27e+01	3.15e+00	≈	5.22e+01	4.38e+00	≈
19	8.83e-01	6.73e-02	9.12e-01	9.13e-02	≈	9.24e-01	8.82e-02	−
20	1.03e+01	5.48e-01	1.03e+01	5.55e-01	≈	1.03e+01	5.32e-01	≈
21	3.07e+02	6.96e+01	3.16e+02	7.10e+01	≈	3.17e+02	7.50e+01	≈
22	5.97e+01	4.10e+01	7.73e+01	3.82e+01	≈	8.97e+01	3.48e+01	−
23	2.99e+03	3.92e+02	3.16e+03	3.76e+02	−	2.97e+03	3.09e+02	≈
24	2.11e+02	9.79e+00	2.04e+02	3.56e+00	+	2.06e+02	5.60e+00	+
25	2.73e+02	9.76e+00	2.65e+02	1.61e+01	+	2.49e+02	6.57e+00	+
26	2.05e+02	2.55e+01	2.02e+02	1.55e+01	≈	2.08e+02	2.87e+01	≈
27	6.13e+02	2.34e+02	3.68e+02	5.31e+01	+	4.10e+02	1.05e+02	+
28	3.00e+02	0.00e+00	2.96e+02	2.80e+01	≈	3.00e+02	0.00e+00	≈

Table 6. Test Results of Benchmarks when D=50 and $D \times 50000$ Evaluations

	JADE		SHADE			RSHADE		
Func.	Mean	Std. dev.	Mean	Std. dev.		Mean	Std. dev.	
1	0.00e+00	0.00e+00	0.00e+00	0.00e+00	≈	0.00e+00	0.00e+00	≈
2	8.86e+01	1.69e+02	1.16e+02	1.21e+02	−	5.12e+01	7.43e+01	≈
3	1.51e+06	3.12e+06	1.29e+06	3.61e+06	≈	9.62e+05	2.37e+06	≈
4	1.37e+03	2.84e+03	0.00e+00	0.00e+00	+	0.00e+00	0.00e+00	+
5	0.00e+00	0.00e+00	0.00e+00	0.00e+00	≈	0.00e+00	0.00e+00	≈
6	4.36e+01	7.95e-01	4.34e+01	0.00e+00	≈	4.36e+01	7.95e-01	≈
7	2.19e+01	9.14e+00	2.13e+01	8.72e+00	≈	1.97e+01	9.36e+00	≈
8	2.10e+01	1.12e-01	2.08e+01	1.07e-01	+	2.08e+01	1.66e-01	+
9	5.35e+01	2.19e+00	5.53e+01	2.12e+00	−	4.06e+01	7.91e+00	+
10	4.08e-02	3.29e-02	9.50e-02	5.64e-02	−	4.51e-02	3.26e-02	≈
11	0.00e+00	0.00e+00	0.00e+00	0.00e+00	≈	0.00e+00	0.00e+00	≈
12	4.70e+01	8.39e+00	5.28e+01	9.95e+00	−	4.84e+01	7.38e+00	≈
13	1.11e+02	2.36e+01	1.32e+02	2.31e+01	−	1.25e+02	2.16e+01	−
14	4.34e-02	2.25e-02	3.70e-02	1.82e-02	≈	4.95e-02	2.63e-02	≈
15	6.18e+03	4.94e+02	6.28e+03	4.49e+02	≈	6.09e+03	4.63e+02	≈
16	9.19e-01	2.08e-01	9.64e-01	1.61e-01	≈	2.92e-01	1.63e-01	+
17	5.08e+01	0.00e+00	5.08e+01	0.00e+00	≈	5.08e+01	0.00e+00	≈
18	9.44e+01	7.31e+00	9.62e+01	6.66e+00	≈	9.55e+01	7.55e+00	≈
19	1.70e+00	1.48e-01	1.81e+00	2.60e-01	−	1.79e+00	2.14e-01	−
20	1.95e+01	5.88e-01	1.90e+01	6.71e-01	+	1.90e+01	7.49e-01	+
21	7.69e+02	4.32e+02	8.27e+02	3.71e+02	≈	7.59e+02	4.13e+02	≈
22	1.28e+01	1.94e+01	9.75e+00	1.13e+00	≈	1.30e+01	8.29e+00	−
23	6.45e+03	5.95e+02	6.96e+03	6.12e+02	−	6.52e+03	7.73e+02	≈
24	2.49e+02	2.05e+01	2.34e+02	1.22e+01	+	2.41e+02	1.22e+01	≈
25	3.56e+02	1.67e+01	3.33e+02	3.13e+01	+	2.99e+02	1.43e+01	+
26	3.31e+02	1.07e+02	2.57e+02	7.39e+01	+	2.96e+02	7.00e+01	+
27	1.33e+03	3.11e+02	8.80e+02	2.80e+02	+	8.48e+02	1.61e+02	+
28	5.75e+02	7.06e+02	4.59e+02	4.23e+02	≈	6.32e+02	8.04e+02	≈

7   Conclusions

In this paper, a new DE variant, viz., RSHADE, was proposed. The motivation for RSHADE was based on the observation that parameter adaptation sometimes stops, which generally causes stagnation. To overcome this phenomenon, we propose utilizing historical information by introducing the array M_r and the three operators Op_1, Op_2, and Op_3. The results of computer simulations were favorable to RSHADE, especially when benchmarks have a larger dimensional size and larger number of evaluations is given.

Although the fundamental advantage has been confirmed in RSHADE, admittedly, there are still many open questions. For instance, although we introduced the three operators, another question arises: how to determine the ratio of the operators. Removing this predetermined ratio could be possible by introducing a certain mechanism that makes the parameters self-adaptive.

References

1. Das, S., Suganthan, P.N.: Differential evolution: A survey of the state-of-the-art. Evolutionary Computation, IEEE Transactions on 15(1), 4–31 (2011)
2. Liang, J.J., Qu, B.Y., Suganthan, P.N.: Alfredo G Hernández-Díaz. Problem definitions and evaluation criteria for the cec 2013 special session on real-parameter optimization. Computational Intelligence Laboratory, Zhengzhou University, Zhengzhou, China and Nanyang Technological University, Singapore, Technical Report, 201212 (2013)
3. Peng, F., Tang, K., Chen, G., Yao, X.: Multi-start jade with knowledge transfer for numerical optimization. In: IEEE Congress on Evolutionary Computation, CEC 2009, pp. 1889–1895. IEEE (2009)
4. Price, K., Storn, R.M., Lampinen, J.A.: Differential evolution: a practical approach to global optimization. Springer (2006)
5. Qin, A.K., Suganthan, P.N.: Self-adaptive differential evolution algorithm for numerical optimization. In: The 2005 IEEE Congress on Evolutionary Computation, vol. 2, pp. 1785–1791. IEEE (2005)
6. Storn, R., Price, K.: Differential evolution–a simple and efficient heuristic for global optimization over continuous spaces. Journal of Global Optimization 11(4), 341–359 (1997)
7. Tanabe, R., Fukunaga, A.: Success-history based parameter adaptation for differential evolution. In: 2013 IEEE Congress on Evolutionary Computation (CEC), pp. 71–78. IEEE (2013)
8. Teo, J.: Exploring dynamic self-adaptive populations in differential evolution. Soft Computing 10(8), 673–686 (2006)
9. Zhang, J., Sanderson, A.C.: Jade: adaptive differential evolution with optional external archive. IEEE Transactions on Evolutionary Computation 13(5), 945–958 (2009)

Multi-agent Control System with Intelligent Optimization for Building Energy Management

Ee May Kan, Khaing Yadanar, Ngee Hoo Ling, Yvonne Soh, and Naing Lin

[1] School of Engineering, Nanyang Polytechnic,
180, Ang Mo Kio Ave 8, Singapore 569830
[2] School of Electrical & Electronic Engineering, Nanyang Technological University,
50 Nanyang Ave, 639798, Singapore
{kan_ee_may,ling_ngee_hoo}@nyp.edu.sg
khaing004@e.ntu.edu.sg
Yvonne_soh@sgbc.sg
Naing.Lin@advantech.com

Abstract. Smart, sustainable and energy-efficient buildings have recently become a trend for intelligent and green building industry. One of the challenges for such a building is to minimize power consumption without compromising the occupant comfort. It is possible to sustain high comfort level with minimal energy consumption through intelligent multi-agent control and optimization. Our work focuses on a new algorithmic aspect of a multi-agent control system that makes buildings more intelligent, resource efficient and comfortable for occupants. A multi-agent control system with intelligent optimization is proposed to optimize HVAC (heating, ventilation and air-conditioning) processes using computational intelligence approaches. The optimum solutions generated through the optimization engine and task scheduling for each agent will be used to sustain a high level of occupant comfort with minimal power consumption, increase energy utilization efficiency, and consequently reduce energy cost. Experimental results and comparisons with other evolutionary approaches demonstrate the overall performance and potential benefits of the new framework.

Keywords: Multi-agent control, optimization, computational intelligence.

1 Introduction

Strategies towards sustainable performance with occupants have become increasingly important. Optimal indoor conditions help to enhance occupant health, productivity and experience. There have been multiple goals and provisions for the improvement of occupant comfort and energy savings [1-3]. The improvement of the indoor environment comfort demands more energy consumption and the building operations require high energy efficiency to reduce energy consumption. Thus, one of the most important issues on energy-efficient buildings is to optimize the requirements of the occupants' comfort and power consumption effectively. It is possible to sustain high

© Springer International Publishing Switzerland 2015

H. Handa et al. (eds.), *Proc. of the 18th Asia Pacific Symp. on Intell. & Evol. Systems – Vol. 2*,
Proceedings in Adaptation, Learning and Optimization 2, DOI: 10.1007/978-3-319-13356-0_40

comfort level with minimal energy consumption through intelligent control of building energy management systems. Intelligent control of the thermal comfort, visual comfort and air quality comfort are vital for both energy efficiency and occupant's quality of living [1].

These three basic factors determine the occupants' quality of lives in a building environment. Thus, the basic control objective for the multi-agent control system is to sustain the occupant comfort level while minimizing the energy footprint of buildings. Our focus in this work is on computational intelligence approach [4] to optimize multi-objective HVAC (heating, ventilation and air-conditioning) processes. There are several important components considered in the proposed approach, for instance, the representation of the multi-objective optimization problem, the fitness function as well as the genetic search operators. With proper and domain-specific representation, we can easily apply our proposed approach to the multi-objective optimization problem. We take into consideration the building operation constraints while optimizing the HVAC (heating, ventilation and air-conditioning) processes. The rest of the paper is organized as follows. Details on the formulation of the problem are presented in Section 2. Section 3 and Section 4 present the solution and implementation, respectively, of the proposed approach substantiated with simulation results. Section 5 shows the results compared with the existing evolutionary approach with brief discussions. Lastly, we summarize the main contributions of this study in Section 6, and enlist several recommendations for the future research.

2   Problem Formulation

The multi-objective optimization problem for building energy management can be formulated in terms of a energy cost function subject to constraints of the building operations. We consider the multi-objective problem as an optimization problem with the conflicting objectives of minimizing the power consumption, maximizing the air quality comfort and maximizing the thermal comfort. Intelligent control of the thermal comfort and air quality comfort are vital for both energy efficiency and occupant's quality of living. The multi-objective optimization problem is defined as follows:

$$f = \delta_1[1 - (\frac{e_T}{T_{set}})^2] + \delta_2[1 - (\frac{e_L}{L_{set}})^2] + \delta_3[1 - (\frac{e_A}{A_{set}})^2] \tag{1}$$

where f represents the overall customer comfort level, which falls into[0,1] . It is the control goal to be maximized; δ_1 , δ_2 and δ_3 are the users-defined factors, which indicate the importance of comfort factors. δ_1 , δ_2 and δ_3 fall into [0,1] , and $\delta_1 + \delta_2 + \delta_3 = 1$. e is the difference between set point and actual sensor measurement; T_{set} , L_{set} and A_{set} are the set points of temperature, illumination and air quality, respectivelyThe goal of a multi-objective optimization algorithm is to identify solutions in the Pareto optimal set. However, identifying the entire Pareto optimal set, for many multi-objective problems, is practically impossible due to its size. In addition, for many problems, especially for combinatorial optimization problems, proof of solution optimality is computationally infeasible. Therefore, a practical approach to multi-objective optimization is to investigate a set of solutions (the best-known Pareto set) that represent the Pareto optimal set as well as possible. With these concerns in mind, a multi-objective optimization approach should achieve the following three conflicting goals:

1. The best-known Pareto front should be as close as possible to the true Pareto front. Ideally, the best-known Pareto set should be a subset of the Pareto optimal set.
2. Solutions in the best-known Pareto set should be uniformly distributed and diverse over of the Pareto front in order to provide the decision-maker a true picture of trade-offs.
3. The best-known Pareto front should capture the whole spectrum of the Pareto front. This requires investigating solutions at the extreme ends of the objective function space.

For a given computational time limit, the first goal is best served by focusing the search on a particular region of the Pareto front. On the contrary, the second goal demands the search effort to be uniformly distributed over the Pareto front. The third goal aims at extending the Pareto front at both ends, exploring new extreme solutions.

3   Solution Methodology

3.1   Multi-objective Optimization

We consider a general multi-objective optimization problem as minimizing a function $f(x)$ depending on q equality and p inequality constraints.

$$\text{min.} \ f(x) = \{f_1(x)f_2(x) \dots f_m(x)\}^T \tag{2}$$

$$x \in D$$

$$where \ x \in R^n, f_i \colon R^n \to R \ and$$

$$D = \begin{cases} x \in R^n \ : \ l_i \leq x \leq u_i, \forall i = 1, \dots, n \\ \qquad g_j(x) \geq 0, \forall j = 1, \dots, p \\ \qquad h_k(x) = 0, \forall k = 1, \dots, q \end{cases} \tag{3}$$

where D indicates a feasible search space; R^n indicates a sequence of real numbers in n-dimensional space; $x = \{x_1 x_2 \dots x_n\}^T$ denotes a sequence of n-dimensional decision variables (integer, continuous or discrete); l_i and u_i represent the lower bound and upper bound of i-th decision variable, respectively; and lastly m denotes the number of objectives. The optimization problem yields Pareto optimal solutions by optimizing the vector function simultaneously. In our study, we consider Pareto front as a collection of non-dominated (Pareto optimal) solutions. A solution x^* is signified as dominated by another solution x, if x is superior than x^* pertaining to all objectives.

3.2   The Proposed Algorithm

The proposed algorithm involves initialization, evaluation and exploration of population by integrating optimization operators with Pareto-dominance criteria. In our proposed algorithm, we first evaluate the particles (potential solutions) and check for dominance relation among the population. Next, we search for non-dominated solutions and store them as potential candidates in order to lead the search particles. It is worth noting that the variable size of the sequence of potential candidates is to improve the computational efficiency of the algorithm during optimization. Additionally, we make use of crowding distance assignment operator to trim down the size of the candidates if it exceeds the allowable size. Lastly, we exploit an efficient evaluation strategy for evaluating the search space. By combining these operators, the algorithm is competent to sustain diversity in the population and successfully explore towards true Pareto optimal fronts. The main steps of the procedure are described as follow:

#1 The algorithm begins with the generation of potential solutions to form an initial population. The iteration counter t is set to 0.
   i. Apply real numbers in the specific range of decision variable randomly to initialize the existing location of the i-th particle x_i;
   ii. Apply uniformly distributed random number in [0, 1] to initialize every particle velocity vector v_i ;
   iii. Assess every particle in the population; Place the individual best location P_i to x_i.
#2 Classify non-dominated solutions among the particles and save those non-dominated solutions as waypoint candidates.
#3 $t = t + 1$.
#4 Iterate the loop as follows:
   i. Choose randomly a global best particle guide P_g for the ith particle among waypoint candidates
   ii. Compute the new v_i, and the new x_i
   iii. Iterate the loop for every particle in the population
      #5 Assess all the particles in the current population.
#6 Examine the Pareto dominance for every particle in the population. Replace local best solution P_i with current solution if P_i is dominated by the current solution.
#7 Classify non-dominated solutions among the particles and save them as waypoint candidates.
#8 Restrict the size of the sequence of waypoint candidates by applying the crowding distance operator if it exceeds the allowable size.
#9 Implement generic evaluation on a particular number of particles.

#10 Return to step 3 if the termination criterion is not met; else output the set of non-dominated solutions from the sequence of waypoint candidates.

Subsequently, we present the main operators of the proposed procedure in the following sections.

3.2.1 Sequence of Potential Candidates

The performance evaluation and selection of the best global particle guide are vital procedures in a multi-objective optimization approach. They sustain a superior range of non-dominated solutions and influence the convergence capability of the algorithm. In the preceding iteration of the procedure, any of the non-dominated solutions from the sequence of potential candidates can be chosen as a global guide. We propose a performance evaluation method to make sure every particle in the population moves towards the true Pareto optimal region. Moreover, the crowding distance operator is used to perform restriction on the waypoint candidates. This operator ensures the highest crowding distance values of non-dominated solutions are stored as potential candidates. Subsequently, we evaluate and select the best global particle guide from a restricted size of the potential candidates. During optimization, the variable size of the potential candidates manages to reduce the computational time. However the computing requirement for the sorting and crowding value becomes greater when the size increases. Hence in the initial steps of the procedure, we set the size to half of the maximum size of potential candidates; followed by increasing the value in a stepwise manner to explore the solution space effectively. This procedure evaluates distinct particle guides from a restricted number of potential candidates and thus improves the algorithm performance by enabling the particles explore towards the true Pareto optimal region.

3.2.2 Crowding Distance Assignment Operator

The crowding distance assignment operator specifies an estimate of the density of solutions. In a particular solution, the average distance of two neighbouring solutions is the crowding distance value. We compute the crowding distance value by first sorting the solution sequence in ascending order of the objective function values and followed by securing boundary solutions with the smallest and greatest values of the objective function. Subsequently, the final crowding distance value of a solution is computed by adding up the individual crowding distance values in every objective function.

The detailed computation of crowding distance value is presented as follows.

#1 Obtain the size of non-dominated solutions from the sequence of potential candidates

$$l = |S| \quad where \ S = [W_1, W_2, \cdots W_k]$$

#2 Initialize distance.

$$For \ i = 1 \ to \ l$$
$$S[i].dist = 0$$

#3 Calculate the individual crowding distance value.

#4 Sort each solution according to objective function value.

For each objective m

$$S = \text{sort} \ (S, m)$$

#5 Set the boundary points to infinite to guarantee their selection.

$$S \ [1]. \ dist = S \ [l]. \ dist = \infty$$

For $i = 2 \ to \ (l - 1)$

$$S \ [i]. \ dist = S \ [i]. \ dist + \frac{S \ [i + 1]. \ m - S \ [i - 1]. \ m}{f_m^{max} - f_m^{min}}$$

3.2.3 Generic Evaluation Operator

In this work, we apply the generic evaluation operator involving priorities, goals and Pareto sets in which we can place the objectives in levels of priority and impose constraints on each of them. The operator evaluates all non-dominated solutions in the population, determines if one Pareto dominates the other, and ranks them accordingly. In order to set the priority levels of our objectives for the optimization problem, we classify them as follows:

- Constraints that the agent has to fulfil due to its building operations, or user requirements
- Zone-specific energy costs to be optimized

The constraints that the agent must fulfil are set as high-priority levels. On the other hand, the energy costs to be optimized are set at lower priority levels in line with the goal of the optimization. By using this method, we can simply modify the constraints or zone-specific energy costs as required by the different group of users. This improves the performance of multi-objective optimization while extending from conventional optimization algorithm. The evaluation operator works on a specific number of particles and distributes the non-dominated solutions along the true Pareto optimal front. Initially, this operator strives to substitute the non-feasible solutions with the evaluated particles of potential candidates. Subsequently, it attempts to exploit the search space in the sequence of waypoint candidates along the Pareto fronts. The evaluation procedure is described as follows.

#1 Sort the particle fitness function and get the particle index number respectively.

#2 Employ crowding distance assignment operator for the calculation of the density of solutions in the sequence of waypoint candidates. Sort them accordingly and choose one of the least crowded solutions from the waypoint candidates randomly as particle guide.

#3 Evaluate a specific number of particles and rank them accordingly.

3.2.4 Constraint Handling Mechanism

We adopt the constraint handling mechanism to deal with the constrained optimization problems. In this simple and effective method, a solution i is considered as a constrained dominate solution j if it meets any of the conditions as follows:

1. Solution i is deemed as feasible while solution j is not.

2. Both solutions i and j are deemed as non-feasible, although solution i has a smaller overall constraint violation.
3. Both solutions i and j are deemed as feasible when solution i dominates solution j.

3.3   The Multi-agent Control System

The multi-agent control system is classified into different agents based on the distinct functions at different zones. The zone agents monitor the energy flow and responsible for energy management in its specific zone based on the occupant preference. The proposed algorithm is embedded in the central coordinator-agent to optimize the set points. Multiple local controller-agents are used to control the devices which are related to the comfort factors. The main comfort factors considered in this study include environmental temperature, illumination level and indoor air quality. Accordingly, the local controller-agents are classified into the temperature controller-agent, the illumination controller-agent, and the air-quality controller-agent. Through the cooperation of these multiple agents, the control objective which is to maximize the occupant comfort and minimize the energy consumption simultaneously can be achieved. By way of example, we consider the multiple building zones with multiple agents portrayed in Fig. 1.

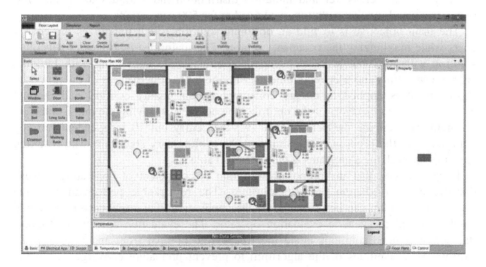

Fig. 1. The multiple zones of a building

Lastly, we demonstrate the efficiency and applicability of the proposed approach in the following sections.

4   Experimental Results

We apply our proposed approach by setting the initial population of the algorithm to 200 and specify the size of non-dominated solutions as 200. We then run the algorithm for 500 iteration steps. In this study, we use set coverage metric (SC) and spacing metric (SP) as performance measures to assess the performance of our proposed approach.

The SC (A, B) metric produces the solution dominance and relative convergence between two solution vectors A and B; calculating the proportion of B, which is weakly dominated by A. The role of SP metric is to assess the distribution of vectors throughout the set of non-dominated solutions and indicate how evenly the generated non-dominated solutions in the approximation set are distributed in the search space. The computational complexity of the algorithm is dominated by the objective function computation, crowding distance computation and the non-dominated comparison of the particles in the population. If there are M objective functions and N number of solutions (particles) in the population, then the objective function computation has $O(MN)$ computational complexity. For comparison purposes, we run the existing evolutionary algorithm by specifying the initial population as 200; probability of crossover as 0.9; and probability of mutation as $1/n$ (n is the size of real variables). Next, we set the crossover and mutation distribution index values to 20 and 100 respectively and run the evolutionary algorithm for 500 generations.

The main objective is to find an optimum solution which minimizes the energy costs while maximizing the occupant comfort. The main idea is to have a generated solution with maximal level of occupant comfort while complying with building energy constraints and user requirements. To show the effectiveness of the proposed algorithm, it is applied to energy consumption, air quality comfort and thermal comfort as three objectives of the optimization engine for solving the multi-objective problem. To examine the performance, we then carry out 10 independent runs by applying both the proposed algorithm and the evolutionary algorithm. The resulting statistics for both the algorithms is tabulated in Table 1.

It can be seen that the proposed algorithm is competent of generating a well-distributed set of Pareto optimal solutions. The remarkable performance of the proposed algorithm can be attributed to the application of the sequence of potential candidates and the generic evaluation operator. The evaluation operator distributes the non-dominated solutions along the true Pareto optimal front and improves the exploratory capabilities of the algorithm to prevent premature convergence. In the preceding iteration of the procedure, any of the non-dominated solutions from the sequence of potential candidates are chosen as a global guide. The selection of the global guide of the particle swarm is a crucial step in a multi-objective algorithm. It affects both the convergence capability of the algorithm as well as the good spread of non-dominated solutions. Our proposed algorithm manages to produce boundary solutions easily in this case.

Table 1. Resulting statistics by the proposed algorithm and the existing evolutionary algorithm for the multi-objective problem. In SC(A,B), A is the proposed algorithm and B is the existing evolutionary algorithm. The best performing algorithm is indicated by bold numbers. From Table 1, it can be seen that the mean value of SC (A, B) is higher than the mean value of SC (B, A) pertaining to set coverage metric. The SC (A, B) metric represents the proportion of B solutions that are weakly dominated by A solutions. Therefore we can conclude that our proposed algorithm is able to perform better than the existing evolutionary algorithm. Likewise, the proposed algorithm has lower spacing metric mean value than the existing evolutionary algorithm; indicating that the proposed algorithm obtains the better distribution of Pareto optimal solutions. A sample of the experimental result corresponding to SC (A, B) median value is presented in Fig. 2 for illustration purposes.

Statistics	Performance metric			
	Set coverage (SC) metric		Spacing (SP) metric	
	SC(A,B)	SC(B,A)	A	B
Best	0.1670	0.0360	164.9022	196.9804
Worst	0.7031	0.5080	236.3021	630.2212
Mean	**0.4466**	0.2302	**206.6202**	402.4579
Variance	0.0349	0.0254	1136.9908	26066.7456
SD	0.1668	0.1426	30.1462	146.5188

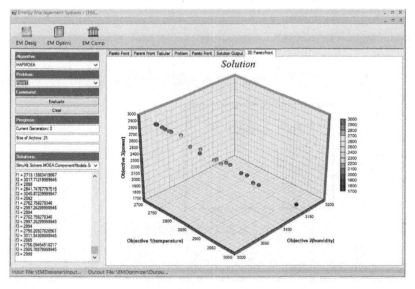

Fig. 2. Non-dominated solutions obtained using the proposed algorithm

5   Results and Discussion

In this section, the optimized solution is going to be compared with the initial layout for the simplest case. The schedules generated from optimization reduce power consumption. The figure below shows the layout simulation before optimization. The energy consumption can be seen in the right panel. The energy measured for this case is 56 with the initial schedule we set randomly. It can be seen that the temperature is not steady and maintained at the desired temperature which is 26°C.

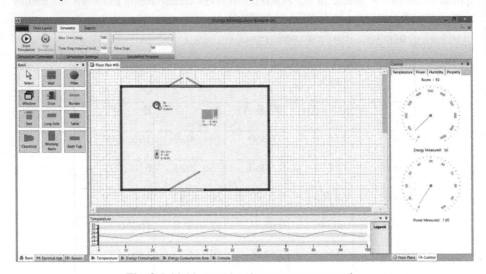

Fig. 3. Initial layout showing power consumption

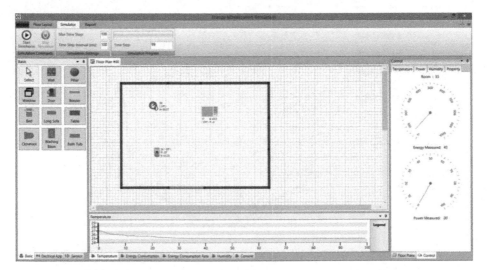

Fig. 4. Optimized layout showing power consumption using the proposed algorithm

The optimized schedule helps to reduce the power consumption while maintaining the desired temperature and humidity. As shown in the figure below, the energy consumption measured is only 43 using the proposed algorithm. The graph at the bottom of the figure shows that the temperature is maintained at 26°C.

As mentioned earlier, there are two types of algorithm applied for this multi-objective optimization process. Using the proposed algorithm, there are 15 solutions generated. User can choose any solution to simulate the optimized zone layout. The optimization process takes longer with the existing multi-objective evolutionary algorithm (MOEA), for e.g. HAPMOEA (Hierarchical Asynchronous Parallel Multi-Objective Evolutionary Algorithm).

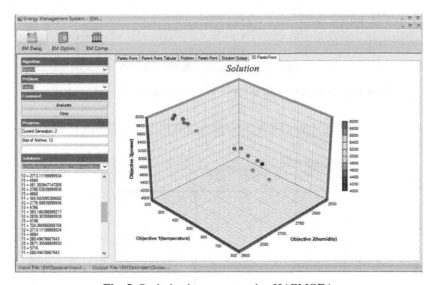

Fig. 5. Optimization process using HAPMOEA

There are 21 solutions generated while the optimization process is carried out with HAPMOEA and the resulted power consumption is about 48.

After the optimization procedure, we may get different solutions which could be better than the initial schedule or worse than the initial one. But certainly, the optimized schedule will be able to maintain the temperature and humidity at certain comfort level. Next, we conduct the experiments with two zones and two air conditionings. Initially, the total power consumption of the two zones is about 152.8 with unsteady temperature and humidity.

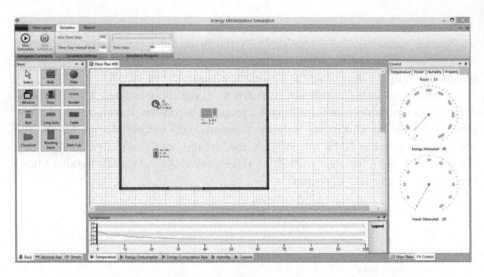

Fig. 6. Optimized layout showing power consumption using HAPMOEA

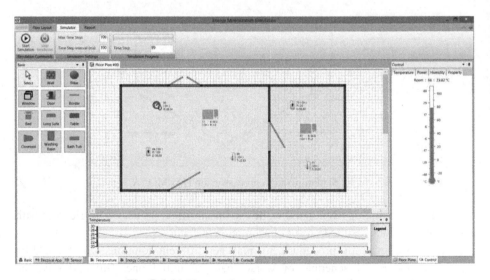

Fig. 7. Initial layout showing power consumption

Subsequently, we optimize the schedule with the proposed algorithm. The total power consumption for two rooms is 79.6 while maintaining desired temperature and humidity.

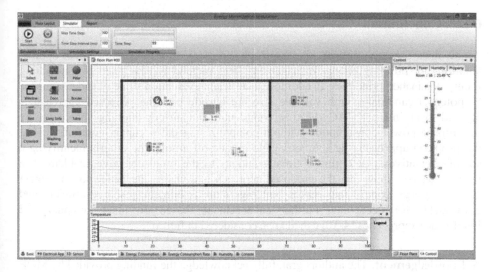

Fig. 8. Optimized layout showing power consumption using the proposed algorithm

The optimization process is then carried out with HAPMOEA. The total power consumption is 82.8 which is higher compared to the solution generated by the proposed algorithm. User can choose the preferred schedule based on their requirement by setting the generation and population size for the optimization process to have more accurate and desired solutions.

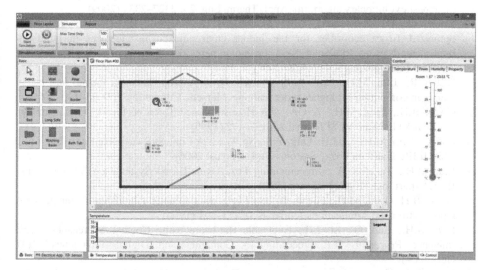

Fig. 9. Optimized layout showing power consumption using HAPMOEA

6   Conclusion

A new multi-objective optimization approach is presented for generating Pareto-optimal solutions for building energy management. This method is developed by integrating Pareto dominance principles. In addition, an efficient evaluation strategy and a sequence of potential candidates with variable size are introduced. The proposed approach is applied to optimize HVAC (heating, ventilation and air-conditioning) processes by minimizing power consumption without compromising the occupant comfort. Using the proposed approach, the solutions yield a trade-off among the criteria identifying a set of alternatives that define optimal solutions to the multi-objective problem. The results obtained show that the proposed approach is a viable alternative for intelligent building energy control that makes buildings more intelligent, resource efficient and comfortable for occupants. Further enhancements based on other computational intelligence approaches [4-10] can also be considered in future.

Acknowledgement. The authors gratefully acknowledge the funding provided by MOE Innovation Fund, Singapore.

References

1. Dounis, A.I., Caraiscos, C.: Advanced control systems engineering for energy and comfort management in a building environment—A review. Renewable Sustainable Energy Rev. 13(6-7), 1246–1261 (2009)
2. Fong, K., Hanby, V., Chow, T.: System optimization for HVAC energy management using the robust evolutionary algorithm. Appl. Therm. Eng. 29, 2327–2334 (2009)
3. Kolokotsa, D., Kalaitzakis, G.S., Stavrakakis, K., Agoris, D.: Genetic algorithms optimized fuzzy controller for the indoor environmental management in buildings implemented using PLC and local operating networks. Engineering Applications of Artificial Intelligence 15, 417–428 (2002)
4. Meuth, R., Lim, M.H., Ong, Y.S., Wunsch, D.C.: A proposition on memes and meta-memes in computing for higher-order learning. Memetic Computing 1(2), 85–100 (2009)
5. Gwee, B.H., Lim, M.H.: A GA with Heuristic-based Decoder for IC Floorplanning. INTEGRATION the VLSI Journal 28, 157–172 (1999)
6. Cao, Q., Lim, M.H., Li, J.H., Ong, Y.S., Ng, W.L.: A Context Switchable Fuzzy Inference Chip. IEEE Trans. on Fuzzy Systems 14(4) (August 2006)
7. Lim, M.H., Takefuji, Y.: Implementing Fuzzy Rule-Based Systems on Silicon Chips. IEEE Expert, pp. 31–45 (February 1990)
8. Gwee, B.H., Lim, M.H.: Polynominoes Tiling by a Genetic Algorithm. Computational Optimization and Applications Journal 6(3), 273–291 (1996)
9. Lim, M.H., Cao, Q., Li, J.H.: Evolvable Hardware using Context Switchable Fuzzy Inference Processor. IEE Proceedings - Computer and Digital Techniques 151(4) (July 2004)
10. Lim, M.H., Yu, Y., Omatu, S.: Extensive Testing of A Hybrid Genetic Algorithm for Solving Quadratic Assignment Problems. Computational Optimization and Applications 23, 47–64 (2002)

A Study on Neighborhood and Temperature in Multi-step Crossover Fusion for Tree Structure

Yoshiko Hanada[1], Koutaro Minami[2], Keiko Ono[3],
Yukiko Orito[4], and Noriaki Muranaka[1]

[1] Faculty of Engineering Science, Kansai University, Osaka, Japan
hanada@kansai-u.ac.jp
[2] Graduate School of Science and Engineering, Kansai University, Japan
[3] Department of Electronics and Informatics, Ryukoku University, Shiga, Japan
[4] Graduate School of Social Sciences, Hiroshima University, Hiroshima, Japan

Abstract. Multi-step crossover fusion (MSXF) is one of promising crossover methods for solving combinatorial optimization problems. MSXF performs multi-step local search from a parent in the direction approaching the other parent. In the transaction process of the local search, a neighborhood solution is stochastically accepted as the next solution, according to Metropolis criterion. To improve the search performance of MSXF, a sophisticated neighborhood structure and an appropriate scheduling of temperature parameter are required. In our previous work, we proposed an improved neighborhood generation method that promotes the recombination of parents' preferable characteristics for tree structures. In this paper, we validate the efficacy of the proposed method and evaluate the temperature scheduling in several symbolic regression problem instances.

Keywords: crossover, local search, distance, metropolis criterion, tree structure.

1 Introduction

Crossover operators or recombination operators have been considered to be the key component of population-based optimization algorithms. Especially in optimizing combinatorial structures such as a graph, it is important to design the operator, with considering problem-specific structures and characteristics. Various kinds of crossover focusing on the inheritance of parents' characteristics have been found in genetic algorithms (GAs) [1,2,3]. For optimization of tree structures found in genetic programming (GP) [4], the design of the crossover operator is also important. A tree structure is a special case of a graph; however it has some difficulties such as a strong dependence among variables due to a self-similarity, which makes the designing of crossover much harder. Several recombination mechanisms to treat tree features have been discussed [5,6,7,8,9]. Since crossovers might bring a drastic change in solutions and break favorable characteristics, in solving combinatorial optimization problems involving complex constraints, the incorporation of local searches into crossover is very effective in order to adjust the structural details of solutions [10].

Multi-step crossover fusion (MSXF) [11] is one of promising crossovers that performs a sequence of stochastic local search based on the Metropolis criterion. The local search gradually moves the offspring from its initial point to the other parent, and

© Springer International Publishing Switzerland 2015
H. Handa et al. (eds.), *Proc. of the 18th Asia Pacific Symp. on Intell. & Evol. Systems – Vol. 2*,
Proceedings in Adaptation, Learning and Optimization 2, DOI: 10.1007/978-3-319-13356-0_41

through the search it can generate a wide variety of solution between parents with keeping their characteristics. The high search performance of MSXF is actualized under the definition of a sophisticated neighborhood structure, and with an appropriate scheduling of temperature parameter. In order to enhance the heredity of parents' characteristic in the local search, an improved neighborhood generation in multi-step crossover fusion that mixes the partial structures of the parents step by step has been proposed [12]. Unlike original manner of MSXF that generates neighborhood solution randomly, the improved manner generates biased neighborhood solutions toward the target solutions.

In our previous work, we proposed a general manner to generate biased neighborhood solution, for tree structures, by defining a pair-wise tree distance [13]. We introduced the proposed manner to MSXF and showed the superior search performance to a conventional crossover. However the search performance in the original framework of MSXF has not been evaluated against the difficulty of the problem instances, and the effectiveness of the proposed manner for neighborhood generation is not clear. A discussion on the appropriate temperature scheduling for the tree structures still also remains. Here, we compare the neighborhood generation methods to show the effectiveness of original MSXF and to validate the efficacy of the proposed one. We adopt several symbolic regression problem instances including arithmetic operators and several numerical functions, and discuss the difference in the search performance against the difficulty of the instance. In addition, we evaluate the temperature scheduling.

2   Multi-step Crossover Fusion

We consider the problem of minimizing a function $f(x)$. First, we begin by recalling the Multi-step Crossover Fusion (MSXF). Let $P(g)$ denote the set of individuals in the gth generation, and $p_i(i = 1, \ldots I)$ denote the ith individual in $P(g)$. In MSXF, the set of individuals that inherit various traits from parents is generated by performing a sequence of stochastic local search from the parent $p_i \in P(g)$ to the parent $p_j \in P(g)$ $(i \neq j)$. In the local search from p_i to p_j, the k_{max} steps local search centering on r_k $(k = 1, \ldots, k_{max})$ is conducted, where the initial point r_1 is p_i. The crossover algorithm of the MSXF is as follows:

M0: Set $R(p_i, p_j) \leftarrow \emptyset$.
M1: Set $k \leftarrow 1$.
M2: Set $R(p_i, p_j) \leftarrow R(p_i, p_j) \cup \{p_i\}$.
M3: Set $r_k \leftarrow p_i$.
M4: If $k \leq k_{max}$ and $r_k \neq p_j$, then perform Steps M5 to M12, otherwise stop.
M5: Set $O(r_k) \leftarrow \emptyset$.
M6: Generate the set of μ candidates $O(r_k)$ randomly in the neighborhood of r_k, where $o_h \in O(r_k)$ $(h = 1, \ldots, \mu)$.
M7: Choose o^* from $O(r_k)$ with probability proportional to $d(o_h, p_j)$, where $d(o_h, p_j)$ is the distance between o_h and p_j.
M8: If $\Delta E > 0^1$ or according to probability proportional to $\exp(-\Delta E/T)$ then perform Step M9 to M10, otherwise return to Step M7.

---

[1] Here, $\Delta E = f(r_k) - f(o^*)$.

M9:   Set $r_{k+1} \leftarrow o^*$.
M10:   Set $R(p_i, p_j) \leftarrow R(p_i, p_j) \cup \{r_{k+1}\}$.
M11:   Set $k \leftarrow k + 1$.
M12: Return to Step M4.

Here, $O(r_k)$ denotes the set of neighborhood solutions of r_k. The set of offspring, $R(p_i, p_j)$, is generated from the parents p_i and p_j. To obtain $R(p_i, p_j)$, at most $k_{max} \times \mu$ solutions are evaluated through the local search. In the Step M7, the closer neighborhood to the parent p_j is more easily selected as the candidate for the transition. The distance $d()$ between two parents should be defined for ease of selection among the neighborhood solutions.

In the original framework of MSXF described above, the neighborhood solutions are generated randomly from the current solution r_k at the Step M6. To promote the recombination of parents' preferable characteristics, the improved manner for generating biased neighborhood solutions toward the parent p_j has been developed [12]. In our previous work, we adopted the improved manner **M6'** below instead of M6 in the algorithm.

M6' :   Generate the set of μ candidates $O(r_k)$ in the neighborhood[2] of r_k,
        where $o_h \in O(r_k)$ $(h = 1, \ldots, \mu)$ and $d(o_h, r_k) < d(o_h, p_j)$.

In this paper, we call a neighborhood solution a *random neighborhood* when M6 is adopted, and call it a *biased neighborhood* when M6' is adopted.

To apply MSXF, we use the following generation alternation model [3,12,14] that is familiar with local search strategies.

G0: Generate P(0) at random.
G1: Set $g \leftarrow 1$.
G2: If $g \leq G$, then perform Steps G3 to G16, otherwise stop.
G3:   Set $h \leftarrow 1$.
G4:   Set $P'(g) \leftarrow P(g)$.
G5:   Select $p_i \in P'(g)$ at random.
G6:   Set $P'(g) \setminus p_i$.
G7:   Set $p_0 \leftarrow p_i$.
G8:   If $h \leq I$, then perform Steps G5 to G14, otherwise stop.
G9:   If $h < I$, then perform Step G10, otherwise set $p_j \leftarrow p_0$ and
        perform Step G12.
G10:   Select $p_j \in P'(g)$ at random.
G11:   Set $P'(g) \leftarrow P'(g) \setminus p_j$.
G12:   Apply crossover (MSXF) to p_i and p_j.
G13:   Set $p^* \leftarrow \text{argmin}_{1 \leq k \leq |R(p_i,p_j)|} f(r_k)$.
G14:   Set $p_h \in P(g) \leftarrow p^*$.
G15:   Set $p_i \leftarrow p_j$.
G16:   Set $h \leftarrow h + 1$.
G17: Return to Step G4.
G18: Set $g \leftarrow g + 1$.
G19: Return to Step G2.

---

[2] Here, we state the most fundamental one. See section 4 for more detail.

3 Definition of Distance for Tree Structure

To apply MSXF, a pairwise distance should be defined. In the original manner of MSXF, the distance is used for the selection of a candidate solution. The improved manner generates neighborhood solutions with keeping the distance condition. In a discrete structure, the distance is often defined based on a difference of gene at the same locus. Since nodes are atomic elements in a tree structure, we define the distance based on the difference of node symbols. In a tree structure, unlike in the case of most GA, the solution size is not fixed, and correspondence of loci is not stable among solutions. Therefore, a kind of position adjustment between trees is required in order to define the locus, the position of node in the tree. We consider the most matching pattern of trees as a trunk to define corresponding nodes in two trees. The most matching pattern is called the largest common subgraph (subtree). Given a pair of trees, the largest common subtree (LSCT, for short) is defined as the isomorphic subtree that has the maximum number of vertices between the two trees. The values stored in the nodes, i.e., nodes' symbols, do not affect whether two subtrees are isomorphic.

Extracted LCSTs depend on the type of tree structure. Here, we treat a rooted tree where one vertex has been designated the root and consider two types of tree; an ordered tree (OT, for short) and an unordered tree (UT, for short). OT and UT are kinds of a directed acyclic connected graph with a fixed root node, and the former considers the appearance position among child nodes while the latter does not. This difference might have a significant effect when evolutionary algorithm such as GP treats a tree structure problem which has function (nonterminal) nodes that work unsymmetrically. For example, in symbolic regression problem, the subtraction is an unsymmetric function of which numerical results, $(a - b)$ and $(b - a)$, are different. OT tends to discriminate two cases in complex trees while they are sometimes considered as structurally equal in UT.

Figures 1 and 2 illustrate LCSTs in OT and UT, respectively. Here, u_* and v_* are nodes in trees A and B. $LSCT_X$ denotes the largest common subtree in tree X. The superscript number of each node means a locus allocated based on the largest common subtree. In the figures, trees consisting of black edges and nodes are LCSTs. Other gray bold edges and nodes constitute the different parts.

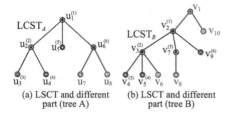

(a) LSCT and different part (tree A)

(b) LSCT and different part (tree B)

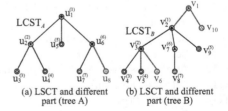

(a) LSCT and different part (tree A)

(b) LSCT and different part (tree B)

Fig. 1. LCSTs between trees A and B (OT): 6 pairs u_1-v_2, u_2-v_3, u_3-v_4, u_4-v_5, u_5-v_7 and u_6-v_9 are corresponding loci, respectively.

Fig. 2. LCSTs between trees A and B (UT): 7 pairs u_1-v_2, u_2-v_3, u_3-v_4, u_4-v_5, u_5-v_9, u_6-v_7 and u_7-v_8 are corresponding loci, respectively.

As shown in the figures, nodes in the tree are divided into two sets; the set of nodes constituting the LCST and the set of nodes included in the remaining part obtained by excluding the LCST from the tree. The corresponding loci can be defined based on matching vertices through the extraction of LCSTs between the two trees. In the figures, (·) means the index of a locus and the nodes with the same index are corresponding. In other remaining parts, no corresponding loci exist.

Here, for the nodes in $LCST_T$, U_T denotes the set of the nodes whose symbol is same as that of the corresponding node, and U'_T denotes the set of the nodes whose symbol differs from that of the corresponding node, respectively, i.e.,

$$LCST_T = G(U_T \cup U'_T, E_{LCST_T}),$$ (1)

where $G()$ means a graph and E_{LCST_T} denotes the set of the edges included in $LCST_T$. In addition, D_T denotes the set of the nodes not included in $LCST_T$, respectively. The distance of trees A and B, $d(A, B)$, is defined as

$$d(A, B) = \frac{1}{2}(|U'_A| + |D_A| + |U'_B| + |D_B|)$$ (2)

where $|\cdot|$ means the number of components included in the set. In the case of the trees shown in Fig. 1 that treat trees as OT, LD_A and OD_A are calculated as $\{u_2, u_3, u_5, u_6\}$ and $\{u_7, u_8\}$, respectively. For the tree B, LD_B and OD_B are calculated as $\{v_3, v_4, v_7, v_9\}$ and $\{v_1, v_6, v_8, v_{10}\}$, respectively. As a consequence, the distance, $d(A, B)$, is $(4 + 2 + 4 + 4)/2 = 7$.

4 Neighborhood Generation Method

In this paper, we compare the biased neighborhood generation and the random neighborhood generation. Here we show the procedures of the generation methods.

4.1 Biased Neighborhood

In our previous work, we adopted the improved neighborhood generation method that considers the distance constraint. The generation methods described below consistently generate biased neighborhoods without destroying the structure of LCST between parents. First, we introduce the following five functions and three fundamental operations to explain the proposed neighborhood generation method.

Functions. Let $N(p)$ be the set of the node included in the tree of an individual p. For all nodes $n \in N(p)$, the five functions are defined as follows.

- **parent(n)**: Return the parent node of the node n
- **child(n)**: Return the set of child nodes of the node n
- **root(n)** : Return the root node of the tree including node n.
- **st(n)**: Return the subtree whose root node is the node n
- **arg(n)**: Return the number of child nodes of the node n

Operations. As in the case of optimizing solutions with tree structures, the number of arguments, child nodes of a functional (nonterminal) node, is defined as constraints. In the symbolic regression problems we discuss in this paper, each of arithmetic operators and other functions has two child nodes or one child node, and constants and variables have no child nodes as terminal nodes.

Here, we introduce the following three types of operation for nodes; *Replace*, *Delete*, *Insert*, to describe the proposed neighborhood generation method. The neighborhood is defined as the trees that are generated by applying these operations during the fixed interval. Given individuals p_u and p_v ($p_u \neq p_v$), for nodes $n^u \in N(p_u)$ and $n^v \in N(p_v)$, three operations are defined below. Note that the constraint in terms of the number of arguments is satisfied after applying these operations.

- **Replace(n^u, n^v)** is an operator to replace the symbol of the node n^u with that of n^v in p_u, and then executes either one of the following two operations if $arg(n^u) \neq arg(n^v)$:
  - In the case of $arg(n^u) > arg(n^v)$: Select the set of nodes $\hat{N} = \{n_k | 1 \leq k \leq arg(n^u) - arg(n^v)\}$ from the set of nodes $child(n^u)$, and delete each subtree $st(n_k)$ ($n_k \in \hat{N}$) from p_u.
  - In the case of $arg(n^u) < arg(n^v)$: Generate the set of terminal nodes $\hat{N} = \{n_k | 1 \leq k \leq arg(n^v) - arg(n^u)\}$ at random, and insert each node of $\hat{N}$ to the node n^u as a child node, in p_u.

- **Delete(n^u)** is an operator to delete the node n^u from p_u, and applicable to only non-terminal nodes[3] except the root node. *Delete(n^u)* executes the following operations:
  1. Select a node from $child(n^u)$, and set it to $n \in N(p_u)$.
  2. Delete the node n^u and subtrees $st(n_k)$ where $n_k \in child(n^u) \setminus \{n\}$, from p_u.
  3. Connect the node n and the $parent(n^u)$.

- **Insert$_{des}$(n^u, n^v)** or **Insert$_{anc}$(n^u, n^v)** is an operator to insert the node n^v or the subtree $st(n^v)$ to a nonterminal node n^u as a child node, or push the nonterminal node n^v on the node n^u as the parent node.
  - *Insert$_{des}$(n^u, n^v)* works as follows:
    1. Select a node from $child(n^u)$, and set it to $n \in N(p_u)$.
    2. Insert n^v or the subtree $st(n^v)$ next to the node n and connect n^v and n^u.
    3. Delete the subtree $st(n)$ from p_u.
  - *Insert$_{anc}$(n^u, n^v)* below is applied if $n_i = root(n_i)$:
    1. Connect n^u to n^v as a child nodes.
    2. Generate the set of terminal nodes $\hat{N} = \{n_k | 1 \leq k \leq arg(n^v) - 1\}$ at random, and insert each node of $\hat{N}$ to the node n^v as a child node, in p_u.

Neighborhood Generation Method. Biased neighborhood solutions $O(r_k)$ is generated from the current solution r_k at each step k of the local search that moves from a

---

[3] *Delete* cannot be applied to terminal nodes due to the constraint about the number of arguments.

parent p_i to a parent p_j. The purpose of the generation manner is to add a few characteristics of the target p_j to r_k step by step. To reduce drastic changes, nodes surrounding LCST or included in LSCT are mainly exchanged, because LSCT is fundament to define the degree of coincidence between two trees.

Here, for tree T, we consider U_T and U'_T that are the set of nodes constituting $LCST_T$. In addition, we focus attention on the node that is a member of D_T and whose adjacent node is included in $LCST_T$, where D_T denotes the set of nodes not included in $LCST_T$. We call such a node a *surrounding node* and define the set of surrounding nodes as D_T^s. In the trees shown in Fig. 1, $D_A^s = \{u_7, u_8\}$ and $D_B^s = \{v_1, v_6, v_8\}$.

In Step M6' of MSXF, the biased neighborhood method generates solutions by the following procedures.

B0: Set $O(r_k) \leftarrow \emptyset$.
B1: If $|O(r_k)| < \mu$, then perform Step B2 to B19, otherwise stop.
B2: Set $r \leftarrow r_k$.
B3: If $distance(r, r_k) \leq distance(p_i, p_j)/k_{max}$, then perform Steps B4 to B17, otherwise perform B18.
B4: Select a node $n \in (U'_r \cup D_r^s \cup D_{p_j}^s)$.
B5: If $n \in U'_r$, then perform Step B6 to B7.
B6: Select $n^j \in LCST_{p_j}$ at the same locus with n.
B7: Apply $Replace(n, n^j)$ to r.
B8: If $n \in D_r^s$, then perform Step B9.
B9: Apply $Delete(n)$ to r.
B10: If $n \in D_{p_j}^s$ then perform Step11.
B11: If $parent(n) \in LCST_{p_j}$, then perform Step B12 to 13, otherwise[4] perform Step B14 to B16.
B12: Select $n^r \in LCST_r$ at the same locus with $parent(n)$.
B13: Apply $Insert_{des}(n^r, parent(n))$ to r.
B14: Select $n^j \in child(n) \cap (U'_{p_j} \cup U_{p_j})$[5].
B15: Select $n^r \in LCST_r$ at the same locus with n^j.
B16: Apply $Insert_{anc}(n^r, n^j)$ to r.
B17: Return to Step B3.
B18: Set $O(r_k) \leftarrow O(r_k) \cup \{r\}$.
B19: Return to Step B1.

4.2   Random Neighborhood

In the original framework of MSXF, the neighborhood solutions are generated randomly from the solution r_k at each step of local search. Here we adopt a simple manner to generate neighborhood by modifying subtrees in tree r_k. In the method, a subtree is selected randomly and replaced with another subtree. The size or the depth of the subtree newly generated is important. If it is excessively large compared to that of the excluded subtree, that might cause the bloat, while too small change in trees cannot search enough

---

[4] One node included in $child(n)$ is a member of $LCST_{p_j}$.
[5] The unique node that satisfies this condition necessarily exists.

in the decision space. Here, we adjust the size, the number of nodes, of subtree by a growing ratio a. When a subtree of which size is r is selected for the elimination, a new subtree are generated within the size of $a \times r$.

Here, $size(T)$ is a function that returns the number of nodes constituting the tree T. In Step M6 of MSXF, the random neighborhood solution set $O(r_k)$ in MSXF is calculated as follows.

R0: Set $O(r_k) \leftarrow \emptyset$.
R1: If $|O(r_k)| < \mu$, then perform Step R2 to R8, otherwise stop.
R2: Set $r \leftarrow r_k$.
R3: Select a node n from $N(r)$ except the root node.
R4: Generate a subtree within the size of $size(st(n)) \times a$, and let n^* be its root node.
R5: Put n^* next to n and connect n^* to $parent(n)$.
R6: Apply $Delete(n)$ to r.
R7: Set $O(r_k) \leftarrow O(r_k) \cup \{r\}$.
R8: Return to Step R1.

5   Numerical Experiments

In this section, we discuss the effect of neighborhood definition on the search performance of MSXF and validate the effectiveness of the biased neighborhood generation. The distance or generated neighborhood solutions depend on the definition of tree. For the biased neighborhood generation, we call a neighborhood *OT neighborhood* when we treat trees as OT, and call it *UT neighborhood* when we treat trees as UT. The effect of the temperature setting in MSXF is then evaluated.

5.1   Problem Domain and Instances

We evaluated the search performance on symbolic regression problem. Here, we considered four kinds of functions. They are estimated by partially using arithmetic operators {addition(+), subtraction(-), multiplication(*) and division(/)} and several numerical functions {modulo(%), sin, cos, exp and ln}. These operators appear as a nonterminal node in trees. The function sin, cos, exp and ln are unary and should have 1 child nodes, and other operators are binary and should have 2 child nodes. In binary operators, addition and multiplication are symmetric functions of which numerical result is stable, regardless of the order between operands, while the results of subtraction, division and modulo are not stable. For the terminal nodes, the variable x and several constants are adopted.

Table 1 shows instances for the experiments. V^{NT} denotes the set of nonterminal nodes (functions) and V^T denotes terminal nodes (the variable and constants). Instances (I) and (II) include only symmetric functions, while instances (III) and (IV) additionally include asymmetric functions. The latter type is difficult than the former type. Instance (I) and (III) can be expressed by only basic arithmetic operators. In these instances, all nonterminal nodes have consistently two child nodes. On the other hand, instances (II) and (IV) consists of arithmetic operators and several numerical functions. Some nonterminal nodes have only one child and that might make the instance complex. The domain of the variable in each instance is shown in Table 1 and the sample points are placed at regular interval in the domain. The fitness function is the sum of error of each sample point, and we assume that the population reaches the optimal solution when the fitness value becomes less than 0.01 for the instances (I) and (III), and less than 2.0 for the instances (II) and (IV).

Table 1. Instances of Symbolic Regression Problem

	function	Node set		Sample points	
		V^{NT}	V^T	Domain	#Points
I	$f_1^{opt} = x^4 + x^3 + x^2 + x$	{+, *}	{x}	[-1, 1]	21
II	$f_2^{opt} = x\sin x(\cos x + 1)$	{+, *, sin, cos, exp, ln}	{x, 0.1, 0.2, 0.3, 0.4, 0.5 0.6, 0.7, 0.8, 0.9, 1.0}	[0, 10]	101
III	$f_3^{opt} = x^3 - x^2$	{+, -, *, /}	{x}	[-1, 1]	21
IV	$f_4^{opt} = x\sin x(\cos x - 1)$	{+, -, *, /, %, sin, cos, } exp, ln }	{x, 0.1, 0.2, 0.3, 0.4, 0.5, 0.6, 0.7, 0.8, 0.9, 1.0}	[0, 10]	101

5.2 Comparison of Neighborhoods

Using the four symbolic regression problem instances, we discuss the performances of MSXFs using the biased neighborhoods and the random neighborhoods. For the comparison, the performance of 1 point crossover (1X) is also shown. In the experiments, the population size was set to 100 and each run was terminated after 40 generations. The generation alternation model described in section was used. The objective of the experiments is to examine the effect of crossovers, therefore neither mutations nor bloat controlling strategies were applied. The number of offspring generated by each pair of parents was set to 100. For MSXF, k_{max} was set to 5 and μ was 20. The initial temperature T_0 was set to 0.1 and updated as $T_{t+1} = 0.1 * T_t$ every 2 generations. To generate random neighborhoods, the growing ratio a was set to 1 and 3. For the random neighborhood generations, the distance definition based on OT was adopted in the local search. Initial solutions were generated randomly but kept the node generation ratio. Here, for the nonterminal nodes, the unary-to-binary generation ratio was set to 0.5:0.5. For the terminal nodes, the variable-to-constant generation ratio was set to 0.8:0.2. All nodes included in each kind were generated with equal probability. The number of nodes included in each initial solution was restricted to less than or equal to 25.

Table 2 shows the success rate (%success), the average of the number of the nodes constituting the obtained tree (size), and the average depth (depth) of trees out of 50 trials. In this table, x1 and x3 mean the MSXF with random neighborhood generation methods of $a=1$ and 3, respectively.

Table 2. Comparison of Neighborhoods

Instance	Crossover		%Success	Size	Depth	Instance	Crossover		%Success	Size	Depth
I	1X		1.00	16.0	6.2	III	1X		0.0	186.2	26.8
	MSXF	OT	1.00	18.3	5.6		MSXF	OT	0.52	30.2	5.8
		UT	1.00	17.0	5.4			UT	0.38	33.1	6.3
		x1	0.96	14.0	4.7			x1	0.04	11.3	4.1
		x3	1.00	19.1	5.8			x3	0.80	43.4	9.2
II	1X		0.5	48.4	16.3	IV	1X		0.50	59.0	17.5
	MSXF	OT	0.84	11.9	5.6		MSXF	OT	0.84	17.0	7.7
		UT	0.78	11.6	5.7			UT	0.56	18.1	8.2
		x1	0.06	5.7	2.7			x1	0.00	6.9	3.3
		x3	0.76	20.2	7.0			x3	0.18	33.5	9.8

From Table 2, except x1, MSXFs are found to be superior to the conventional crossover, 1X, in the success rate. The marked effectiveness can be observed in the depth of constructed tree, i.e., MSXFs can find precise solutions with shallow trees. In MSXF with the random neighborhood, x1, trees do not grow due to too small change, which cannot search enough in the decision space. By enlarging the size of random substitute subtree for an excluded subtree, the growth of tree is observed and the success rate is improved; however that has possibilities to cause the bloat. In this result, we showed only the result of x3. In other settings of a, more than 2, the marked growth is also observed. On the other hand, we confirm that the biased neighborhoods, OT neighborhood and UT neighborhood, can find compact but precise tree by promoting the recombination of parents' preferable characteristic. From the comparison of OT neighborhood and UT neighborhood, we can reconfirm that the former is superior to the latter, regardless of the instance properties. OT neighborhood that considers the symmetric property between child nodes outperforms UT neighborhood even on the instance which does not require unsymmetric functions.

5.3   Comparison in Temperature Setting

The effect of temperature scheduling is discussed by adjusting the initial temperature and the cooling rate. Here we use the instance (IV) and in the experiments, each run was terminated after 100 generations. The initial temperature was varied from 0.01 to 5. Two cooling rates, a rapid cooling (0.1) and a slow cooling (0.8) were evaluated. Other parameters were the same as those described in the previous section.

Figures 3 and 4 show the comparison of the cooling on four neighborhood generations method, in the success rate and the average size of obtained trees. Regardless of the initial temperature and the cooling rate, biased neighborhood generations outperform random neighborhood generations, i.e., the biased one can find optimal solutions with compact trees. The performance in higher initial temperature with the slow cooling tends to decay in the biased neighborhood methods, from the perspective of the success ratio.

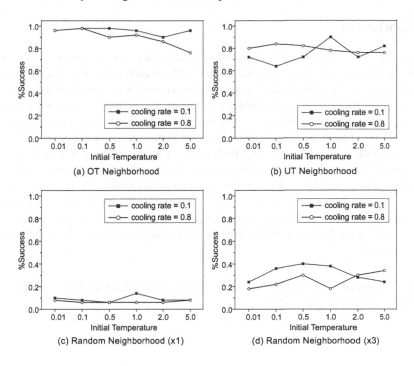

Fig. 3. Comparison of Cooling Scheduling (%Success)

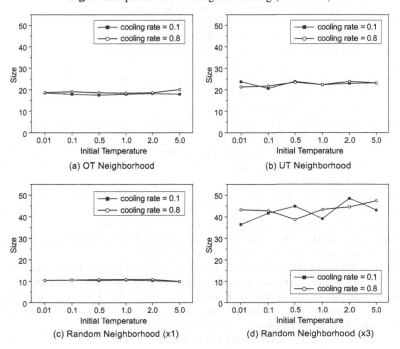

Fig. 4. Comparison of Cooling Scheduling (Size)

6 Conclusion

In this paper, we reconfirmed the positive effectiveness of two proposed biased neighborhood generation in MSXF by comparing it with a random neighborhood generation adopted in typical local searches. The search performance was evaluated in two types of instance of symbolic regression problem; instances that include only symmetric functions and instances that additionally include asymmetric functions. Through the experiments, OT neighborhood that considers the symmetric property between child nodes was found to be superior to UT neighborhood even on the instance which does not require unsymmetric functions. In addition, we confirmed that OT neighborhood outperforms the other neighborhood generation methods, regardless of temperature scheduling. For the examination we chose a simple conventional method to assess the search performance but we should compare them with some state-of-the-arts. In addition, the detailed analyses of behaviors of the proposed method are required. These tasks are left as a future goal.

Acknowledgment. This work was supported by JSPS KAKENHI Grant Numbers 26330290, 25730148, and 26730133.

References

1. Ono, I., Kobayashi, S.: A Genetic Algorithm Taking Account of Characteristics Preservation for Job Shop Scheduling Problems. In: Proc. of the International Conference on Intelligent Autonomous Systems, vol. 5, pp. 711–718 (1998)
2. Sakuma, J., Kobayashi, S.: Extrapolation-Directed Crossover for Job-shop Scheduling Problems: Complementary Combination with JOX. In: Proc. Genetic and Evolutionary Computation Conference 2000, pp. 973–980 (2000)
3. Nagata, Y.: New EAX crossover for large TSP instances. In: Runarsson, T.P., Beyer, H.-G., Burke, E.K., Merelo-Guervós, J.J., Whitley, L.D., Yao, X. (eds.) PPSN IX. LNCS, vol. 4193, pp. 372–381. Springer, Heidelberg (2006)
4. Koza, J.R.: Genetic Programming: On the Programming of Computers by Means of Natural Selection. MIT Press, Cambridge (1992)
5. Francone, F.D., Conrads, M., Banzhaf, M., Nordin, P.: Homologous Crossover in Genetic Programming. In: Proc. Genetic and Evolutionary Computation Conference 1999, vol. 2, pp. 1021–1026 (1999)
6. Poli, R., McPhee, N.F., Rowe, J.E.: Exact Schema Theory and Markov Chain Models for Genetic Programming and Variable-length Genetic Algorithms with Homologous Crossover. Genetic Programming and Evolvable Machines 5(1), 31–70 (2004)
7. Hasegawa, Y., Iba, H.: A Bayesian Network Approach to Program Generation. Proc. of IEEE Trans. Evolutionary Computation 12(6), 750–764 (2008)
8. Beadle, L., Johnson, C.G.: Semantically Driven Crossover in Genetic Programming. In: Proc. IEEE World Congress on Computational Intelligence 2008, pp. 111–116 (2008)
9. Ono, K., Hanada, Y., Shirakawa, K., Kumano, M., Kimura, M.: Depth-dependent crossover in genetic programming with frequent trees. In: Proc. 2012 IEEE International Conference on Systems, Man, and Cybernetics, pp. 359–363 (2012)
10. Freisleben, B., Merz, P.: New Genetic Local Search Operators for the Traveling Salesman Problem. In: Ebeling, W., Rechenberg, I., Voigt, H.-M., Schwefel, H.-P. (eds.) PPSN 1996. LNCS, vol. 1141, pp. 890–899. Springer, Heidelberg (1996)

11. Yamada, T., Nakano, R.: Scheduling by Genetic Local Search with Multi-Step Crossover. In: Ebeling, W., Rechenberg, I., Voigt, H.-M., Schwefel, H.-P. (eds.) PPSN 1996. LNCS, vol. 1141, pp. 960–969. Springer, Heidelberg (1996)
12. Ikeda, K., Kobayashi, S.: Deterministic multi-step crossover fusion: A handy crossover composition for GAs. In: Guervós, J.J.M., Adamidis, P.A., Beyer, H.-G., Fernández-Villacañas, J.-L., Schwefel, H.-P. (eds.) PPSN 2002. LNCS, vol. 2439, pp. 162–171. Springer, Heidelberg (2002)
13. Hanada, Y., Hosokawa, N., Ono, K., Muneyasu, M.: Effectiveness of Multi-step Crossover Fusions in Genetic Programming. In: Proc. IEEE Congress on Evolutionary Computation, pp. 1743–1750 (2012)
14. Hanada, Y., Hiroyasu, T., Miki, M.: Genetic Multi-step Search in Interpolation and Extrapolation domain. In: Proc. of Proc. Genetic and Evolutionary Computation Conference 2007, pp. 1242–1249 (2007)

Considering Reputation in the Selection Strategy of Genetic Programming

Chiao-Jou Lin, Rung-Tzuo Liaw, Chien-Chih Liao, and Chuan-Kang Ting

Department of Computer Science and Information Engineering and
Advanced Institute of Manufacturing with High-tech Innovations
National Chung Cheng University, Taiwan
{lrt101p,lcch97p,ckting}@cs.ccu.edu.tw

Abstract. Genetic programming (GP) is an evolutionary algorithm inspired by biological evolution. GP has shown to be effective to build prediction and classification model with high accuracy. Individuals in GP are evaluated by fitness, which serves as the basis of selection strategy: GP selects individuals for reproducing their offspring based on fitness. In addition to fitness, this study considers the reputation of individuals in the selection strategy of GP. Reputation is commonly used in social networks, where users earn reputation from others through recognized performance or effort. In this study, we define the reputation of an individual according to its potential to produce good offspring. Therefore, selecting parents with high reputation is expected to increase the opportunity for generating good candidate solutions. This study applies the proposed algorithm, called the RepGP, to solve the classification problems. Experimental results on four data sets show that RepGP with certain degrees of consanguinity can outperform two GP algorithms in terms of classification accuracy, precision, and recall.

1 Introduction

Genetic programming (GP) is an evolutionary algorithm known for its effectiveness on machine learning tasks [6]. Similar with other evolutionary algorithms, GP mimics the operators of natural evolution such as selection, crossover, and mutation. However, GP particularly uses high-level building blocks of variable length: their size and complexity can change during evolution. The individuals in GP are usually represented as parse tree, which facilitates GP to deal with rules, programs, and mathematical expressions for various problems. In addition, due to the structure and characteristic of tree, the results of GP are easier to be interpreted. GP has been used in plenty of applications, especially in machine learning tasks such as classification. Many studies have shown that GP is able to build prediction and classification model with high accuracy on real-world applications [10,12,7,8,9,4,3]. Espejo et al. [4] conducted a survey of existing research on GP in classification and showed that GP can help construct useful and accurate classifiers in several ways. In addition to learning of classifiers, GP can be used in data preprocessing, e.g., feature selection and construction. In

© Springer International Publishing Switzerland 2015
H. Handa et al. (eds.), *Proc. of the 18th Asia Pacific Symp. on Intell. & Evol. Systems – Vol. 2,*
Proceedings in Adaptation, Learning and Optimization 2, DOI: 10.1007/978-3-319-13356-0_42

the credit industry, GP is employed to build the credit scoring models to recognize whether the loan customers are creditworthy or not in the transactions of banks [7,8,1]. These studies show that GP performs better than artificial neuron network, decision tree, and logistic regression.

Fitness plays an important role in GP because it serves as the basis of selection and guides the search in GP. In this study, we propose using reputation as another metric for evaluating individuals. In the light of reputation, the evaluation of an individual is determined by the performance of its related individuals in the population. In human society, the reputation of a person acquired from others is generally more reliable than one's own promotion. Additionally, in social networks, users can be ranked according to their interaction with others on a community question/answer portal, such as Yahoo! Answers. Users raise their levels by providing high-quality answers. As for social media, the number of followers of a user implies one's popularity. Therefore, each unit can be assessed and gain reputation from others according to its performance among all units. Reputation can be used to rank units or determine the competition winner.

This study applies the notion of reputation in the selection strategy of GP. Specifically, reputation is defined as the ability to produce good offspring in GP. Individuals having high reputation represent that they hold high potential for producing strong offspring. Considering that parent selection aims to choose individuals promising for generating good offspring, we replace fitness with reputation as the evaluation metric in the parent selection of GP. For this, we present a method to calculate the reputation of an individual and introduce the degree of consanguinity as a parameter for reputation. This study examines the effectiveness of the proposed GP on the classification task. Four data sets including 2-label and 3-label data are adopted in the experiments. According to the experimental results, GP using reputation can outperform the original GP in test accuracy.

The remainder of this paper is organized as follows. Section 2 reviews the related work on reputation. The proposed GP is described in Section 3. Section 4 presents the experimental results. Finally, we draw conclusions of this work in Section 5.

2   Related Work

Reputation is a rating gained from others in appreciation of one's effort or performance. In the social media and networks, the rating behavior is common on question/answering portals and forums, where participants gain their reputation through providing satisfactory answers. Zhang et al. [13] tested some network-based ranking algorithms to find out users with high expertise on the Java Forum. They considered the number of questioners and answerers and claimed that the rank of answerers should be promoted. The expertise scores are propagated through the online community. Ipeirotis et al. [5] evaluated the quality of workers on Amazon Mechanical Turk, which is a crowdsourcing platform for people to offer money for solutions. Identifying correct answers and high quality workers is an important task for avoiding spammers and ineffective workers. The EM algorithm is used to estimate the quality of a worker by comparing the submitted answers with correct answers judged by several workers.

Another common issue on community question/answering portals is to rank the quality of answers. Agichtein et al. [2] proposed a ranking system to automatically assess the answers on Yahoo! Answers. They classified the quality of answers into three levels and the results are used to identify the high-quality content on Yahoo! Answers. The relationships between users, answers, and questions are built to analyze their interaction. The ranking problem is formulated as a binary classification problem. The evaluation results indicate that the proposed ranking system assesses the answer quality as correctly as humans do. Su et al. [11] analyzed the data quality, throughput, and user behavior to collect the human-reviewed data at Internet-scale. They further conducted experiments on Yahoo! Answers and concluded that human-reviewed data on question/answer forum achieve high similarity with the assessment of answers and questions made by human.

Motivated by the above-stated utility of reputation in social networks, this study proposes using reputation in place of fitness in the parent selection to improve the performance of GP in classification. More details about the proposed GP are described below.

3   Methodology

This study presents a GP using reputation, called RepGP, to solve the classification problem. More specifically, GP is used as a learning algorithm for training classifiers to classify instances. In RepGP, parents are selected according to their reputation rather than fitness. This section introduces GP for classification and then describes the proposed RepGP in details.

3.1   Genetic Programming for Classification

Following the principle of natural evolution, GP applies parent selection, recombination (crossover), and survival selection to evolve candidate solutions into the optima. Figure 1 shows the flow chart of using GP in classification. The training set is used to evaluate the individuals; that is, the accuracy of a classifier (represented by an individual in GP) on the training data serves as the fitness value of the individual. In the fitness evaluation, each individual classifies the training data and obtains the ratio of correctly classified instances. The fitness of an individual x is thus defined by the accuracy of classification, i.e.,

$$f(x) = \frac{\text{number of instances correctly classified by } x}{\text{number of instances in the training set}}. \tag{1}$$

After the evolutionary process, GP obtains a population of classifiers. Considering these classifiers made up of different parameters and structures, we adopt them as multiple classifiers and use a weighted voting to establish the

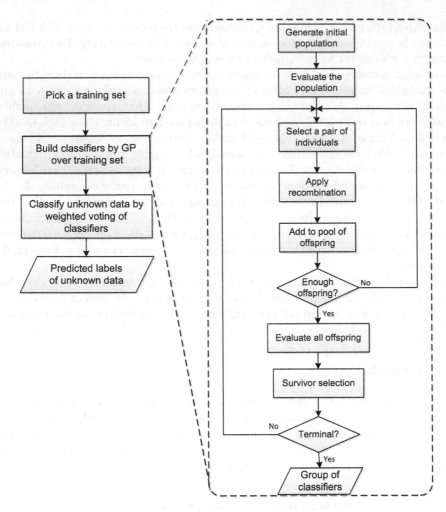

Fig. 1. Flow chart of classification using GP

final classifier for the proposed GP. In this study, we select only the top 20% of individuals in the population to participate in the voting. The weight of a classifier is determined by the proportion of its fitness value to the sum of all voters' fitness.

3.2   Reputation in Parent Selection

This study applies the notion of reputation in the parent selection of GP. The reputation of an individual is calculated after survival selection. Different from conventional GP, the proposed RepGP selects parents according to reputation instead of fitness. Generally, an individual with higher reputation has a higher probability to be selected as a parent for reproduction. In this study, we regard reputation as the potential for producing good offspring. An individual earns reputation from its descendants that can survive after survival selection. In this regard, the reputation is based on the consanguinity of survival individuals: an individual that has more descendants surviving will receive more reputation. This study introduces the degree of consanguinity ρ to limit the scope of reputation propagation. For example, parents are of first-degree consanguinity and grandparents are of second-degree consanguinity.

Here we utilize the thread structure to demonstrate the reputation defined in this study. An individual gives reputation to its ancestry, limited by the parameter ρ (degree of consanguinity). In addition, the amount of reputation obtained is dependent upon the ratio of genes inherited from ancestries. For simplicity, this study adopts the expectation $\frac{1}{2}$ as the ratio; that is, each individual is assumed to inherit $\frac{1}{2^k}$ of genes from its k_{th}-degree ancestry. Figure 2 illustrates the inheritance ratios in a family. The solid lines indicate direct inheritance, while the dashed lines indicate indirect inheritance.

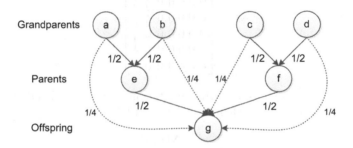

Fig. 2. Example of inheritance ratios

The reputation value is determined by the inheritance ratio. Noteworthily, the sum of all proportions of reputation issued from a specific individual is stipulated to be one. Figure 3 shows the proportions of transferred reputation in the example of Figure 2. According to the inheritance ratio, individual g transfers $\frac{1}{4}$ reputation to its parents e and f and $\frac{1}{8}$ to its grandparents a and b, whereas individual e transfers $\frac{1}{2}$ reputation to its parents a and b.

To calculate reputation, we define a reputation matrix $\mathbf{M} = [m_{ij}]_{N \times N}$ to record the proportions of reputation transferred to each individual, where N denotes the population size and element m_{ij} stands for the proportion of reputation transferred from individual j to individual i. Note that each row of $\mathbf{M}$ gives the total proportion of reputation that individual i receives from its descendants, while each column of $\mathbf{M}$ gives the total proportion of reputation that

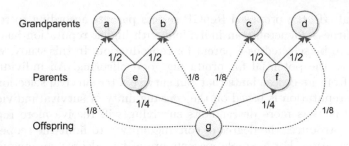

Fig. 3. Example of proportions of reputation transferred

individual j transfers to its ancestries. As above stated, the proportions of reputation transferred from an individual should be summed to one; therefore, the sum of each non-empty column is one, i.e., $\sum_{j=1}^{N} m_{ij} = 1$ if $\exists j, m_{ij} \neq 0$. For example, the reputation matrix for Fig. 3 is

$$\mathbf{M} = \begin{bmatrix} 0 & 0 & 0 & 0 & \frac{1}{2} & 0 & \frac{1}{8} \\ 0 & 0 & 0 & 0 & \frac{1}{2} & 0 & \frac{1}{8} \\ 0 & 0 & 0 & 0 & 0 & \frac{1}{2} & \frac{1}{8} \\ 0 & 0 & 0 & 0 & 0 & \frac{1}{2} & \frac{1}{8} \\ 0 & 0 & 0 & 0 & 0 & 0 & \frac{1}{4} \\ 0 & 0 & 0 & 0 & 0 & 0 & \frac{1}{4} \\ 0 & 0 & 0 & 0 & 0 & 0 & 0 \end{bmatrix}.$$

Let $\boldsymbol{r} = (r_1, \ldots, r_N)^{\top}$ denote the reputation values of all individuals. The reputation after transfer can then be calculated by

$$\boldsymbol{r}' = \mathbf{M}\boldsymbol{r}.$$

4   Experimental Results

This study conducts experiments to examine the performance of RepGP in classification problems. Two standard GP methods, i.e., GP1 using (μ, λ) and GP2 using $(\mu + \lambda)$ survival selection, are included in the performance comparison. Note that RepGP can only use $(\mu + \lambda)$ survival selection in that it requires the relationship of ancestry and posterity. Table 1 summarizes the parameter setting for the three algorithms used in the experiments. This study experiments with different degrees of consanguinity to examine its effect on RepGP performance. Four data sets from UCI Machine Learning Repository are used in the experiments: Australian Credit data (ACD) and German Credit data (GCD) are of two labels; Iris data (ID) and Wine data (WD) are of three labels. Table 2 lists the properties of these data sets. Each algorithm is tested using 10-fold cross-validation.

First, we investigate the accuracy of classification on the test data for the three algorithms. Accuracy measures the number of correctly classified instances

Table 1. Parameters setting

Parameters	GP1	GP2	RepGP
Population size		10000	
Parent selection		Greedy over-selection	
Parent selection strategy	Fitness	Fitness	Reputation
Crossover		One-point crossover with $p_c = 1.0$	
Survival selection	(μ, λ)	$(\mu + \lambda)$	$(\mu + \lambda)$
Survival selection strategy		Fitness	
Generation		500	
Cross-validation		10-fold cross-validation	
Degree of consanguinity (ρ)	-	-	1, 2, 3, 5

Table 2. Property of data sets

Data set	Volume	#Features	#Labels	#Label1	#Label2	#Label3
ACD	690	14	2	387	303	-
GCD	1000	20	2	700	300	-
ID	150	4	3	50	50	50
WD	178	13	3	59	71	48

against the number of total instances in the test set. According to Table 3, RepGP with certain degrees of consanguinity can outperform GP1 and GP2 on ACD, GCD and ID. This preferable outcome shows the capability of reputation to improve GP in classification. Although RepGP can achieve better results, the trend of best consanguinity degree is inconclusive.

Table 3. Accuracy (%) of GP1, GP2, and RepGP(ρ) with different degrees of consanguinity (ρ) on the four test data sets

	ACD	GCD	ID	WD
GP1	84.90	72.60	93.26	95.56
GP2	84.78	73.70	94.89	**95.58**
RepGP(1)	85.07	72.90	96.44	93.82
RepGP(2)	**86.52**	74.30	96.67	94.19
RepGP(3)	84.78	72.10	**98.78**	94.76
RepGP(5)	83.19	**74.40**	96.44	94.76

Considering that accuracy may be misleading if the data set is unbalanced, we further examine the precision and recall of the three algorithms. Precision measures the number of correctly classified positive instances against the number of instances classified as positive. High precision is desirable for reducing

Table 4. Precision (%) of GP1, GP2, and RepGP(ρ) with different degrees of consanguinity (ρ) on the four test data sets

	ACD	GCD	ID	WD
GP1	82.26	**79.58**	92.76	93.22
GP2	80.79	58.45	92.62	**97.09**
RepGP(1)	82.28	56.94	92.95	91.94
RepGP(2)	**85.43**	59.47	**95.30**	93.41
RepGP(3)	83.00	54.88	92.81	94.48
RepGP(5)	79.38	59.57	94.08	93.99

Table 5. Recall (%) of GP1, GP2, and RepGP(ρ) with different degrees of consanguinity (ρ) on the four test data sets

	ACD	GCD	ID	WD
GP1	83.06	40.33	94.00	87.30
GP2	**86.32**	42.67	92.00	88.83
RepGP(1)	84.69	39.67	**96.67**	90.48
RepGP(2)	84.04	45.00	94.67	89.95
RepGP(3)	82.94	39.33	94.67	90.48
RepGP(5)	84.04	**45.67**	95.33	**91.01**

the number of negative instances that are incorrectly predicted as positive instances. On the other hand, recall measures the number of correctly classified positive instances against the number of positive instances. High recall is desirable for reducing the number of positive examples that are incorrectly predicted as negative examples. Tables 4 and 5 show the precision and recall of the three algorithms on the four data sets. The results show that RepGP can achieve best precision on ACD and ID and best recall on GCD, ID, and WD. These outcomes reconfirm the advantage of reputation in GP. However, the trend of consanguinity degree's effect on precision and recall is inconclusive as well. In comparison of Tables 3 and 4, high accuracy does not imply high precision and recall.

Figure 4 shows the anytime behavior of the three algorithms in terms of training accuracy during evolution. Note that training accuracy is used as the fitness value for the three GP algorithms. The figure shows that GP1 converges fastest with best training accuracy on three data sets; however, Table 3 indicates that GP1 is generally inferior to GP2 and RepGP in test accuracy. This outcome reveals GP1 may suffer from overfitting. By contrast, RepGP generally converges slower on the four data sets but achieves better test accuracy than GP1 and GP2.

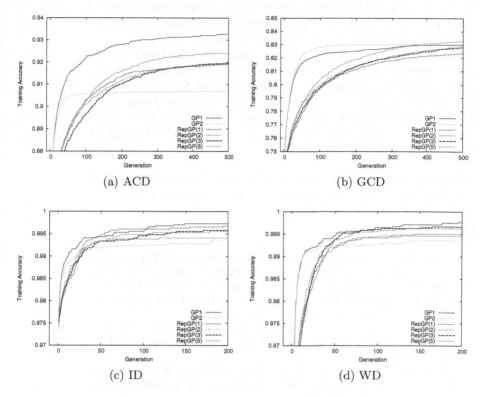

(a) ACD

(b) GCD

(c) ID

(d) WD

Fig. 4. Variation of training accuracy during evolution

5   Conclusions

This study proposes using reputation in place of fitness in the parent selection strategy of GP. A significant difference from fitness is that the reputation of an individual is gained from others. This study calculates reputation based on the consanguinity between individuals. More specifically, the individuals with more descendants surviving can gain more reputation, which reflects the individuals' potential for producing good offspring. The proposed RepGP therefore selects the individuals with high reputation as parents for reproduction. In calculating reputation, this study considers the consanguinity and uses the genealogy to construct the reputation matrix. The reputation of an individual can then be computed by the matrix. In addition, the degree of consanguinity is introduced to control the scope of reputation transfer.

Several experiments were conducted to examine the performance of RepGP in comparison with two GP algorithms. The results show that RepGP using certain degrees of consanguinity can outperform GP using (μ, λ) and $(\mu + \lambda)$ survival selection in terms of test accuracy, precision, and recall. The results also reveal that, although RepGP has a slower convergence and lower accuracy on the training data, it can achieve higher accuracy on the test data than the other two GP algorithms. These preferable outcomes validate the effectiveness of reputation on improving GP in classification.

Some directions remain for future work. First, the trend of consanguinity degree's effect on the RepGP performance is inconclusive. Determining or controlling the consanguinity degree is a promising task for enhancing RepGP. Second, the proposed reputation is applied to replace fitness in parent selection and is independent of the algorithmic framework of GP. Thus, incorporating the parent selection using reputation into some state-of-the-art GP is another promising direction to explore.

Acknowledgment. This work was supported by Ministry of Education, Taiwan, and Metal Industries Research and Development Centre.

References

1. Abdou, H.A.: Genetic programming for credit scoring: The case of egyptian public sector banks. Expert Systems with Applications 36(9), 11402–11417 (2009)
2. Agichtein, E., Castillo, C., Donato, D., Gionis, A., Mishne, G.: Finding high-quality content in social media. In: Proceedings of the International Conference on Web Search and Web Data Mining, pp. 183–194 (2008)
3. Al-Madi, N., Ludwig, S.A.: Improving genetic programming classification for binary and multiclass datasets. In: Center for Information-Development Management, pp. 166–173 (2013)
4. Espejo, P.G., Ventura, S., Herrera, F.: A survey on the application of genetic programming to classification. IEEE Transactions on Systems, Man, and Cybernetics, Part C 40(2), 121–144 (2010)
5. Ipeirotis, P.G., Provost, F., Wang, J.: Quality management on amazon mechanical turk. In: Proceedings of the ACM SIGKDD Workshop on Human Computation, HCOMP 2010, pp. 64–67 (2010)
6. Koza, J.R.: Genetic programming as a means for programming computers by natural selection. Statistics and Computing 4(2), 87–112 (1994)
7. Ong, C.S., Huang, J.J., Tzeng, G.H.: Building credit scoring models using genetic programming. Expert Systems with Applications 29(1), 41–47 (2005)
8. Ong, C.S., Huang, J.J., Tzeng, G.H.: Two-stage genetic programming (2sgp) for the credit scoring model. Applied Mathematics and Computation 174(2), 1039–1053 (2006)
9. Riekert, M., Malan, K., Engelbrecht, A.P.: Adaptive genetic programming for dynamic classification problems. In: IEEE Congress on Evolutionary Computation, pp. 674–681 (2009)
10. Savic, D.A., Walters, G.A., Davidson, J.W.: A genetic programming approach to rainfall-runoff modelling. Water Resources Management 13(3), 219–231 (1999)
11. Su, Q., Pavlov, D., Chow, J.H., Baker, W.C.: Internet-scale collection of human-reviewed data. In: Proceedings of the 16th International Conference on World Wide Web, pp. 231–240 (2007)
12. Whigham, P.A., Crapper, P.F.: Modelling rainfall-runoff using genetic programming. Mathematical and Computer Modelling 33(6-7), 707–721 (2001)
13. Zhang, J., Ackerman, M.S., Adamic, L.: Expertise networks in online communities: structure and algorithms. In: Proceedings of the 16th International Conference on World Wide Web, pp. 221–230 (2007)

Nature-Inspired Chemical Reaction Optimisation Algorithm for Handling Nurse Rostering Problem

Yahya Z. Arajy and Salwani Abdullah

Data Mining and Optimisation Research Group (DMO),
Centre for Artificial Intelligence Technology,
Universiti Kebangsaan Malaysia, 43600 Bangi Selangor, Malaysia
yahya.araj@gmail.com, salwani@ukm.edu.my

Abstract. The optimisation of the nurse rostering problem is chosen in this work seeking to improve the organization of hospital duties and to elevate health care by enhancing the quality of the decision-making process. Nurse rostering is a difficult and complex problem with a large number of demands and requirements that conflict with hospital workload constraints in terms of employee work regulations and personal preferences. We propose a variable population-based metaheuristic algorithm, the chemical reaction optimisation (CRO), to solve the NRP at the First International Nurse Rostering Competition (2010). The CRO algorithm features an adaptive search procedure that systematically controls the selection between an intensive search strategy and diversification search based on specific criteria to reach the best solution. Computational results were measured with three complexity levels as a total of 30 variant instances based on real-world constraints.

Keywords: nurse rostering problem, chemical reaction optimisation, real-world problems.

1 Introduction

Health care is an important factor in real-life practice. The nurse rostering problem (NRP) is listed as one of the scheduling problems encountered in a real-world modern hospital. Nurse rostering involves planning a periodical schedule for nurses by assigning daily duties for each nurse, which integrates the requirement to satisfy a number of constraints. All hospital demands (hard constraints) must be fulfilled (i.e., assigned all shifts required for a number of nurses), while accommodating the various requirements of the nurses (i.e., workload and personal demands) to obtain a feasible schedule (roster) that practically works. NRP must be solved by effectively utilizing limited resources to enhance hospital efficiency, while considering the well-being and job satisfaction of the nurses. This arrangement can elevate health care by improving the quality of the decision-making process. The importance of health care, as well as the complex NP-hard problem [1] that involves various requirements and real-world constraints, provides a scientific challenge for researchers. NRP is one of the most intensively explored topics. Over the last 45 years, research on the effect of different

© Springer International Publishing Switzerland 2015
H. Handa et al. (eds.), *Proc. of the 18th Asia Pacific Symp. on Intell. & Evol. Systems – Vol. 2*,
Proceedings in Adaptation, Learning and Optimization 2, DOI: 10.1007/978-3-319-13356-0_43

techniques and approaches on NRP has aimed to understand and solve this problem efficiently. These include the studies of Burke et al. [1], Ernst et al. [2], and Cheang et al. [3].

One of the first methods implemented to solve NRP is integer and mathematical programming. Warner and Prawda [4] presented a mathematical programming model for scheduling of nurses in a hospital. Thornton and Sattar [5] revisited nurse rostering and integer programming. Millar and Kiragu [6] attempted cyclic and non-cyclic scheduling using network programming. Moz and Pato [7, 8] applied an integer multi-commodity flow model, developed mathematical ILP formulations with additional constraints in a multi-level acyclical network, and used CPLEX solver. These techniques were used to guarantee the production of the most stable and desirable solutions. However, these techniques failed when utilized to solve a large number of constraints with a large number of instances. A significant deterioration in the quality of the solutions was observed when a vast search space problem (e.g., nurse rostering) was involved.

Researchers have applied heuristic/metaheuristic algorithms (e.g., evolutionary algorithms, VNS, simulated annealing, scattered search, and tabu search) in their attempts to solve NRP. Brusco and Jacobs [9] generated a cyclic schedule for continuously operating organizations. They combined simulated annealing metaheuristic with basic local search heuristic and applied the model on hospitals and other labor sectors continuously. Burke et al. [10] integrated the tabu search method with commercial techniques to solve NRP in Belgian hospitals. Burke et al. [11] also proposed a hybridization method to improve the quality of solutions. They applied a variable neighbourhood search method to search for new solutions by using heuristic ordering for repairing the schedule along with backtracking. The technique depended on two neighbourhood structures that successively assigned shifts to a free duty nurse and swapped shifts between two assigned nurses. Bellanti et al. [12] introduced a tabu search procedure and an iterated local search to solve NRP. Both hard and soft constraints are violated during the process, and various neighbourhood structures are operated on partial solutions. This procedure assists the search in working intensively to satisfy all constraints with each move; thus, avoiding the generation of infeasible rosters. Burke et al. [13] developed two hybrid tabu search algorithms, with diversification and greedy shuffling heuristics as presented in Bellanti et al. [12], who used the algorithms on a number of neighbourhoods to assign work shifts for a free duty nurse, swap shifts between two assigned nurses, and replace parts of the working schedule of nurses. The diversification method can be used to escape the current search space and allow a wild search in case the solution becomes trapped in local optima.

The present work focuses on the latest NRPs presented in the First International Nurse Rostering Competition (INRC-2010) organized by the International Series of conferences on the Practice and Theory of Automated Timetabling (PATAT-2010). INRC-2010 provided datasets with various complexity levels that incorporated a significant number of common real-world constraints [14].

A number of approaches have been anticipated by the participants of the competition in relation to this challenging and interesting problem. We briefly review the methods proposed by the INRC-2010 competition finalists.

Burke and Curtoiso [15] proposed two algorithms to solve NRP. The first algorithm, a variable depth search, is an ejection chain-based method. The algorithm was tested on short-term instances (sprint; 10-Seconds-timeframe) for the 2010 nurse rostering completion data. The algorithm initially uses a hill climbing strategy and employs different neighbourhoods. Burke and Curtoiso applied a dynamic programming algorithm in the constructive phase to build an individual schedule for each nurse. The second algorithm, a branch and price method, was applied based on the complexity of instances (medium and long). The algorithm embeds the pricing problem as a resource constrained shorted path problem. Similarly, Burke and Curtoiso applied the same technique as the first algorithm and solved NRP using a dynamic programming approach.

Valouxis et al. [16] attempted to solve NRP by applying a systematic two-phase approach. They partitioned the problem into sub-problems of manageable computational sizes. Integer mathematical programming was used to solve each sub-problem individually. Valouxis et al. systematically separated the daily workload assignment for each nurse in the first phase and scheduled specific shifts for the assigned nurse in the second phase. They also incorporated additional local optimisation search techniques to search across combinations of partial schedules for nurses.

Nonobe [17] used a metaheuristic algorithm by employing a general-purpose problem solver based on a tabu search approach for constraint problems (COP). In the present work, we only reformulated NRP as COP and implemented user-defined constraints. The constraint weights and tabu tenure was dynamically controlled during the search to enhance performance. The approach was applied on three tracks of the competition.

Bilgin et al. [18] developed a single approach to tackle three sets of NRP compaction tracks. They presented a hyper-heuristic approach combined with a greedy shuffle heuristic. NRP was modelled as a 0–1 matrix with random nurse assignments in such a way that all solutions were feasible. The proposal depended on the heuristic selection method and the move acceptance criterion (applied with simulated annealing). Bilgin et al. initially attempted to solve NRP with integer linear programming (ILP) using IBM CPLEX. The results obtained from different instances were varied in terms of complexity; in some areas, the hyper-heuristic approach outperformed the ILP (and vice versa).

Martin Josef [19] presented a variable neighbourhood search for personal nurse rostering. The technique consisted of two phases. The first phase was an initial constructive approach that met all shift requirements with guaranteed feasible solutions. A principle called most-constrained-first was used to assign nurses in the first phase. The second was an iterative improvement phase based on a VNS local search. After each round of VNS, a neighbourhood structure called perturb structure was applied as a diversified technique to escape the local optima trap.

In the present work, a nature-inspired chemical reaction optimisation (CRO) algorithm with four neighbourhood structures are employed (coded as *N1_move*, *N2_swap*, *N3_block-swap* and *N4_taskreplacement*) with tuned parameters that are based on our preliminary experiments to deal with NRP. The proposed algorithm is tested on INRC-2010 datasets.

The paper is organized as follows: Section 2 presents the problem description and formulation. Section 3 describes the CRO algorithm. Section 4 provides the experimental results. Section 5 presents the concluding remarks.

2   Problem Description and Formulation

NRP focuses on the accurate generation of a valid roster, which is represented by assigning shifts for each nurse [14]. The final roster must satisfy the various personal preferences and work regulations of nurses in the form of soft constraints. The terms used in this work are described below:

- **Roster:** a plan formulated for a number of days for one hospital ward.
- **Shift types:** a time frame for which a nurse with a certain skill is required.
- **Employees:** refer to the number of required nurses provided for each day and shift type.
- **Schedule constraints**: a large number of items (i.e., the min/max amount of work, weekends, and night shifts) to which the schedules (sometimes called work patterns) of each employee are usually subjected to.

2.1   Constraints

The nurse rostering problem considered in this work involves of assigning duties to employees in practice with a given set of constraints namely Hard and Soft Constraints. Usually, two types of constraints are defined as:

1- Hard constraints: represent the requirements that must be satisfied to guarantee the usability of the roster. Hard constraints are addressed by feasible solutions. Hospital and law requirements are generally considered as hard constraints. State or contract-based law requirements include the limited working hours and number of shifts allowed. Hospital requirements specify the compulsory coverage requested for each day to maintain the needed level of care quality. When the hard constraints are satisfied, a feasible (useable) roster is obtained.

2- Soft constraints: increase the practical roster quality. A feasible roster not only fulfills actual hospital needs, but also produces a satisfied workforce that can efficiently meet care demands. Soft constraints can be diverse. Common soft constraints include the desired preferences of each nurse (e.g., requests for free days, preference for excluding certain shift types for specific days, and demands for day rest after night shifts). Nurses are bound by contracts with a set of assignments. Each contract is characterized by a number of regulations that should be fulfilled. Different contracts have different numbers of rules and penalty values. Some soft constraints have special demands (i.e., unwanted patterns). Unwanted patterns are shift sequences that unwanted to take place on certain days during the work routine. Other soft constraints are only applicable for employees with high skill levels (e.g., head nurse). With the exception of the head nurse, some constraints are occasionally sacrificed to improve the roster quality of other nurses. After all, the actual quality of a solution is measured by soft constraint violations. Table 1 shows common real-world constraints for NRP.

Table 1. Description of constraints

Constraints	Description
H1	All demanded shifts must be assigned during the schedule planning
H2	A nurse can only work one shift per day (i.e., no nurse is allowed to work two shifts during the same day)
S1-2	The maximum/minimum number of shifts that should be assigned to nurse s
S3-4	The maximum/minimum number of consecutive working days set for nurse s
S5-6	The maximum/minimum number of consecutive free days set for nurse s
S7	The night shift type
S8-9	The maximum/minimum number of consecutive working weekends set for nurse s
S10	The maximum number of working weekends for nurse s
S11	Complete weekends; if nurse s is assigned to work during a weekend day, then nurse s should work for the entire weekend
S12	Identical shift types during the weekend; nurse s should be assigned the same shift T on the working weekend days
S13-14	Request On (off) if nurse s requests (not) to work any shift at day d
S15-16	Request On (off) if nurse s requests (not) to work a specific shift T at specific day d
S17	Alternative skill; certain shifts can only be set for a nurse if the nurse has all the required skills for that assignment
S18	The set of unwanted patterns of nurse s

We present a symbolic definition of the problem below to introduce hard and soft constraints:

- A set D of days, during which nurses are scheduled: $|D| = D$. D usually consists of four weeks (i.e., $D = 28$);
- A set S of nurses, each associated with a set of available skills and works under exactly one contract: $|S| = S$;
- A set T of shifts, each characterized by a set of required skills: $|T| = T$.

The constraints in Table 1 are variations of the penalty values based on contracts. Thus, each individual solution can have a different penalty value according to the working contract even if the nurses share the same soft constraints.

The evaluation function $f(X)$ for this problem is intended to sum up all penalties associated with the violation of soft constraints in the planning period as defined in Formula (1):

$$f(x) = \sum_{s=1}^{S} \sum_{n=1}^{18} y_{s,n} \cdot f_{s,n} \qquad (1)$$

where X is the current solution and $y_{s,n}$ is the weight of the soft constraint S_n for nurse s regulated by the contract of nurse s. Each soft constraint (i.e., S1 to S18) has a different penalty value based on the contracts. If one of the soft constraints is disabled based on the contract regulation, then the value of $y_{s,n}$ will be set as 0.

Thus, the objective is to find a feasible solution X' such that $f(X') \leq f(X)$ for all X in the feasible search space. Further mathematical clarifications are presented in the studies of Lü et al.[20] and Haspeslagh et al.[14].

2.2   Solution Space and Initial Solution

NRP is modelled as a 2D 0–1 matrix, in which the columns represent the shifts arranged per day and the rows represent the nurses. Figure 1 shows that a nurse is assigned to a particular shift on a specific day if the resultant matrix intersection element is 1. The initial solution is the selection of a nurse and assigning a shift that satisfies some of the soft constraints. In case of failure, a shift is randomly assigned to another nurse based on feasibility. The coverage should be met and no nurse can work more than one shift per day. Feasibility is maintained during the subsequent search by considering only assignment moves within the same column. No assignment can be removed without making a new one within the same column.

Fig. 1. Planning roster

2.3   Neighbourhood Structures

Our local search procedure uses four distinct neighbourhood structures: *single-move, single-swap, block-swap, and N4_taskreplacment.*

- ***N1_move:*** A shift is randomly transferred from a selected nurse on duty to any other free nurse on the same day (Fig. 2).
- ***N2_swap:*** For a single day, two shifts assigned to a pair of nurses are swapped on the same day, unless they are pre-assigned with the same shift type.

- *N3_block-swap*: For a selected number of days in a row (e.g., 2 ☐ 4 days), selected shifts assigned to a pair of nurses are swapped on the same day (Fig. 3).
- *N4_taskreplacement:* a form of random crossover, where a random number of shifts for random days is selected and then swapped between two selected nurses.

During any kind of movement, feasibility is always maintained with no violation of the hard constraints.

Fig. 2. Move neighbourhood structure

Fig. 3. Swap neighbourhood structure

3   CRO Approach

Chemical Reaction Optimisation (CRO) is a population-based swarm intelligent algorithm that mimics the natural activities of chemical reactions. CRO imitates the actions and changes accrued on the molecules during the reaction, where it traces the energy transformation for each molecule [21]. In metaheuristic terms, a molecule represents a solution. Each molecule possesses two kinds of energies: potential energy (*PE*) and kinetic energy (*KE*). In this study, *PE* represents the objective function value of the solution, whereas *KE* symbolizes the ability of a molecule to escape from a local minimum. *KE* allows molecules to move to a high potential state and, thus, the chance for favorable structures during future changes is enhanced. Therefore, the higher the *KE* of the molecule, the higher the possibility of the molecule possessing a new neighbouring molecule with a higher *PE* [21]. When the chemical reaction ends, a molecule that possesses minimum or free energy is produced. The product is known as a stable molecule, which implies a solution with low objective function values. The main difference between CRO and other swarm intelligent algorithms is the ability of CRO to auto-adjust the population size during the search process.

The formulation of *PE* is as follows:

$$PE_\omega = f(\omega). \tag{2}$$

Given a molecule ω, the new neighbouring molecule of ω is denoted by ω'. The acceptance criteria allow the excited molecule to be replaced by 'ω:

$$PE_\omega + KE_\omega \geq PE_{\omega'}. \tag{3}$$

3.1   Elementary Reactions of CRO

A sequence of collisions occurs among molecules in a chemical reaction. Molecules collide with the walls of the container or with one another. The CRO algorithm has four elementary reactions: on-wall collision, decomposition, inter-molecular collision, and synthesis. These fundamental reactions are grouped based on the number of molecules involved in the operation. The on-wall collision and decomposition reactions are single molecular reactions that are activated when the molecule hits the container wall, whereas the inter-molecular collision and synthesis reactions are multi-molecule reactions triggered by the collision of more than one molecule.

Intensification and diversification are generally the main mechanisms in the search process. Intensification explores the closely local search space (neighbouring area). Whenever the process fails to determine any neighbouring solution with low energy (objective function value) for a period of time, diversification allows the diversion to a relatively different area to continue the search. In CRO, on-wall ineffective collision and inter-molecular collision are categorized under intensification, whereas decomposition and synthesis are classified under diversification. The CRO algorithm is detailed in the study of Lam et al.[21]. CRO has been used in different domains and can obtain promising results [22–25]. These fundamental collisions can be described as follow:

1-   On-wall collision: An on-wall collision reaction is triggered when a molecule hits the container wall and bounces back. The attributes of ω change because of the collision. Consequently, a new molecule (ω') s created. After the on-wall ineffective collision, ω loses a fraction of *KE* to the buffer. Thus, the *KE* for ω' is represented as follows:

$$KE_{\omega'} \geq (PE_\omega + KE_\omega - PE_{\omega'}) \times q, \tag{4}$$

where $q \in [KELossRate, 1]$ and $(1-q)$ represent the fraction of *KE* lost to the environment after the on-wall collision. *KELossRate* is a system parameter that limits the maximum percentage of *KE* lost at a time. The lost energy is stored in a central buffer; the energy will be utilized later in the decomposition process. The update of the buffer is represented as follows:

$$buffer = buffer + (PE_\omega + KE_\omega - PE_{\omega'}) \times (1 - q). \tag{5}$$

2- Decomposition: A decomposition occurs when a molecule hits the container wall and then decomposes into two or more pieces. This reaction produces molecules that have different structures from the original one. The KE of the resultant molecules is shown as follows:

$$KE_{\omega1'} = (PE_\omega + KE_\omega - PE_{\omega1'} - PE_{\omega2'} + buffer) \times k1 \times k2, \tag{6}$$

$$KE_{\omega2'} = (PE_\omega + KE_\omega - PE_{\omega'} - PE_{\omega2'} + buffer) \times k3 \times k4. \tag{7}$$

The condition for accepting the new solution is denoted as follows:

$$PE_\omega + KE_\omega + buffer \geq PE_{\omega1'} + PE_{\omega2'}, \tag{8}$$

where $k1$, $k2$, $k3$, and $k4$ are random numbers uniformly and independently generated from interval [0–1]. Subsequently, the buffer is updated by the following formulation:

$$buffer = PE_\omega + KE_\omega - PE_{\omega1} - PE_{\omega2} + buffer - KE_{\omega1} - KE_{\omega2}. \tag{9}$$

3- Inter-molecular collision: An inter-molecular collision occurs when two or more molecules collide with each other and then bounce away. This reaction is similar to on-wall collision and produces a similar energy that changes the effect on the involved molecules. The accepting condition for inter-molecular collision is as follows:

$$PE_{\omega1} + KE_{\omega1} + PE_{\omega2} + KE_{\omega2} \geq PE_{\omega1'} + PE_{\omega2'}. \tag{10}$$

The KE of the two new molecules is derived as follows:

$$KE_{\omega1'} = (PE_{\omega1} + KE_{\omega1} + PE_{\omega2} + KE_{\omega2} - PE_{\omega1'} - PE_{\omega2'}) \times k, \tag{11}$$

$$KE_{\omega2'} = (PE_{\omega1} + KE_{\omega1} + PE_{\omega2} + KE_{\omega2} - PE_{\omega1'} - PE_{\omega2'}) \times (1 - k). \tag{12}$$

where k is a random number uniformly and independently generated from interval [0-1].

4- Synthesis: A synthesis occurs when a new molecule is generated as an outcome of the collision and combination of two molecules. For example, the molecular structures of two original molecules $\omega1$ and $\omega2$ collide with each other and a new molecule ω' bounces back. Thus, the newly produced molecule ω' has a high chance of escaping the trap of a local optima in the subsequent elementary reactions associated with it.

The accepting criteria for the synthesis reaction are as follows:

$$PE_{\omega 1} + KE_{\omega 1} + PE_{\omega 2} + KE_{\omega 2} \geq PE_{\omega'}. \tag{13}$$

The KE for the resultant molecule ω' is derived as follows:

$$KE_{\omega'} = PE_{\omega 1} + KE_{\omega 1} + PE_{\omega 2} + KE_{\omega 2} - PE_{\omega'}. \tag{14}$$

3.2   CRO for NRP

The CRO algorithm is a population-based metaheuristic that can dynamically adjust the population size. The algorithm has fundamental search elements that can divert the solution once it is trapped in local optima (e.g., exploitation and exploration). CRO has already been applied to NRP in an earlier study [26]. However, the application didn't show promising results.

Moreover, the environmental setup was not sufficiently explained and the neighbourhood structures used were not mentioned at all. We believe that CRO can compete with well-known approaches, if provided with a well-tuned parameters and appropriate structures. Thus, we formulated a new experimental setup using varies neighbourhood structures as presented in Section 2.3.

We embedded problem encoding, neighbourhood structures, initial population approach, and local search methods to make the basic CRO suitable for solving NRP. Named as CRO–NRP, it is shown in the following pseudo code:

```
Algorithm CRO-NRP
Input: parameter values
\\ Initialization
Set PopSize, KELossRate, MoleColl, buffer, InitialKE, α,
and β
Create PopSize number of molecules

Set initial Objective function values for all molecules
as a PE
\\ Iterations
while the stopping criteria not met do
Generate b ε [0, 1]
    if b > MoleColl then
        Randomly select one molecule Mω
            if Decomposition criterion α met then
              Trigger Decomposition
            else
              Trigger OnwallIneffectiveCollision
```

```
        end if
    else
      Randomly select two molecules M_{ω₁} and M_{ω₂}
        if Synthesis criterion β met then
          Trigger Synthesis
        else
          Trigger IntermolecularIneffectiveCollision
        end if
    end if
Check for any new minimum solution
end while
\\ The final stage
Output the best solution found and its objective function
value
```

As an initial step of the setup, we visualized a container with a number of chemical substances inside in the form of molecules, with each molecule representing a different solution. The container size is called *PopSize* and represents the population size. The collection of molecules (solution) is based on random generations. Each solution has *PE and KE*. Subsequently, we initialized the settings of the algorithm and assigned values to the algorithmic parameters, including *KELossRate, MoleColl, buffer, InitialKE, α*, and *β*. *KELossRate* is used to control the amount of *KE* to be transferred to the buffer in an on-wall collision. *MoleColl* is the fraction of a number of intermolecular reactions among all possible reactions. *InitialKE* is the initial *KE* assigned to each new solution. *α* and *β* are the decomposition and synthesis acceptance parameters that control the degree of diversification.

Subsequently, we evaluated the *PE* value of each solution in the population and set the number of iterations needed. In our case, the process is terminated based on the time frame given (10 Seconds.).

One elementary reaction occurs in each iteration. A random number r in interval [0-1] is generated and then compared with the *MoleColl*. Consequently, we can determine whether a uni-molecular or multi-molecular reaction process is selected. If $r >$ *MoleColl*, then one molecule $M_ω$ from the population is randomly selected. We check if the decomposition condition is met and then perform the decomposition operation to produce $M_{ω1}$ and $M_{ω2}$ as an outcome of $M_ω$. Otherwise, the on-wall ineffective collision is performed, which produces only $M_ω'$. However, if $r <$ *MoleColl*, then a random selection of $M_{ω1}$ and $M_{ω2}$ is performed and the synthesis criterion is checked to execute the synthesis operator and create $M_ω'$. Otherwise, an intermolecular collision occurs and produces $M_{ω1'}$ and $M_{ω2'}$. Once an elementary reaction operation is successfully accepted as a new solution, the new solution takes its place in the population by removing the old one. At the end of each iteration, we trace any recently created superior solution. By adding two new solutions and removing only one (e.g., decomposition) and vice versa (e.g., synthesis), a new adjustment in original population size is observed. Finally, we output the best results when the process is terminated. Figure 4 illustrates clearly the visualization of the process. Further details on the implementation of CRO can be found in related studies [21–23, 25].

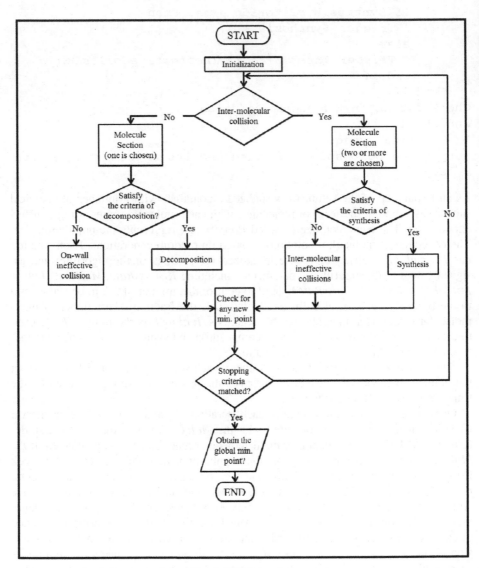

Fig. 4. CRO flowchart

4   Experiment Results

The excremental environment for the proposed method was implemented in Java, and simulations were performed on 2.8 GHz CPU with 4 GB RAM. The tested datasets can be downloaded from https://www.kuleuven-kulak.be/nrpcompetition/instances-results. For each dataset, the algorithm was run for 10 Seconds based on the competition rules time. Each instance was tested 30 times.

The details of the parameter values conducted in the experiment are presented below. The parameters were tuned with a simple iterated search technique and without any special tuning tool.

- PopSize=5
- KELossRate= 0.08
- MoleColl=0.08
- InitialKE =198000
- α =90000
- β = 10

4.1   Comparison Results

Table 2 shows the computational statistics of the proposed CRO–NRP. The second column shows the best known results for these datasets [20]. The rest of the table presents the best results obtained by CRO–NRP, the average function value, the best time needed to obtain the results, and the number of iterations.

CRO–NRP can obtain one best result on the **sprint_hidden04** dataset and tie with 17 datasets with other best known results in literature. Thus, CRO–NRP can reach most of the lower bound values under the competition time provided.

We conducted a comparison of population-based algorithms, including the first version of CRO_V1 and genetic algorithm (GA) as proposed by Zheng et al.[26] and the harmony search algorithm proposed by Awadallah et al.[27]. Table 3 shows that all results obtained by CRO-NRP outperformed all other metaheuristic population methods over all 30 instances, except for some authors who have unreported results represented as "-" in the table. A good combination of neighbourhood structures with the original ability of the algorithm is essential for adapting single and multi-solutions during the process in addition to intensification and diversification strategies, all of which can contribute to a fast and stable search mechanism.

Table 2. CRO computational results

Dataset	Best known	CRO-NRP	Avr.	T_best (sec)	Itr
sprint01	56	56	56.03	0.324	18441
sprint02	58	58	58.03	0.597	36225
sprint03	51	51	51.2	0.548	33427
sprint04	59	59	59.2	0.57	35968
sprint05	58	58	58	0.316	18744
sprint06	54	54	54	0.412	25507
sprint07	56	56	56	0.463	29114
sprint08	56	56	56	0.416	26109
sprint09	55	55	55.1	0.504	31652
sprint10	52	52	52.069	0.298	17671
sprint_late01	37	40	41.44	0.679	24494
sprint_late02	42	43	44.89	0.464	20947
sprint_late03	48	48	50.58	1.46	56416
sprint_late04	75	77	91.79	1.38	56085
sprint_late05	44	44	45.75	0.824	33280
sprint_late06	42	42	42.24	0.29	27901
sprint_late07	42	48	52.27	0.315	24436
sprint_late08	17	17	20.7931	0.112	11030
sprint_late09	17	17	20.79	0.146	15640
sprint_late10	43	51	55.48	0.221	20987
sprint_hidden01	33	33	37.27	0.459	24659
sprint_hidden02	32	32	35.82	0.518	28903
sprint_hidden03	62	64	67.517	0.637	24035
sprint_hidden04	67	**66**	68.586	0.699	38980
sprint_hidden05	59	60	64.34	0.515	19189
sprint_hidden06	134	147	184.759	0.749	41287
sprint_hidden07	153	171	207.172	0.425	22733
sprint_hidden08	209	224	263.034	0.676	26801
sprint_hidden09	338	356	378.069	0.67	26368
sprint_hidden10	306	334	362.414	0.441	15401

Table 3. Comparison with other methods

Dataset	BKS	CRO–NRP	CRO_V1	GA	HSA
sprint01	56	56	75	72	61
sprint02	58	58	75	75	63
sprint03	51	51	69	69	56
sprint04	59	59	80	76	66
sprint05	58	58	-	-	63
sprint06	54	54	-	-	60
sprint07	56	56	-	-	64
sprint08	56	56	-	-	61
sprint09	55	55	-	-	61
sprint10	52	52	-	-	60
sprint_late01	37	40	52	50	56
sprint_late02	42	43	60	61	61
sprint_late03	48	48	59	58	68
sprint_late04	75	77	-	-	136
sprint_late05	44	44	-	-	61
sprint_late06	42	42	-	-	50
sprint_late07	42	48	-	-	70
sprint_late08	17	17	-	-	20
sprint_late09	17	17	-	-	23
sprint_late10	43	51	-	-	77
sprint_hidden01	33	33	62	58	57
sprint_hidden02	32	32	76	72	55
sprint_hidden03	62	64	98	93	90
sprint_hidden04	67	**66**	87	89	94
sprint_hidden05	59	60	-	-	81
sprint_hidden06	134	147	-	-	238
sprint_hidden07	153	171	-	-	288
sprint_hidden08	209	224	-	-	317
sprint_hidden09	338	356	-	-	606
sprint_hidden10	306	334	-	-	416

5   Conclusion

In this study, we discussed the use of the Chemical Reaction Optimisation (CRO) algorithm to solve nurse rostering problem. CRO is a population-based nature-inspired algorithm that copy the way in of the natural activities of chemical reactions.

The overall goal was to tweak the algorithm using a combination of four neighbour-hood structures. A total of 30 well-known datasets from INRC-2010 demonstrated the strength of the algorithm. Depending on the excited elements of the algorithm (i.e., adjusting the population size using exploitation and exploration techniques) one lower bound result was obtained and a match of 17 instances for the best known solutions in literature was identified. Our future work aims to examine different datasets with various complexities. The solutions can be further improved by enhancing the neighbourhood structures through advanced parameter-tuning methods.

References

1. Burke, E.K., De Causmaecker, P., Vanden Berghe, G.: Novel meta-heuristic approaches to nurse rostering problems in Belgian hospitals. In: Handbook of Scheduling: Algorithms, Models and Performance Analysis, pp. 44.41–44.18 (2004)
2. Ernst, A.T., Jiang, H., Krishnamoorthy, M., Sier, D.: Staff scheduling and rostering: A review of applications, methods and models. European Journal of Operational Research 153, 3–27 (2004)
3. Cheang, B., Li, H., Lim, A., Rodrigues, B.: Nurse rostering problems——a bibliographic survey. European Journal of Operational Research 151, 447–460 (2003)
4. Warner, D.M., Prawda, J.: A mathematical programming model for scheduling nursing personnel in a hospital. Management Science 19, 411–422 (1972)
5. Thornton, J., Sattar, A.: Nurse rostering and integer programming revisited. In: International Conference on Computational Intelligence and Multimedia Applications, pp. 49–58 (1997)
6. Millar, H.H., Kiragu, M.: Cyclic and non-cyclic scheduling of 12 h shift nurses by network programming. European Journal of Operational Research 104, 582–592 (1998)
7. Moz, M., Pato, M.V.: Solving the problem of rerostering nurse schedules with hard constraints: new multicommodity flow models. Annals of Operations Research 128, 179–197 (2004)
8. Moz, M., Vaz Pato, M.: A genetic algorithm approach to a nurse rerostering problem. Computers & Operations Research 34, 667–691 (2007)
9. Brusco, M.J., Jacobs, L.W.: Cost analysis of alternative formulations for personnel scheduling in continuously operating organizations. European Journal of Operational Research 86, 249–261 (1995)
10. Burke, E.K., De Causmaecker, P., Vanden Berghe, G.: A hybrid tabu search algorithm for the nurse rostering problem. In: McKay, B., Yao, X., Newton, C.S., Kim, J.-H., Furuhashi, T. (eds.) SEAL 1998. LNCS (LNAI), vol. 1585, pp. 187–194. Springer, Heidelberg (1999)
11. Burke, E., De Causmaecker, P., Petrovic, S., Berghe, G.V.: Variable neighborhood search for nurse rostering problems. In: Metaheuristics: Computer Decision-making, pp. 153–172. Springer (2004)
12. Bellanti, F., Carello, G., Della Croce, F., Tadei, R.: A greedy-based neighborhood search approach to a nurse rostering problem. European Journal of Operational Research 153, 28–40 (2004)
13. Burke, E.K., Curtois, T., Post, G., Qu, R., Veltman, B.: A hybrid heuristic ordering and variable neighbourhood search for the nurse rostering problem. European Journal of Operational Research 188, 330–341 (2008)

14. Haspeslagh, S., De Causmaecker, P., Schaerf, A., Stølevik, M.: The first international nurse rostering competition 2010. Annals of Operations Research 1–16 (2012)
15. Burke, E.K., Curtois, T.: New approaches to nurse rostering benchmark instances. European Journal of Operational Research 237, 71–81 (2014)
16. Valouxis, C., Gogos, C., Goulas, G., Alefragis, P., Housos, E.: A systematic two phase approach for the nurse rostering problem. European Journal of Operational Research 219, 425–433 (2012)
17. Nonobe, K.: INRC2010: An approach using a general constraint optimization solver. In: The First International Nurse Rostering Competition (INRC 2010) (2010)
18. Bilgin, B., De Causmaecker, P., Rossie, B., Berghe, G.V.: Local search neighbourhoods for dealing with a novel nurse rostering model. Annals of Operations Research 194, 33–57 (2012)
19. Martin, J.G.: Personnel rostering by means of variable neighborhood search. In: Operations Research Proceedings 2010, pp. 219–224. Springer (2011)
20. Lü, Z., Hao, J.-K.: Adaptive neighborhood search for nurse rostering. European Journal of Operational Research 218, 865–876 (2012)
21. Lam, A.Y., Li, V.O.: Chemical reaction optimization: A tutorial. Memetic Computing 4, 3–17 (2012)
22. Lam, A.Y., Li, V.O.: Chemical-reaction-inspired metaheuristic for optimization. IEEE Transactions on Evolutionary Computation 14(3), 381–399 (2010)
23. Lam, A.Y., Xu, J., Li, V.O.: Chemical reaction optimization for population transition in peer-to-peer live streaming. In: 2010 IEEE Congress on Evolutionary Computation (CEC), pp. 1–8. IEEE (2010)
24. Nguyen, T.T., Li, Z., Zhang, S., Truong, T.K.: A hybrid algorithm based on particle swarm and chemical reaction optimization. Expert Systems with Applications 41, 2134–2143 (2014)
25. Xu, J., Lam, A.Y., Li, V.O.: Chemical reaction optimization for task scheduling in grid computing. IEEE Transactions on Parallel and Distributed Systems 22, 1624–1631 (2011)
26. Zheng, Z., Gong, X.: Chemical Reaction Optimization for Nurse Rostering Problem. In: Frontier and Future Development of Information Technology in Medicine and Education, pp. 3275–3279. Springer (2014)
27. Awadallah, M.A., Khader, A.T., Al-Betar, M.A., Bolaji, A.L.A.: Nurse scheduling using harmony search. In: 2011 Sixth International Conference on Bio-Inspired Computing: Theories and Applications (BIC-TA), pp. 58–63. IEEE (2011)

Multi Objective Optimization for Route Planning and Fleet Assignment in Regular and Non-regular Flights

Takahiro Jinba[*], Tomohiro Harada, Hiroyuki Sato, and Keiki Takadama

The University of Electro-Communications, Chofu, Japan
{jinba,harada}@cas.hc.uec.ac.jp
{sato,keiki}@hc.uec.ac.jp

Abstract. To optimize the flight schedule that consists of (1) the regular flight operated on the same day and time through one year and (2) the non-regular flight operated on the different day and time according to month, this paper proposes a new multi-objective fleet assignment method. To investigate the effectiveness of our method, this paper applies it to a test problem of Japanese domestic airport network optimization for two months, off-peak and peak months, using a real-world data. The following extensions are introduced: (1) our method can not only obtain a flight network for each month simultaneously, but also can find a network that has an equivalent profit given by the conventional method; (2) in peak month, our method can find a network that has higher profit than the conventional method; and (3) our method can find a network that has a well-balanced profit between off-peak and peak months.

Keywords: Multi-objective optimization, Regular and non-regular flights optimization, Evolutionary algorithm, Fleet assignment problem.

1 Introduction

In order to generate the flight network for a different demand of passengers in each season, airline companies consider the *regular* and the *non-regular* flights separately [9]. The regular flight is operated on the schedule and the same airplane assignment through one year, while the non-regular flights are operated on the different schedule and airplane assignment according to operated month. Since the current rate of the non-regular flights in whole flights is approximately 30% [10] and the number of the non-regular flight is increasing every year, the responsibility of the airline company is increasing. For these reason, the optimization for the regular and the non-regular flight network is required. Our previous work [8] proposed the optimization method that can consider the common components (*i.e.*, the regular flights) that are optimized from the viewpoint of all objective functions (all months) and the specialized components (*i.e.*, the non-regular flights) that are optimized from the viewpoint of single objective function (one month) at the same time. However, when this previous work is applied to the flight scheduling problem by combining a conventional route plan-

---

[*] Corresponding author.

ning and fleet assignment method (Priority-based GA for Fleet Assignment Problem (FAP) with Connection Network Model [5]) which optimizes a flight's route and assignment of aircraft's type by considering a passenger's demand in just one season, cannot assign them in each month (*i.e.*, the non-regular flights do not considered) because the conventional route planning and fleet assignment method does not consider a lot of month.

To tackle this problem, this paper proposes a new multi-objective fleet assignment method for the flight network optimization that assigns aircraft for both the regular and non-regular flights. Specifically, our method extends the PriGA for FAP with Connection Network Model [5] to optimize the flight's route and assignment of aircraft's type for the regular and non-regular flight by considering passenger's demand in each month. This paper aims to investigate the effectiveness of our method through an experiment using a real-world data (*i.e.*, the transportation cost of aircraft, the sales of airline and the passenger's demand).

This paper is organized as follows. Section 2 introduces three mechanisms of (1) NSGA-II [4], (2) the multi-objective optimization for regular and non-regular flights [8], and (3) PriGA that is a one of network optimization methods. Section 3 describes the mechanism of our proposed methods. Section 4 explains a test problem. Section 5 tests our methods on the test problem introduced in section 4. Section 6 discusses the experimental results. Finally, our conclusions and future works are presented in section 7.

2   Conventional Methods

2.1   NSGA-II (Elitist Non-dominated Sorting Genetic Algorithm)

Elitist Non-dominated Sorting Genetic Algorithm (NSGA-II [4]) is a one of major multiple objective optimization (MOO) method based on evolutionary algorithm (EA), which has the following characteristics: (1) the non-dominated sort to sort solutions according to the order of the Pareto front, (2) the crowding distance calculation to maintain a diversity of solutions, (3) the tournament selection to select solutions having higher fitness in a high possibility as parents, and (4) the archive function to keep solutions having high fitness to the next generation.

Non-dominated Sort
The left side in Fig. 1 shows the concept of the non-dominated sort of each solution in two objectives optimization. In this figure, the vertical and horizontal axes indicate the objective functions $f1(x)$ and $f2(x)$, respectively, and the plots indicates solutions. If each objective function value of a solution X are higher than each one of a solution Y in all objective functions, the solution X dominates the solution Y. As shown in this figure, each solution is classified to the rank 1, 2, or 3 where (1) the solutions in the rank 1 are not dominated by any other solutions; (2) the solutions in the rank 2 are not dominated by others except for the solutions in the rank 1; and (3) the solutions in the rank 3 are not dominated by others except for the solutions in the ranks 1 and 2. Iteratively, the solutions in the rank n are not dominated by others except for the solutions in the ranks 1, 2 … and, n-1. The best solutions are ranked at 1.

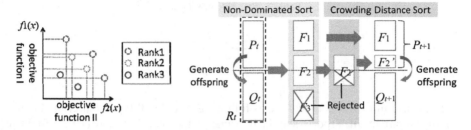

Fig. 1. Concept of the non-dominated sort (left) and schematic of NSGA-II procedure (right)

The right side in Fig. 1 shows the overview of the generation in NSGA-II. In this figure, P_t indicates the parent population archived as solutions having high fitness in the generation t, Q_t indicates the offspring population generated by through genetic operation to P_t, F_1, F_2 ... indicate the population in the first, second non-dominated rank, and F_2' indicates the population selected from F_2 by the crowding tournament selection operator. NSGA-II evolves solutions as following: (1) the population R_t is generated by combining the parent population P_t and the offspring population Q_t; (2) All solutions in the population R_t are sorted by the non-dominated sort and classified into each rank; (3) P_{t+1} is selected from R_t according to the non-dominated rank without exceeding the population size N; (4) The crowding distance sort is conducted for the population in the next rank to add the solutions having high diversity to the P_{t+1}; and (5) The new offspring population Q_{t+1} is generated through the genetic operations (*i.e.*, the tournament selection, crossover and mutation) from the population P_{t+1}.

2.2   Route Optimization: PriGA

Connection Network Model

The fleet assignment problem (FAP) [5] consists of determining the aircraft type to assign to each flight leg in order to minimize the total costs while satisfying aircraft routing and availability constraints. To represent the FAP, connection network model (hereafter we call it as the CNM) was proposed [2]. The purposes of the CNM are: (1) satisfying passenger's demand of each flight leg; and (2) to make a daily flight schedule having the minimum total costs. Note that the CNM does not decide the time of the flight schedule but decides a flight route, the number of aircraft, and an aircraft assignment for each flight route.

The preconditions of this model are as following: (1) a daily OD (Origin-Destination) table that indicates the passenger's demand of each flight leg is given; (2) the preparing cost of aircraft is given depending on departure airport; and (3) the number of aircraft owned by an airline is unlimited. Table.1 shows an example of an OD table. In this table, row indicates a departure airport number and column indicates a destination airport number respectively, and each cell indicate the number of passenger's demand from airport i to j (hereafter we call it as the OD value). A passenger's demand indicates the number of persons who wants to use airway. For example, table 1 indicates that the number of persons who wants to go from airport A to B is 30 persons.

Table 1. Example of OD table

$i \diagdown^j$	A	B
A	0	30
B	50	0

The constraints of this model are as following: (A) in all type of aircraft, the number of aircraft in all airports should be same at the morning; and (B) all flight leg should satisfy a passenger's demand. In the constraints, (A) is constraint to repeat the planned daily flight schedule. The cost of flight schedule is defined as the summation of the transportation cost (*i.e.*, an airline has to spend the transportation charges when the passengers transport to the destination) and preparing cost of aircraft. The preparing cost arises when an aircraft departs a first airport.

Priority-Based GA for FAP with Connection Network Model

Priority-based GA (PriGA) is usually applied to Vehicle Routing Problem (VRP) [3] and the ship routing problem [6]. PriGA can generate a lot of loop type routes. In the PriGA for FAP with CNM [5], a flight network is represented by a chromosome with integer string as shown in Fig. 2. This example has five airports, and each airport is mutually connected. Each gene locus in this figure indicates each flight line, the first horizontal line (*i.e.*, St.1 ... and St.5) indicates the airport number of a departure place, the second line indicates the index, the third line indicates the airport number of a destination place, and the last line indicates the values of chromosome. The gene length corresponds to the number of flight lines. The values of chromosome represent a priority of a flight line, and a high value of chromosome represents a high priority. Note that the values of chromosome are not overlapped. The flight network that is represented by this gene is evolved according to the conduct of the genetic operations.

From station:	St.1				St.2				St.3				St.4				St.5			
i :	1	2	3	4	5	6	7	8	9	10	11	12	13	14	15	16	17	18	19	20
To station :	2	3	4	5	1	3	4	5	1	2	4	5	1	2	3	5	1	2	3	4
Chromosome :	12	16	11	10	19	17	9	18	20	2	8	3	7	1	14	5	15	13	6	4

Fig. 2. Route representation of PriGA for FAP with connection network model

The algorithm of the decoding procedure of PriGA for FAP with CNM is shown as following: (1) select an index i having the highest priority from set of not selected indexes; (2) set the departure i^{th} airport as the departure airport O_i of a route from the first line; (3) set the destination i^{th} airport as D_i, and add the flight leg from O_i to D_i to the route; (4) if the route does not return to the airport O_i, select an index i having the highest priority from set of not selected indexes and have the departure airport number "D_i", and return to step 3; and (5) repeat step 1 to step 4 until all indexes are selected.

Fig. 3 shows the mechanism of PriGA for FAP with CNM. This example shows a flight network having three airports (the gene length is six). The decoded flight network from the gene in this figure has two routes.

The difference from a simplified GA is three points as following: (1) Each individual indicates a lot of flight route; (2) In an individual, the values of gene are not overlapped; and (3) The profit of flight network is regarded as the fitness of individual. By this mechanism, routes having higher profit are explored. The algorithm of PriGA for FAP with CNM is shown as following: (1) generate predetermined amount of individuals randomly as the initial population. Note that individuals are initialized having no overlapped value in gene; (2) decode each gene of individual to a flight network and evaluate them; (3) select parent individuals depending on the fitness; (4) generate offspring by the crossover operation; (5) conduct the mutation operation with the predetermined mutation rate; (6) exchange individuals that have lower fitness in the population with generated offspring; and (7) repeat step 2 to step 6 until the predetermined termination condition is satisfied.

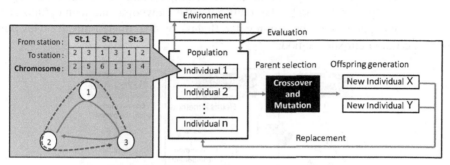

Fig. 3. Mechanism of PriGA for FAP with CNM

There are many kinds of the crossover and mutation in the genetic operations of PriGA. This paper employs following two operations.

Position-Based Crossover
The left side in Fig. 4 shows the procedure of position-based crossover operator employed in PriGA, and this procedure is as following: (1) Randomly select the set of gene locus from the parent 1; (2) Generate a proto-child by copying the selected gene locus on these positions into the corresponding positions of it; (3) Delete the gene locus already selected from the parent 2; and (4) Place the gene locus into the empty positions of the proto-child from left to right according to the order of the sequence to generate one offspring.

Swap Mutation
The right side in Fig. 4 shows the procedure of swap mutation operator that exchanges randomly selected two points gene value.

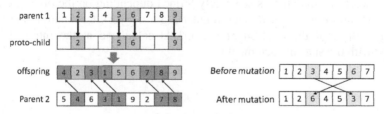

Fig. 4. Position-based crossover (left side) and swap mutation (right side)

2.3   Multi-objective Optimization for Regular and Non-regular Flights

Some of the MOO problems in the real world require to optimize most objects from the multi-objective viewpoint (hereafter we call it as the *common* components) but to optimize others from the single-objective viewpoint (hereafter we call it as the *special* components). In the flight network optimization problem, the regular and non-regular flights correspond to the common and special components respectively. To optimize the flight network composed of the regular and non-regular flights, we proposed the MOO method for the regular and the non-regular flights [8] based on the NSGA-II. Our previous method firstly optimizes the common components (*i.e.*, the regular flights in the flight network) by the multi-objective viewpoint, and then optimizes the special components (*i.e.*, the non-regular flights) by the single-objective viewpoint using a certain heuristic method.

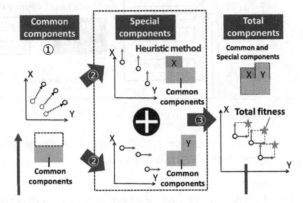

Fig. 5. Schematic of the MOO method for flight network procedure

Fig. 5 shows this conventional method for flight network, where the vertical and horizontal axes represent the two objective functions for the profit of an airline company of the months X and Y, and the plots indicate the solutions (flight network). Concretely, as shown in this figure, (1) the regular flight network is formed by the multi-objective viewpoint; (2) the non-regular flight network is formed by a certain heuristic method from the single-objective viewpoint of months X and Y, respectively. In detail, the non-regular flight network for month X is formed depending on the objective function value (*i.e.*, the profit of airline) of month X, and the non-regular flight network for month Y is separately formed depending on the objective function value of month Y; (3) after the non-regular flight network formulation, the objective function values of the total flight network is calculated at the intersection point (marked with the star) of each month.

3   Proposed Methods

3.1   Fleet Assignment Method Based on Minimum and Maximum Passenger's Demand

This proposed method is based on the conventional fleet assignment method [5] that assigns type of aircraft to each flight leg by considering just one month. In the case of the fleet assignment by considering a lot of month, daily OD tables according to each month are given, and aircraft should be assigned to satisfy OD tables of all month. For example, when considering two months (*e.g.*, a peak season and an off-peak season), two OD tables are given. However, the conventional fleet assignment method cannot assign aircraft of a lot of month at the same time (*i.e.*, the non-regular flights do not considered).

To overcome this problem, we propose a new fleet assignment method that assigns aircraft for both the regular and non-regular flights to a flight route for each month. Concretely, we propose two fleet assignment methods, one decides the fleet assignment based on a month having maximum OD, and another decides it based on a month having minimum OD. Note that the way of route planning of aircraft is same as the PriGA mechanism introduced in section 2.2 [5], and the proposed method assigns type of aircraft to each flight leg in the decoded order by PriGA after route planning. To explain the proposed methods, we show a flight route example in the off-peak and peak months as shown in Fig. 6. In this figure, the left and right side indicates a flight routes in the off-peak and peak months generated by the PriGA respectively, and the nodes, the arrows, and the values indicate an airport, a flight route, and a passenger's demand of an airway respectively. The flight route departs from the airport A, and returns to the airport A via the airports B and C.

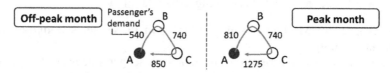

Fig. 6. Flight route example in off-peak and peak month (A-B-C-A)

Fleet Assignment Method Based on Minimum Passenger's Demand: MIN Assignment Method

Fleet assignment method based on a *minimum* passenger's demand (hereafter we call it as the *MIN assignment method*) assigns the aircraft to each airway of flight network focusing on a month which has the minimum passenger's demand. The procedure of the MIN assignment method is as following: (1) calculate the minimum passenger's demand among all months of each flight route and define the maximum value of them as MaxFlow; (2) assign the type and the number of aircraft for each flight route in the regular flights according to MaxFlow; (3) decide a non-regular flight route to satisfy a demand of each month that is not satisfied by the regular flight; and (4) assign the type and the number of aircraft for the non-regular flights. Note that step 1 and 2 are same as the conventional fleet assignment method [5].

In the step 1, a minimum passenger's demands are calculated according to the comparison of passenger's demands among all month (*e.g.*, between the off-peak and peak month in the case of two months) in each airway included in the flight route, and MaxFlow is defined as the maximum value in the minimum passenger's demands. Fig. 7 shows the calculation example of the MaxFlow for the MIN assign method. In this example, the minimum passenger's demands are calculated as shown in the right side of Fig. 7 and the MaxFlow is calculated as 850.

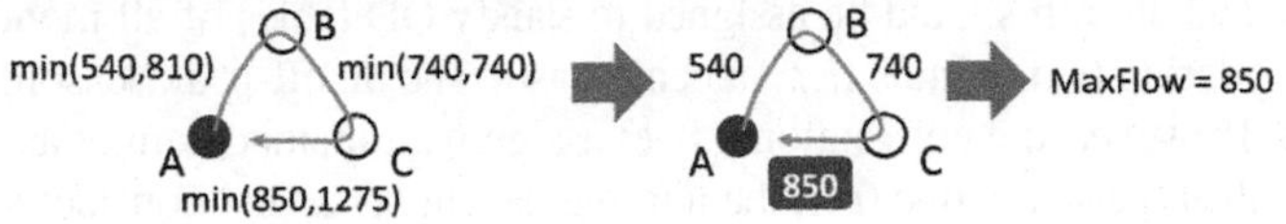

Fig. 7. Calculation example of MaxFlow for the MIN assignment method

In the step 2, the type and the number of aircraft for the regular flight are assigned to the flight route according to MaxFlow that is calculated in the step 1. Specifically, an aircraft assignment that has the minimum cost is selected from all available combination of aircraft that satisfy MaxFlow and some constraints (*e.g.*, the number of aircraft). In the step 3, the non-regular flights are assigned to satisfy a demand of each month that is not satisfied by the regular flight. At first, a flight route of the non-regular flight is decided. To simplify the problem of the route planning for the non-regular flights, the non-regular flights are assigned to a simple round trip route. Note that if the demand of all airways is already satisfied, the non-regular flight is not assigned. In the step 4, the MaxFlow of the route for the non-regular flight is calculated in the same way as the regular flights, and the type and the number of aircraft for the non-regular flight are assigned according to the MaxFlow.

Fleet Assignment Method Based on Maximum Passenger's Demand: MAX Assignment Method

Fleet assignment method based on a *maximum* passenger's demand (hereafter we call it as the *MAX assignment method*) assigns the aircraft to each airway of flight network focusing on a month which has the maximum passenger's demand. The procedure of the MAX assignment method is as following: (1) calculate the maximum passenger's demand among all months of each flight route and define the maximum value of them as MaxFlow; (2) assign the type and the number of aircraft for each flight route in the regular flights according to MaxFlow; and (3) change excessively assigned regular flights regarding the demand into non-regular flights.

In the step 1, a maximum passenger's demands are calculated according to the comparison of passenger's demand among all months (*e.g.*, between the off-peak and peak month in the case of two months) in each airway included in the flight route, and MaxFlow is calculated in the same way as the MIN assignment method. Fig. 8 shows the calculation example of the MaxFlow for the MAX assign method. In this example, the maximum passenger's demands are calculated as shown in the right side of Fig. 8 and the MaxFlow is calculated as 1275.

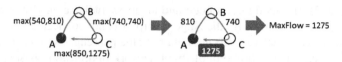

Fig. 8. Calculation example of MaxFlow for the MAX assignment method

The step 2 is same as the MIN assignment method. In the step 3, since the regular flights are assigned focusing the month which has the maximum demand in the MAX assignment method, an excessively assigned regular flight for the month that has the lower demand (*e.g.*, the off-peak month) is changed into a non-regular flight for other month (*e.g.*, the peak month). The route, the type, and the number of the non-regular flights that is changed from the regular flight are inherited. If there is no excessively assigned regular flight in all flight routes, the non-regular flight is not assigned.

3.2 Fleet Assignment Method Based on Weighting Factor

MIN and MAX assignment method explained in section 3.1 assign the aircraft focusing on the demand of either off-peak or peak month. A profit of flight network generated by the MIN assignment method becomes high in the off-peak month, while one generated by the MAX assignment method becomes high in the peak month. For this reason, it is difficult to find a solution around the center of Pareto. And these proposed methods are assign the regular flight to satisfy the demand for just one month and then assign the non-regular one for other month. For this reason, these proposed methods cannot assign a different non-regular flight in the off-peak and peak month.

To tackle this problem, we propose a fleet assignment method based on a weighting factor. Since this proposed method assigns the regular flights using the weighting factor, this method can assign a different non-regular flight in each month and can find a solution around the center of Pareto (*i.e.*, the balance of profit in the off-peak and peak month is good). This method is based on the MIN or MAX assignment method. This method adjusts MaxFlow by multiplying the weighting factor after calculating MaxFlow by the same procedure as the step 1 of the MIN and MAX assignment method, and assigns the regular flights slightly lower regarding the passenger's demand. The later procedure of this method is the same as the one of the MIN and MAX assignment method. The weighting factor is set from 0.5 to 1.0. If the weighting factor is less than 0.5, a solution becomes infeasible because the rate of the non-regular flights increases. Note that if non-regular flights that are same type in all month are assigned, these flights are regarded as the regular flight because these are common in all month.

4 Test Problem: Domestic Airport Network Based on Connection Network Model

To investigate the effectiveness of the proposed methods, we introduce a test problem that consists of a five Japanese domestic airports network based on the CNM (as shown Fig. 9). In this figure, a node and an arrow indicate an airport and an airway, respectively. The reason why the Nagoya airport is not connected with except Fukuoka is lack of these real-world flight data (*i.e.*, the passenger's demand). The influence of above is that the test problem of flight network optimization is simplified. However, it is difficult to find the optimal solution because still the search space is large (*i.e.*, since this test problem has 14 airway, the search space is 14! combinations).

Fig. 9. Flight network of the test problem

In this paper, we optimize the flight network by considering two months (*i.e.*, an off-peak season (April) and an peak season (August)). Table 2 shows daily OD tables for the off-peak (left) and peak month (right), respectively. In this table, row and column indicate a departure airport and a destination one respectively, and this table is based on a data of ANA's website in Japan [1]. If an OD value is zero, no flight exists between these two airports.

Table 2. Daily OD tables in off-peak (left) and peak month (right)

	Fukuoka	Osaka	Nagoya	Tokyo	Sapporo		Fukuoka	Osaka	Nagoya	Tokyo	Sapporo
Fukuoka	0	764	102	4235	96	Fukuoka	0	774	161	4790	360
Osaka	764	0	0	3603	177	Osaka	774	0	0	3788	585
Nagoya	102	0	0	0	0	Nagoya	161	0	0	0	0
Tokyo	4235	3603	0	0	3491	Tokyo	4790	3788	0	0	6211
Sapporo	96	177	0	3491	0	Sapporo	360	585	0	6211	0

This test problem has three types of aircraft as shown in Table 3, and simplifies the number of type of aircraft to three types, small, medium and large aircrafts. A boarding rate for each month determined by a data of ANA's website in Japan [1] is considered by multiplying to the number of seats of each aircraft. For example, if the number of seats of an aircraft X is 100 and the boarding rate is 0.5, the capacity of the aircraft X is defined as up to 50 persons. The boarding rate is different according to month, and in this test problem, we set the boarding rate in the off-peak and the peak month as 0.56 and 0.67 respectively. We define a count of flight (as shown in Table 2) based on the real timetable of ANA as a constraint. Each type of aircraft has to be assigned less than this constraint flight count, and if the flight counts of some flight routes exceed the constraint, they are treated as the infeasible solution.

Table 3. Aircraft's data (the number of seat and constraints of flight's count)

Type	Large	Medium	Small
Seats	495	267	143
Count(April)	79	29	35
Count(August)	82	28	34

5 Experiment

5.1 Experiment Cases

We conduct ten experimental cases shown in Table 4. In Table 4, row indicates each experiment cases and column indicates the experiment cases number, an assignment method and a weighting factor.

Table 4. Experiment cases

Case	Assignment method	Weighting factor	Case	Assignment method	Weighting factor
1	fixing (Max)	fixing (1.0)	6	vary	fixing (0.7)
2	fixing (Min)	fixing (1.0)	7	vary	fixing (0.6)
3	vary	fixing (1.0)	8	vary	fixing (0.5)
4	vary	fixing (0.9)	9	vary	vary (interval: 0.1)
5	vary	fixing (0.8)	10	vary	vary (interval: 0.05)

Both the assignment method and the weighting factor have two patterns "fixing" and "vary". The "fixing" assignment method corresponds to the MIN or MAX assignment method, and the "fixing" weighting factor tests a constant weighting factor value. On the other hand, in the case of "vary" assignment method the individual of PriGA has additional information respect to the assignment method, the MIN or MAX assignment method, while in the case of "vary" weighting factor, the individual of PriGA also has additional information respect to the value of the weight that is represented as a discrete value from 0.5 to 1.0. In cases 9 and 10, the interval of the discrete value of the weighting factor is set as the parameter. In the case of "vary", both the assignment method and the weighting factor of the offspring are inherited from one parent selected at random, and mutate at random by the probability of 30%.

5.2 Evaluation Criteria and Parameter Setting

The profit of a daily flight schedule in the off-peak and the peak months is employed as the evaluation criteria to evaluate the proposed methods. We define the profit (*i.e.*, the fitness) of a flight network generated by the proposed methods as a value by subtracting cost of airline (*i.e.*, the transportation cost) from sales of it. The transportation costs of each airway and each type of aircraft are defined by a computation expression of the cost of fuel in the relevant study [6]. The sales of airline is defined as a value which the number of passengers multiplied by a fare according to each airway, and the fare is defined based on the ANA's website in Japan [1]. Note that we not consider a discounted service of the fare.

Table 5 shows the experimental parameters. In detail, a population size as the number of individuals in population is set as 200, the number of generations is set as 3000, the position-based crossover as the crossover operation and the swap mutation as the mutation operation are employed, their rate are respectively set as 0.5 and 0.5, and the tournament selection is employed to select parent.

Table 5. Experimental parameters

Population size	Generation	Crossover	Mutation	Crossover rate	Mutation rate	Parent selection
200	3000	position-based crossover	swap mutation	0.5	0.5	tournament selection

5.3   Results

Fig. 10 shows solutions optimized by the proposed methods, where the horizontal and vertical axes indicate the profit of the off-peak and the peak months respectively and the plots indicate a generated flight network by each experimental case. In this figure, these plots are classified by the background color according to the MIN or MAX assignment method. Solutions (A) and (B) are compared by the next section. From Fig. 10, the case 10, where the "vary" assignment method and the "vary" weighting factor with interval of 0.05 is employed, found the solutions that dominate solutions of other cases. In Fig. 10, the dotted lines indicate the profit of the flight network in each month given by separately optimized solutions with single objective PriGA

The proposed methods cannot only optimize the flight network of each month simultaneously, but also can obtain the equivalent profit in comparison with one given by single objective optimization. In the peak month, the proposed methods can find the solution that has higher profit than the conventional one.

Focusing on the distribution of the solutions according to each assignment method, from Fig. 10, the solutions of the MIN assignment method are distributed into the right side of this figure (*i.e.*, these solutions have the high profit in the off-peak month). This is because the MIN assignment method precedes the fleet assignment for the off-peak month. While from the same reason, the solutions of the MAX one are distributed into the upper side of this figure (*i.e.*, these solutions have the high profit in the peak month).

Focusing on the distribution of the solutions according to each value of the weighting factor, the solutions of the experiment cases with the fixed weighting factor are distributed not only into the both ends of the Pareto (*i.e.*, these solutions are specialized for just one month), but also into the center of the Pareto (*i.e.*, these solutions have good balance profit). This result reveals that generated solutions differ depending on the variation of the weighting factor.

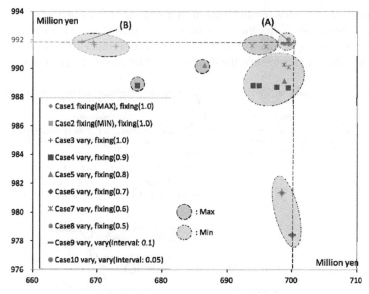

Fig. 10. Solution of each experiment cases

6   Discussion

6.1   Effect of Weighting Factor

Fig. 11 shows the flight network optimized by the proposed method (case 10) that has the highest sum profit of the off-peak and the peak months (*i.e.*, the solution (A) in Fig. 10). In these figures, the left figure indicates the regular flight network and the upper right figure and the lower right figure indicates the non-regular one of the off-peak and the peak months respectively. Each node and each arrow indicate an airport and a flight route respectively. In the experimental cases that the weighting factor is less than 1.0, since the regular flights are assigned slightly lower regarding the passenger's demand and a different non-regular ones are assigned to each month to satisfy the passenger's demand that is not satisfied by the regular flights, the proposed method increases the diversity (*i.e.*, representation of PriGA) of solutions. Specifically, focusing on the flight route between Tokyo and Fukuoka, the one small airplane and two large airplanes are assigned in the off-peak month as the non-regular flights while the one medium airplane is assigned in the peak month.

6.2   Exploration of High Profit Solutions

We consider that the proposed method can find the solutions that have higher profit than conventional PriGA in the peak month. We compare the best solution optimized by the proposed method (*i.e.*, the solution (A) in Fig. 11) and the solution optimized by the conventional one from the viewpoint of the just one peak month (*i.e.*, the solution (B) in Fig. 10). In this experiment, both the proposed method and the conventional one regard a solution that does not satisfy the passenger's demand as the infeasible solution, and a feasible solution optimized by these methods satisfies absolutely all passenger's demand. For this reason, the number of passengers (*i.e.*, the sales) of all solution is equal and the difference of the profit accrues from the costs.

Table 6. Comparison of flight routes in the proposed method and the conventional one

Route number	Transportation cost (yen)		Fleet assignment		Influence
	proposed method solution (A)	conventional method solution (B)	proposed method solution (A)	conventional method solution (B)	
Route 2	7,519,044	6,919,212	Small 1, Medium 1, Large 1	Large 2	High
Route 4	25,152,352	23,165,390	Small 2, Medium 2, Large 10	Medium 1, Large 11	Low
Route 7	5,737,776	8,473,266	Medium 1, Large 2	Small 3, Medium 3	Middle
Total	38,409,172	38,557,868	Small 3, Medium 4, Large 13	Small 3, Medium 4, Large 13	

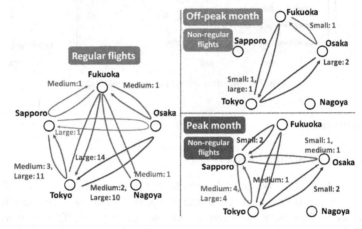

Fig. 11. Flight network optimized by the proposed method (the solution (A) in Fig. 10)

Table 6 shows the cost and the type of assigned aircraft of a flight route. Note that in this table flight routes that have different the fleet assignment in the solution (A) and (B) are shown. In this table, the route number indicates the decoded order by PriGA and the influence indicates a degree of the cost difference in the case that the high cost aircraft (*i.e.*, the medium or small type) are assigned instead of the low cost one (*i.e.*, the large type). From this influence, in the route 2, 4, and 7, it is indicated that the medium and large type of aircraft are assigned to the lower influence route 4 and 7, while the large type of aircraft is assigned to the high influence route 2. Since the solution (A) optimized by proposed method assigns aircraft for the regular flights slightly lower regarding the passenger's demand by using the weighting factor, the large type remains until the last route is assigned the aircraft. The large type is distributed well-balance over each route, and the cost of whole network becomes optimal. While the conventional method cannot represent the solution (A) because the conventional method assigns the aircraft which have the minimum cost to the route in decoded order by PriGA. In the solution (B), since the low cost aircraft (*i.e.*, the large) is exhausted in the initial phase of assignment, the routes in the initial phase are optimal assignment. However, the whole network is not optimal assignment. Specifically, in the route 7 in table 6, since the solution (B) optimized by the conventional method has already exceeded the constraint count of the large type, the large type cannot be assigned.

7   Conclusion

This paper proposes a new multi-objective fleet assignment method for the flight network optimization that assigns aircraft for both the regular and non-regular flights. To investigate the effectiveness of our method, this paper applies it to a test problem of Japanese domestic airport network optimization for two months, off-peak and peak months, using a real-world data. The intensive experiments have revealed the following implications: (1) our method can not only obtain a flight network for each month simultaneously, but also can find a network that has an equivalent profit given by the conventional method; (2) in peak month, our method can find a network that has higher profit than the conventional method; and (3) our method can find a network that has a well-balanced profit between off-peak and peak months by using the weighting factor.

The following future works must be done in the near future: (1) a consideration of a cost besides the transportation cost; (2) a flight network optimization by considering the regular flight rate; and (3) a proposal of a flight schedule optimization by considering a departure time.

References

1. ANA Japan, http://www.ana.co.jp/
2. Abara, J.: Applying integer linear programming to the fleet assignment problem. Interfaces 19, 20–28 (1989)
3. Dantzig, G.B., Ramser, J.H.: The truck dispatching problem. Management Science, 80–91 (1959)
4. Deb, K.: Multi-objective Optimization using Evolutionary Algorithms, Wiley (2001)
5. Gen, M., Cheng, R., Lin, L.: Network models and optimization: Multi-objective genetic algorithm approach. Springer (2008)
6. Inoue, H.: Simulation Analysis of Airline Network with Various Aircraft. Journal of Economics, 53-73 (2012)
7. Iseya, S., Takadama, K.: Evolutionary spoke network based on demand and operation. In: 2010 Second World Congress on Nature and Biologically Inspired Computing (NaBIC), pp. 678–683. IEEE (2010)
8. Jimba, T., Kitagawa, H., Azuma, E., Sato, K., Sato, H., Hattori, H., Takadama, K.: Towards Network Optimization of Regular and Non-regular Flights. In: The 16th International Symposium on Intelligent and Evolutionary Systems (IES2012), pp. 124–128 (December, 2012)
9. Sato, M., Adachi, N.: Airline Schedule Planning by Genetic Algorithms. The Japanese Society for Artificial Intelligence 16(6), 493–500 (2001)
10. Sato, M., Matumoto, S., Teramoto, Y., Adachi, N.: The Fully Dated Airline Crew Pairing by Genetic Algorithms. The Japanese Society for Artificial Intelligence 16(3), 324–332 (2011) (in Japanese)

Distance Based Locally Informed Particle Swarm Optimizer with Dynamic Population Size

Nandar Lynn and Ponnuthurai Nagaratnam Suganthan

School of Electrical & Electronics Engineering
Nanyang Technological University
nandar001@ntu.edu.sg, epnsugan@ntu.edu.sg

Abstract. In this paper, distance-based locally informed particle swarm optimizer is introduced with dynamic population size and constant neighborhood size. The population is reduced from a larger size for exploration and a smaller size for exploitation. The simulation study is conducted extensively on the shifted and rotated benchmark problems. The results show that the proposed algorithm outperforms the current state-of-art algorithms in solving shifted and rotated unimodal and multimodal problems.

Keywords: distance-based fully informed particle swarm optimizer, neighborhood size, dynamic population size, exploration, exploitation.

1   Introduction

Particle Swarm Optimization (PSO), inspired from the social behavior of fish schooling and bird flocking, is a population-based optimization algorithm [1,2,3]. In PSO, each particle in the swarm represents a potential solution and they collaborate with each other to find the optimum over the search space. Imitating the swarm social behavior, the particles retain their best solution and share this promising solution with other particles in the swarm. Thus, in PSO, each particle is attracted to the best solution found by itself and another best solution found by any neighbor particle in the swarm.

Based on a particle's neighborhood size, there are two main variants in PSO [4], namely global PSO and local PSO. If the entire population is defined as each particle's neighborhood, it is called global PSO. In global PSO, a particle finds the optimal solution over the search space while learning from its own previous best and the best solution among the entire swarm as follows:

$$V_i^d = w * V_i^d + c_1 * rand1_i^d * \left(pbest_i^d - X_i^d\right) + c_2 * rand2_i^d * \left(gbest^d - X_i^d\right) \quad (1)$$

$$X_i^d = X_i^d + V_i^d \quad (2)$$

Where, i=1, 2,...., N and N is population size. d= 1, 2,..., D and D is the number of parameters to be optimized in an optimization problem. X_i^d and V_i^d represent the position and velocity of ith particle in the population. $pbest_i^d$ is the best position found by

© Springer International Publishing Switzerland 2015

H. Handa et al. (eds.), *Proc. of the 18th Asia Pacific Symp. on Intell. & Evol. Systems – Vol. 2*,
Proceedings in Adaptation, Learning and Optimization 2, DOI: 10.1007/978-3-319-13356-0_45

578   N. Lynn and P.N. Suganthan

i^{th} particle itself and *gbest* is the best position found by the entire swarm neighborhood. *w* is inertia weight which balance the global and local search abilities of PSO. *w* can be either constant or linear/non-linear function of time. c_1 and c_2 are acceleration coefficients and $rand1_i^d$ and $rand2_i^d$ are two uniformly distributed random numbers in the range of [0,1].

Unlike global PSO, if a particle in the swarm has a small social neighborhood with only few other particles, it is known as local PSO. In local PSO, The velocity of a particle is updated using the local best solution *lbest* found within its small local neighborhood as in the following equation (4) and the position is updated using the same equation (2).

$$V_i^d = wV_i^d + c_1 * rand1_i^d * (pbest_i^d - X_i^d) + c_2 * rand2_i^d * (lbest_i^d - X_i^d) \qquad (3)$$

Where, $lbest(lbest_i^1, lbest_i^2, lbest_i^3, \ldots, lbest_i^D)$ is the local best neighborhood position found by ith particle within its neighborhood. The global PSO converges fast, however it can be trapped in a local optimum while local PSO searches the optimum solution with slower convergence and more accurately [4,5]. In order to improve PSO, different types of neighborhood topologies (such as Ring, Star, Mesh and von Neumann topologies) had been studied in [6,7,8]. In [8], Kennedy and Mendes mentioned that the performance of the individual particle depended on the topology used and also the way they interacted with their neighbors. Suganthan introduced PSO with increasing neighborhood particles in which local neighborhood size was dynamically increased and global best (*gbest*) was replaced by local best (*lbest*) [9]. In FDR-PSO [10], not just following the best position found so far by any particle, the particle's velocity update is also influenced by best nearest neighbor selected based on fitness distance ratio. In unified PSO (UPSO) [11], global best and local neighborhood best positions were used in order to combine exploration and exploitation properties of global and local PSOs. In fully informed PSO (FIPS) [12], instead of following local best performer, the individual particle in the swarm is fully informed by the neighbors' best positions. Star, ring, four clusters, pyramid and square neighborhood topologies were studied in FIPS. In order to increase local search ability of PSO, distance-based locally informed PSO (LIPS) is proposed in [13]. Instead of using *gbest* to guide the search of a particle, each particle is locally informed by its nearest neighbor particles measured in term of Euclidean distance and local search ability is enhanced. In this paper, LIPS with decreasing population size is proposed to improve the local search performance of LIPS on solving shifted and rotated unimodal and multimodal problems. CEC 2005 shifted and rotated benchmark problems are used to evaluate the performance of modified LIPS algorithm. LIPS's framework is briefly described in section II. Modified LIPS with decreasing population size is introduced in section III and its performance on shifted and rotated multimodal problems are discussed in section IV. Finally, the paper is concluded in section V.

2   Distance-Based Locally Informed PSO (LIPS)

In PSO, each individual particle is influenced by *gbest* and even if it is far from the global optimal solution, may allocate in the local optimum and occur premature convergence to the local optima for the multimodal problems. To avoid this condition,

LIPS uses a set of neighbor particles as a source of influence for each individual particle in the swarm. Thus, the behavior of a particle is affected by its local neighborhood. The topology structure can affect searching the optimum by defining the neighborhood such as in ring topology, each particle only interacted with its immediate neighbors on its right and left according to particle indices. These immediate neighbors may be from different regions and it can affect exploitation process at later search and can lead to slow convergence.

In LIPS, rather than using neighborhood topologies, the particles select their nearest neighborhood in terms of Euclidean distance and share the local information from its nearest neighbor to find the optimum over the search space. Similar to FIPS [12], LIPS also uses the personal *pbest* of its nearest neighbors and the velocity of a particle is updated using the formula as follows:

$$V_i^d = \chi * (V_i^d + \varphi(P_i^d - X_i^d)) \tag{4}$$

Where,

$$P_i = \frac{\sum_{j=1}^{nsize}(\varphi_j * nbest_j)/nsize}{\varphi} \tag{5}$$

$$\varphi_j \sim U\left(0, \frac{4.1}{nsize}\right) \tag{6}$$

$$\varphi = \sum_{j=1}^{nsize} \varphi_j \tag{7}$$

$nbest_j$ is the jth nearest neighborhood to i^{th} particle's *pbest*. *nsize* is the neighborhood size and it is dynamically increased from 2 to 5. χ is defined as construction coefficient and $\chi = 0.7298$. φ_j is random positive number drawn from uniform distribution in the range of (0, 4.1/*nize*). φ is defined as acceleration weight and summation of φ_j. The position is updated using the same equation as in (2). Euclidean distance based neighborhood selection is $O(N^2)$, where N is the number of particles of the current population. Using the Euclidean distance based neighborhood selection, the neighbors are from the same region and it improves local search ability and fine tuning of the algorithm. Besides, using the information of all neighborhood particles, it can improve exploitation process especially at the later stage and lead to fast convergence and high accuracy. These two advantages can help LIPS to solve shifted and rotated multimodal problems.

3   LIPS with Dynamic Population Size

In this section, LIPS is modified with dynamic population size and constant neighborhood size. Population size is directly related to the search space. In PSO algorithms, the particles are initialized with a random within the search space. At the early stage, the particles will explore over the search space to find the possible regions containing the optimum. At the later stage when the good region is identified, the particles will fine tune the region to converge to the optimum.Thus, PSO algorithm requires the large population size in the beginning to explore globally all over the search space and the small population size is needed for local fine tuning in the end of the search. In LIPS, the population size is fixed. However large population size is not necessary in the end of the search process as the search space become narrower than in the beginning.

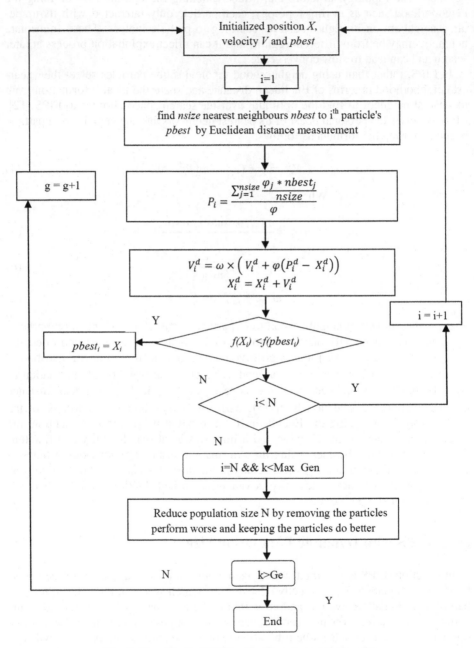

Fig. 1. Flow Chart of LIPS with dynamic population size

In this paper, LIPS with dynamic population size is proposed to enhance local search ability of LIPS. The population size is dynamically decreased by reducing the particles perform worse and by keeping the particles perform better. The population size will be dynamically decreased until 90 percent of the generations and at the last 10 percent of the generations, the fixed population size is used to maintain diversity and not to trap in the local optimum as shown in Fig. 2.

In the proposed algorithm, the velocity is updated using the LIPS equation (4) and the position is updated using the equation (2). The population size is dynamically decreased from 100 to 20 until 90% FES and population size 20 is used for the last 10% FES. Rather than using dynamic neighborhood size, the constant neighborhood size is used in the proposed algorithm. The neighborhood size, *nsize* 3 is used in this paper.

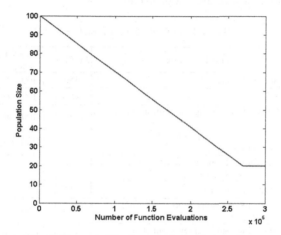

Fig. 2. Dynamically decreased population size

4   Simulation Study

4.1   CEC 2005 Test Functions

In order to evaluate the performance of modified LIPS algorithm, the 14 cases of CEC 2005 test functions in [17] are used in this paper. These 14 cases of CEC 2005 problems can be grouped into two categories, unimodal and multimodal problems. The search range, *f_bias F(x*)* and acceptance criteria of test functions are described in Table I.

4.2   Experiment Results and Discussion

The performance of the proposed algorithm is evaluated by comparing with following PSO algorithms:

- Distance-based locally informed PSO (LIPS)
- Fully-informed PSO (FIPS)
- Unified PSO (UPSO)

In FIPS, ring topology is used as the neighborhood topology and in LIPS, Euclidean distance based neighborhood selection is used to select the particles' neighbors. In both algorithms, the information from all the neighbors is used for updating velocity of every particle in the swarm. UPSO is combined of global and local PSO variants and cyclic topology is used for local variant. In order to compare fairly, average population size 60 of dynamically decreased population size of proposed algorithm is used for LIPS, FIPS and UPSO and their parameter settings are described in Table. 2. For the performance test, the experiment is conducted for 30 run times for each test function and the number of function evaluation 300000 FES is used in each run for each function.

The proposed algorithm is tuned with different neighborhood *nsize* 2 and 3 and firstly compared with original LIPS. The comparison of the two algorithms is shown in Table 3. As seen in comparison Table. 3, the proposed algorithm performs better than original LIPS on most of the test problems.

Table 1. CEC 2005 Test Functions

	Test Functions	$F(x^*)$	Search range	Criteria
Unimodal Test Functions	F_1: Shifted Sphere	-450	$[-100,100]^D$	1e-6
	F_2: Shifted Schwefel'sProblem 1.2	-450	$[-100,100]^D$	1e-6
	F_3: Shifted Rotated High Conditioned Elliptic	-450	$[-100,100]^D$	1e-6
	F_4:Shifted Schwefel'sProblem 1.2 with Noise in Fitness	-450	$[-100,100]^D$	1e-6
	F_5: Schwefel's Problem 2.6 with Global Optimum on Bounds	-310	$[-100,100]^D$	1e-6
Multi-modal Test Functions	F_6: Shifted Rosenbrock's	390	$[-100,100]^D$	1e-2
	F_7: Shifted RotatedGriewank's Function without Bounds	-180	$[-600,600]^D$	1e-2
	F_8: Shifted Rotated Ackley's Function with Global Optimumon Bounds	-140	$[-32,32]^D$	1e-2
	F_9: Shifted Rastrigin's	-330	$[-5,5]^D$	1e-2
	F_{10}: Shifted RotatedRastrigin's	-330	$[-5,5]^D$	1e-2
	F_{11}: Shifted RotatedWeierstrass	90	$[-0.5,0.5]^D$	1e-2
	F_{12}: Schwefel's Problem 2.13	-460	$[-\pi,\pi]^D$	1e-2
	F_{13}: Expanded ExtendedGriewank's plus Rosenbroack's Function (F_8F_2)	-130	$[-3,1]^D$	1e-2
	F_{14}: Shifted Rotated Expanded Scaffer's F_6	-300	$[-100,100]^D$	1e-2

Table 2. Parameter Settings

Algorithm	Constriction Coefficients χ	Acceleration coefficients c_1, c_2	Population size	Dimension (D)	Reference
FIPS	0.7298	~	60	30	[12]
UPSO	0.7298	1.49554	60	30	[11]
LIPS	0.7298	~	60	30	[13]
Modified LIPS	0.7298	~	100~20	30	

Table 3. Comparison of LIPS and modified LIPS with different neighborhood size

Function Type	F	LIPS	Modified LIPS*(2)	Modified LIPS*(3)
		ErrorMeans (F(x)-F(x*))		
Unimodal Test Functions	F_1	0	0	0
	F_2	307.12	**33.93**	**0.56**
	F_3	1.03E+07	**8.37E+06**	**2.05E+06**
	F_4	2.21E+04	2.62E+04	**2.01E+04**
	F_5	7.64E+03	7.60E+03	**6.93E+03**
Multi-modal Test Functions	F_6	82.02	**27.01**	**53.66**
	F_7	30.22	**22.83**	45.01
	F_8	20.91	20.92	20.96
	F_9	52.52	**40.59**	**40.11**
	F_{10}	71.75	**56.75**	**44.58**
	F_{11}	22.69	**21.89**	**18.88**
	F_{12}	8.85E+03	**2.24E+03**	**1.26E+03**
	F_{13}	5.06	4.03	**2.84**
	F_{14}	12.53	**12.30**	**12.00**

For both *nsize* 2 and 3, modified LIPS algorithm outperforms on unimodal functions F_2 and F_3 and on multimodal functions F_6 and F_9 to F_{14}. In functions F_1 and F_8, the proposed algorithm achieves the same performance as the state-of-art methods. Similarly, the proposed algorithm with *nsize* 3 provides better result on F_4 and F_5 and the algorithm with *nsize* 2 offers better result on F_6. Therefore, for overall, the proposed LIPS with dynamic population size obtained better performance than original LIPS on 12 out of 14 test functions and same results on the rest 2 functions.

The simulation results of the proposed algorithm with *nsize* 3 are compared with other PSO algorithms and the comparison of the experiment results are shown in Table. 4. The convergence characteristics graphs are also analyzed for unimodal and multimodal functions and are shown in Fig.2. Following [17], the median performance of 30 runs is used to analyze the convergence performance of the algorithms. As highlighted in Table. 3, UPSO performs best on unimodal functions F_2 and F_3 and on multimodal function F_7. FIPS obtain the best result on unimodal functions F_4 and F_5 and on multimodal function F_6. The proposed algorithm outperforms best on multimodal functions F_9 to F_{14} and second best on unimodal functions F_2 and F_3. These results can be clearly observed in Fig.2. Among multimodal functions, the convergence graphs of the functions F_9 to F_{12} are presented in Fig. 3 (b), (c), (d) and (e) respectively. The convergence analysis shows that LIPS with dynamic population size performs better than original LIPS and outperforms on most of multimodal problems.

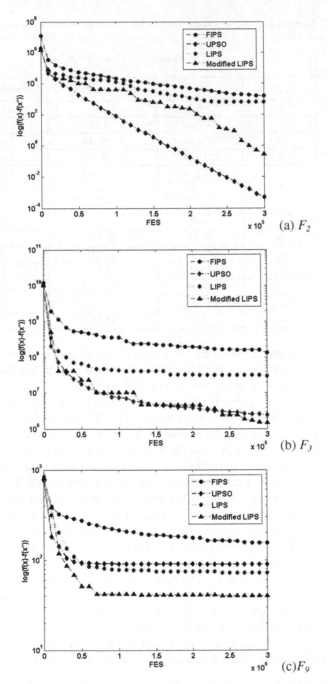

Fig. 3. Median convergence characteristics graphs of 30 dimensional CEC 2005 benchmark functions:(a) F_2: shifted schw fel's problem 1.2, (b)F_3: shifted rotated high conditioned elltic function, (c) F_9: shifted rastrigin's function, (d)F_{10}: shifted rotated rastrigin's function, (e)F_{11}:, shifted rotated weierstrass function (f). F_{12}: schwefel's problem 2.13

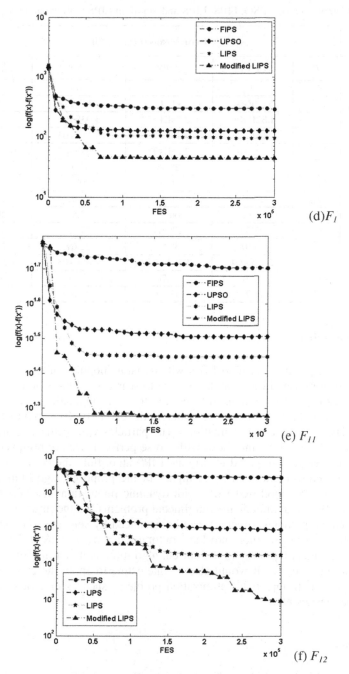

(d)F_1

(e) F_{11}

(f) F_{12}

Fig. 3. (*continued*)

Table 4. Comparison of UPSO, FIPS, LIPS and modified LIPS with dynamic population

Function Type	F	ErrorMeans ($F(x)$-$F(x^*)$)			
		UPSO	FIPS	LIPS	Modified LIPS*(3)
Unimodal Test Functions	F_1	0	0	0	0
	F_2	**5.83E-04**	394.17	307.12	0.56
	F_3	**1.82E+06**	2.79E+07	1.03E+07	2.05E+06
	F_4	6.65E+03	**2.20E+03**	2.21E+04	2.01E+04
	F_5	6.02E+03	**2.35E+03**	7.64E+03	6.93E+03
Multi-modal Test Functions	F_6	52.14	**28.48**	82.02	53.66
	F_7	**0.01**	0.07	30.22	45.01
	F_8	20.95	20.94	20.91	20.96
	F_9	81.62	66.97	52.52	**40.11**
	F_{10}	93.39	179.60	71.75	**44.58**
	F_{11}	30.01	38.51	22.69	**18.88**
	F_{12}	2.48E+04	9.91E+05	8.85E+03	**1.26E+03**
	F_{13}	4.86	12.70	5.06	**2.84**
	F_{14}	12.75	13.19	12.53	**12.00**

5   Conclusion

In this paper, we present modified LIPS with dynamic population size. In the beginning, the algorithm used large population size to search over the whole search space and the population size is dynamically reduced to small population size to exploit in the good region in the end. In order to maintain the diversity, fixed population size is used at the last 10% function evaluations. The particles with better performance are kept to continue the search and those with worse performance are removed from the population. Therefore, compared to original LIPS algorithm, the computational time is reduced and convergence ability is higher in the proposed algorithm.Simulation results also show that modified LIPS with dynamic population size performs better than original LIPS on unimodal and multimodal problems. As the future work, instead of finding nearest particles, diversity can be enhanced by selecting random particles and the probability to get stuck into local optimum can be decreased. The next is to introduce hierarchical topology to the current algorithm as it can improve the performance of the local search. It would also be interesting to apply the current approach for solving the multi-objective optimization problems such as electrical power systems, optimal design, etc.

References

1. Kennedy, J., Eberhart, R.C.: Particle swarm optimization. IEEE International Conference on Neural Networks 4, 1942–1948 (1995)
2. Eberhart, R.C., Kennedy, J.: A new optimizer using particle swarm theory. In: Sixth International Symposium on Micro Machine and Human Science, pp. 39–43 (1995)

3. Shi, Y., Eberhart, R.C.: A modified particle swarm optimizer. IEEE Congress on Evolutionary Computation, 69–73 (1998)
4. Eberhart, R.C., Simpson, P., Dobbins, R.: Computational Intelligence PC Tools. Academic Press, Boston (1996)
5. Kennedy, J., Eberhart, R.C., Shi, Y.: Swarm Intelligence. Morgan Kaufmann/Academic Press, San Francisco (2001)
6. Kennedy, J.: Small worlds and mega-minds: Effects of neighborhoodtopology on particle swarm performance. In: Proceedings of Congress on Evolutionary Computation, pp. 1931–1938 (1999)
7. Kennedy, J., Mendes, R.: Population structure and particle swarm performance. In: IEEE Congress on Evolutionary Computation, Honolulu, HI, pp. 1671–1676 (2002)
8. Kennedy, J., Mendes, R., Neighborhood, R.: topologies in fully informedand best-of-neighborhood particle swarm. IEEE Transaction on System, Man and Cybernetics Part – C 36(4), 515–519 (2006)
9. Suganthan, P.N.: Particle swarm optimizer with neighborhood operator. In: Proceedings of Congress on Evolutionary Computation, Washington, DC, pp. 1958–1962 (1999)
10. Peram, T., Veeramachaneni, K., Mohan, C.K.: Fitness-distance-ratiobased particle swarm optimization. In: Proceedings of Swarm Intelligent Symposium, pp. 174–181 (2003)
11. Parsopoulos, K.E., Vrahatis, M.N.: UPSO: Aunified particle swarm optimization scheme. Lecture Series on Computer and Computational Sciences, vol. 1, pp. 863–867 (2004)
12. Mendes, R., Kennedy, J., Neves, J.: The fully informed particleswarm: Simpler, maybe better. IEEE Transaction on Evolutionary Computation 8, 204–210 (2004)
13. Qu, B., Suganthan, P.N., Das, S.: A distance-based locally informed particle swarm model for multimodal optimization. IEEE Transaction on Evolutionary Computation 17(3), 387–402 (2013)
14. Clerc, M., Kennedy, J.: The particle swarm-explosion, stability, andconvergence in a multidimensional complex space. IEEE Transaction on Evolutionary Computation 6(1), 58–73 (2002)
15. Wolpert, D.H., Macready, W.G.: No free lunch theorems for optimization. IEEE Transaction on Evolutionary Computation 1, 67–82 (1997)
16. Liu, L., Yang, S., Wang, D.: Particle swarm optimization with composite particles in dynamic environments. IEEE Transaction on System, Man and Cybernetics Part- B 40(6), 1634–1648 (2010)
17. Suganthan, P.N., Hansen, N., Liang, J.J., Deb, K., Chen, Y.-P., Auger, A., Tiwari, S.: Problem definitions and evaluation criteria for the CEC2005 special session on real-parameter optimization. In: Proceedings of Congress on Evolutionary Computation, pp. 1–50 (2005)

Improve Robustness and Resilience
of Networked Supply Chains

Robert de Souza and Rong Zhou

The Logistics Institute – Asia Pacific, National University of Singapore
tlihead@nus.edu.sg, tlizr@nus.edu.sg

Abstract. With increased risk events and complex supply chain networks in recent years, it is imperative for a company to enhance the robustness and
resilience of its supply chain in fighting with any risk event. The current paper
defines the concepts of supply chain robustness and resilience and also identifies levers to enhance them. Subsequently, some key solutions are presented in
each of four phases of supply chain risk management, e.g. risk identification,
assessment, mitigation and monitoring. Finally, a master facilitative control
tower is proposed as the platform to provide global information for independent
local control towers to enable early risk detection and collaboration in risk
mitigation.

Keywords: supply chain risk management, control tower, master facilitative
control tower.

1   Introduction

Supply chain risk management has stayed at the top of the corporate agenda, given the
propensity for an ever-increasing number of disasters and disruptions to business in
the past decade. It is imperative that a company should understand its disruption profile prior to or at least upon a risk event, the available disruption reduction levers for
mitigation, and the efficient and effective ways supply chain risk management can
contribute to cost avoidance. Compounding the situation is that supply chains intersect other supply chains, sharing resources and members (nodes), in complex patterns.

This paper documents the progress of our current research in supply chain risks,
their identification and propagation, and the subsequent visualization and monitoring.
The intent is to improve the understanding and resolution of the challenges in building
robust and resilient networked supply chains. Generally, after the occurrence of a risk
event, the disruption profile of the attacked company can undergo two stages: decay
and recovery (Figure 1).

The decay stage shows the company's robustness level upon the attack of a risk
event while the recovery stage shows its resilience level in bringing performance back
to its original level. Figure 2 is the disruption profile of a robust and resilient company. On the contrary, Figure 3 reveals that a company is fragile in facing a disaster and
lacks resilience in recovering from the disruption.

© Springer International Publishing Switzerland 2015   589
H. Handa et al. (eds.), *Proc. of the 18th Asia Pacific Symp. on Intell. & Evol. Systems – Vol. 2*,
Proceedings in Adaptation, Learning and Optimization 2, DOI: 10.1007/978-3-319-13356-0_46

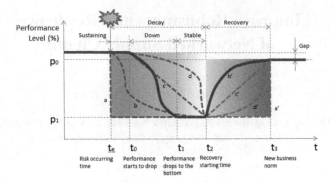

Fig. 1. Disruption profile - a generic decay and recovery curve

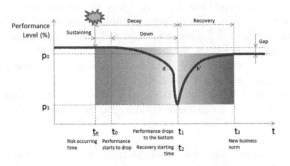

Fig. 2. Disruption profile of robustness and resilience

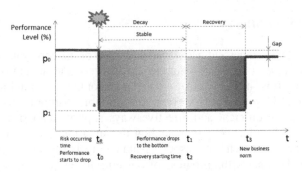

Fig. 3. Disruption profile of no robustness and resilience

Robustness is thus defined as the ability of the supply chain (or its entity) to withstand a disruption without adapting its initial stable configuration and Resilience is the property that enables a supply chain to resume its original shape or position after being disrupted. We present a series of levers that could be unilaterally or multilaterally applied according to the generic decay and recovery pattern evident or anticipated. Essentially, there are four disruption reduction levers (Figure 4): act to either delay the disruption impact (1), or act to reduce the disruption duration (2), or speed up the recovery (3), and contain as far as possible the disruption severity (4). One or more of these levers is necessarily engaged for a strategic response.

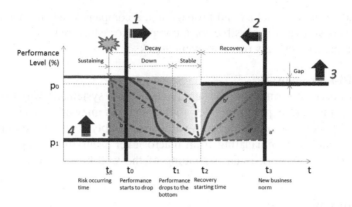

Fig. 4. Disruption reduction levers

Though an understanding of the disruption profile is key, just as important is a methodological response. In order to manage supply chain risks, a company should thus be able to systematically identify potential risks/disruptions embedded in its supply chain, assess their impacts, build in mid- or long-term preparedness or mitigation strategies, and monitor risks.

In section two, we summarize some of our current research work in each phase of supply chain risk management: risk identification, risk assessment, risk mitigation, and risk monitoring to enhance robustness and resilience of supply chains. We also introduce the concept of master facilitative control tower in section three to support supply chain risk management in complex networks. Section four concludes the work.

It should be noted that this paper does not intend to give a complete framework of supply chain risk management. Instead, it summarized the ongoing work that has been done under the research project titled: the master facilitative control tower for risk management of complex supply chains.

2   Improve Robustness and Resilience in the Four Phases of Supply Chain Management

As mentioned in the above section that there are four phases in supply chain risk management: risk identification, assessment, mitigation, and monitoring/alerting. Each phase is important in enhancing the robustness and resilience of a supply chain.

2.1   Risk Identification

Risk identification is the key and the first step for risk management in a supply chain. Supply chain partners need to systematically identify the potential risks for either focal companies or the overall network. Thus, a risk framework should be proposed as a stracturalized check list for a company to go through its risk factors. Furthermore, as a supply chain is generatlly a network, risks may propagate from one node to another one. Thus risks through propagation should also be considered. Finally, after understanding the risk factors and propagation, a company needs to identify the most risky partners so that it is possible for a company to reduce risk impact by switching to safer partners before any real risk event attacks.

2.1.1   Risk Framework

Companies need to scrutinize their supply chains in a systematic way to find the potential risks they are exposed to. We provide a risk framework considering three levels of risk factors: those arising from within one supply chain, those impacting with an industry, and those emanating from macro level events impacting several industries [1]. This framework is an expanding of the well-known work of Chopra and Sodhi [2].

2.1.2   Risk Propagation

As supply chain entities are linked together in networks, the disruption in one location may propagate to other locations through financial, information, or material flows in an unexpected manner. The characteristics of propagation include the time delay and risk amplification. Thus, a minor risk in the up-stream may cause serious risks to entities in the downstream of the supply chain. The mechanics and paths of the propagation should be investigated for risk analysis and evaluation.

Zhou et al. [3] model the propagation of the supply delay risk. The analysis of several scenarios demonstrates that inventory levels, lead times, and risk duration are key factors determining the behaviour of risk propagation. A new time-based Inoperability Input-Output Model is developed to quantify the propagation of disruptions in a supply chain occurring at any node. Companies can use the method to have a quantitative estimation of the impact of the disruptions, such as the percentage of non-deliveries due to the disruptions. With this visibility of propagation index, companies then can take mitigating action to counter the disruptions [4].

2.1.3   Risk Exposure Index

It is important for a firm to understand its risk exposure level within the supply chain and that contributed by its partners so that it can choose (1) the best partners before any risk actually occurs and (2) alternative partners post-disruption, without compromising or necessarily increasing the network's risk exposure index.

Simchi-Levi proposed a method using simulation to calculate risk exposure index [5]. We postulate a statistical methodology for determining a risk exposure index for the supply chain as a whole as well as its constituents. Such an approach would be applicable to a comprehensive strategic supply chain diagnosis in view of risk-averse/prone members. Firstly, a set of risk factors are identified. Secondly, the proposed framework measures the degree of risk exposure of a focal company. The degree of risk exposure on the focal company can, in turn, be extended to the entire supply chain (network). With the risk exposure index, a firm can now more confidently determine both, its operational levers and the interaction levels of activities with its partners so that the potential risk exposure would be managed in advance, pre- and post-disruption.

2.2   Risk Assessment

Supply chain risk assessment helps a company to understand the potential risks and their consequences so that it can prioritize resources for risk mitigation and management. In assessing the impact of certain disruptions to a company or a supply chain or a network, we propose simulation and VaR (Value at Risk) approaches for investigative analysis.

Simulation enables companies to analyze and evaluate their systems' performance in the occurrence of risks. A simulation model has to be built for the company based on its operational data. Generally, the following data are required.

- The emerging pattern of the disruption profile/distribution
- The time of disruption
- The duration of the disruption
- Operational data e.g. inventory level, product unit cost, demand, etc.

With greater risks and vulnerabilities and increasing complexities in supply chains, how to provide responsive and effective "What if" analysis of the impact of disruptions and mitigation policies in complex supply chains becomes an issue of high attention from both practice and academia. VaR is commonly used for risk assessment by providing a quantitative analysis of risk and it is defined as a threshold value such that for a certain probability value, the amount of loss on the portfolio does not exceed this value within a specified time. For industry, the risk can be quantified using VaR based on "losses" which includes both tangible and intangible costs for a disruption [6].

2.3   Risk Mitigation

With the risks now identified and prioritized, risk mitigation is the next step to making supply chains more robust and resilient through the adoption of preferred mitigation policies. Supply chain mitigation policies can be categorized into different types based on the risk issues including supply chain structure, visibility, resilience and buffer issues, etc. We have analyzed some typical issues and designed related mitigation policies in inventory control, sourcing strategy, resource pooling, and postponement.

2.3.1   Inventory Control Policy

Vulnerabilities manifest as supply lead time fluctuations or demand loss or uncertainty, play major roles in the increase of inventory cost. Inappropriate inventory policies, which do not take into account these risks and modify their parameters, may incur a huge cost in backorders or inventory levels at the onset of a disruption. An efficient and effective inventory ordering policy should consider these factors efficiently and effectively while determining policy parameters [7].

A new inventory policy, which adjusts its parameters according to the variation in demand, lead time or other cost parameters, is postulated. The policy – a 'time-adjusted' (r, Q) policy determines the r (reorder point) and Q (order quantity) values at the start of each defined time period based on the new demand forecast for order placement, rather than using the same r and Q values.

In testing this policy with the existing inventory policy of a global MNC, the performance of the policies is proven to be more responsive and effective numerically under various demand distributions. The time-adjusted (r, Q) policy generally outperforms the manufacturer's existing policy under identified disruptions. The following figures show the impact on cost variations under lead-time variability and demand uncertainty, respectively, when demand is assumed to be normally distributed.

2.3.2   Emergency Sourcing Strategies

A supplier's capacity is affected when a risk event occurs and it fails to deliver the requested products in time due to its reduced capacity. As a result, the manufacturer may face some penalties because of its failure to supply the products to customers in time. One way to mitigate the problem is to order some emergency replenishment from alternate suppliers or from the same supplier but from a different location, albeit at a higher price. But what price should the manufacturer pay for the emergency orders?

We consider only those risk events, which have low probability but high impact (LPHI). An exponential distribution is used to represent the reduction in supplier capacity. Based on this stochastic capacity, we develop an expression for the total inventory cost for the manufacturer when it places the order without considering capacity reduction. The cost includes lost sales because of its failure to supply customers due to the non-availability of parts. We also determine the inventory cost when emergency orders are placed to minimize the number of stock outs. Comparing the total inventory cost, we derive the necessary conditions for placing an emergency order. This study provides the necessary conditions under which the manufacturer should place an emergency order to minimize the total cost [8].

2.3.3   Optimal Allocation of Safety Stock

Demand uncertainty is one of the major vulnerabilities in supply chains. It is important to have a holistic approach that systematically allocates the optimal safety stock level for every node in the supply chain and, at the same time, coordinate the production plans of manufacturers and replenishment policies of distribution centers/retailers.

We use an evolutionary optimization and agent-based simulation at the upper layer to produce the series of inventory positions for all agents for a period of time, and use a look-ahead Model Prediction Control (MPC) algorithm to produce operational inventory positions based on actual demand information but with reference to the inventory positions.

The solution can meet the actual demand better and dampen demand fluctuation in the supply chain. The solution can also maintain good Value-Added performance at the same time with acceptable customer service level at the operational level. This approach is in a two-layer structure: an upper layer module for performance optimization and a lower-layer module for addressing demand uncertainty.

The integrated inventory control algorithms designed are more effective and efficient than existing approaches in mitigating demand uncertainty in the supply chain, while allowing for various inventories related constraints to be evaluated.

2.4   Risk Monitoring

After identifying the key risks, carefully choosing the right (networked) supply chain stakeholders, implementing mitigation strategies to improve robustness and resilience, then a company still needs to monitor its operational processes for timely risk detection and early quick fixes. An essential element in achieving agility in supply chains is visibility (Christopher and Peck 2004). Thus, there is an immediate need for the visualization of impacts of these decisions, more so, when the supply chain is complex.

Supply chain visualization tools for risk tracking and analyzing are important for companies as the cost of being unprepared and the lack of visibility for potential and ongoing disruptions could be immense.

The supply chain networks and their key entities can be mapped and visualized based on their physical locations. The visualization platform has implemented the following functions:

- A visualization dashboard
- Tracking of disruption events associated with the geographical locations of the supply chain nodes
- Display of temporal, spatial and connectivity patterns
- Scenario analysis to prepare the plan for supply chain risk mitigation

As the result, we have built the supply chain visualization platform (RiskVis) [9] [10] to enable the timely capture of data from the other modules addressing the respective challenges including risk identification and mitigation and network-based analysis, modeling and simulation. These collect, monitor and analyze critical items such as:

- Inbound and outbound logistics: it can help to detect shipment delays and understand the movement of material/parts/products along the supply chain.
- Inventory level: It can keep obsolescent inventory at minimal levels while providing enough buffer for unforeseen events.
- Order fulfillment and manufacturing operations.
- Risks such as natural disasters that might affect part of the supply chain.

3   Improve Robustness and Resilience through the Master Facilitative Control Tower

The concept of supply chain control tower is to provide a more flexible and agile control structure to co-ordinates and manages the flow of goods and information, from supplier to inbound logistics, manufacturing, outbound logistics to the end customer. A control tower offers a centralized view of planning and execution systems and a consolidated platform for enabling rapid recognition and faster response to change. Thus it can facilitate communication among four phases of supply chain risk management as illustrated in Fig. 5.

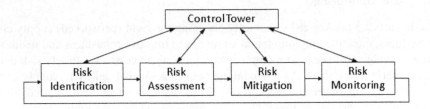

Fig. 5. Control tower and supply chain risk management

However, as today's supply chain networks become complex and one node may play a part simultaneously in several supply chains, thus the concept of the master facilitative control tower (MFCT) is developed to design and develop risk tracking methodologies crossing different intersecting supply chain networks (Fig. 6).

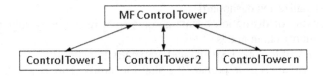

Fig. 6. The concept of the master facilitative control tower

A MFCT can provide global information to independent control towers as a neutral platform. Each control tower may register its public information to the MFCT and through the synergy of information, the MFCT is able to have a global view across many supply chain networks. Some main features of the MFCT are as follows.

- Provide a management platform for risk information sharing and infusion.
- Facilitate collaboration in producing integrated and quick fix highly responsive solutions.
- Provide analysis and insights about cascading risk propagation across integrated supply chain networks
- Provide a visualization about connectedness in supply chains and the identification of key linkages in and across the networks
- Provide for an integrated response based on the challenges inherent in risk identification, assessment and mitigation, and monitoring.

4   Conclusions and Future Research

In this paper, we first identify levers to enhance supply chain robustness and resilience, which are illustrated in a few figures. Then we present some of our main solutions in each of four phases of supply chain risk management that can materialize the robustness and resilience. Finally, in order to feed those solutions with timely and accurate information to pinpoint the source of risk, assess the impact, and identify the right mitigation strategies, the concept of a master facilitative control tower is posited as the grand challenge.

Acknowledgements. This work was supported by the Science and Engineering Research Council (SERC) of the Agency for Science, Technology and Research (A*STAR) of Singapore under Grant 1224204048.

References

1. Zhou, R., de Souza, R., Goh, M.: Supply chain risks and frameworks. In: Tan, Y.W., Sim, T., de Souza, R. (eds.) Managing logistics and supply chain challenges – Singapore insights and perspectives, pp. 435–448. Singapore, Cengage (2013)
2. Chopra, S., Sodhi, M.S.: Managing risk to avoid supply-chain breakdown. Mit Sloan Management Review 46(1), 53–62 (2004)
3. Zhou, R., Goh, M., der Souza, R.: Modelling the propagation of delay risks in a supply chain. In: 24th Annual POMS Conference, Denver, United States (2013)
4. Li, Z.P., et al.: An Extended Risk Matrix Approach for Supply Chain Risk Assessment. In: IEEE International Conference on Industrial Engineering and Engineering Management (IEEM), Bangkok (2013) (accepted)
5. SCDigest. Risk Exposure Index Starting to Gain Traction, Change Supply Chain Thinking, David Simchi-Levi Says (2012), http://www.scdigest.com/assets/on_target/13-04-24-1.php?CID=6971
6. Zhang, A.N., Wagner, S.M., Goh, M., Terhorst, M., Ma, B.: Quantifying Supply Chain Disruption Risk Using VaR. In: IEEE International Conference on Industrial Engineering and Engineering Management (IEEM), Hong Kong, pp. 272–277 (2012)
7. Ghosh, S., Goh, M., de Souza, R.: A time-adjusted (r, Q) policy for supply chain risk management. Journal of Operational Research Society (2014) (Submitted)
8. Ghosh, S., Goh, M., de Souza, R.: Emergency ordering during catastrophic risks (working paper) (2014)
9. Goh, R.S.M., Wang, Z.X., Yin, X.F., Fu, X.J., Ponnambalam, L., Lu, S.F., Li, X.R.: RiskVis: Supply Chain Visualization with Risk Management and Real-time Monitoring. In: The 9th Annual IEEE International Conference on Automation Science and Engineering (IEEE CASE 2013), Madison Wisconsin, USA (2013)
10. Wang, Z.X., Goh, R.S.M., Yin, X.F., Ponnambalam, L., Fu, X.J., Lu, S.F.: Understanding the Effects of Natural Disasters as Risks in Supply Chain Management: A Data Analytics and Visualization Approach. In: 2013 Singapore Management University Summer Institue, Analytics for Business, Consumer and Social Insights (BCSI) Area of Excellence (AoE), Singapore, August 3-5 (2014)

An Efficient Representation for Genetic-Fuzzy Mining of Association Rules

Chuan-Kang Ting, Ting-Chen Wang, and Rung-Tzuo Liaw

Department of Computer Science and Information Engineering and
Advanced Institute of Manufacturing with High-tech Innovations
National Chung Cheng University, Taiwan
{ckting,wtc102p,lrt101p}@cs.ccu.edu.tw

Abstract. Data mining is a blooming area in information science. Mining association rules aims to find the relationship among items in the databases and has become one of the most important data mining technologies. Previous study shows the capability of genetic algorithm (GA) to find the membership functions for fuzzy data mining. However, the chromosome representation cannot avoid the occurrence of inappropriate arrangement of membership functions, resulting in inefficiency of GA in searching for the optimal membership functions. This study proposes a novel representation that takes advantage of the structure information of membership functions to deal with the issue. In the light of overlap and coverage, we propose two heuristics for appropriate arrangement of membership functions. The experimental results show that GA using the proposed representation can achieve high fitness and suitability. The results also indicate that the two heuristics help to well exploit the structure information and therefore enhance GA in terms of solution quality and convergence speed on fuzzy association rules mining.

1 Introduction

With the rapid increase in the amount and size of data, data mining emerges to leverage and transform the information hidden in the data into explicit knowledge [5,6,14]. Data mining technologies include description, classification, estimation, clustering, and association rules. Description means the observation and explanation by human beings, which is the primordial way of data mining. Classification categorizes data into finite outcomes to which the decision maker prefers. A well-known method is the support vector machine [3]. Estimation approximates the possible value of the desired outcome, which is usually a real-valued number. Approximation model or surrogate is a well-developed and promising approach for estimation [13]. Clustering groups the data with high similarity and explains the differences between groups. A common method is the K-means clustering [16]. Association rules mining attempts to find the relationship among items in the database.

This study focuses on mining association rules through genetic fuzzy system. The aim of mining association rules is to find the relationship among items, which is also known as homogeneous group or affinity group. A noted case is the prediction of customer's behavior in Walmart. The Apriori algorithm is the most famous approach for mining association rules. It establishes association rules depending on a user-defined minimum confidence. Srikant and Agrawal [15] considered Boolean association rules and proposed a method for quantitative association rules, which can deal with data with quantitative value or categories. The proposed method is similar to the Apriori algorithm, whereas the former requires data discretization as a pre-process. Rather than discretization, Hong et al. [10,11,12] employed fuzzy logic and fuzzy sets to analogue the value of original data and accordingly proposed the fuzzy transaction data mining algorithm (FTDA). The quantitative values of data are transferred into fuzzy values by membership functions. This fuzzy data mining method holds three advantages: extension, tolerance, and suitability for nonlinear systems. The FTDA uses fuzzy sets for each item's membership functions and establishes association rules by these membership functions. The results are known as fuzzy association rules.

Although FTDA enables fuzzy data mining, the genetic algorithm (GA) used in FTDA is inefficient in finding optimal setting for the membership functions. In addition, FTDA fails to consider the relationship between membership functions. This study aims to address these issues and improve GA on fuzzy data mining. Specifically, we propose a novel chromosome representation for the GA. The representation considers the structure of membership functions and their relationship. Based on the representation, two heuristics are presented to eliminate inappropriate arrangement of membership functions and thus reduce the search space. Experiments are conducted to examine the performance of the proposed GA in comparison to FTDA.

This paper is organized as follows. Section 2 introduces association rules and fuzzy association rules. Section 3 describes the proposed GA. Experimental results are presented in Section 4. Finally, we draw conclusions and recommend the directions for future work in Section 5.

2   Fuzzy Association Rules

Association rules can be generally expressed in the following mathematical form:

$$X \rightarrow Y,$$

where X and Y are sets of items. Taking the famous Walmart example "If buying *bread* and buying *cheese*, then buying *milk*," this association rule can be written as

$$bread, cheese \rightarrow milk.$$

Given an itemset $I = \{i_1, \ldots, i_m\}$ with m items and a database D with n transactions $D = \{T_1, \ldots, T_n\}$, where each $T_i \subseteq I$ is assigned with a unique ID. An association rule $X \rightarrow Y$ represents the possibility of $X \subseteq T_i$ then $Y \subseteq T_i$. Two metrics are used to quantify the importance of an association rule: support and confidence.

Definition 1 (Support). *The support of association rule $X \rightarrow Y$ is defined by*

$$\text{Support} (X \rightarrow Y) = \mathcal{P} (X \cap Y).$$

Support calculates the probability that X and Y coexist.

Definition 2 (Confidence). *The confidence of association rule $X \rightarrow Y$ is defined by*

$$\text{Confidence} (X \rightarrow Y) = \mathcal{P} (X|Y) = \frac{\mathcal{P} (X \cap Y)}{\mathcal{P} (Y)}.$$

Confidence calculates the probability that X exists given Y exists.

The Apriori algorithm [1,2] is widely used to mine association rules. This method first selects the large itemsets $L = \{L_1, \ldots, L_m\}$ from all candidate itemsets $C = \{C_1 \ldots, C_m\}$ based on the user-defined minimum support $support_{\min}$. Then the algorithm establishes association rules according to the user-defined minimum confidence $confidence_{\min}$. Since the size of candidate itemsets C is exponential to the number of items m in the itemset I, the Apriori algorithms has an exponential time complexity $O(2^m)$.

Hong et al. [10] proposed the FTDA to construct fuzzy association rules based on the fuzzy support and fuzzy confidence of the set of membership functions for all m items in the database D. Let $\Upsilon_{j,k}$ denote the fuzzy region of k-th membership function for item I_j. The fuzzy value $f_{j,k}^{(i)}$ for fuzzy region $\Upsilon_{j,k}$ is transformed from $v_j^{(i)}$, i.e., the quantity of the j-th item in the i-th transaction in database D.

Definition 3. *The fuzzy support of region $\Upsilon_{j,k}$ is defined by*

$$\text{FuzzySupport} (\Upsilon_{j,k}) = \frac{1}{n} \sum_{i=1}^{n} f_{j,k}^{(i)}.$$

The large 1-itemset L_1 is determined by checking whether $f_{j,k}^{(i)} \geq support_{\min}$ for $\Upsilon_{j,k}$. This step is similar to the Apriori algorithm. In checking the fuzzy support of a set of membership functions $MF = \{mf_1, \ldots mf_p\}$, the fuzzy value of MF is given by the intersection operator, i.e.,

$$f_{MF}^{(i)} = \bigcap_{k=1}^{p} f_{mf_k}^{(i)}.$$

In fuzzy logic, the intersection operator is commonly implemented by taking the minimum. Accordingly, the fuzzy value of MF is

$$f_{MF}^{(i)} = \min_{1 \leq k \leq p} f_{mf_k}^{(i)}.$$

Definition 4 (Fuzzy Support). *The fuzzy support of large itemsets MF is defined by*

$$\text{FuzzySupport} (MF) = \frac{1}{n} \sum_{i=1}^{n} f_{MF}^{(i)}.$$

If $f_{MF}^{(i)}$ is greater than the minimum support $support_{min}$, then the set MF is added to the large p-itemsets L_p. The selection of large itemsets continues until $L_p = \emptyset$. Next, the fuzzy association rules are established from the large itemsets according to their fuzzy confidence. For a large p-itemset $MF = \{mf_1, \ldots mf_p\}$ in L_p, the candidate rules have the following form:

$$mf_1 \wedge \cdots \wedge mf_{k-1} \wedge mf_{k+1} \wedge \cdots \wedge mf_p \to mf_q \quad \text{for } 1 \leq k \leq p.$$

Here we denote the above rule by MF_k .

Definition 5 (Fuzzy Confidence). *The fuzzy confidence of a candidate rule MF_k is defined by*

$$\text{FuzzyConfidence}\,(MF_k) = \frac{\sum_{i=1}^{n} f_{MF}^{(i)}}{\sum_{i=1}^{n} \left(f_{mf_1}^{(i)} \wedge \cdots f_{mf_{k-1}}^{(i)} \wedge f_{mf_{k+1}}^{(i)} \wedge \cdots \wedge f_{mf_p}^{(i)} \right)}.$$

If the fuzzy confidence of a candidate rule is greater than the user-defined minimum confidence, i.e., FuzzyConfidence(MF_k) $\geq confidence_{min}$, then this rule is established as a fuzzy association rule. Note that all candidate rules generated from the large itemsets L should be checked for fuzzy association rules.

The set of membership functions plays an essential role in the FTDA because the establishment of fuzzy association rules is subject to the set of membership functions. The fuzzy regions define the explanation of item's quantity and therefore influence the fuzzy association rules obtained. Hence, the main goal in mining fuzzy association rules is to find the set of membership functions for all items in the database that are beneficial for establishing fuzzy association rules.

3   Genetic Algorithm for Fuzzy Association Rules Mining

Inspired from Darwin's evolutionary theory, GA evolves a population of candidate solutions into the optima [9]. In GA, the candidate solutions are represented as chromosomes. The parent selection, crossover, mutation and survival selection of GA mimic the evolutionary operators in nature [4,7,9]. The parent selection operator picks chromosomes from the population as the parents for reproduction. The offspring are generated by crossover of parents; afterward, GA performs mutation on the offspring for increasing diversity. The reproduction process is repeated until the subpopulation is filled. Then the survival selection determines the chromosomes that can survive into the next generation.

This study focuses on the design of representation for GA to efficiently mine fuzzy association rules. More details about the proposed GA are described below.

3.1   Representation

In genetic fuzzy mining of association rules, GA is used to find suitable membership functions for a given item. The shape of membership function is commonly set to be triangular for simplicity. A corresponding fuzzy region is parameterized

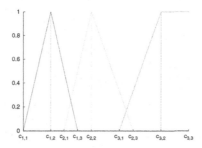

Fig. 1. An example of three membership functions

by the three vertices of triangle. Hong et al. [10] used a real-valued string as a chromosome for representing membership functions. For an item with ℓ linguistic terms, a chromosome is composed of 3ℓ parameters to determine the ℓ membership functions. Let $c_{i,j}$ denote the j-th parameter of i-th membership function for a given item. Two constraints are imposed for the legality of membership functions:

$$c_{i,1} \leq c_{i,2} \leq c_{i,3} \tag{1}$$

$$c_{1,2} \leq c_{2,2} \leq \cdots \leq c_{l,2} \tag{2}$$

The first constraint holds the triangular shape and the second constraint ensures the order of linguistic terms. Figure 1 illustrates an arrangement of membership functions that satisfy the above constraints. As a chromosome violates the constraints, its parameters are sorted and reordered to fix this issue.

Although the above representation can maintain the shape and order of membership functions, it may still generate inappropriate membership functions in terms of overlap and coverage. Coverage denotes the range covered by all membership functions for an item. In fuzzy systems, coverage reflects the explanation capability of the membership functions. Overlap is measured by the same coverage between all pairs of membership functions; therefore, high overlap brings about trivial linguistic terms. For example, the membership functions in Fig. 2 are inappropriate in that the overlap of the left two membership functions is too high and the coverage is incomplete due to the gap between the right two membership functions. This study proposes introducing the structure information of membership functions into the representation to address the issue of inappropriate arrangement of membership functions. The proposed chromosome representation consists of two parts: 1) parameters and 2) structure type of the membership functions. For three membership functions, there are totally 93 structure types that satisfy the constraints (1) and (2) on shape and order. Figures 3 and 4 illustrate these 93 structures. An integer is used in the chromosome representation to indicate the structure type.

Moreover, the structure type facilitates development of heuristics for enhancing the performance GA. In this study, we propose two heuristics for elimination

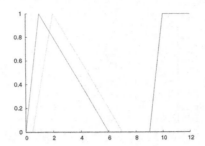

Fig. 2. An illustration of inappropriate membership functions

of inappropriate arrangement of membership functions:

$$c_{i,3} \geq c_{i+1,1} \tag{3}$$

$$c_{i,3} \leq c_{i+1,3} \tag{4}$$

The heuristics use two inequalities (3) and (4) to secure the coverage and moderately decrease the overlap, respectively. According to these two heuristics, 12 structure types are selected from the 93 structure types. These 12 structure types guarantee appropriate arrangement of membership functions considering coverage and overlap.

The proposed representation using the above heuristics has two major advantages. First, it can eliminate inappropriate structures for the membership functions. Second, the two heuristics decrease the number of structure types and thus reduce the search space of GA. These advantages help to improve GA in convergence speed and solution quality.

3.2   Genetic Operators

The genetic operators in GA include parent selection, crossover, mutation, and survival selection. For parent selection, this study adopts the well-known k-tournament selection, which randomly picks k individuals from the population and selects the best among the k individuals as a parent. Performing tournament selection twice obtains a pair of parents for reproduction.

This study employs the crossover and mutation operators commonly used in genetic fuzzy data mining [8,10], to wit, max-min-arithmetical (MMA) crossover and creep mutation. The MMA crossover generates four candidates and selects the best two of them as offspring. The four candidates are generated in different ways: two are produced by the whole arithmetic crossover, one is the maximum and one is the minimum of the two parents. The whole arithmetic crossover conducts a linear combination of two parents. As for mutation, the parameters are performed with creep mutation with a bound ε, whereas the structure type is mutated by random resetting.

The $(\mu + \lambda)$ survival selection is adopted, in which both parent and offspring populations are considered for survival.

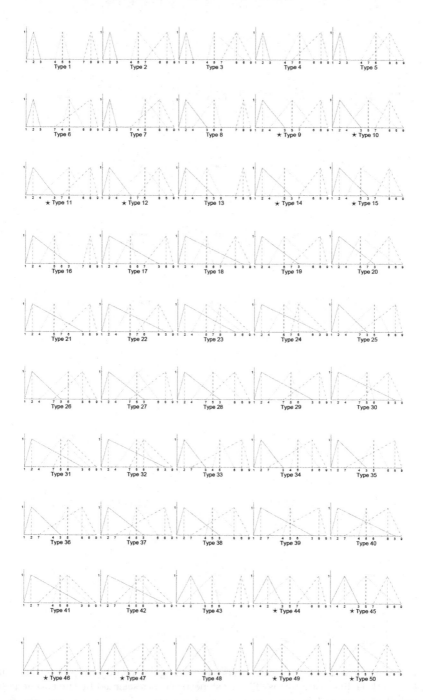

Fig. 3. The 93 structure types for triangular membership functions corresponding to three linguistic terms. The 12 structure types selected by the two heuristics are marked with asterisk.

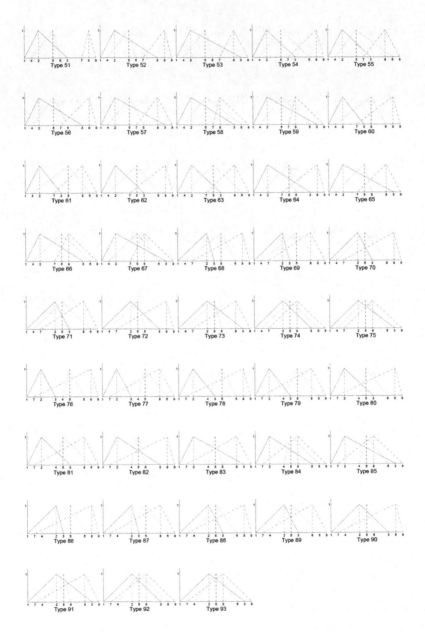

Fig. 4. (Continued) The 93 structure types for triangular membership functions corresponding to three linguistic terms. The 12 structure types selected by the two heuristics are marked with asterisk.

3.3   Fitness Evaluation

In this study, we apply the fitness function proposed by Hong et al. for mining fuzzy association rules [10]. For a given item, the fitness of a chromosome is determined by three factors, i.e., overlap, coverage, and fuzzy support. Let Υ_i denote the fuzzy region of i-th membership function for a given item.

Definition 6 (Overlap Factor). *The overlap factor of chromosome C_k is defined by*

$$\text{Overlap}\,(C_k) = \sum_{i<j} \left(\max\left(\text{ovlratio}\,(\Upsilon_i, \Upsilon_j), 1\right) - 1\right)$$

with

$$\text{ovlratio}\,(\Upsilon_i, \Upsilon_j) = \frac{\text{The same coverage of } \Upsilon_i \text{ and } \Upsilon_j}{\min\,(c_{i,3} - c_{i,2}, c_{j,2} - c_{j,1})}.$$

Overlap considers the same coverage between all pairs of membership functions. Definition 6 gives the overlap factor, where the ratio of overlap ovlratio (Υ_i, Υ_j) is determined by the proportion of the same coverage between two membership functions and is normalized by the smaller of $(c_{i,3} - c_{i,2})$ and $(c_{j,2} - c_{j,1})$, which stand for the right side of the left membership function and the left side of the right membership function, respectively. The overlap factor is nonnegative and its best value is zero. Note that the zero overlap factor does not imply non-overlap of all pairs of membership functions but indicates the overlap is adequate according to the definition of overlap ratio.

Definition 7 (Coverage Factor). *The coverage factor of chromosome C_k is defined by*

$$\text{Coverage}\,(C_k) = \frac{\max\,(I)}{\text{range}\,(\Upsilon_1, \ldots, \Upsilon_l)}.$$

Coverage factor is inversely proportional to the range covered by all membership functions. A small coverage factor implies that the given item is well represented. The best coverage factor is 1 for full coverage of the item's quantity.

The fitness evaluation considers the fuzzy support and the suitability, where the latter is defined by the sum of overlap and coverage factors. Formally, the fitness value of chromosome C_k is computed by

$$f\,(C_k) = \frac{\sum_{X \in L_1} \text{FuzzySupport}\,(X)}{\text{Overlap}\,(C_k) + \text{Coverage}\,(C_k)}, \tag{5}$$

where L_1 denotes the large 1-itemset obtained from the membership functions set C_k. Note that the use of only L_1 without $L_{>1}$ is to balance the computation time and quality in calculating fuzzy support [10].

The fitness function considers fuzzy support as well as the arrangement of membership functions. According to the definitions of overlap and coverage, the proposed two heuristics for structure types guarantee that the coverage factor is at least $\frac{\max(I)}{c_{l,3} - c_{1,1}}$. The heuristics help to eliminate the gap between membership functions and avoid strong overlap.

Table 1. Parameter setting

Parameter	Value
Representation	Real-valued
Parent selection	2-tournament
Crossover	MMA ($d = 0.35$, $p_c = 0.8$)
Mutation	Creep ($\varepsilon = 3$, $p_m = 0.01$)
Survival selection	$\mu + \lambda$
Population size	50
#Generations	500

4   Experimental Results

In this study, we conduct a series of experiments to examine the effectiveness of the proposed representation. The test algorithms include the original GA [10], GA using new representation (GA_{93}), and GA using new representation and two heuristics (GA_{12}). The subscript of GA_{93} and GA_{12} accounts for the number of structure types explored in the GA. Table 1 summarizes the parameter setting for the three GAs, which follows the setting in [10] for comparison. The minimum support is set to 0.04. The test data consists of 10000 transactions and 64 items [10]. Each experiment runs 30 independent trials considering the stochastic nature of GA.

Figure 5 shows the anytime behavior of GA, GA_{93}, and GA_{12} on item 1. The results indicate that GA_{93} and GA_{12} converge faster and obtain better solution quality than GA does. Moreover, GA_{12} achieves the fastest convergence speed and highest fitness. Figure 6 depicts the membership functions obtained from GA, GA_{93}, and GA_{12} on item 1 after 500 generations. All the three algorithms result in good coverage; however, GA_{93} and GA_{12} acquire better overlap. We further compare the mean fitness values of all 64 items obtained from the three algorithms. According to Fig. 7 and Table 2, GA_{12} outperforms GA and GA_{93} on most of the items, while both GA_{93} and GA_{12} perform generally better than GA does. These preferable outcomes validate that the proposed representation can improve the effectiveness and efficiency of GA in mining fuzzy association rules. In addition, the two heuristics help GA to efficiently explore appropriate arrangements of membership function and further enhance the performance of GA.

Figures 8 and 9 present the fuzzy support and suitability for GA, GA_{93}, and GA_{12} on all 64 items. The results indicate that GA_{93} and GA_{12} lead to better suitability but lower fuzzy support than GA does. Likewise, GA_{12} obtains higher suitability and yet lower fuzzy support than GA_{93}. The results reveal the trade-off between fuzzy support and suitability. The use of structure information helps to maintain the suitability of membership functions while pursuing high fuzzy support.

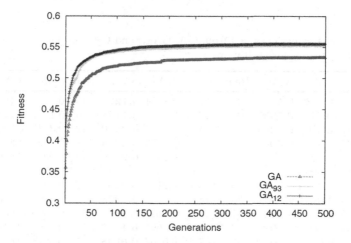

Fig. 5. Progress of the mean best fitness for GA, GA_{93}, and GA_{12} on item 1

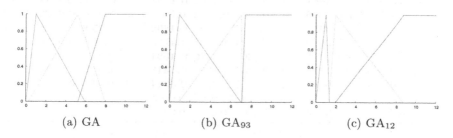

(a) GA   (b) GA_{93} (c) GA_{12}

Fig. 6. Membership functions obtained by GA, GA_{93} and GA_{12} on item 1

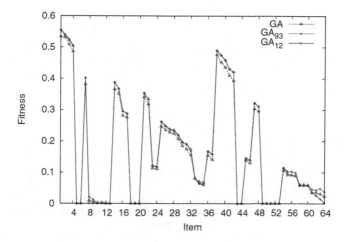

Fig. 7. Fitness values obtained from GA, GA_{93}, and GA_{12} on all 64 items after 500 generations

Table 2. Fitness values obtained from GA, GA_{93}, and GA_{12} on all 64 items. Boldface marks the best fitness for each item

Item	GA	GA_{93}	GA_{12}	Item	GA	GA_{93}	GA_{12}	Item	GA	GA_{93}	GA_{12}
1	0.5334	0.5530	**0.5555**	23	0.1132	0.1194	**0.1227**	45	0.1395	0.1447	**0.1469**
2	0.5315	0.5379	**0.5397**	24	0.1106	0.1177	**0.1184**	46	0.1307	**0.1415**	0.1389
3	0.5087	0.5209	**0.5254**	25	0.2475	0.2617	**0.2630**	47	0.3037	0.3203	**0.3228**
4	0.4862	0.5033	**0.5048**	26	0.2357	0.2475	**0.2489**	48	0.2960	**0.3109**	0.3103
5	0.0000	0.0000	0.0000	27	0.2288	0.2362	**0.2396**	49	0.0000	0.0000	0.0000
6	0.0000	0.0000	0.0000	28	0.2242	0.2331	**0.2360**	50	0.0000	0.0000	0.0000
7	0.3813	0.4014	**0.4015**	29	0.2070	0.2183	**0.2200**	51	0.0000	0.0000	0.0000
8	0.0088	**0.0212**	0.0000	30	0.1855	0.1952	**0.1978**	52	0.0000	0.0000	0.0000
9	0.0056	**0.0124**	0.0000	31	0.1747	**0.1915**	0.1894	53	0.0000	0.0000	0.0000
10	**0.0021**	0.0000	0.0000	32	0.1559	0.1691	**0.1747**	54	0.1069	0.1140	**0.1159**
11	0.0015	**0.0036**	0.0000	33	0.0797	0.0807	**0.0838**	55	0.0918	0.0972	**0.1030**
12	0.0000	**0.0017**	0.0000	34	**0.0701**	0.0629	0.0655	56	0.0914	0.0919	**0.1026**
13	0.0000	0.0000	0.0000	35	**0.0670**	0.0597	0.0603	57	0.0886	0.0907	**0.0989**
14	0.3660	0.3877	**0.3881**	36	0.1549	0.1662	**0.1687**	58	**0.0606**	0.0569	0.0590
15	0.3517	0.3664	**0.3669**	37	0.1408	**0.1591**	0.1581	59	0.0605	**0.0622**	0.0566
16	0.2816	0.2953	**0.2955**	38	0.4756	0.4894	**0.4899**	60	**0.0590**	0.0580	0.0585
17	0.2762	0.2873	**0.2884**	39	0.4523	0.4738	**0.4753**	61	0.0357	**0.0460**	0.0355
18	0.0000	0.0000	0.0000	40	0.4361	0.4576	**0.4591**	62	0.0359	**0.0436**	0.0266
19	0.0000	0.0000	0.0000	41	0.4100	0.4297	**0.4312**	63	0.0313	**0.0482**	0.0154
20	0.0000	0.0000	0.0000	42	0.3931	0.4209	**0.4211**	64	0.0235	**0.0383**	0.0019
21	0.3411	0.3528	**0.3538**	43	0.0000	0.0000	0.0000				
22	0.3177	0.3339	**0.3343**	44	0.0000	0.0000	0.0000				

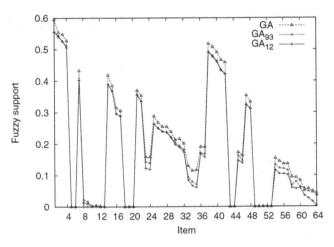

Fig. 8. Fuzzy support for GA, GA_{93}, and GA_{12} after 500 generations

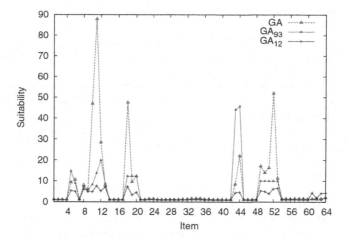

Fig. 9. Suitability (overlap plus coverage) for GA, GA$_{93}$, and GA$_{12}$ after 500 generations

5 Conclusions

This study proposes a new representation that considers the structure type of membership functions for mining fuzzy association rules. In addition, we present two heuristics for securing the coverage and moderating the overlap of membership functions. The heuristics are applied to eliminate inappropriate arrangement of membership function. Precisely, for three membership functions, the two heuristics reduce the number of structure types from 93 to 12; that is, 81 inappropriate structure types are pruned away.

The experimental results show that the proposed representation is capable of improving GA in fitness and suitability of membership functions. In particular, GA using the new representation with the two heuristics can achieve better solution quality and faster convergence speed than other two test GAs. The preferable outcomes validate that the utility of the proposed representation. In addition, the two heuristics help GA to efficiently explore appropriate arrangements of membership functions and further promote the efficiency of GA in mining fuzzy association rules.

Some directions remain for future work. Memetic algorithms are powerful problem solvers, which combines global search with local enhancement. Incorporating local search on the structure type is a promising direction for improving this work. In addition, design of local enhancement operator for parameters of membership functions is another way to enhance genetic-fuzzy data mining.

Acknowledgment. The authors would like to thank Shiau-Huei Su for her preliminary work and Tzung-Pei Hong and Chun-Hao Chen for the test data and comments. This work was supported by Ministry of Education, Taiwan.

References

1. Agrawal, R., Imielinski, T., Swami, A.N.: Mining association rules between sets of items in large databases. In: Proceedings of the 1993 ACM SIGMOD International Conference on Management of Data (1993)
2. Agrawal, R., Srikant, R.: Fast algorithms for mining association rules. In: Procddings of 20th International Conference on Very Large Data Bases (VLDB), pp. 487–499. Morgan Kaufmann (1994)
3. Chang, C.-C., Lin, C.-J.: Libsvm: a library for support vector machines. ACM Transactions on Intelligent Systems and Technology 2(27), 1–27 (2011)
4. Eiben, A.E., Smith, J.E.: Introduction to Evolutionary Computing. In: Natural Computing. Springer, Berlin (2003)
5. Fayyad, U., Piatetsky-Shapiro, G., Smyth, P.: From data mining to knowledge discovery in databases. AI Magazine 17, 37–54 (1996)
6. Fayyad, U.M., Piatetsky-Shapiro, G., Smyth, P.: Advances in knowledge discovery and data mining. In: Fayyad, U.M., Piatetsky-Shapiro, G., Smyth, P., Uthurusamy, R. (eds.) From Data Mining to Knowledge Discovery: An Overview, pp. 1–34. American Association for Artificial Intelligence (1996)
7. Goldberg, D.E.: Genetic Algorithms in Search, Optimization and Machine Learning. Addison-Wesley (1989)
8. Herrera, F., Lozano, M., Verdegay, J.L.: Fuzzy connectives based crossover operators to model genetic algorithms population diversity. Fuzzy Sets System 92(1), 21–30 (1997)
9. Holland, J.: Adaptation in Natural and Artificial Systems. University of Michigan Press (1975)
10. Hong, T.-P., Chen, C.-H., Lee, Y.-C., Wu, Y.-L.: Genetic-fuzzy data mining with divide-and-conquer strategy. IEEE Transaction on Evolutionary Computation 12(2), 252–265 (2008)
11. Hong, T.-P., Chen, C.-H., Wu, Y.-L., Lee, Y.-C.: A GA-based fuzzy mining approach to achieve a trade-off between number of rules and suitability of membership functions. Soft Computing 10(11), 1091–1101 (2006)
12. Hong, T.-P., Kuo, C.-S., Chi, S.-C.: Mining association rules from quantitative data. Intelligent Data Analysis 3(5), 363–376 (1999)
13. Jin, Y.: Surrogate-assisted evolutionary computation: Recent advances and future challenges. Swarm and Evolutionary Computation 1, 61–70 (2011)
14. Piatesky-Shapiro, G., Brachman, R., Klösgen, W., Simoudis, E.: An overview of issues in developing industrial data mining and knowledge discovery applications. In: Proceedings of Knowledge Discovering and Data Mining, pp. 89–95 (1996)
15. Srikant, R., Agrawal, R.: Mining quantitative association rules in large relational tables. SIGMOD Record (ACM Special Interest Group on Management of Data) 25(2), 1–12 (1996)
16. Wagstaff, K., Cardie, C., Rogers, S., Schroedl, S.: Constrained K-means clustering with background knowledge. In: Proceedings 18th International Conference on Machine Learning, pp. 577–584 (2001)

Medical Image Management System with Automatic Image Feature Tag Adding Functions

Tomoyuki Hiroyasu[1], Yuji Nishimura[2], and Utako Yamamoto[1]

[1] Faculty of Life and Medical Sciences, Doshisha University,
Tataramiyakodani, 1-3, Kyotanabe-shi, Kyoto, Japan
{tomo,utako}@mis.doshisha.ac.jp
[2] Graduate School of Life and Medical Sciences, Doshisha University,
Tataramiyakodani, 1-3, Kyotanabe-shi, Kyoto, Japan
ynishimura@mis.doshisha.ac.jp

Abstract. In this paper, the new medical image management system is proposed. In this system, the system saves image data at any time and extracts feature quantity of each image automatically. At the same time, the information on feature quantity is stored in the metadata of the image. Since image processing is performed automatically, the users do not have burdens for adding feature information. To examine the validity of the proposed system, jpeg pictures which have Exchangeable image file format (Exif) data are stored and the image features are extracted. In the evaluation experiment, the experiment of system of operation is conducted and it is checked whether the system operates normally. At the same time, required time to extract features and write to the metadata is measure and evaluated.

Keywords: Medical Image, Object Detect, Exchangeable Image File, Digital Imaging and Communications in Medicine.

1 Introduction

In recent years, even though cancer care and diagnostic technic progress, the number of cancer death is increasing. At the same time, multiple primaries that cancer generates to multiple organs become the problem [1–5]. There are the Simultaneous multiple primaries and the Non-synchronous multiple primaries in the multiple primaries. Simultaneous multiple primaries is the cancer that occurs in the multiple organs within two months. Non-synchronous multiple primaries is the cancer that occurs in multiple organs after it passes for more than two months. The time of multiple primaries is different from person to person. Therefore, the early detection of the cancer is very important [6–9]. Currently, there is a diagnostic imaging as an effective method of diagnosis for finding cancer. The medical image used at the time of diagnostic imaging is image of Digital Imaging and Communications in Medicine (DICOM) standard. The DICOM image can include information such as patient information and photography information in metadata, and is outputted from diagnostic imaging equipment, such as

© Springer International Publishing Switzerland 2015   613
H. Handa et al. (eds.), *Proc. of the 18th Asia Pacific Symp. on Intell. & Evol. Systems – Vol. 2*,
Proceedings in Adaptation, Learning and Optimization 2, DOI: 10.1007/978-3-319-13356-0_48

Computed Tomography (CT) and Magnetic Resonance Imaging (MRI). However, usually, only the part that needs DICOM image for diagnosis is generated and only this part is paid attention. In this procedure, it is impossible to obtain cancer information complication. That means there is a possibility cancer occurs by the part that is not observed. The early cancer may be overlooked and it may be discovered in the last stage. In order to prevent oversight of the early cancer, it is necessary to conduct advanced image analysis by the medical specialist [10–13]. Therefore, a system that conducts image analysis automatically and enables prevention of the cancer oversight is required [14, 15]. We propose a system that adds the analysis result to the metadata of the DICOM image automatically. Since DICOM image can store several types of information in the metadata, image feature values which are analyzed automatically are added into tag information. A user only uploads images to a proposal system, and image analysis is conducted automatically on back of the system. Each process is performed as batch processing, a user need not to wait until process is completed. Because of this operation, a user is free from getting image analysis knowledge and the mitigation of a burden of a user is attained. Furthermore, by adding analysis results to the metadata of image and checking several types of diseases from several points, sick oversight prevention is assisted.

After, implemented the system, we conducted the evaluation experiment for confirming the usefulness of the proposed system. In order to use a proposal system by actual environment, user interface was created so that user could access the system. At the same time, the system that uses Joint Photographic Experts Group (JPEG) image containing not DICOM image but Exchangeable Image File (Exif) data by problem of personal information or security was constructed. Exif is the image standard same as DICOM and photography environment information is stored [16]. When the proposed system treats Exif correctly, it is also made clear that the system treats DICOM properly. In order to conduct image analysis, image processing using OpenCV is performed [17, 18]. In the evaluation experiment, the information of the person reflected from the image was extracted using OpenCV and this information is added to the Exif tag. To confirm the effectiveness of the system, the time for adding tag information and the time for searching are examined.

2   Proposed System: Medical Image Management System with Automatic Image Feature Tag Adding Functions

In this research, medical image management system with automatic image feature tag adding functions is proposed. This system processes the medical image data saved at server at any time and extract of feature quantity. Feature quantity information is stored in the metadata of an image. Proposal system consists of following several servers.

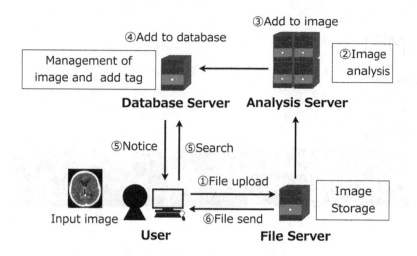

Fig. 1. Configuration of the proposed system

- File server
  Images are uploaded to this server and images are stored.
- Analysis server
  Server extracts new information from image inputted by image analysis automatically, and adds it to metadata of the image.
- Database server
  This server creates database of analysis result and image metadata, and manages added information and image.

With the proposed system, users enable to search and notice from analysis result or metadata. The block diagram of proposal system is shown in Fig. 1. The process of the proposal system is described as follows:

1. A user uploads image to the file server.
2. Image analysis is started simultaneously with extraction of metadata from the image.
3. The analysis result is added to metadata of image as tag information.
4. The tag information added to the extracted metadata and metadata is inserted in a database.
5. A user can obtain other information by search and notice of result which were added.

Input images are analyzed with OpenCV library. OpenCV is an image processing API which is developed by Intel and offers several types of image processing algorithms [17].

For managing images, such as search operation, system does not treat images themselves but tag information.

Fig. 2. Interface of the system

Usually, when input image and analysis image manage, management is performed using the metadata of image. Therefore, server creates the database of analysis result obtained by OpenCV and image. In other words, Database server manages image management and storage server manages image preservation.

3   Implementation of Proposal System and Evaluation Experiment

3.1   Implementation of Proposal System

In order to check the usefulness of the proposal system, we implemented the proposal system. The implemented system is a web application system and user can use the system through the Internet.

The interface of the web application is shown in Fig. 2.

The prosed system consists of the following servers.

- File server
- Analysis server
- Database server
- Web server

These four servers are located in the same LAN. Table. 1 shows the specifications of servers.

Table 1. SPECIFICATIONS OF A SERVER

OS	debian7.5 64bit
Kernel	3.2.0-4-amd64
CPU	Quad-Core AMD Opteron(tm) Processor 2356 1.2GHz
Core	4
Memory	8GB

The following functions are implemented.

– Uploading image to system
– Automatic image analysis by OpenCV
– Addition of analysis result to metadata of image
– Notification to user and search from metadata and analysis results

The block diagram of a system is shown in Fig. 3.

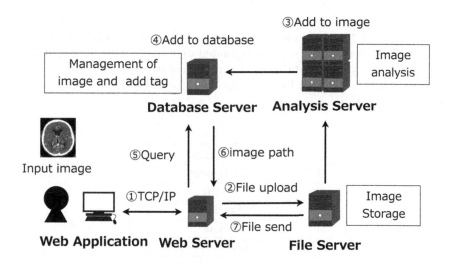

Fig. 3. Configuration of implemented system

The user uploads image through the interface implemented on the web server. The input images are saved at the file server. The system transmits image to analysis server and conducts image analysis by OpenCV there. Then information on analysis result is stored in the metadata of the image and inserts metadata and analysis result of image in database.

3.2   Adding Image Information to Exif Tag

Here, the function where information of analysis results of images is added to Exif tag automatically is implemented. Exif standard which stores several types of image information such as photography information is an extension of JPEG file format. JPEG has a structure shown in the Fig. 4.

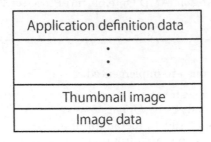

Fig. 4. Structure of JPEG

Application definition data is identified as two bytes values called maker and this data has values which are corresponded to markers. Exif information is one of application definition data. Exif information is stored into JPEG image as an application maker (APP1). Fig. 5 shows the structure of Exif information which is embedded in APP1 marker.

Exif Header
TIFF Header
0th IFD
Exif IFD
1st IFD
JPEG Thumbnail

Fig. 5. Structure of Exif

Exif information is defined based on TIFF (Tagged Image File Format) form. For storing metadata, a directory structure called IFD (Image File Directory) is used. The structure of IFD is shown in the Fig. 6.

Each IFD consists of tag number, data type, number of data and offset of data substance. In the proposed system, the additional information is added to tag information after 32768. There is a private field in TIFF which is one of APP1 markers and 32768th and more number is assigned to this field. For our additional tag, we implemented IFD as gUndifinedh type which can store various data type.

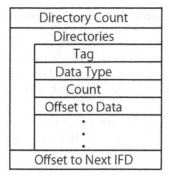

Fig. 6. Structure of IFD

3.3 Experimental Assessment

We conducted the evaluation experiment to check the usefulness of implemented system. In this experiment, instead of DICOM images, images containing Exif data are operated. Both Exif and Dicom have meta data and image itself. Thus, the discussions of usefulness of the proposed system in this experiment have validity for the situation using DICOM images. In the experiment, pictures where there are five people are existed are used. The size of image is 2816x2112 pixels. The image was transmitted to the implemented system through the interface. The time between staring uploading images and ending adding tag information is measured. At the same time, the time of searching images is also measured. The number of people on the picture is added to metadata. This number is detected As added information to metadata, the number of people on the picture is stored. This information is derived through the process to extract face, nose, and mouth. This process is the same as a DICOM image used at medical scene.

The human information is reflected from several types of images. These results are added to metadata of image. The system detects cancer automatically from the information of metadata. Through this process, not only the focused disease is observed but other diseases which are not focused can be discovered. Thus, this system prevents from disease oversight. In the experiment, the process performs for 10, 50 and 100 sheets respectively. Time of search of 10, 50, and 100 sheets are measured in the evaluation experiment. A setup of the tag added to metadata is shown in Table. 2CTable. 3CTable. 4.

Table 2. FACE TAG

Field Name	Face
Tag Number	32769
Type	Undefined

Table 3. NOSE TAG

Field Name	Nose
Tag Number	32770
Type	Undefined

Table 4. MOUTH TAG

Field Name	Mouth
Tag Number	32771
Type	Undefined

3.4   Extraction of Personal Information by Image Analysis

In this system, to extract personal information, the detector contained in OpenCV is used. This detector uses the Haar-Like feature and is performing object recognition [19] [20] [21]. Haar-Like feature is a brightness value of the local domain that combined two kinds of square areas, black and white. The search window for feature quantity extraction is shown in the Fig. 7. This feature has robustness to lighting conditions and noizes. People's face can be recognized by denoting relative distribution of light and darkness by Haar-Like feature quantity for light and darkness of eyes or a nose.

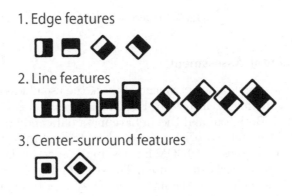

Fig. 7. Search window of Haar-Like feature extraction

4   Result

Fig. 8 shows that image was analyzed automatically when the image was inputted into the implemented system, and result of image analysis was added to Exif of the image.

This result show that the implemented system works well. In Fig. 9, the results of processing time are shown. Fig. 10 shows the searching time. The example of image of recognition result are shown in Fig. 11CFig. 12CFig. 13.

Fig. 9 shows that the operation of 100 sheets takes around 100 seconds. In this time, object was detected and the information is added to metadata. From Fig. 10, searching operation of 100 sheets takes less than 40 seconds. These results described that the implemented system is worth using as a real system. Fig. 9 shows that operation time of adding information to tag is increasing lineally with along to the number of images. This means that when the system treats huge number of images it takes a huge time. During diagnosis, several type of information should be offered in a short time. To conquer this defect, it is necessary to introduce the mechanism to shorten the operation time such as a parallel processing algorithm.

Fig. 8. Metadata addition of analysis results

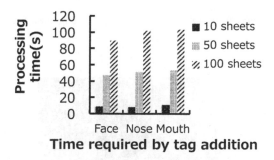

Fig. 9. processing time

In the proposed system, it is assumed that several types of image features are added to the tag automatically. For example, an early detection of cancer can be performed by analyzing a lot of images and adding onset probability data of several types of cancer. Therefore, it is necessary to conduct a large-scale evaluation experiment treating a vast quantity of data.

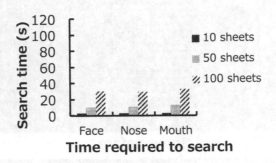

Time required to search

Fig. 10. search time

Fig. 11. Face detection

Fig. 12. Nose detection

Fig. 13. Mouth detection

5   Conclusion and Future Works

In this paper, the system which adds analysis result to the metadata of image automatically is proposed. This system processes the medical image data saved at servers at any time and extract of feature quantity. Then information on feature quantity is stored in the metadata of the image. As for this system, since image processing is performed automatically, the user does not have any burden for extracting this information. Moreover, information is added to metadata and this information is checked from several points of views periodically. This process is expected to prevent from disease oversight.

In this paper, the implemented system was introduced. In this system, instead of using DICOM images but images containing Exif data are used for the reason of personal information. Exif is standard for images and it has information tags of images. This construction is the same as DICOM. Through the evaluation experiment, the implemented system is fully useful and the expected specifications are implemented. In the future research, it is necessary to conduct the large-scale evaluation experiment treating a vast quantity of the data.

References

1. Kosak, T., Miwa, K., Yonemura, Y., et al.: A clinicopathologic study on multiple gastric cancers with special reference to distal gestrectomy Cancer. Cancer 65, 2602–2605 (1990)
2. Moertel, C.G., Gargen, J.A., Soule, E.D.: Multiple gastric cancers; review of the literature and study of 42 cases. Gastroenterology 32, 1095–1103 (1957)
3. Yamagiwa, H., Yoshimura, H., Matsuzaki, O., Ishihara, A.: Pathological study of multiple gastric carcinoma. Acta Pathol Jpn 30, 421–426 (1980)
4. Kurita, H.: A study of clinico-epidemiology on multiple gastric cancer. Nagoya Med. J. 22, 239–249 (1977)
5. Noguchi, Y., Ohta, H., Takagi, K., et al.: Synchronous multiple early gastric carcinoma: A study of 178 cases. World J. Surg. 9, 786–793 (1985)
6. Pepe, M.S., Etzioni, R., Feng, Z., et al.: Phases of Biomarker Development for Early Detection of Cancer. Journal of the National Cancer Institute 93, 1054–1061 (2013)
7. Henson, D.E., Srivastava, S., Kramer, B.S.: Molecular and genetic targets in early detection. Curr. Opin. Oncol. 11, 419–425 (1999)
8. Srivastava, S., Kramer, B.S.: Early detection cancer research network. Lab Invest 80, 1147–1148 (2000)
9. Greenwald, P.: New directions in cancer control. Johns Hopkins Med. J. 151, 209–213 (1982)
10. Fu, K.S., Mui, J.K.: A survey on image segmentation. Pattern Recognition 13, 3–16 (1981)
11. Strom, J., Cosman, P.C.: Medical image compression with lossless regions of interest. Signal Processing 59, 155–172 (1997)
12. Mori, K., Urano, A., Hasegawa, J., et al.: Virtualized endoscope system – an application of virtual reality technology to diagnostic aid. IEICE Transactions on Information and Systems E79-D, 809–819 (1996)

13. Mori, K., Deguchi, D., Sugiyama, J., et al.: Tracking of a bronchoscope using epipolar geometry analysis and intensity-based image registration of real and virtual endoscopic images. In: Special Issue on Medical Image Computing and Computer-Assisted Intervention - MICCAI 2001, vol. 6, pp. 321–336 (2002)
14. Hiroyasu, T., Uehori, K., Yamamoto, U., Tanaka, M.: Construction of an Interactive System Aims to Extract Expert Knowledge about the Condition Cultured Corneal Endothelial Cells 65, 1805–1810 (2013)
15. Choobineh, J., Vokurka, R.J., Vadi, L.: A prototype expert system for the evaluation and selection of potential suppliers 16, 106–127 (1980)
16. Kosak, T., Miwa, K., Yonemura, Y., et al.: Using extended file information (EXIF) file headers in digital evidence analysis. International Journal of Digital Evidence, Economic Crime Institute (ECI) 2, 1–5 (2004)
17. OpenCV: Open source Computer Vision library, http://opencv.org
18. Bradski, G., Kaehler, A.: Learning OpenCV: Computer Vision with the OpenCV Library. O'Reilly Media Inc. (2008)
19. Viola, P., Jones, M.J.: Rapid Object Detection using a Boosted Cascade of Simple Features 1, 511–518 (2001)
20. Lienhart, R., Maydt, J.: An extended set of Haar-like features for rapid object detection 1, 900–903 (2002)
21. Li, S.Z., Zhu, L., Zhang, Z., Blake, A., Zhang, H., Shum, H.-Y.: Statistical learning of multi-view face detection. In: Heyden, A., Sparr, G., Nielsen, M., Johansen, P. (eds.) ECCV 2002, Part IV. LNCS, vol. 2353, pp. 67–81. Springer, Heidelberg (2002)

The Impact of the Malfunction of a Sector in Supply Chain on the Ordering Policy of Each Sector

Hiroshi Sato[*], Tomohiro Shirakawa, Masao Kubo, and Akira Namatame

Department of Computer Science, National Defense Academy of Japan
1-10-20 Hashirimizu, Yokosuka, Kanagawa 239-8686 JAPAN
{hsato,shirakawa,masaok,nama}@nda.ac.jp

Abstract. Maintaining an effective management of supply chain even in the disasters is becoming crucial issue for manufacturing industry, because the contemporary supply chain networks become global and complex and the behaviors of the network are hardly predictable. The Beer Game is a simple but very useful example of supply chain management. The game consists of four sectors: factory, distributor, wholesaler, and retailer and deliver the beer to the customers. Many business schools adopt it to learn the key point of supply chain. In this study, computer agents play the game instead of humans. Agents are evolved with a genetic algorithm. We examine how the agents handle the game especially when some part of supply chain exhibit malfunction. Through simulations, we confirmed that effective ordering strategies are different between sectors.

Keywords: The Beer Game, Supply Chain Management, Genetic Algorithm.

1 Introduction

Supply chain is a flow of a product from factory to consumer [1]. Nowadays, the supply chain becomes global and complex network. This network produces effective productivity in peacetime, but the function of the supply chain will lost when a disaster occurs. For example, a large number of supply chains are forced to stop their operation after the 2011 Tohoku earthquake and tsunami, and 2011 Thailand floods [2]. The research of resilience of supply chain is an urgent issue in order to establish stable supply of products. Many cases such as mentioned above are studied, but it is difficult to draw general knowledge from these cases because each case has each own properties.

This paper takes up the Beer Game [3] as the benchmark of the supply chain management. This game is a very simple model of supply chain, but it can reproduce the important issue such as bullwhip effects [4][5]. Human players usually play the beer game, however, in this study the computer agents play the game. We analyze the performance of computer agents in the supply chain especially in the situation where some part of the network break down because of the disasters such as earthquake. Our purpose is to get the simple rules of thumb such as "Should we focus on the flow, or

---

[*] Corresponding author.

should we focus on stock?" because the simpler the rules become, the easier the managers of each sector handle the situation.

2   The Beer Game

2.1   Outline of the Game

The Beer Game is invented in 1960's in the business school of Massachusetts Institute of Technology. It has been used as an educational tool and a research tool. The supply chain in the beer game consists of four sectors: "factory", "distributor", "wholesaler", and "retailer" (Fig. 1). These four players make a team. The goal of the team is to minimize the total cost of operation. Every turn, the players decide how much case of beer they will order. The turn of the game is called "week" and the total cost is the sum of inventory fee and the loss of backorder. During the play of the game, players cannot communicate with each other. Only they can see is the flow of the beer in the supply chain.

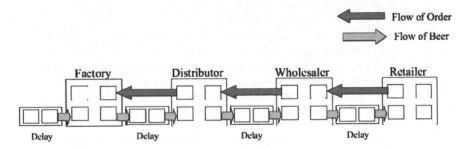

Fig. 1. The game board of Beer game

The players have to estimate the demand of the future and to make an appropriate order based on their strategies. This uncertainty makes the mismatch of demand and supply and unstableness in the supply chain [6].

2.2   The Problems in Beer Game

Bullwhip effect is the most interesting and catastrophic phenomenon where we can see the very large swings of inventories in response to the change of the demand. Fig. 2 shows the typical situation of Bullwhip effect.

The cause of this oscillation comes from the psychological factors of the player such as misperception and panic and the operational factors such as forecast error and lead-time variety.

The previous studies on how to prevent the occurrence of the bullwhip effect are classified into the two categories. The studies in the first category focus on the detection of the point where the demand of the customer changes [7][8], and those in the second category study the appropriate policies in each sector [6]. These studies assume there is no failure or accident during the operation. However, our concern is how we can find resilient strategies even when the disaster strikes the supply chain.

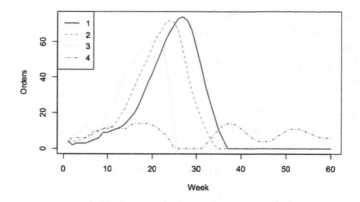

Fig. 2. Typical situation of Bull whip effect. The legend in the figure shows the layer of the supply chain: 1-> factory, 2-> distributor, 3 -> wholesale, 4 -> retailer. Small change of the layer 4 (blue) causes big change to layer 1 (black) in supply chain

3   Implementing the Beer Game by Agent Modeling

In this study, we adopt the Strozzi's model [6] for the implementing the Beer Game by computer agent. There are four layers of agents in the Beer Game: factory, distributor, wholesaler and retailer. Basically, all agents have the same structure. Fig. 3 shows agent at i-th layer.

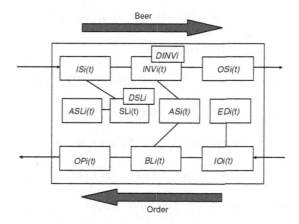

Fig. 3. Basic structure of the agent on i-th layer at time t in the Beer Game

The following are the state variables of the agent.

- $ISi(t)$: incoming shipment at time t,
- $INVi(t)$: inventory of the beer at time t,
- $DINVi$: desired inventory (constant),

- *OSi(t)*: outgoing shipment at time t,
- *ASLi(t)*: adjustment of supply chain at time t,
- *SLi(t)*: supply chain at time t,
- *DSLi*: desired supply chain (constant).
- *ASi(t)*: adjustment of stock at time t,
- *EDi(t)*: expected demand at time t,
- *OPi(t)*: orders placed at time t,
- *BLi(t)*: backlog of orders at time t,
- *IOi(t)*: incoming orders at time t,

Transition of the states is determined by the following equations.

$$INV_i(t) = \begin{cases} INV_i(t-1) + IS_i(t-1) - BL_i(t-1) - IO_i(t-1) \\ if \quad INV_i(t-1) + IS_i(t-1) \geq BL_i(t-1) + IO_i(t-1) \\ 0 \quad otherwise \end{cases} \tag{1}$$

$$IS_t(t) = OS_{i-1}(t-1) \tag{2}$$

$$BL_i(t) = \begin{cases} BL_i(t-1) + IO_i(t-1) - INV_i(t-1) - IS_i(t-1) \\ if \quad BL_i(t-1) + IO_i(t-1) \geq INV_i(t-1) + IS_i(t-1) \\ 0 \quad otherwise \end{cases} \tag{3}$$

$$OS_t(t) = \min\{INV_i(t-1) + IS_{i-1}(t-1) + IO_i(t-1)\} \tag{4}$$

$$ED_i(t) = \theta \cdot IO_i(t-1) + (1-\theta) \cdot ED_i(t-1) \tag{5}$$

Here, $\theta(0 \leq \theta \leq 1)$ is a control parameter,

$$OP_t(t) = \max\{0, \theta \cdot IO_i(t-1) + (1-\theta) \cdot ED_i(t-1\} \tag{6}$$

$$ASL_i(t) = \alpha_{SL}(DSL - SL_i(t)) \tag{7}$$

$$AS_i(t) = \alpha_S(DINV_i(t) - INV_i(t) + BL_i(t) \tag{8}$$

$$SL_t = WIS_i(t) + IO_{i-1}(t) + BL_{i-1}(t) + OS_{i-1}(t) \tag{9}$$

Let $\beta = \alpha_{SL} / \alpha_s$ and $Q = DINV + \beta \cdot DSL$, then we get:

$$OP_i(t) = \max\{0, ED_i(t) + \alpha_S(Q - INV_i(t) + BL_i(t) - \beta \cdot SL_i(t))\} \tag{10}$$

We assume $\alpha_s \leq \alpha_{SL}$ and $\beta \leq 1$. (α_s, β) represents the characteristic of agents: If $\alpha_s > \beta$, the agent pays attention to inventory and $\alpha_s < \beta$, the agent pays attention to supply chain. In this study, we examine the evolution of (α_s, β) using genetic algorithm.

4   Genetic Algorithm

In order to search the appropriate characteristic parameter (α_s, β) of each agent, we use genetic algorithm (GA). The followings are the procedures of the GA.

4.1   Fitness Function

As fitness function of GA, we use the cost function of the Beer Game. The objective is to minimize the cost. There are two types of cost. The first one is the cost of inventory (0.5\$ per case per week). The second one is the penalty of back log (2\$ per case per week). Equation (x) is the fitness function.

$$Fitness = \sum_{t=1}^{T}(2\sum_{i=1}^{m}BL_i(t)+0.5\sum_{i=1}^{m}INV_i(t)) \tag{11}$$

where, T is the total number of weeks and m is the number of sectors.

4.2   Coding and Crossover Operator

We adopt real number coding instead of binary coding. As crossover operator, we use REX [9]. REX is a multi-parents crossover and it is known for its high performance. Let $\mathbf{x}^i$ is the representation of individual that is a real-coded vector of n dimension and k is the number of parents for the crossover. The children are

$$\mathbf{x}_c = \mathbf{x}_g + \sum_{i}^{n+k} \varphi(0,\sigma^2)(\mathbf{x}_i - \mathbf{x}_g) \tag{12}$$

where $\varphi(0,\sigma^2)$ is random number chosen from the probability distributions with average 0 and variance σ^2. In this study, we use normal distribution as $\varphi(0,\sigma^2)$.

4.3   Generation Alteration Model

The performance of crossover strongly depends on generation alteration model. In this study, we use JGG [10]. The procedure of JGG is as follows:

1. [Initialization] Create m individuals as initial population and evaluate them.
2. [Selection 1] Select n_p parents from population by random sampling without replacement.
3. [Reproduction] Repeat crossover to parents and generate n_c children
4. [Selection 2] Select best n_p individual from children and replace with parents.
5. Go to 2.

JGG requests sufficiently large number of children is generated during crossover.

5   Simulation Results

As shown in Section 3, the ordering policy of each sector can be characterized by α_s and β. In this paper, we optimize these parameters of each four agents using genetic algorithm as shown in Section 4.

The purpose of this simulation is to investigate how the effective policy of each agent in the Beer Game will be when we change the situation. There are two types of change: one is extension of the supply chain. In original Beer Game, the length of supply chain is 4. In this study, we extend it to 8 and 12. The other change is malfunction of the sector can be occurred. In this study, sector in layer 2 breaks down during 21st week and 25th week in the period of 60 weeks. Demand of customer is the same as the original beer game (First four week: 4, after that: 8).

Fig. 4~6 show the best and average fitness during the evolution. The values are the average of 10 trials. In all length of supply chain, malfunction makes the situation (= fitness) worse.

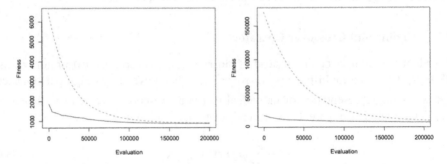

Fig. 4. Best and average fitness during evolution. The length of the supply chain is 4. Normal condition (left) and malfunction condition (right)

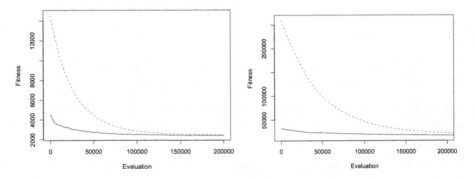

Fig. 5. Best and average fitness during evolution. The length of the supply chain is 8. Normal condition (left) and malfunction condition (right)

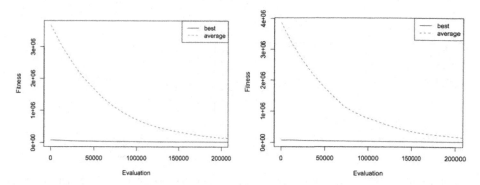

Fig. 6. Best and average fitness during evolution. The length of the supply chain is 12. Normal condition (left) and malfunction condition (right).

Fig. 7~9 show the one of best solution found by genetic algorithm. In small length of supply chain, bullwhip effect can be suppressed well. For example, please compare Fig. 2 and Fig. 7 (left). The orders placed by agents became a half. However, in large length of supply chain, we can see the large overshoot around 40th week.

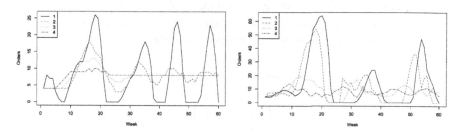

Fig. 7. Change of the orders in the Beer Game played by the best solution of GA. The length of supply chain: 4, normal condition (left) and malfunction condition (right). Lower numbers correspond to upper layer in the chain.

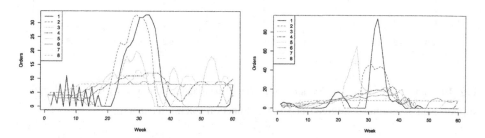

Fig. 8. Change of the orders in the Beer Game played by the best solution of GA. The length of supply chain: 8, normal condition (left) and malfunction condition (right). Lower numbers correspond to upper layer in the chain.

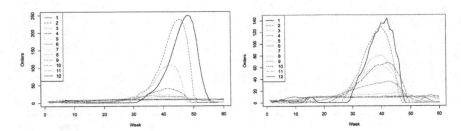

Fig. 9. Change of the orders in the Beer Game played by the best solution of GA. The length of supply chain: 12, normal condition (left) and malfunction condition (right). Lower numbers correspond to upper layer in the chain.

Fig. 10~12 show the results of the evolved parameters of each agent. Each point corresponds to the each setting of damage. If the point is located at lower right, this means that the agent is sensitive to inventory, and if the point is located at upper left, this means that the agent is sensitive to flow. We can see that the agents in upstream (factory and distributor) are sensitive to inventory and the agent in downstream (retailer) is sensitive to flow. This trend is constant when we change the length of the supply chain and when we introduce malfunction of sector.

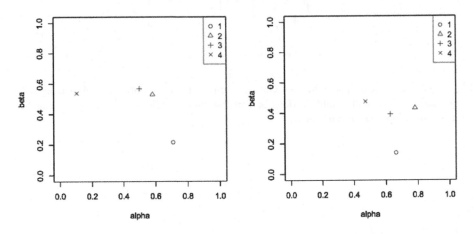

Fig. 10. Parameters of ordering policies (α_s, β) of best agents obtained by GA. The length of supply chain: 4, normal condition (left) and malfunction condition (right).

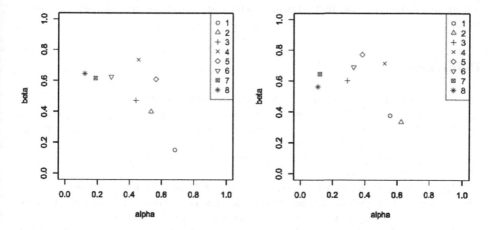

Fig. 11. Parameters of ordering policies (α_s, β) of best agents obtained by GA. The length of supply chain: 8, normal condition (left) and malfunction condition (right).

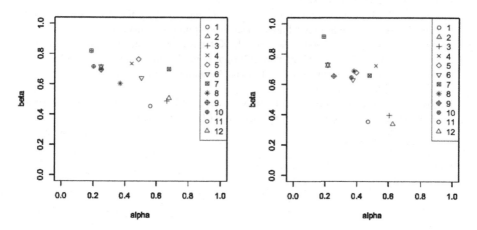

Fig. 11. Parameters of ordering policies (α_s, β) of best agents obtained by GA. The length of supply chain: 12, normal condition (left) and malfunction condition (right).

6   Conclusion

In this paper, we investigated the effective ordering policy of the supply chain using the Beer Game. Genetic algorithm is used to obtain the optimal policy. We introduce two changes in original Beer Game. One is the extension of the length of supply chain. The other is the malfunction of sector during the simulation. From the simulation results, we obtained the fact that the sectors in the upstream should be sensitive to the inventory and the sectors in downstream should be sensitive to flow. This trend never changed when we introduce the changes stated above. Getting more precise policy of each sector and using more complex supply chain network are our future work.

References

1. Hugos, M.H.: Essentials of Supply Chain Management third edition, Wiley (2011)
2. MacKenzie, C.A., Santos, J.R., Barker, K.: Measuring changes in international production from a disruption: Case study of the Japanese earthquake and tsunami. International Journal of Production Economics 138(2), 293–302 (2012)
3. Sterman, J.D.: Flight Simulator for Management Education, http://web.mit.edu/jsterman/www/SDG/beergame.html
4. Jarmain, W.E.: Problems in Industrial Dynamics. MIT Press, Cambridge (1963)
5. Sterman, J.D.: Modeling managerial behaviour: mis-perceptions of feedback in a dynamic decision making experiment Management Science 35, 321–339 (1989)
6. Strozzi, F., Bosch, J., Zaldivar, J.: M., Beer game order policy optimization under changing customer demand. Decision Support Systems 42(4), 2153–2163 (2007)
7. O'Donnell, T., Maguire, L., McIvor, R., Humphreys, P.: Minimizing the bullwhip effect in a supply chain using genetic algorithms. International Journal of Production Research 44(8), 1523–1543 (2006)
8. Steven, O., Kimbrough, D.J.: Wu, Fang Zhong, Computers Play the Beer Game: Can Artificial Agents Man-age Supply Chains? Decision Support Systems 33(3), 323–333 (2002)
9. Akimoto, Y., Hasada, R., Sakuma, J., Ono, I., Kobayashi, S.: Generation Alternation Model for Real-coded GA Using Multi-Parent Proposal and Evaluation of Just Generation Gap (JGG). In: SICE Symposium on Decentralized Autonomous Systems, vol. 19, pp. 341–346 (2007) (in Japanese)
10. Kobayashi, S.: The Frontiers of Real-Coded Genetic Algorithms, Transactions of Japanese Society for Artificial Intelligence 24(1), 147–162 (2009) (in Japanese)

Offline Design of Interactive Evolutionary Algorithms with Different Genetic Operators at Each Generation

Hisao Ishibuchi, Takahiko Sudo, Koji Ueba, and Yusuke Nojima

Department of Computer Science and Intelligent Systems, Graduate School of Engineering,
Osaka Prefecture University, 1-1 Gakuen-cho, Naka-ku, Sakai, Osaka 599-8531, Japan
{hisaoi,nojima}@cs.osakafu-u.ac.jp,
{takahiko.sudo,koji.ueba}@ci.cs.osakafu-u.ac.jp

Abstract. In interactive evolutionary computation (IEC), each solution is evaluated by human users. This means that only a small number of solutions can be evaluated during the execution of an IEC algorithm. In some application fields of IEC such as hearing aid design and music composition, only a single solution is evaluated at a time. Moreover, accurate and precise fitness evaluation is not easy for human users. Based on these discussions, we formulated an IEC model with the minimum requirements for human users' fitness evaluation ability: They can evaluate only a single solution at a time, they can memorize only a single previous solution, and their evaluation result on the current solution is whether it is better than the previous solution or not. In this paper, we propose a meta-level approach to the design of interactive algorithms for our IEC model. The novelty of our approach is to use a different operator to generate each solution. An IEC algorithm is coded by a string of operators to generate solutions such as random creation, mutation and crossover. The string length is the same as the number of solutions to be generated. Each string (i.e., each IEC algorithm) is evaluated through simulations in our IEC model. A population of strings is evolved by a meta-level evolutionary algorithm. We perform experimental studies using a well-known test function as a surrogate of a human user. For a simple test function, we obtain (1+1)ES-style algorithms where the next solution is generated by mutation from the best solution among the examined ones. For a complicated test function, we obtain $(\mu+1)$ES-style algorithms with crossover, mutation and random creation.

Keywords: Interactive evolutionary computation, interactive algorithms, algorithm design, meta-level evolutionary algorithms, hyper-heuristics.

1 Introduction

Interactive evolutionary computation (IEC) is characterized by subjective fitness evaluation by a human user [13]. No explicit formulation of a fitness function is assumed in IEC. A number of successful applications of IEC have been reported in the literature [1]-[3], [9]-[11]. IEC algorithms are usually based on discretized subjective evaluation by human users (e.g., 1: very bad, 2: bad, 3: average, 4: good, 5: very

H. Handa et al. (eds.), *Proc. of the 18th Asia Pacific Symp. on Intell. & Evol. Systems – Vol. 2*,
Proceedings in Adaptation, Learning and Optimization 2, DOI: 10.1007/978-3-319-13356-0_50

good). In this case, it is assumed that human users can divide a population of solutions into different classes. However, this is not always easy for human users. Some IEC algorithms are based on much simpler evaluation mechanisms such as a pair-wise comparison [5], [15] where two solutions are compared with each other. In this case, it is assumed that two solutions can be evaluated simultaneously. However, this is not always the case. In some application tasks of IEC such as hearing aid design [14] and music composition [4], we can evaluate only a single solution at a time.

The simplest solution evaluation by human users is to examine only a single solution. The examined current solution is compared with the previous one. The evaluation result is whether the current solution is better than the previous one or not.

Based on these discussions, we formulated an IEC model with the minimum requirement for human users' fitness evaluation ability by assuming the following [7]:

(i)   Human users can evaluate only a single solution at a time.
(ii)   Human users can memorize only a single previous solution. After a new solution is evaluated, their memory is replaced with the evaluated one.
(iii)   Human users can evaluate the current solution in comparison with the previous solution in their memory. The evaluation result is whether the current solution is better than the previous one or not.
(iv)   Human users can perform a pre-specified number of evaluations in total.
(v)   Human users want to have a single final solution after a pre-specified number of evaluations.

The point in the design of IEC algorithms is to decrease the burden of human users in fitness evaluation [12]. Our IEC model was formulated to minimize the burden in fitness evaluation. It can be also used for decreasing the burden in entering evaluation results through a computer keyboard. One idea is to read a human user's response through his/her facial expression or brain wave activity. It is very difficult for automated recognition systems to assign a human user's reaction into one of multiple evaluations (e.g., from 1 to 5). However, the recognition task in our IEC model is simple: to assign a human user's reaction into one of two cases: the current solution is better than the previous one or not. The use of our IEC model together with an automated recognition system is an interesting but totally different future research topic.

The main feature of our IEC model is the necessity of solution re-evaluation before choosing a single final solution. Some solutions are needed to be evaluated several times to identify a single final solution. This is often the case in our everyday life. For example, we usually examine some pairs of glasses several times to compare them with each other before buying a single pair. It is very difficult for us to choose a single best solution after evaluating a number of solutions just once. Let us explain this feature using a simple example with the following five solutions:

Example 1: (Worst) $x^C \prec x^B \prec x^A \prec x^E \prec x^D$ (Best)

In this example, $x \prec y$ means that a solution y is preferred to (i.e., better than) another solution x. Among the five solutions, x^C is the worst while x^D is the best.

Let us assume that the five solutions x^A, x^B, x^C, x^D and x^E are evaluated in this order. First x^A is shown to a human user. Next x^B is evaluated in comparison with x^A.

The evaluation result is "x^A is better than x^B (i.e., $x^B \prec x^A$)". Then x^C is evaluated as $x^C \prec x^B$. After the evaluation of the three solutions, we can say that x^A is the best from the evaluation results $x^B \prec x^A$ and $x^C \prec x^B$. Then x^D is evaluated as $x^C \prec x^D$. After the evaluation of x^D, we cannot say which is the best between x^A and x^D. Finally x^E is evaluated as $x^E \prec x^D$. It is clear from this evaluation result that x^E is not the best. However, we cannot still say which is the best between x^A and x^D. The comparison between x^A and x^D is needed to identify the best solution. This example explains the necessity of solution re-evaluation to identify the best solution.

Let us assume that x^A is re-evaluated after x^D (i.e., before x^E is evaluated). That is, x^A is re-evaluated in comparison with x^D. The evaluation result is $x^A \prec x^D$. From this result, we can say that x^D is the best among x^A, x^B, x^C and x^D. Then x^E is evaluated in comparison with x^A (since x^A is re-evaluated after x^D). The evaluation result is $x^A \prec x^E$. After the evaluation of x^E, we cannot say which is the best between x^D and x^E. Thus x^D is re-evaluated after x^E. The evaluation result is $x^E \prec x^D$. From this result, we can say that x^D is the best. In this case, the five solutions are evaluated in the order of $x^A x^B x^C x^D x^A x^E x^D$.

The re-evaluation of x^A and x^D can be performed after the evaluation of the five solutions. For example, x^A can be re-evaluated after x^E. The re-evaluation result is $x^A \prec x^E$. From this result, we can say that x^D is the best (without re-evaluating x^D) from $x^B \prec x^A$, $x^C \prec x^B$, $x^C \prec x^D$, $x^E \prec x^D$ and $x^A \prec x^E$. That is, the best solution can be identified after the evaluation of the five solutions in the order of $x^A x^B x^C x^D x^E x^A$. This example shows that the total number of evaluations (including re-evaluations) to identify the best solution depends on the evaluation order. This suggests the necessity of an appropriate re-evaluation schedule in IEC algorithms.

In this paper, we discuss the design of interactive algorithms for our IEC model. This paper is organized as follows. In Section 2, we explain a (1+1)ES-style IEC algorithm [7] where a single best solution is always stored. In Section 3, we explain how the (1+1)ES-style IEC algorithm can be generalized to a $(\mu+1)$ES-style IEC algorithm [8] where μ is the upper limit of candidate solutions stored during the execution of the algorithm. In Section 4, we propose a meta-level approach to the design of IEC algorithms. A single IEC algorithm is coded by a string of operators where a different operator is used to generate each solution to be evaluated. Re-evaluation is also used in our meta-level approach as an operator to specify the next solution to be evaluated. In Section 5, we report experimental results where some well-known test functions are used as surrogates of human users to evaluate the current solution in comparison with the previous one. Finally this paper is concluded in Section 6. We also suggest some future research topics in Section 6.

2   Our (1+1)ES-Style IEC Algorithm

Since our IEC model is based on the comparison of the current solution with the previous one, a (1+1)ES-style IEC algorithm seems to be a reasonable choice. In the standard (1+1)ES-style algorithm, the best solution among the examined ones is always stored and used to generate a new solution. The generated solution is compared with the stored solution. If the new solution is better, the stored solution is replaced with the new one. A straightforward implementation of such a generation update mechanism under our IES model is to re-evaluate the best solution whenever the current solution is not the best solution (i.e., whenever the best solution is not stored in the human user's memory [7]).

Using Example 1 in Section 1 with the five solutions, we explain such a straightforward implementation. In Example 1, x^A and x^B are evaluated in this order where x^A is better than x^B (i.e., $x^B \prec x^A$). After the evaluation of x^B, the current solution x^B is not the best among the examined two solutions. Thus the previous solution x^A is re-evaluated. After the re-evaluation of x^A, the best solution is the current solution (i.e., x^A). Then the next solution x^C is evaluated. The evaluation result is $x^C \prec x^A$. Since the current solution x^C is not the best among the examined three solutions, the previous solution x^A is re-evaluated again. After that, the next solution x^D is evaluated in comparison with x^A as $x^A \prec x^D$. Since the current solution x^D is the best among the examined four solutions, the next solution x^E is evaluated. The evaluation result is $x^E \prec x^D$, which means that the current solution x^E is not the best. Thus the previous solution x^D is re-evaluated before a new solution is generated. The five solutions are examined in the straightforward implementation in the order of $x^A x^B x^A x^C x^A x^D x^E x^D$. This implementation corresponds to a human user who always wants to have the best solution among the examined ones in his/her memory through re-evaluation. However, as shown by the order $x^A x^B x^A x^C x^A x^D x^E x^D$, this implementation is not efficient.

A little bit more efficient implementation [7] of a (1+1)ES-style IEC algorithm is to re-evaluate a solution only when we cannot identify a single best solution. Let us explain this implementation using the same example. After the evaluation of x^B as $x^B \prec x^A$, the current solution x^B is not the best solution. However, we can say that x^A is the best. Thus we can use x^A to generate a new solution. It should be noted that IEC algorithms can remember all the examined solutions whereas human users can remember only a single solution (x^B in the current situation). Then the next solution x^C is evaluated in comparison with x^B as $x^C \prec x^B$. We can still say that x^A is the best (whereas the current solution x^C is not the best). Thus we evaluate the next solution x^D. The evaluation result is $x^C \prec x^D$. Now, we cannot say which is the best between x^A and x^D. Thus we re-evaluate x^A. From the re-evaluation result $x^A \prec x^D$, we can say that x^D is the best. Then the next solution x^E is evaluated in comparison with the previous solution x^A. The evaluation result is $x^A \prec x^E$. We cannot say which is the best between x^D and x^E. Thus we re-evaluate x^D. From the evaluation result $x^E \prec x^D$, we can say that x^D is the best. The evaluation order is $x^A x^B x^C x^D x^A x^E x^D$, which is shorter than $x^A x^B x^A x^C x^A x^D x^E x^D$ in the case of the straightforward implementation.

The efficient implementation can be also explained using the concept of candidate solutions for the best solution among the examined ones. Let us denote the set of candidate solutions by S. In the above-mentioned example with the evaluation order $x^A x^B x^C x^D x^A x^E x^D$, S is not changed during the evaluation of x^A, x^B and x^C (i.e., $S = \{x^A\}$). After the evaluation of x^D, x^D is added to S (i.e., $S = \{x^A, x^D\}$). Then x^A is re-evaluated to decrease the number of candidate solutions in S (i.e., $S = \{x^D\}$). Since S includes only a single candidate solution, the next solution x^E is evaluated. Based on the evaluation result $x^A \prec x^E$, x^E is added to S (i.e., $S = \{x^D, x^E\}$). Then x^D is re-evaluated. After the re-evaluation of x^D, $S = \{x^D\}$. As explained using this example, the number of candidate solutions in S is decreased from two to one by re-evaluating the older candidate solution in S whenever it is increased from one to two.

One important requirement in our IEC model is that a single final solution should be obtained after a pre-specified number of evaluations (i.e., assumption (v) in Section 1). In the above-mentioned example, the best solution x^D is identified after the seven evaluations in the order of $x^A x^B x^C x^D x^A x^E x^D$. If the total number of evaluations is limited to six, we stop the execution of our IEC algorithm after the sixth evaluation (i.e., after the evaluation of x^E). As a result, we have a candidate solution set S with two candidates: $S = \{x^D, x^E\}$. This poses an additional problem: which solution should be used as a final solution when S includes multiple candidates? To avoid this situation, we handle the assumption (v) in Section 1 as the following requirement:

(v) Only a single final solution should be included in the candidate solution set S when the execution of IEC algorithms is terminated.

Let x_T and S_T be the evaluated solution at the T-th evaluation and the candidate solution set after the T-th evaluation, respectively. We assume that T is the upper limit on the total number of evaluations. In this case, the requirement (v) means $|S_T| = 1$. When $|S_{T-1}| = 2$, the older candidate solution in S_{T-1} is always re-evaluated to decrease the number of candidate solutions. Thus $|S_T| = 1$ is satisfied. When $|S_{T-1}| = 1$, there are two cases: (a) the solution x_{T-1} at the $(T-1)$th evaluation is included in S_{T-1} (i.e., $S_{T-1} = \{x_{T-1}\}$), and (b) x_{T-1} is not included in S_{T-1} (i.e., $S_{T-1} \neq \{x_{T-1}\}$). In the case of (a), we can examine a new solution x_T at the T-th evaluation without increasing the number of candidate solutions. This is because the final solution in S_T is the better one between x_{T-1} and x_T. However, in the case of (b), the evaluation of a new solution x_T increases the number of candidate solutions when x_T is evaluated as being better than x_{T-1}. Thus, we do not evaluate any new solution at the T-th evaluation in the case of $S_{T-1} \neq \{x_{T-1}\}$. These discussions are summarized as follows:

Termination Rules in Our (1+1)ES-style IEC Algorithm:
If $|S_{T-1}| = 2$, the older candidate solution is re-evaluated at the last evaluation.
If $S_{T-1} = \{x_{T-1}\}$, a new solution is generated from x_{T-1} to be evaluated.
If $|S_{T-1}| = 1$ and $S_{T-1} \neq \{x_{T-1}\}$, no solution is evaluated at the last evaluation.

Except for the last evaluation, the following simple rule is always used:

Solution Selection Rules in Our (1+1)ES-style IEC Algorithm ($t + 1 < T$):
If $|S_t| = 2$, the older candidate solution in S_t is re-evaluated at the $(t+1)$th evaluation.
If $|S_t| = 1$, a new solution x_{t+1} is generated from the candidate solution in S_t.

3   Our (μ+1)ES-Style IEC Algorithm

In our (1+1)ES-style IEC algorithm, the size of the candidate solution set S can be viewed as one. By increasing its size from one to μ, we can implement a (μ+1)ES-style IEC algorithm [8]. This is to generalize a single-point search to a multi-point search. However, this generalization is not always useful especially when the number of evaluations is severely limited. Actually, experimental results in our former study [8] showed that the increase in the size of S (i.e., the value of μ) degraded the performance of the $(\mu+1)$ES-style IEC algorithm. We will show its potential advantages over the $(1+1)$ES-style algorithm by computational experiments later in this paper.

The basic idea in the implementation of the $(\mu+1)$ES-style IEC algorithm is to maintain the number of candidate solutions as being equal to μ by re-evaluating candidate solutions. Let us assume that the number of candidate solutions is increased from μ to $(\mu+1)$ after the evaluation of x_t at the t-th evaluation. In this case, the candidate solution set S_t after the t-th evaluation can be written as $S_t = \{x_{i_1}, x_{i_2}, ..., x_{i_\mu}, x_t\}$ where i_j's are μ indexes ($1 \leq i_1 < i_2 < ... < i_\mu < t$). We can always decrease the number of candidate solutions from $(\mu+1)$ to μ by re-evaluating a candidate solution x_{ij} (excluding x_t) at the $(t+1)$th evaluation to compare x_{ij} with x_t. This is because one of x_{ij} and x_t is always removed from the candidate solution set depending on the evaluation result of x_{ij} in comparison with x_t. Basic rules for solution selection in our $(\mu+1)$ES-style IEC algorithm can be written as follows:

Basic Solution Selection Rules in Our $(\mu+1)$ES-style IEC Algorithm:

If $|S_t| = \mu+1$, a candidate solution (excluding x_t) is randomly selected from S_t. The selected candidate solution is re-evaluated at the $(t+1)$th evaluation.
If $|S_t| \leq \mu$, a new solution is generated from a randomly selected candidate solution (including x_t). The generated solution is evaluated at the $(t+1)$th evaluation.

In this manner, we can always decrease the number of candidate solutions in S_t whenever it is increased from μ to $(\mu+1)$. In addition to these solution selection rules, we need archive management rules in order to decrease the number of candidate solutions from μ to one (i.e., $|S_T| = 1$ should be satisfied). The archive management rules correspond to the termination rules for our $(1+1)$ES-style algorithm in Section 2.

Due to the page limitation, we cannot explain how to derive our archive management rules in detail. We only explain basic properties with respect to the number of candidate solutions. Our archive management rules are derived from those properties.

First, let us explain the increase in the number of candidate solutions after the evaluation of a new solution x_{t+1} at the $(t+1)$th evaluation. When the current solution x_t at the t-th evaluation is included in S_t, $|S_{t+1}| = |S_t|$ always holds. This is because (i) if x_t is better than x_{t+1}, x_t simply remains in S_{t+1}, and (ii) if x_{t+1} is better than x_t, x_t is replaced with x_{t+1}. When the current solution x_t is not included in S_t, $|S_{t+1}| = |S_t|$ still holds if x_t is better than x_{t+1}. In this case, $S_{t+1} = S_t$ (i.e., there is no change in the candidate solution set). However, if x_{t+1} is better than x_t, x_{t+1} is added to S_{t+1} without removing any candidate solution from S_{t+1}. Thus $|S_{t+1}| = |S_t| + 1$ holds.

Next, let us explain how $|S_t|$ can be decreased by the re-evaluation of a candidate solution x_i at the $(t+1)$th evaluation (i.e., by comparing x_i with x_t). When the current solution x_t is included in S_t, $|S_{t+1}| = |S_t| - 1$ always holds. This is because one of x_i and x_t is always removed from S_{t+1} depending on the evaluation result of x_i in comparison with x_t. When the current solution x_t is not included in S_t, $|S_{t+1}| = |S_t| - 1$ still holds if x_t is better than x_i. In this case, x_i is removed from S_{t+1}. However, x_i is evaluated as being better than x_t, $|S_{t+1}| = |S_t|$ holds. That is, no candidate solution is removed after the re-evaluation of x_i. In this case, the re-evaluated solution x_i at the

$(t+1)$th evaluation continues to be included in S_{t+1}. This means that the current solution at the $(t+1)$th evaluation is included in S_{t+1}. Thus we can remove one candidate solution by re-examining a different candidate solution at the $(t+2)$th evaluation. These discussions show that we can always remove at least one candidate solution by performing solution re-examination twice. This means that we can always decrease the number of candidate solutions from μ to one (i.e., we can remove $(\mu-1)$ candidate solutions) by performing solution re-examination $2(\mu-1)$ times.

Using these properties, we can derive the following archive maintenance rules for our $(\mu+1)$ES-style IEC algorithm (detailed derivation procedures are omitted):

Archive Maintenance Rules in Our $(\mu+1)$ES-style IEC Algorithm:
A new solution is generated for the $(t+1)$th evaluation from a randomly selected candidate solution in S_t (including x_t) in the following two cases:
(i) x_t is included in S_t, $|S_t| \leq (T-t+1)/2$, and $|S_t| \leq \mu$,
(ii) x_t is not included in S_t, $|S_t| \leq (T-t)/2$, and $|S_t| \leq \mu$.
Otherwise, a candidate solution is randomly selected from S_t (excluding x_t) and re-evaluated at the $(t+1)$th evaluation.

The selection of the next solution is governed by these archive management rules in our $(\mu+1)$ES-style IEC algorithm in this paper. In our former work [8], we derived a simple rule $|S_t| \leq (T-t)/2$ without considering the two cases separately depending on the inclusion of x_t in S_t. Our archive management rules can be viewed as being an improved version from our former work [8]. Our $(\mu+1)$ES-style IEC algorithm in Section 2 is obtained by specifying μ as $\mu=1$ in our archive management rules.

4   Meta-Level Approach to the Design of IEC Algorithms

In IEC algorithms, usually the number of evaluations is severely limited (e.g., 200 evaluations). An important issue in this situation is how to generate each solution to be evaluated. In standard fitness function-based evolutionary algorithms, hundreds of solutions are often included in a population (e.g., 200 solutions). Such a population is evolved over thousands of generations (e.g., 2000 generations). As a result, hundreds of thousands of solutions are evaluated. This is totally different from IEC algorithms. Such a large difference in the number of solution evaluations may suggest the necessity of a different algorithm design scheme for IEC algorithms from standard evolutionary algorithms. Moreover, since re-evaluation of solutions is needed in our IEC model, we may need a totally different algorithm design scheme. Motivated by these discussions, we propose a meta-level approach to the design of IEC algorithms.

In meta-level optimization, a single solution is an IEC algorithm. Each solution is coded by a string. Let us assume that T is the upper limit on the number of evaluations in our IEC model. In this case, an IEC algorithm is used to generate T solutions to be evaluated. In our meta-level approach, we assume that a different operator is used to generate each solution. Thus an IEC algorithm in our meta-level approach can be represented by a string of T operators. Each operator shows how to generate a solution to be evaluated. More specifically, let us denote a meta-level solution (i.e., an IEC algorithm) by a string $\boldsymbol{\tau}$ of length T as $\boldsymbol{\tau} = \tau_1 \tau_2 \ldots \tau_T$ where τ_t shows how to generate the t-th solution x_t to be evaluated in an IEC algorithm at the t-th evaluation.

As each element τ_t of a string $\boldsymbol{\tau}$ (i.e., as each operator τ_t in an IEC algorithm $\boldsymbol{\tau}$), we use one of the following six operators in our meta-level approach:

Operator 0: Re-evaluation (if inapplicable, random creation is used),
Operator 1: Re-evaluation (if inapplicable, mutation is used),
Operator 2: Random creation,
Operator 3: Crossover (if inapplicable, random creation is used),
Operator 4: Crossover (if inapplicable, mutation is used),
Operator 5: Mutation,

where re-evaluation means the random selection of a candidate solution from S_t (excluding the current solution x_t) to be evaluated at the the $(t+1)$th evaluation. If there is no candidate solution except for x_t (i.e., $S_t = \{x_t\}$), re-evaluation is not applicable. In this case, random creation is used in Operator 0 while mutation is used in Operator 1. Mutation is applied to a randomly selected candidate solution (including x_t). Except for the generation of the first solution, mutation is always applicable since we have at least one candidate solution. The first solution is always generated by random creation (since all of the other operators are inapplicable to generate the first solution). Crossover is applied to two candidate solutions that are randomly selected from S_t. If the number of candidate solutions is one or zero, crossover is not applicable. In this case, random creation is used in Operator 3 while mutation is used in Operator 4.

We use six integers (0 to 5) to show the six operators. This means that an IEC algorithm is coded by an integer string $\boldsymbol{\tau}$ of length T with an alphabet $\{0, 1, 2, 3, 4, 5\}$: $\boldsymbol{\tau} = \tau_1 \tau_2 \ldots \tau_T$ where $\tau_t \in \{0, 1, 2, 3, 4, 5\}$ for $t = 1, 2, \ldots, T$. Thus the search space size is 6^T. We use a simple evolutionary algorithm with the following components:

- Random creation of initial strings (i.e., randomly generated initial population),
- Binary tournament selection for choosing a pair of parents,
- Uniform crossover,
- Mutation (the current value is replaced with a randomly specified integer),
- $(\mu + 1)$ES-style generation update mechanism to construct the next population.

The main feature of our meta-level optimization is its fitness evaluation. Since each string is an IEC algorithm, its performance is used as its fitness. However, its actual execution using a human user is impractical (since hundreds of thousands of algorithms are evaluated). We use a test function as a surrogate of a human user in each IEC algorithm. The current solution is compared with the previous one using the test function. The IEC algorithm uses the evaluation result (i.e., whether the current solution is better than the previous one or not). The value of the test function at the obtained solution by the IEC algorithm is used as its fitness value. In our computational experiments, the average result over 100 runs of the IEC algorithm on the test function is used as its fitness value.

When a string $\boldsymbol{\tau}$ (i.e., an IEC algorithm $\boldsymbol{\tau}$) is evaluated, our archive management rules are also used to have a single final solution. More specifically, a string $\boldsymbol{\tau}$ is used to generate the next solution for the $(t+1)$th evaluation in the following two cases:

(i) x_t is included in S_t and $|S_t| \leq (T - t + 1)/2$,

(ii) x_t is not included in S_t and $|S_t| \leq (T - t)/2$.

Otherwise, a candidate solution is randomly selected from S_t (excluding x_t) and re-evaluated at the the $(t+1)$th evaluation.

In our meta-level approach, we do not use the parameter μ to pre-specify the size of the candidate solution set. The number of candidate solutions is indirectly adjusted by the choice of solution generation operators in τ except for the final stage of the IEC search under the control of our archive management rules.

5 Experimental Results

First we report experimental results of our $(1+1)$ES-style and $(\mu+1)$ES-style IEC algorithms. As surrogates of human users, we use the following two test functions with 50 decision variables $x = (x_1, x_2, ..., x_{50})$ where $-5.12 \le x_i \le 5.12$ ($i = 1, 2, ..., 50$):

Sphere function: $F_{\text{Sphere}}(x) = \sum_{i=1}^{n} x_i^2$.

Rastrigin function: $F_{\text{Rastrigin}}(x) = 10n + \sum_{i=1}^{n}(x_i^2 - 10\cos(2\pi x_i))$.

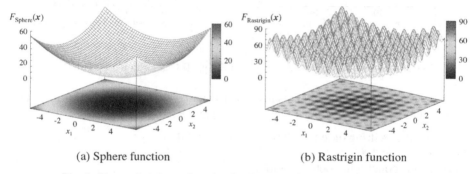

(a) Sphere function   (b) Rastrigin function

Fig. 1. Shape of each test function for the case of two decision variables

The shape of each test function is shown for the case of two decision variables in Fig. 1. As we can see from Fig. 1, Sphere function is a simple unimodal function whereas Rastrigin function is a complicated function with many local minima.

In our $(1+1)$ES-style and $(\mu+1)$ES-style IEC algorithms, initial solutions are randomly generated using a uniform distribution over the possible range of each variable $[-5.12, 5.12]$. A new solution is generated from a candidate solution using a polynomial mutation operator with $P_m = 1$ and $\eta_m = 20$ (for details, see [6]). Each algorithm is terminated after 200 evaluations (i.e., $T = 200$).

We apply our $(\mu+1)$ES-style IEC algorithm to each test problem 10,000 times for $\mu = 1, 2, 3, 4$. Average experimental results are shown in Fig. 2. We can see from Fig. 2 (a) that the increase in the size of the candidate solution set (i.e., the value of μ) deteriorates the performance of our $(\mu+1)$ES-style IEC algorithm. The same observation was obtained in our former study [8] where we used a knapsack problem as a surrogate of a human user. This observation suggests that single-point search is more efficient than multi-point search in our IEC model when we use a simple test function (e.g., Sphere function) as a surrogate of a human user. However, we cannot observe such a clear advantage of single-point search in Fig. 2 (b). This is because Rastrigin function is a highly nonlinear complicated function. One may think that multi-point search should be more efficient than single-point search for complicated test functions such as Rastrigin function. This may be true when more computation load is available. In Fig. 2, each run is terminated after 200 solution evaluations.

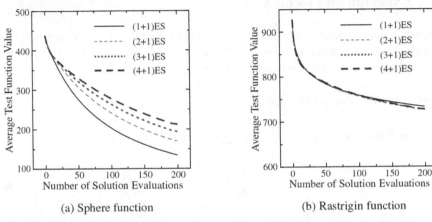

(a) Sphere function   (b) Rastrigin function

Fig. 2. Experimental results of our $(\mu + 1)$ES-style IEC algorithm for $\mu = 1, 2, 3, 4$.

Next we report experimental results of our meta-level optimization. Our meta-level optimization is performed under the following parameter specifications:

Coding: Integer string of length 200 with $\{0, 1, 2, 3, 4, 5\}$,
Population size: 100,
Termination condition: 1000 generations,
Crossover: Uniform crossover with the crossover probability 1.0,
Mutation: Random generation of an integer value with the mutation probability 1/200,
Fitness evaluation: Average performance of 100 computer simulations.

We apply our meta-level approach to each test function ten times. In order to evaluate the performance of the obtained algorithms, we apply them to each test problem 10,000 times. The average performance over 100,000 runs (i.e., $10 \times 10,000$ runs) is compared with that of our (1+1)ES-style IEC algorithm in Fig. 3. In Fig. 4, we show the average number of candidate solutions over 100,000 runs.

In Fig. 3 (a), our meta-level optimization result is slightly inferior to our (1+1)ES-style IEC algorithm. This observation suggests again that single-point search is efficient for simple test function. Actually, the average number of candidate solutions in the obtained algorithm in Fig. 4 (a) is almost always less than 2. In the 100,000 runs on Sphere function in Fig. 3 (a) and Fig. 4 (a), we monitor how each solution is actually generated in the obtained algorithm. We obtain the following results: 55.7% (mutation), 8.3% (crossover), 6.8% (random creation), and 29.2% (re-evaluation). We can see that mutation is mainly used for creating new solution. The corresponding results by the (1+1)ES-style algorithm in Fig. 3 (a) are as follows: 71.1% (mutation), 0.0% (crossover), 0.5% (random creation), and 28.4% (re-evaluation).

In Fig. 3 (b), we can observe a clear performance improvement by our meta-level approach for Rastrigin function. From Fig. 4 (b), it seems that the obtained algorithm iterates global multi-point search using many candidate solutions and focused search using a few candidate solutions several times. This seems to be a reasonable search strategy for complicated test functions. In Fig. 3 (b) and Fig. 4 (b), solutions are generated in the obtained algorithm as follows: 11.4% (mutation), 41.9% (crossover), 15.3% (random creation), and 31.4% (re-evaluation). Crossover and random creation are frequently used in the obtained algorithm. This may be the reason for its higher performance in Fig. 3 (b) than our $(\mu+1)$ES-style IEC algorithm in Fig. 2 (b).

We also examine four different resolutions of coding: one operator is used to generate a single solution (string length 200 as we have explained), two solutions (string length 100), five solutions (string length 40), and ten solutions (string length 20). However, no clear differences are observed in their experimental results.

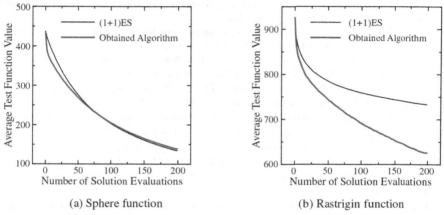

(a) Sphere function   (b) Rastrigin function

Fig. 3. Performance comparison between meta-level optimization and (1+1)ES-style algorithm.

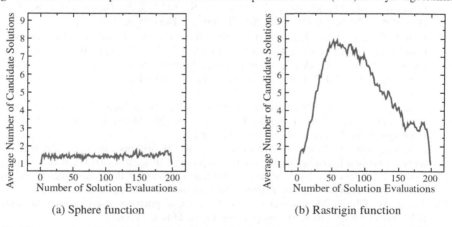

(a) Sphere function   (b) Rastrigin function

Fig. 3. Average number of candidate solutions in the execution of the obtained algorithms

6   Conclusion

We proposed a meta-level approach to the design of IEC algorithms. The main feature of our approach is that a different operator is used to generate each solution. An IEC algorithm is coded as a string of operators where the string length is the same as the number of solutions to be generated. We obtained promising experimental results where an efficient multi-point search algorithm was designed for a complicated test function. The designed algorithm seemed to adjust the diversity-convergence balance automatically over 200 evaluations. Future research topics include the choice of an appropriate surrogate function and the use of multiple surrogate functions.

References

1. Arevalillo-Herráez, M., Ferri, F.J., Moreno-Picot, S.: Distance-based Relevance Feedback using a Hybrid Interactive Genetic Algorithm for Image Retrieval. Applied Soft Computing 11(2), 1782–1791 (2011)
2. Cho, S.B.: Towards Creative Evolutionary Systems with Interactive Genetic Algorithm. Applied Intelligence 16(2), 129–138 (2002)
3. Cho, S.B.: Emotional Image and Musical Information Retrieval with Interactive Genetic Algorithm. Proceedings of the IEEE 92(4), 702–711 (2004)
4. Fernández, J.D., Vico, F.: AI Methods in Algorithmic Composition: A Comprehensive Survey. Journal of Artificial Intelligence Research 48, 513–582 (2013)
5. Fukumoto, M., Inoue, M., Imai, J.: User's Favorite Scent Design using Paired Comparison-based Interactive Differential Evolution. In: Proc. of CEC 2010, pp. 4519–4524 (2010)
6. Hamdan, M.: On the Disruption-level of Polynomial Mutation for Evolutionary Multi-objective Optimisation Algorithms. Computing and Informatics 29(5), 783–800 (2010)
7. Ishibuchi, H., Hoshino, K., Nojima, Y.: Problem Formulation of Interactive Evolutionary Computation with Minimum Requirement for Human User's Fitness Evaluation Ability. In: Proc. of 16th Asia Pacific Symposium on Intelligent and Evolutionary Systems, pp. 52–57 (2012)
8. Ishibuchi, H., Sudo, T., Nojima, Y.: Archive Management in Interactive Evolutionary Computation with Minimum Requirement for Human User's Fitness Evaluation Ability. In: Rutkowski, L., Korytkowski, M., Scherer, R., Tadeusiewicz, R., Zadeh, L.A., Zurada, J.M. (eds.) ICAISC 2014, Part I. LNCS, vol. 8467, pp. 360–371. Springer, Heidelberg (2014)
9. Kim, H.S., Cho, S.B.: Application of Interactive Genetic Algorithm to Fashion Design. Engineering Applications of Artificial Intelligence 13(6), 635–644 (2000)
10. Lai, C.C., Chen, Y.C.: A User-oriented Image Retrieval System based on Interactive Genetic Algorithm. IEEE Trans. on Instrumentation and Measurement 60(10), 3318–3325 (2011)
11. Lameijer, E.W., Kok, J.N., Bäck, T., Ijzerman, A.P.: The Molecule Evaluator. An Interactive Evolutionary Algorithm for the Design of Drug-like Molecules. Journal of Chemical Information and Modeling 46(2), 545–552 (2006)
12. Sun, X., Gong, D., Zhang, W.: Interactive Genetic Algorithms with Large Population and Semi-supervised Learning. Applied Soft Computing 12(9), 3004–3013 (2012)
13. Takagi, H.: Interactive Evolutionary Computation: Fusion of the Capabilities of EC Optimization and Human Evaluation. Proceedings of the IEEE 89(9), 1275–1296 (2001)
14. Takagi, H., Ohsaki, M.: Interactive Evolutionary Computation-based Hearing Aid Fitting. IEEE Trans. on Evolutionary Computation 11(3), 414–427 (2007)
15. Takagi, H., Pallez, D.: Paired Comparisons-based Interactive Differential Evolution. In: Proc. of NaBIC 2009, pp. 475–480 (2009)

Memetic Algorithms of Graph-Based Estimation of Distribution Algorithms

Kenta Maezawa and Hisashi Handa

Kindai University, Higashi-Osaka 577-8502, JAPAN,
handa@info.kindai.ac.jp,
http://www.info.kindai.ac.jp/~handa/hisashi

Abstract. This paper constitutes Memetic Algorithms of the graph-based Estimation of Distribution algorithms which have been proposed by us previously. The graph-based EDA employs graph-kernels to estimate the probabilistic distributions of individuals, i.e., graphs. In this paper, a greedy-search is introduced into the graph-based EDA. The experimental results on edge-max and edge-min problems examined in this paper show the effectiveness of the proposed method, i.e., the Memetic Algorithms of the graph-based EDA.

Keywords: Memetic Algorithm, Estimation of Distribution Algorithms, Graph Kernel.

1 Introduction

This paper constitutes a Memetic Algorithms for the graph-based Estimation of Distribution Algorithms [1]. In order to cope with rugged fitness landscapes, the graph-based Estimation of Distribution Algorithms employ kernel density estimation by using the Graph kernels. The sampling of offspring is established by Gibbs sampler. In the sampling phase, a Greedy method is incorporated into the Graph-based Estimation of Distribution Algorithms.

The organization of the remainder of the paper is as follows: the next section explains the graph kernels used in this paper. Section 3 introduces the graph-based Estimation of Distribution Algorithms by referring to selection, estimation, sampling, and Memetic Algorithms. Preliminary experiments on two test functions, edge-max problems and edge-min problems, are examined in Section 4.

2 Graph Kernels

2.1 The Difficulties in Graph-Related Problems for EAs

Suppose that the adjacency matrix representation of the genotype is used for undirected graph problems. Moreover, we assume that we only concern with the topology of edges now. Figures 1 (a) and (b) show that they are different genotypes although they yield the same phenotype, i.e., topology. Such reflected/rotated solutions in the adjacency matrixes coding yield multi-modal fitness functions even if original problems have the uni-modal landscape.

© Springer International Publishing Switzerland 2015
H. Handa et al. (eds.), *Proc. of the 18th Asia Pacific Symp. on Intell. & Evol. Systems – Vol. 2*,
Proceedings in Adaptation, Learning and Optimization 2, DOI: 10.1007/978-3-319-13356-0_51

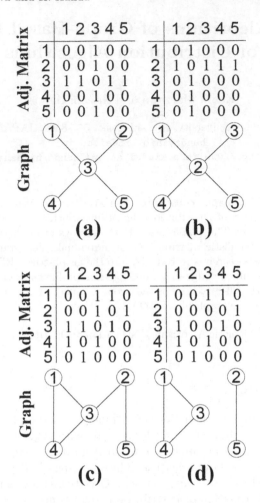

Fig. 1. An Example of the difficulties in Graph-related problems

On the other hand, Figures 1(c) and (d) are of similar genotypes: the difference is the edge between nodes 2 and 3. However, they yield different graphs: The graph by Figure 1 (d) indicates a bipartite graph. In most cases, these solutions cause different performance. Such difference may be significantly small if the edge between nodes 1 and 4 is removed. That is, graph-related problems tend to have high dependency so that the fitness landscape is quite rugged.

2.2   Shortest-Path Kernel on Graphs

Kernel methods are one of main study areas in Machine Learning. Recently, the kernel functions for graphs have been proposed [12]. The shortest-path graph kernel proposed by Borgwardt and Keiegel [13] is used in this paper: First, all the pairs-shortest-paths in two graphs G, and G' are calculated by using Floyd-

Warshall algorithms. The shortest-path kernel is defined by comparing all paris of shortest path lengths from G, and G':

$$k(G, G') = \sum_{v_i, v_j \in G} \sum_{v'_i, v'_j \in G'} k_{length}(d(v_i, v_j), d(v'_i, v'_j)),$$

where $d(v_i, v_j)$ denotes the path distance of the shortest path between nodes v_i, v_j. k_{length} is a kernel that compares the lengths of two shortest paths. In the case of a delta kernel [13],

$$k_{length}(d(v_i, v_j), d(v'_i, v'_j)) = \begin{cases} 1 & \text{if} \ \ d(v_i, v_j) = d(v'_i, v'_j) \\ 0 & \text{otherwise.} \end{cases}$$

Let $h(G)$ be a histogram of distances for all the shortest paths in a graph G. Each bin of $h(G)$ can be written as follows:

$$h(D, G) = |\{v_i, v_j \in G | d(v_i, v_j) = D\}|.$$

In the case of the delta kernel, the shortest-path kernel $k(G, G')$ can be rewritten as follows:

$$k(G, G') = \sum_D h(D, G) \cdot h(D, G'). \tag{1}$$

Figure 2 illustrates how graph kernels work. The graphs in this figure are the same as the ones in Figure 1. "Dist. Matrix" in Figure 2 represents the distance matrixes of shortest-path between corresponding nodes. "Histogram" means the histogram of distances mentioned in the previous paragraph. As depicted in this figure, rotated/reflected solutions indicate the same histograms. Meanwhile, solutions in Figure 2 (c), (d), which are similar in the adjacency matrix representation, show quite different histograms.

3 The Proposed Method

3.1 Overview

The proposed method is based upon Estimation of Distribution Algorithms. Estimation of Distribution Algorithms are a class of evolutionary algorithms which adopt the probabilistic models to reproduce individuals in the next generation, instead of conventional crossover and mutation operations. This probabilistic model is estimated from the genetic information of selected individuals in the current generation. The proposed method employs non-parametric probabilistic model: kernel density estimation.

The procedure of the proposed method is the same as the one of the conventional EDAs.

1. Firstly, the N selected individuals are selected from the population in the previous generation

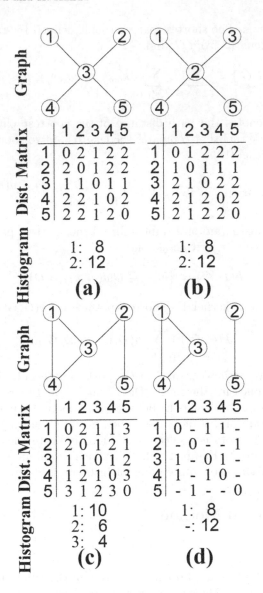

Fig. 2. Short-path graph kernels

2. Secondly, the probabilistic model is estimated from the genetic information of the selected individuals
3. A new population whose size is M is then sampled by using the estimated probabilistic model.
4. Finally, the new population is evaluated.
5. Steps 1-4 are iterated until stopping criterion is reached.

One of the distinguished points of the proposed method in comparison with conventional EDAs is the use of graph in genotype of individuals. For implementation, we do not care the data structure of graphs. We can use either of the adjacency matrix data structure or the adjacency list data structure. The following subsections introduces "selection," "estimation," and "sampling." Moreover, in "sampling" subsection, a greedy search method is also described.

3.2   Selection of Better Individuals

The comma-strategy and the plus-strategy in Evolution Strategies are examined in this paper.

comma-strategy Suppose that μ parents generate λ offspring. The μ survivals, i.e., the parents in the next generation, are selected from offspring. That is, the best μ individuals among offspring are selected.
plus-strategy In the case of the plus-strategy, the parents of the next generation is constituted by selecting μ individuals from the parents and the offspring in the current generation.

3.3   Estimation of Individual Distributions

Suppose that N denotes the number of the selected individuals at each generation, and the selected individuals are represented by $G_1, G_2, \ldots, G_N$. A kernel density function $f(G)$ can be defined as follows:

$$f(G) = \frac{1}{N \cdot p_n} \sum_i k(G, G_i) \tag{2}$$

where k denotes the kernel function in equation (1), and p_n stands for a normalizing parameter which normalizes the magnitude of the kernel function to 1. By using equation (1), equation (2) can be rewritten as follows:

$$f(G) = \frac{1}{N \cdot p_n} \sum_i \sum_D h(D, G) \cdot h(D, G_i)$$

$$= \sum_D h(D, G) \cdot \frac{1}{N \cdot p_n} \sum_i h(D, G_i)$$

Therefore, the probability density estimation for the selected individuals in the proposed method is to sum up $h(D, G_i)$ of all the selected individuals for each possible distance D. Hence, the procedure of this is summarized as follows:

1. The distance matrixes of shortest-path for all the selected individuals are calculated.
2. A histogram of distances for all the selected individuals is constituted.

3.4   Sampling of New Individuals

Gibbs sampler is adopted to generate offspring:

1. A graph G' is copied from one of selected individuals.
2. Two nodes in the G' are randomly selected.
3. Constitute a new graph G'' by adding or by removing the edge between two nodes selected in the previous step. IF the edge exists in the graph G', the edge is removed in G''. Otherwise, the edge is added in G''.
4. The new graph G'' is accepted with the following probability p_{accept}:

$$p_{accept} = \frac{f(G'')}{f(G') + f(G'')},$$

   where $f()$ indicate the probability density function explained in the previous subsection.
5. If the new graph is accepted in the previous step, The graph G' is substituted into The graph $G'($
6. If the number of iterations is less than an upper bound I, go back to 2.

We examine two kinds of the number of parents used to generate offspring.

individual-based N in equation (2) is set to be 1. Hence, each parent in Step 1 in above enumeration is used to constitute probabilistic model. It is just like Evolutionary Strategies in the case of real-valued function optimization.
group-based N in equation (2) is set to be the number of selected individuals μ. That is, all the selected individuals is used for the probabilistic model. This is similar to conventional EDA.

3.5   Constitution of a Memetic Algorithm for the Proposed Method

We employ a Greedy method in the procedure in section 3.4. As mentioned in section 3.4, the procedure is iterated I times. The first half of I, the procedure in section 3.4 is carried out. The second half of I, the probability p_{accept} in the step 4 is changed as follows:

$$p_{accept} = \begin{cases} 1 & f(G'') > f(G') \\ 0 & \text{Otherwise} \end{cases}$$

4   Experiments

4.1   Test Problems

This paper examined two simple problems: Edge-max problems and Edge-min problems.

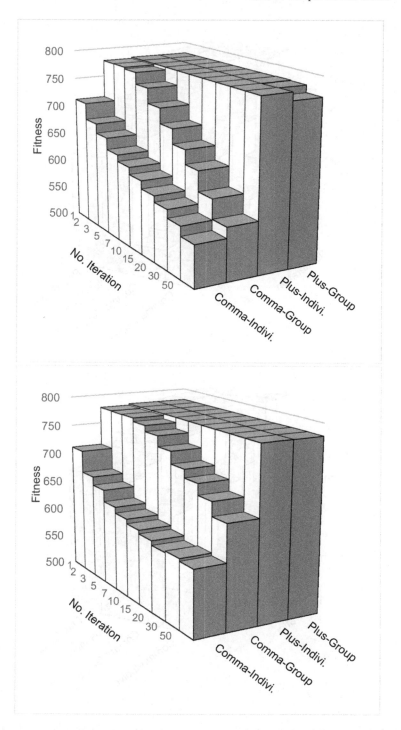

Fig. 3. Experimental results: Edge-Max problems; Original(UPPER), and Memetic Algorithms(LOWER)

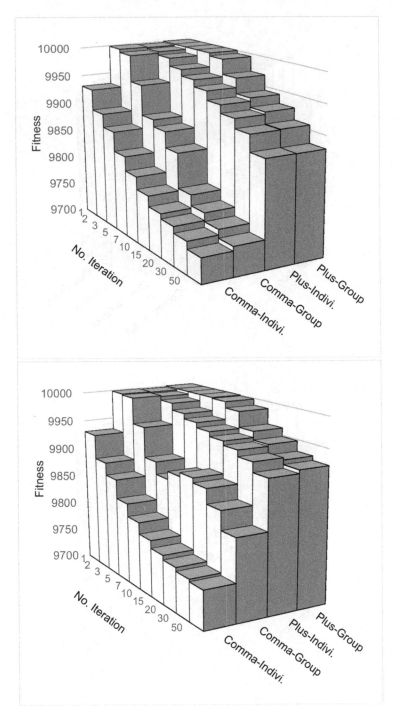

Fig. 4. Experimental results: Edge-Min problems; Original(UPPER), and Memetic Algorithms(LOWER)

Edge-max problems The edge-max problems are just like one-max problems used in binary Genetic Algorithms and binary Estimation of Distribution Algorithms. The objective of the edge-max problems is to find out a complete graph. The fitness F_{emax} of the Edge-max problems is defined as follows:

$$F_{emax}(G) = \text{(the number of edges in } G),$$

where $F_{emax}(G)$ should be maximized.

Edge-min problems As contrasted with the Edge-max problems, the solution of the Edge-min problems is of a graph with no edges. The fitness F_{emin} of the edge-min problems is defined as follows:

$$F_{emin}(G) = 10000 - \text{(the number of edges in } G).$$

This fitness function F_{emin} should also be maximized.

4.2 Experimental Results

First, the experimental settings are described: The number of nodes in the Edge-max and the Edge-min problems are set to be 40. Therefore, the best fitness for F_{emax} is $40 \times 39/2 = 780$. The number of parents μ is set to be 100. The number of generated offspring is set to be 200. The number of generations is set to be 1500. The number of iterations in the Gibbs sampling is set to be either of 1, 2, 3, 5, 7, 10, 20, 30, or 50.

Figure 3 shows the experimental results for the Edge-Max problems. The upper graph shows the results of the original proposed method as in [1]. The lower graph shows the results of the Memetic Algorithms. We examined four algorithms for the original proposed method and the MA: "Comma-individuals," "Comma-group," "Plus-individuals," and "Plus-group." As you can see, the MA with "Plus-group" show the best performance for almost all the experiments in these figures.

Figure 4 shows the experimental results for the Edge-Min problems. The Edge-Min problems seem to be more difficult than the Edge-Max problems. The use of the Greedy Method improves the performance of the proposed method.

5   Conclusions

This paper constitutes a Memetic Algorithm for the graph-related problems. The graph-based EDA employs graphs as the genotype of the EAs. The proposed method adopted graph kernels in the kernel density estimation in order to estimate the distribution of individuals, i.e., graphs. The Gibbs sampler is used to sample offspring from the estimated probabilistic model. In the iteration in the Gibbs sampler, the Greedy method is incorporated in order to constitute a Memetic Algorithm. The experimental results on the Edge-Max and the Edge-Min problems elucidate the effectiveness of the Memetic Algorithms for the graph-based Estimation of Distribution Algorithms, which are proposed in this paper.

Acknowledgment. This work was partially supported by the Grant-in-Aid for Scientific Research (C) and the Grant-in-Aid for Young Scientists (B) of MEXT, Japan (26330291, 23700267).

References

1. Handa, H.: Use of graph kernels in Estimation of Distribution Algorithms. In: Proc. 2012 IEEE Congress on Evolutionary Computation (2012)
2. Larrañaga, P., Lozano, J.A.: Estimation of Distribution Algorithms. Kluwer Academic Publishers (2003)
3. Pelikan, M., Goldberg, D.E., Cantú-paz, E.: BOA: The bayesian optimization algorithm. In: Proc. of 1999 Genetic and Evolutionary Computation Conf., pp. 525–532 (1999)
4. Bosman, P.A.N., Grahl, J., Thierens, D.: Enhancing the performance of maximum–likelihood gaussian eDAs using anticipated mean shift. In: Rudolph, G., Jansen, T., Lucas, S., Poloni, C., Beume, N. (eds.) PPSN 2008. LNCS, vol. 5199, pp. 133–143. Springer, Heidelberg (2008)
5. Handa, H.: EDA-RL: estimation of distribution algorithms for reinforcement learning problems. In: Proc. 2009 ACM Genetic and evolutionary computation, pp. 405–412 (2009)
6. Alba, E., Chicano, F.: ACOhg: Dealing with huge graphs. In: Proc. 2007 ACM Genetic and Evolutionary Conference, pp. 10–17 (2007)
7. Chicano, F., Alba, E.: Ant colony optimization with partial order reduction for discovering safety property violations in concurrent models. Information Processing Letters 106(6), 221–231 (2008)
8. McDermott, J., O'Reilly, U.-M.: An executable graph representation for evolutionary generative music. In: Proc. the 2011 ACM Genetic and Evolutionary Conference, pp. 403–412 (2011)
9. Lewis, T.E., Magoulas, G.D.: Strategies to minimise the total run time of cyclic graph based genetic programming with GPUs. In: Proc. the 2009 ACM Genetic and Evolutionary Conference, pp. 1379–1386 (2009)
10. Shirakawa, S., Nagao, T.: Graph structured program evolution with automatically defined nodes. In: Proc. the 2009 ACM Genetic and Evolutionary Conference, pp. 1107–1115 (2009)
11. Mabu, S., Hirasawa, K., Hu, J.: A Graph-Based Evolutionary Algorithm: Genetic Network Programming (GNP) and Its Extension Using Reinforcement Learning. Evolutionary Computation 15(3), 369–398 (2007)
12. Kashima, H., Tsuda, K., Inokuchi, A.: Marginalized Kernels Between Labeled Graphs. In: Proc. 20th International Conference on Machine Learning, pp. 321–328 (2003)
13. Borgwardt, K.M., Kriegel, H.-P.: Shortest-path kernels on graphs. In: Proc. 5th International Conference Data Mining (2005)

Extreme Learning Machine
for Active RFID Location Classification

Felis Dwiyasa and Meng-Hiot Lim

School of Electrical and Electronic Engineering,
Nanyang Technological University,
50 Nanyang Avenue, Singapore 639798
http://www.ntu.edu.sg

Abstract. This paper is a preliminary work which seeks the possibilities of using Extreme Learning Machine (ELM) for location classification. We gathered signal strength data from Radio Frequency Identification (RFID) tags and fed the data into the ELM to find in which room a tag is located. We also investigated ELM configuration that results best accuracy for solving our classification problem in terms of the number of training data, regularization factor, the number of time samples, and the number of hidden neurons. Given the problem is to identify in which room a tag is located among 6 rooms by using 2 readers, we achieved 87 percent accuracy with 1 sample, regularization factor $C = 2^{30}$, 5 percent training data, and 100 hidden neurons configuration. In simulation-based testing, we found that ELM classification performance is better than LANDMARC performance and comparable with WPL and ELM regression performance with nearest-room coordinate conversion.

Keywords: ELM, classification, signal strength, RFID.

1 Introduction

Radio Frequency Identification (RFID) is a wireless communication technology which is intended for identification, positioning and tracking purpose. Its application includes product tracking in manufacturing, parking access, inventory control, personnel tracking, and many more.

RFID tags are devices which are identified and tracked, whereas readers are the devices which do the identification and tracking. Based on the type of tag's power source, RFID can be classified into three types: passive, active and semi-active. Passive tags use the reader's emitted signals for powering up and responding, active tags use an on-board power source for powering up and responding, whereas semi-active tags use an on-board power source for powering up and use the reader's emitted signal for responding [7].

Passive tags are low-cost-battery-less tags which can only be detected when the distance from the reader is at most 3 meters. Active RFID tags are more expensive and requires internal battery, but they have the largest access range among all types of RFID. Because active RFID signal can reach up to hundreds

© Springer International Publishing Switzerland 2015   657
H. Handa et al. (eds.), *Proc. of the 18th Asia Pacific Symp. on Intell. & Evol. Systems – Vol. 2*,
Proceedings in Adaptation, Learning and Optimization 2, DOI: 10.1007/978-3-319-13356-0_52

of meters, the tags do not have to be physically placed very close to the readers in order to be identified and tracked. Therefore, active RFID tags are very suitable to be used in applications that need more automation and flexibility. Due to this advantage, active type RFID is chosen to be discussed throughout this paper.

There has been numerous approaches in location system, including triangulation, scene analysis, and proximity [3]. As a form of scene analysis, location fingerprinting takes signal strength from reference tags to characterize the location and match it with the data of the current tag, either by offline database matching [1] [6] or online database matching such as in LANDMARC [8] and its derivatives [12] [11]. The matching process in LANDMARC approach is done by calculating euclidean distance to the nearest reference tags [8]. The major drawback of the LANDMARC approach is that its performance is affected by the density of reference tags.

There are also some works employing various neural network architectures for location fingerprinting, including Support Vector Machine (SVM) [2], Multi Layer Perceptrons (MLP), and Radial Basis Function (RBF) [9]. The main advantage of neural networks is that they provide weights calculation method that requires less intervention and adjustment.

As one of the neural network variants, Extreme Learning Machine (ELM) introduced by Huang et. al. [5] has been known to have learning speed up to thousand times faster than conventional neural networks. ELM has also shown good accuracy in solving various regression and classification problems [4].

ELM has been used in active RFID location system by solving it as a regression problem and showing its superiority over the LANDMARC and Weighted Path Loss (WPL) method [13]. However, there does not appear to be research studies which apply the ELM in location system by viewing it as a classification problem. Therefore, in this preliminary work we investigate whether the ELM can be applied in location system as a multi-class classifier, where the ELM is required to identify in which room a tag is located. We explored various ELM parameters such as regularization factor and hidden layer size to find a configuration that produces the best classification accuracy.

2   Extreme Learning Machine

ELM is a neural network algorithm which uses single hidden layer feed-forward network architecture as shown in Fig. 1. The main advantage of ELM is that its training time is much faster than conventional learning method. Instead of optimizing all weights in the hidden layers and output layer, ELM assigns random weights to its hidden layers and solves the optimal weights for the output layer only [5].

ELM has been tested on many data sets which involve regression and classification problems. Despite its fast training time, the accuracy of ELM has been found to be comparable with the accuracy of other neural network architectures [4] [10] [14].

ELM uses the same algorithm for solving both regression and classification problems. However, there is an additional step required to prepare training data

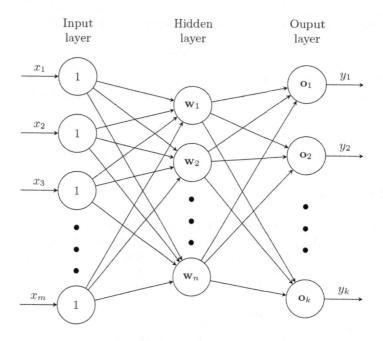

Fig. 1. Extreme Learning Machine Architecture

for ELM classifier. In ELM classifier, the number of output neurons k represents the number of classes to be distinguished. The k unique output class in the training set needs to be expanded to k binary data representing the k output neurons in such a way that only one output neuron is active for each class. For example, if we have three classes $[A, B, C]$ need to be classified, binary data $[1, 0, 0]$, $[0, 1, 0]$, and $[0, 0, 1]$ correspond with class A, B, and C respectively.

Let m be the number of input neurons, n be the number of hidden neurons, and k be the number of output neurons. Given $p \times m$ training input matrix $\mathbf{S}$ and $p \times k$ output vector $\mathbf{T}$, where p is the number of data in training set, the ELM classification algorithm during training stage is as follow:

1. Initialize $m \times n$ hidden neuron weights matrix $\mathbf{W}$ with random weights.
2. Compute $p \times n$ hidden neuron outputs $\mathbf{H}$ as given by

$$\mathbf{H} = \phi(\mathbf{SW}) \qquad (1)$$

where $\phi(.)$ is an activation function operator.

3. Calculate $n \times k$ output neurons weights matrix $\mathbf{V}$ as

$$\mathbf{O} = \left(\frac{\mathbf{I}}{C} + \mathbf{H}^T\mathbf{H} \right)^{-1} \mathbf{H}^T\mathbf{T} \qquad (2)$$

where $\mathbf{I}$ is an $n \times n$ identity matrix and C is regularization factor coefficient which is used to avoid matrix inversion problem [4].

At testing stage, the calculation of $q \times k$ output data $\mathbf{Y}$ is given by

$$\mathbf{Y} = \phi(\mathbf{XW})\mathbf{O} \qquad (3)$$

where $\mathbf{X}$ is $q \times m$ input data.

If the testing data is fed one by one, the equivalent formula for Eq. 3 is

$$\mathbf{y}(t) = \phi(\mathbf{x}(t)\mathbf{W})\mathbf{O} \qquad (4)$$

where $\mathbf{y}(t)$ is $1 \times k$ output vector and $\mathbf{x}(t)$ is $1 \times m$ input vector at time t.

3   Methodology

This work includes an experimental testing and a simulation-based testing. The experimental testing is aimed to check whether our proposed method is applicable in real-world situation, whereas the simulation-based testing is used to compare our proposed method and existing methods.

3.1   Experimental Testing

We collected data by using active RFID system[1] which are shown in Fig. 2. The RFID devices send and receive radio frequency signal at 2.4 GHz center frequency.

Fig. 2. RFID Reader (left) and Tag (right)

Tags were configured to transmit ping signal every 3 seconds. When a reader senses a ping signal from a tag, it measures the received signal strength and represents it as Link Quality Indicator (LQI). The signal strength data are then communicated across readers and recorded by the computer connected to one of the readers.

---

[1] Readers and tags are from 1Rwave LLP.

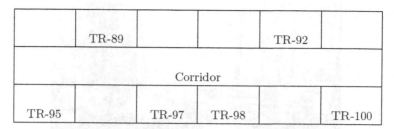

Fig. 3. Room Layout during Experimental Testing

Experiment was conducted in a university environment that consists of several small-sized classrooms as shown in Fig. 3. The size of each room is 5.8 m × 7.8 m. Some tables, chairs, and electronics are presents in the rooms. There is a corridor separating opposite rooms. Including the corridor, rooms and walls the total area is approximately 13.6 m × 47.4 m.

Two readers and 16 tags were used during the experiment. The readers were equipped by omni-directional external antennas. Each reader was mounted on a table in front of the room as shown in Fig. 4a and tags were placed facing up in the middle of the room. Tags were mounted on top of chair's writing pads as shown in Fig. 4b. One of the readers was connected to a computer through a serial cable.

We divided the 16 tags into 2 groups of 8 tags. The first group of tags was placed at one room while the second group was placed at another room. Readers were placed in TR-95 and TR-100.

After collecting data for 10 minutes, we swapped the tags and collected data for another 10 minutes. Then we selected another pair of rooms and repeated the same procedure.

Moving average was performed to every set of data for every 1 minute. This moving average process was intended to reduce data fluctuation as well as to fill missing data that sometimes occur in our experimental data.

After applying moving average, we constructed the signal strength data from all readers along with the tag identification number as training data and testing data to be fed into ELM. Each data consists of id, ss_1, ss_2 where id represents tag identification number, ss_1 is signal strength data tuple from reader 1, and ss_1 is signal strength data tuple from reader 2. The length of ss_1 and ss_2 is the number of time samples we are using.

With input range normalized to [0, 1] as recommended by [5], we varied the number of training data, regularization factor, the number of time samples, and the number of hidden neurons as shown in Table 1. The test case are designed to observe the best value of ELM adjustable parameters. Sigmoid function was the only activation function used in this test.

Training data are randomly selected from the data sets and therefore simulation needs to be repeated in order to exclude the random effect from the accuracy result. For each test case, we repeated the ELM simulation 50 times and calculated the average of testing accuracy.

(a) Mounting of readers on tables

(b) Mounting of tags on writing pads

Fig. 4. Experimental Setup

Table 1. Test Cases Specification

Test case	Percentage of training data	Regularization factor (C)	The number of time samples	The number of hidden neurons
1	0.5%, 1%, 5%, 10%, 20%, ..., 90%	2^{30}	1	5, 10, 100, 500, 1000
2	10%	2^{-10} to 2^{30}	1	100, 1000
3	10%	2^{10}	1, 2, 3, ..., 20	5, 10, 100, 500, 1000
4	10%	2^{30}	1, 2, 3, ..., 20	5, 10, 100, 500, 1000

3.2   Simulation-Based Testing

We used simulation-based testing to compare the performance of our ELM classifier algorithm with the performance of LANDMARC's k-nearest neighbors, Weighted Path Loss (WPL), and ELM regression algorithm.

To model the distances into signal strength, we used a simplified ITU indoor propagation model where reference path loss coefficient and path loss exponent are obtained from empirical curve fitting [13] as follow

$$PL = -52.40 - 10 * 3.58 * log(d) \tag{5}$$

where d is the distance between transmitter and receiver. To simplify the propagation model, we do not include wall attenuation and object diffraction because they are highly related to materials types and placements. Instead, we added white Gaussian noise component to the path loss model as shown in Fig. 5. The path loss noise is ranging from 0 dB to 10 dB. The 0-dB path loss noise refers to an ideal condition where no noise is present.

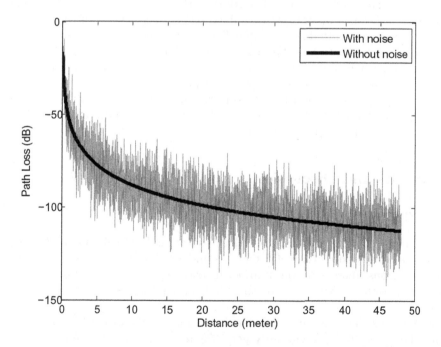

Fig. 5. Path Loss Model

The room layout for the simulation-based testing were made to resemble the 6-room layout used in experimental testing shown in Fig. 3. We added reference tags and two additional readers to support the requirement of LANDMARC k-nearest neighbors and Weighted Path Loss (WPL) algorithm. The tracked location are spread randomly in the 6 rooms. One reference tag is placed at the center of every room. In total we used 4 readers, 6 reference tags, and 6000 tracked location as shown in Fig. 6.

With the given setting, we took ideal condition for both LANDMARC and WPL. In LANDMARC approach we assume that the signal strength values from all readers and all reference tags are available. As for WPL algorithm, we assume that the radio propagation parameters are perfectly known.

Apart from LANDMARC and WPL, we also measured the performance of ELM regression. ELM regression and ELM classification basically take the same input. The difference is in the internal processing and the output neurons. ELM regression uses two output neurons which is X and Y coordinates. ELM classification uses six output neurons, where each neuron corresponds to each room.

The outputs of LANDMARC, WPL, and ELM regression are estimated coordinates. In order to be compared with the ELM classification, those outputs need to be converted into the estimated rooms. Here we introduce two types of coordinate-conversion method. The first method is exact conversion which set the room boundary exactly as it is. Any coordinates that fall into the boundary of room i are classified as room i. Coordinates that do not belong to any defined room boundaries are classified as undefined places. The output of the converter is marked as correct only if it is not undefined and it matches the actual location. The second method is to calculate the distances of each coordinate to the center of each room, where the nearest room which is the room having the minimum distance is then selected as the converter output.

Both ELM regression and ELM classification used sigmoid activation function, regularization factor $C = 2^{30}$, 1 time sample, 600 training data, and 100 hidden neurons in this simulation-based testing.

4   Results and Discussion

4.1   Experimental Result

As shown in Fig. 7, we obtained better accuracy for 100, 500, and 1000 hidden neurons as compared to that of 5 neurons and 10 neurons. Except for the case using 5 neurons, increasing the number of training data from below 5 percent of the total data to 5 percent of the total data produces significantly better accuracy.

Fig. 8 shows that regularization factor, C, has a significant effect to the accuracy. For $C = 2^{-10}$ we can only get 53% accuracy for 1000 neurons, which is even much lower when using 100 neurons. Accuracy is monotonically incrementing when the C factor is incremented. The accuracy improvement starts to diminish around $C = 2^{25}$ and $C = 2^{28}$ for 1000 neurons and 100 neurons respectively.

We also found that using one sample gives better accuracy than using time series. As shown in Fig. 9, for $C = 2^{10}$ the accuracy of time series is slightly worse than the accuracy of 1–2 samples data. The performance decline is more

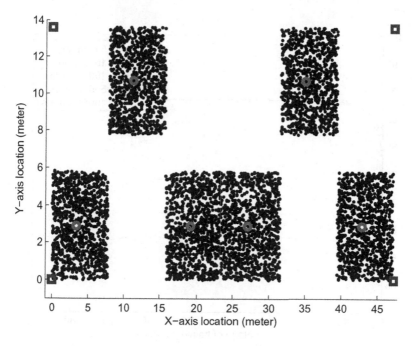

Fig. 6. Simulated Room Layout. Squares represent readers, circles represent reference tags, and dots represent tracked location

observable in Fig. 10 where for $C = 2^{30}$ the accuracy difference between one sample and time series is about 30 percent.

The results show that the best accuracy was obtained when we used 1 sample, regularization factor $C = 2^{30}$, at least 5 percent training data, and at least 100 hidden neurons.

4.2   Simulation-Based Result

Our simulation results as shown in Fig. 11 and Fig. 12 show that our ELM classifier surpass the LANDMARC approach regardless of the path loss noise and coordinate conversion method. This result might be caused by the lack of density of the reference tags we used which was only 1 reference tag for each room.

As for the WLP and ELM regression, we noticed different behavior in different coordinate conversion methods. When the coordinates from the WLP and ELM regression are converted by finding the nearest room, their performance are comparable to our proposed ELM classification. However, their accuracy become much worse than ELM classification when their coordinates are converted by using exact room boundaries. This phenomenon indicates that the estimated coordinates produced by WLP and ELM regression sometimes do not fall into the six rooms defined. It may fall into the corridor area or adjacent rooms.

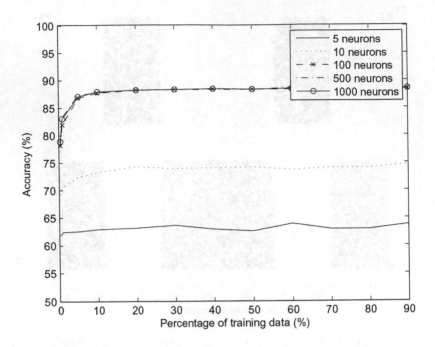

Fig. 7. Test Case #1: Performance versus the Percentage of Training Data

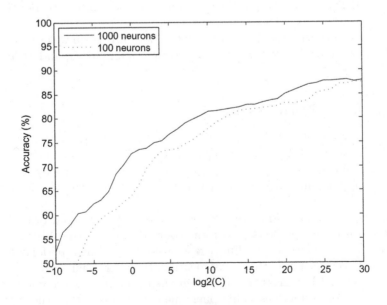

Fig. 8. Test Case #2: Performance versus Regularization Factor

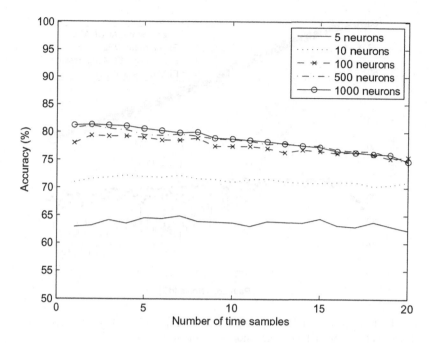

Fig. 9. Test Case #4: Performance versus the Number of Time Samples, $C = 2^{10}$

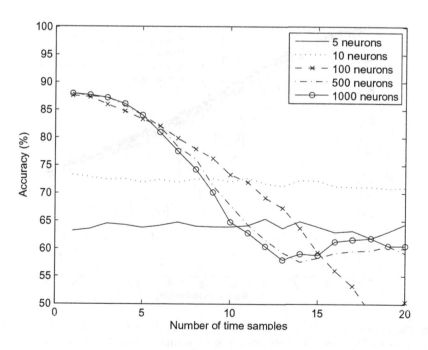

Fig. 10. Test Case #4: Performance versus the Number of Time Samples, $C = 2^{30}$

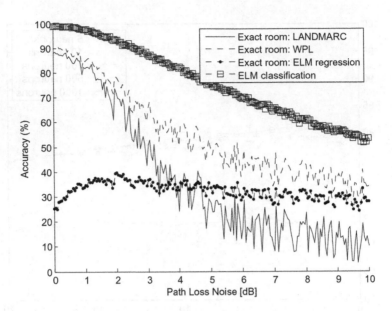

Fig. 11. Performance of ELM Classification Compared to LANDMARC, Weighted Path Loss (WPL) and ELM Regression if Coordinates are Converted into Rooms by Using Exact Room Boundaries

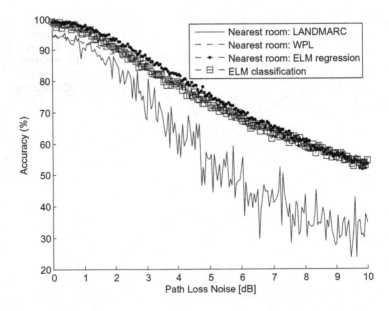

Fig. 12. Performance of ELM Classification Compared to LANDMARC, Weighted Path Loss (WPL) and ELM Regression if Coordinates are Converted into Rooms by Finding The Nearest Room

Although the WPL performance and ELM performance seem to be comparable, we should note that WPL highly depends on the estimation of path loss parameters which may not be accurately estimated in practical situation.

5   Conclusion and Future Works

This work shows that location detection in our 6-room experimental test can be predicted by using ELM classification with 87 percent accuracy. Using 5 percent of data as training data was sufficient to achieve good accuracy.

Increasing the number of hidden neurons may help the accuracy, but there is a saturation point for accuracy improvement. In our experimental data, the improvement is significant only until the number of hidden neurons reaches 100. Beyond that point, there is only slight improvement observed. The same thing applies to the regularization factor C which experiences better accuracy for higher C, where the accuracy starts to saturate when C reaches 2^{25} and 2^{28} for the number of hidden neurons 1000 and 100.

We also found that feeding time series into the ELM is not a good way to approach this problem. It was found that time series produces lower accuracy instead of higher accuracy for any regularization factor we have tested.

Our simulation-based testing proves that ELM classification performs better than LANDMARC in case of room classification. Its performance is also comparable with the performance of WPL and ELM regression. Compared to WPL approach, ELM requires less parameter adjustment because it does not depend on the estimates of radio propagation parameters.

This preliminary work can be expanded in several different ways. We may try to classify more rooms instead of just 6 rooms with more readers and bigger coverage. Because the RFID tags in the experimental testing were always put in the middle of the room, we may also try to put the tags near a wall to investigate whether room boundaries could affect the classification accuracy. Dynamic environment, which may be created by moving some furnitures during the experiment or letting people walk around the devices, can be another scenario to be explored.

Moreover, due to its quick training time, we think ELM is very practical to be used for online training. Therefore it will be interesting to investigate if using reference tags and ELM online training can produce better accuracy, especially in dynamic environment.

References

1. Bahl, P., Padmanabhan, V.N.: RADAR: An in-building rf-based user location and tracking system. In: Nineteenth Annual Joint Conference of the IEEE Computer and Communications Societies (INFOCOM), vol. 2, pp. 775–784. IEEE (2000)
2. Brunato, M., Battiti, R.: Statistical learning theory for location fingerprinting in wireless LANs. Computer Networks 47(6), 825–845 (2005)
3. Hightower, J., Borriello, G.: Location systems for ubiquitous computing. Computer 34(8), 57–66 (2001)

4. Huang, G.B., Zhou, H., Ding, X., Zhang, R.: Extreme learning machine for regression and multiclass classification. IEEE Transactions on Systems, Man, and Cybernetics, Part B: Cybernetics 42(2), 513–529 (2012)
5. Huang, G.B., Zhu, Q.Y., Siew, C.K.: Extreme learning machine: a new learning scheme of feedforward neural networks. In: IEEE International Joint Conference on Neural Networks, vol. 2, pp. 985–990. IEEE (2004)
6. Kaemarungsi, K., Krishnamurthy, P.: Modeling of indoor positioning systems based on location fingerprinting. In: Twenty-third Annual Joint Conference of the IEEE Computer and Communications Societies (INFOCOM), vol. 2, pp. 1012–1022. IEEE (2004)
7. Lahiri, S.: RFID sourcebook. IBM press (2005)
8. Ni, L.M., Liu, Y., Lau, Y.C., Patil, A.P.: Ultra low-power RFID tag with precision localization using ir-uwb. In: 2011 IEEE MTT-S International Microwave Symposium Digest (MTT), pp. 1–4 (2011)
9. Shareef, A., Zhu, Y., Musavi, M.: Localization using neural networks in wireless sensor networks. In: Proceedings of the 1st International Conference on MOBILe Wireless MiddleWARE, Operating Systems, and Applications, p. 4. ICST (Institute for Computer Sciences, Social-Informatics and Telecommunications Engineering) (2008)
10. Wang, D., Huang, G.B.: Protein sequence classification using extreme learning machine. In: IEEE International Joint Conference on Neural Networks (IJCNN), vol. 3, pp. 1406–1411. IEEE (2005)
11. Yang, Z., Liu, Y.: Quality of trilateration: Confidence-based iterative localization. IEEE Transactions on Parallel and Distributed Systems 21(5), 631–640 (2010)
12. Zhao, Y., Liu, Y., Ni, L.M.: VIRE: Active RFID-based localization using virtual reference elimination. In: International Conference on Parallel Processing (ICPP), pp. 56–56. IEEE (2007)
13. Zou, H., Wang, H., Xie, L., Jia, Q.S.: An RFID indoor positioning system by using weighted path loss and extreme learning machine. In: 1st IEEE International Conference on Cyber-Physical Systems, Networks, and Applications (CPSNA), pp. 66–71. IEEE (2013)
14. Zou, H., Xie, L., Jia, Q.S., Wang, H.: An integrative weighted path loss and extreme learning machine approach to RFID based indoor positioning. In: 4th IEEE International Conference on Indoor Positioning and Indoor Navigation (IPIN), pp. 1–5. IEEE (2013)

An Experimental Study of Correlation between Text and Image Similarity by Information Fusion Approach

Ruoxu Ren[1,2], Chee Khiang Pang[2], Li Ma[1], and Partha Dutta[1]

[1] Rolls-Royce Singapore Pte. Ltd, Advanced Technology Centre, Singapore
{Ruoxu.Ren,Li.Ma,Partha.Dutta}@Rolls-Royce.com
[2] National University of Singapore, Singapore
{ren.ruoxu,justinpang}@nus.edu.sg

Abstract. The goal of this experimental study is to explore two problems that are often encountered in *image retrieval* tasks. The first is whether text similarity consistently implies image similarity. If the answer is negative, the second question is to find the conditions for the implication to hold. In order to answer the questions, a subset of Wikipedia articles are selected as the data for exploration. By using text mining techniques and an open source image recognizer, text and image semantics are extracted from the articles and their correlations are studied to answer the two questions. The experimental results show that there are conditions for the implication to hold, which are discussed in this paper.

Keywords: Multimedia Information Fusion, Image Retrieval, Latent Semantic Analysis, Term Frequency–Inverse Document Frequency, Cosine Similarity

1 Introduction

Image Retrieval (ImR) is a technique for efficient image browsing, retrieving and searching from a collection of images. With the appearance of World Wide Web (WWW), considerable attention has been received for ImR from large–scale articles, given a multimedia user query. The user query includes both text and sample images to describe user's interest. In ImR, the set of articles is modelled as a list of text–image tuples, where image and text are from the same article in each tuple. Due to the rich semantics of text and the high computational cost of retrieving information from images, the traditional approach for ImR only takes text queries into account. This approach calculates the similarity scores of query text to texts in the dataset given a specific similarity measure. Similarities of images are ranked according to the scores.

One limitation of the traditional method is that it may give false positive results due to the homographs in the text. Homographs are words whose spelling are identical but differ in meanings. Such as 'apple' can refer to either a fruit or brand name. An example for false positive results of ImR due to homographs is given in Figure 1.

© Springer International Publishing Switzerland 2015
H. Handa et al. (eds.), *Proc. of the 18th Asia Pacific Symp. on Intell. & Evol. Systems – Vol. 2*,
Proceedings in Adaptation, Learning and Optimization 2, DOI: 10.1007/978-3-319-13356-0_53

Due to the increasing computational power of computers, retrieving image information nowadays has become less expensive compared to that in last decade. Researchers have proposed methods [1,2,3,4,5,6,7,8] that include both text and image information to alleviate the false positive cases. For example in Late Semantic Fusion (LSF) [1], false positive results are reduced by adding a re-ordering step to the top-ranked images returned by text similarity. The re–ordering further filters out irrelevant images by comparing image similarities. Late Semantic Fusion has been reported to improve the ImR performance in certain tasks [1].

On the other hand, the ImR results can also be challenged by the false negative case. Images that are relevant to a query can be incorrectly filtered out due to the lack of similarity between their corresponding texts and the query. An example is shown in Figure 2, where the query 'tile roof' fails to retrieve any relevant images.

Given the challenges described above, it is important to study two related problems in image retrieval. Firstly, whether text similarity consistently implies image similarity. Secondly, if the implication does not always hold, under which conditions it is violated. In this paper, we propose an experimental study for the two problems.

The organization of the paper is as follows. Section 2 presents the problem formulation. The mathematical preliminaries are shown in Section 3. A detailed description of the experiment and the discussion of results are given in Section 4. Finally, Section 5 concludes the paper.

Fig. 1. An example of text homographs giving different types of images, which may contribute to false positive ImR results: Images are retrieved by query **'beetle'**. The results include images of the insect **'beetle'** as well as the car **'beetle'**.

(a) Images retrieved by the query **'tile roof'**

(b) A relevant image is not retrieved by **'tile roof'**

Fig. 2. An example of false negative case of ImR: Given text query **'tile roof'**, traditional ImR returns images in 2(a). However, a relevant image in 2(b) is overlooked because the similarity between its related text and the query is not strong enough.

2   Problem Formulation

To answer the two questions discussed in Section 1, the distance metrics for texts as well as the distance measure for images need to be defined. In this paper, Term Frequency-Inverse Document Frequency (TF–IDF) [9] and Latent Semantic Analysis (LSA) [10] are explored for text representation. Overfeat [11] is chosen for image representation. As to distance metric, Cosine and Euclidean distances are explored for both text and image. The details are presented Section 3.

During the first step of our applied method, images and text information are analysed in parallel. On the one hand, text-image tuples are grouped by text similarities. This is achieved by a clustering method based on a specific text distance metric. On the other hand, the Nearest Neighbours (NNs) of each image are calculated based on image representation and an image distance metric. After the text data groups and image NNs are obtained, similarities of images in the same group are inspected. If the images in the same group are consistently more similar to each other than to those from a different group, it indicates that the similarity of text strongly implies the similarity of images. If this does not hold for a group, we further investigate to see whether there is an underlying pattern. The flowchart of the experimental study is shown in Figure 3. The details will be discussed in the next section.

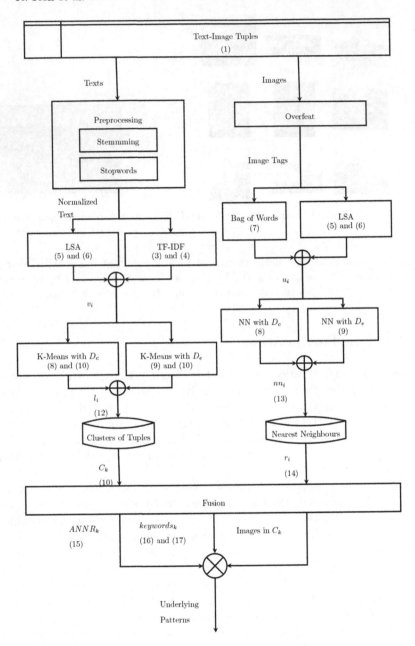

Fig. 3. The steps used in the experiment to explore the connection of text and image similarities. Numbers in the flowchart stand for equation number in Section 3.

3   Mathematical Preliminaries

The detailed mathematical preliminaries are shown in this section. The representations for texts and images, distance metrics, clustering method and measure for strength of implication are discussed.

As introduced in Section 2, text-image tuples are used in this experiment. The images are indexed from 1 to N. The ith image is denoted as I_i and the text from the same article is denoted as T_i. The tuples are ordered pairs of texts and images. The ith pair is defined as

$$P_i = (T_i, I_i). \tag{1}$$

It is worth mentioning that there can be multiple images that share the same text in the original documents. A subset of texts are duplicated to make text-image tuple as one-to-one mapping.

3.1   Representations for Texts

As stated in Section 2, for the representation of text both TF–IDF and LSA are explored. In order to obtain TF–IDF feature, two steps are carried out. In the first step, raw text features are extracted using bag-of-words model from normalized text. Suppose that there are in total m distinctive words in all the texts and they are denoted as $w_1, w_2, \cdots, w_m$. Denote the set of words as W, where $W = \{w_i | w_i \in [1, m]\}$. Let $tf(w, T_k)$ represents the frequency of a word w that appear in T_k. The raw feature of T_k is calculated as

$$t(T_k) = (tf(w_1, T_k), tf(w_2, T_k), \cdots, tf(w_m, T_k)). \tag{2}$$

In the second step, the raw features are transformed by TF–IDF, as shown in the following:

$$tfidf(w_i, T_k) = tf(w_i, T_k) * log\left(\frac{N}{df(w_i)}\right), \tag{3}$$

where $df(w_i)$ is the number of texts that w_i appears.
The TF–IDF representation for text is then calculated as:

$$TFIDF(T_k) = (tfidf(w_1, T_k), tfidf(w_2, T_k), \cdots, tfidf(w_m, T_k)). \tag{4}$$

To obtain text representation by LSA, a singular value decomposition is performed on the TF–IDF feature as:

$$X = U\Sigma V^T, \tag{5}$$

where the X is an $m \times N$ matrix whose k^{th} column is $TFIDF(T_k)$. U is an $m \times m$ orthogonal matrix, Σ is an $m \times N$ diagonal matrix whose entries are the singular values of X and V is $N \times N$ orthogonal matrix. The rank k approximation of X is calculated using

$$X_k = U_k \Sigma_k V_k^T, \tag{6}$$

This approximation is derived by keeping the largest k singular values in Σ. The corresponding rows of U, Σ and V form U_k, Σ_k and V_k, respectively. The LSA representation of T_k is the kth column of X_k. The final representation of texts is denoted as v_i, where v_i can be derived either from (4) or (6).

3.2   Representations for Images

For images, the representations are extracted using Overfeat [11]. An example of Overfeat output is shown in Figure 4. For each image, Overfeat generates top N semantic tags and the associated confidence levels. The number of tags is user-defined. In this experiment, the top one hundred tags are selected for each image, due to the fact that the confidence level of the 100th tag is approximately 0. To obtain the image representaion, Overfeat tags are collected for all images during the first step. After obtaining the tags, the representation of images are constructed using bag-of-words model. The vocabulary is the set of words that appear in Overfeat tags. For instance, each image in Figure 4 consists of five tags. Suppose there are only three images in the data set as shown in Figure 4. By combining the tags, the vocabulary consist of the fifteen text segments. The image feature is the vector of its associated confidence levels. For example, the image feature of the left image in Figure 4 becomes $[0.503962, 0.387938, 0.306734, 0.0121338, 0.00479733, 0, 0, 0, 0, 0, 0, 0, 0, 0, 0]$. Suppose the confidence level given by Overfeat for word o_j and I_k is $p(o_j, I_k)$ and there are in total s distinctive Overfeat words. The representation of images are calculated using

$$t(I_k) = (p(o_1, I_k), p(o_2, I_k), \cdots, p(o_s, I_k)). \tag{7}$$

Similar to the text representation, LSA is also applied on the representation of images as (5) and (6). The LSA feature serves as another candidate for image representation. The final representation of images is denoted as u_i, where u_i is either from (6) or (7) .

3.3   Distance Metric

Based on the representations discussed in Section 3.1 and 3.2, there are two distance metrics which are explored for both images and texts in this experiment. They are Euclidean Distance and Cosine Distance as shown in the following two equations.

$$\text{Euclidean Distance: } D_e(\boldsymbol{a}, \boldsymbol{b}) = \sqrt{\sum_{i=1}^{s}(a_i - b_i)^2}, \tag{8}$$

where $\boldsymbol{a} = (a_1, a_2, \cdots, a_s)$ and $\boldsymbol{b} = (b_1, b_2, \cdots, b_s)$.

$$\text{Cosine Distance: } D_c(\boldsymbol{a}, \boldsymbol{b}) = 1 - \frac{\boldsymbol{a}^T \boldsymbol{b}}{\|\boldsymbol{a}\|\|\boldsymbol{b}\|}, \tag{9}$$

where $\| \cdot \|$ denotes the l^2 norm.

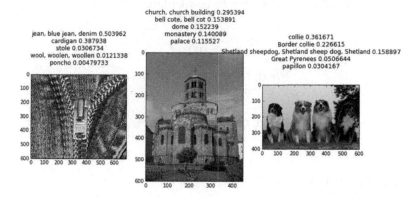

Fig. 4. An example of image representations based on **Overfeat** results. In the representation each image is a distribution of fixed set of text tags.

3.4 K-Means Clustering

Given a set of text representations $(v_1, v_2, \cdots, v_N)$, we perform the clustering using the method which K-Means selects centroids by minimizing the within-cluster sum of square and assigns each vector to the closest centroid. Let the clusters be $C = \{C_1, \cdots, C_m\}$.

$$\underset{C}{\mathrm{argmin}} \sum_{i=1}^{k} \sum_{v_j \in C_i} D(v_j, \mu_i), \qquad (10)$$

where μ_i is the mean of points in C_i and $D(\cdot, \cdot)$ is either $D_c(\cdot, \cdot)$ or $D_e(\cdot, \cdot)$.

3.5 Measurement for Strength of Implication

In order to obtain the strength of implication, the Average Nearest Neighbour Ratio (ANNR) is calculated in this experiment as follows. The clusters obtained from Section 3.4 are numbered from 1 to M. C_k stands for cluster k, which is

$$C_k = \{P_{k_1}, P_{k_2}, P_{k_3}, \cdots, P_{k_s}\}. \qquad (11)$$

The label of P_i is l_i, defined as

$$l_i = k, \text{ if } P_i \in C_k. \qquad (12)$$

One hundred Nearest Neighbours of each tuple are selected for analysis using the corresponding image information. Suppose the Nearest Neighbours of P_i is denoted as

$$nn_i = \{P_{i_1}, P_{i_2}, .., P_{i_{100}}\}. \qquad (13)$$

Nearest Neighbour Ratio (NNR) for P_i is calculated as

$$r_i = \frac{|C_{l_i} \cap nn_i|}{100},$$ (14)

where $|\cdot|$ is the cardinality.
The ANNR for cluster C_k is calculated as

$$ANNR_k = \frac{1}{|C_k|} \sum_{P_i \in C_k} r_i .$$ (15)

4   Experimental Results

This section discusses a specific application on which the proposed experimental study method is used. A detailed analysis of the experimental results is also presented.

4.1   Data Description

The data for this experiment is from ImageCELF 2010 – 2011 Wikipedia Collection [13] [14] [16], which contains almost 240,000 images and their source articles. The articles are from three different languages, which are English, German and French. In the experimental study, 34,617 images are used as well as the corresponding English articles. We choose the dataset based on two considerations. Firstly, Wikipedia articles are well-edited for single topics. Secondly, Wikipedia articles are rich in correlated texts and images. An example of the data is shown in Figure 5.

4.2   Text-Image Fusion

As discussed in Section 3.5 and Figure 3, images and texts are analysed independently during the first step. Texts are grouped by text features and one hundred NNs of each image are retrieved. At the fusion step, the clusters of texts and the NNs are combined in the ANNR calculation. The ANNR performs as an index for the strength of implication from text similarity to image similarity. To find the underlying patterns of clusters with both high ANNR and low ANNR, keywords and randomly selected images of these clusters are studied. The choice of representations and distance metrics, as well as the analysis of results are shown in the next subsection.

4.3   Results and Discussion

Representations and Distance Metrics. From Section 3, representations and distance metrics of texts and images need to be defined. Three kinds of combinations are explored for both image and text. The first combination is sparse

```
|name                     = Piano
|image                    = Grand piano and upright piano.jpg
|image_capt               = A grand piano (left) and an upright
piano (right)
|background               = keyboard
|hornbostel_sachs         = 314.122-4-8
|hornbostel_sachs_desc    = Simple [[chordophone]] with
[[Musical keyboard|keyboard]] sounded by hammers
|inventors                = [[Bartolomeo Cristofori]]
|developed                = Early 18th century
|range                    = [[Image:Range of piano.svg|200px|
center]]
}}
The '''piano''' (an abbreviation of '''pianoforte''') is a
[[musical instrument]] played using a [[musical keyboard|
keyboard]].<ref>{{cite web|url=http://
www.oxforddictionaries.com/definition/english/pianoforte?
q=pianoforte|title=Definition of "pianoforte" in the Oxford
Dictionary.|publisher=Oxford University Press}}</ref> It is
widely used in [[Classical music|classical]] and [[jazz]]
music for solo [[performance]]s, [[musical ensemble|
ensemble]] use, [[chamber music]], [[accompaniment]] and for
[[musical composition|composing]] and [[rehearsal]].
Although the piano is not portable and often expensive, its
versatility and ubiquity have made it one of the world's
most familiar musical instruments.|
```

(a) Text   (b) Image

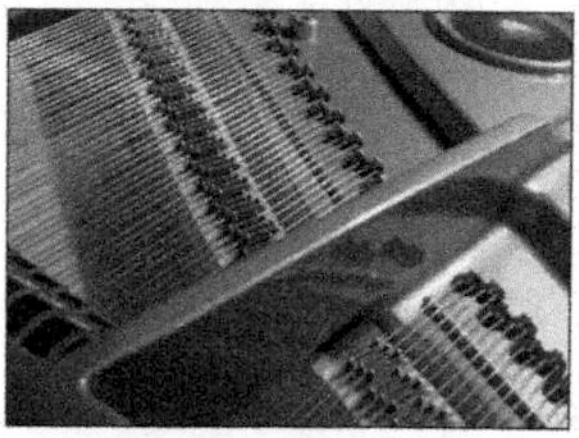

Fig. 5. An example of image and text data in the Wikipedia article dataset. 5(a) and 5(b) are texts (in mark-up format) and image from the same article.

representation (TF–IDF for text and bag-of-words for images) and Cosine distance. The second combination is LSA representation and Euclidean distance. The last one is LSA representation and Cosine distance. The appropriate combination with best performance is selected, in terms of reflecting the similarity between image and text. The analysis results for determining an appropriate combination for text are shown in Figure 6 and Figure 7. By comparing Figure 6(a) and 6(b), LSA and Cosine distance is better than TF–IDF and Cosine distance combination. The TF–IDF is of high dimension and the majority of distances are close to 1, which leads to loss of accuracy in computation. By comparing Figure 6(b) and 6(c), it is not clear that which distance metric is suitable. When the distance metrics are visualized as shown in Figure 7, it can be seen that Cosine distance is better than Euclidean distance. Due to the distribution of LSA representation, Cosine distance which groups documents by the angles between them is better than Euclidean which calculates the absolute distances. Therefore, LSA and Cosine distance is selected for text.

The reasoning for choosing appropriate image representation and distance metric are similar to that for text. Due to the page limitation of this paper, the details are omitted. Similarly to text, LSA representation and Cosine Distance is chosen for image.

Once the representations and distance metrics are determined, the next step is to cluster texts according to their similarity scores. The K-Means clustering is performed using the package in Scikit-Learn [17]. The semantics of each clusters are observed from their keywords. The keywords are selected in terms of highest TF–IDF scores. Some examples of keywords are shown in Table 1. The number of clusters are chosen by manual inspection based on the homogeneity of the keywords in each cluster.

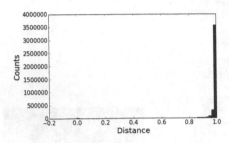 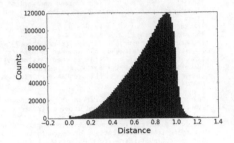

(a) Histogram of distance scores from TF–IDF representation and Cosine distance

(b) Histogram of distance scores from LSA representation and Cosine distance

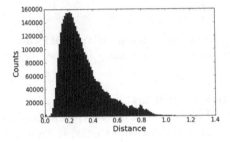

(c) Histogram of distance scores from LSA representation and Euclidean distance

Fig. 6. The distribution of pairwise text distances in a sample (1000 articles that are randomly selected) set of Wikipedia data. (a), (b) and (c) are histograms of pairwise distances of different combinations of text representations and distance metrics. From the diagram, TF–IDF and CD is hard to measure because of its high skewness.

Analysis of Underlying Patterns. After imposing the cluster structure on the data based on text similarities, the ANNR, text keywords and randomly selected images are combined for analysis. There are four observations in this experiment, which are discussed in this subsection.

The ANNR of each cluster is plotted in Figure 8. For better visualization, the clusters are ordered from small to large in terms of ANNR score. There are forty clusters and they are labelled from zero to thirty-nine. From the figure, ANNR ranges from as low as 0.02 to as high as 0.5. We are interested in exploring the underlying patterns for clusters where high and low ANNR are observed. Cluster 0 to 5 and cluster 35 to 39 are explored in this experiment.

In the case of high ANNR, there are two observations as shown in Figure 9. One observation is that, the images are of similar objects, where there is no

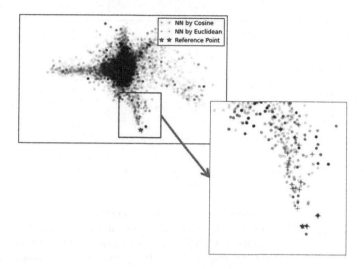

Fig. 7. Distribution of text representations in the 2D projected space given by the first 2 principal components of LSA, where NN stands for Nearest Neighbours. From the diagram it is obvious that the points are distributed in a structured way where cosine distance gives better measurement of proximity than Euclidean distance.

Table 1. Example of Text Keywords inside Some Clusters

Cluster	Keywords
1	centuri, king, greek, templ, roman, church, sculptur, di, ancient, bc
2	film, award, actor, star, movi, best, actress, hollywood, role, disney
3	olymp, summer, game, medal, athlet, beij, gold, ioc, won, stadium
4	rail, train, car, locomot, passeng, railway, freight, coupl, oper, servic

ambiguity in image concepts. For example in Figure 9(a), the images are mostly about the racing car. In Figure 9(b), the objects all belong to the category of food. Another observation is that the text keywords well-describe the images in the two clusters, which means the concept of text and that of image are closely related.

Observation 1: In cases where implication from text similarity to image similarity is strong, there are no ambiguities in image concepts.

Observation 2: In cases where implication from text similarity to image similarity is strong, the concept of text and images are strongly correlated.

In another case of low ANNR, two patterns are found and two examples are shown in Figure 10. For the first kind of pattern, what Overfeat detects is correct, although different from what text conveys. An example cluster is shown

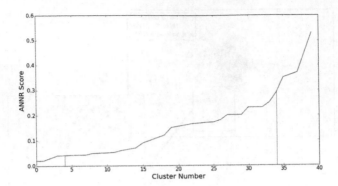

Fig. 8. ANNR scores for different clusters(sorted in ascending order). From the results, the median of the ANNR scores of different clusters is around 0.2. However, there are some clusters that have much larger ANNR scores as well as clusters that have near-zero ANNR scores. It implies that the connection between the text and image similarities vary in different clusters.

in Figure 10(a). Overfeat does not detect the images as oil paintings. It recognizes the objects inside the paintings instead. For example, what Overfeat detects for the middle image in the first row in Figure 10(a) are 'ram', 'tup','ox' and 'bighorn'. For the first pattern, it can be seen that the image has multiple meanings. If users query for 'bighorn', the image is incorrectly filtered out due to the lack of semantic correlation for its text to the query. Another pattern is that text and image from the same article are weakly correlated. An example is shown in Figure 10(b). For the left image in the first row, what Overfeat detects are 'Egyptian cat', 'lynx, catamount' and 'tabby, tabby cat'. By referring to the source article as shown in Figure 11, the text is about an image processing technique and the cat image is used as an example for illustration purpose. In fact, any image can be used for demonstration and there is weak correlation between text and image. As a consequence, text information degrades the performance of ImR.

Observation 3: The implication from text similarity to image similarity is weak, if images have multiple concepts.

Observation 4: The concept of text and image from the same article can be uncorrelated.

Therefore for Wikipedia data, if there are no ambiguities in the meanings of images, text similarity usually implies image similarity. However in some cases, even if the meanings of images are clear, the implication does not hold. As shown in the the example in Figure 11, text information and image information are weakly correlated, due to the nature of text topic. There is another case that the implication does not hold, where images have multiple meanings and what text conveys is not necessarily what users need.

(a) Cluster keywords: 'formula', 'grand', 'race', 'monaco', 'driver', 'car', 'team', 'championship', 'belgian', 'brazilian'

(b) Cluster keywords: 'wine', 'fruit', 'chees', 'cook', 'food', 'milk', 'tea', 'sweet', 'dish', 'tradit'

Fig. 9. Two clusters that have high ANNR scores in the experiment: the most common words (keywords) and sample images of the two clusters are presented in (a) and (b) respectively. From the result, the keywords in each cluster are similar to each other. The high ANNR scores observed in these clusters are due to the strong connection between text and image similarities.

(a) Cluster keywords: 'paint', 'portrait', 'painter', 'artist', 'oil', 'art', 'galleri', 'isbn', 'del', 'da'

(b) Cluster keywords: 'color', 'light', 'red', 'green', 'blue', 'pigment', 'yellow', 'eye', 'imag', 'purpl'

Fig. 10. Two clusters that have near-zero ANNR scores in the experiment: the most common words (keywords) and sample images of the two clusters are presented in (a) and (b) respectively. From the result, the keywords in each cluster are similar to each other. The low ANNR scores observed in these clusters are due to mismatch between the diversity of images and the relative uniformity of their texts.

```
'''Dither''' is an intentionally applied form of [[noise]] used to
randomize [[quantization error]], preventing large-scale patterns such
as [[color banding]] in images. Dither is routinely used in processing
of both digital audio and digital video data, and is often one of the
last analog stages of audio production to [[compact disc]].

A typical use of dither is: given an image in grey-scale, convert it
to black and white, such that the density of black dots in the new
image approximates the average level of grey in the original image.
[[File:1 bit.png|frame|A [[grayscale]] image represented in 1 bit
[[black-and-white]] space with dithering]]

== Etymology ==
[[Image:Didder def 1707.png|right|thumb|250px|The first dictionary
definition of "didder," from [[Thomas Blount (lexicographer)|Thomas
Blount]], ''Glossographia Anglicana Nova...'',<ref group=note>Full
title: ''Glossographia Anglicana Nova: Or, A Dictionary, Interpreting
Such Hard Words of whatever Language, as are at present used in the
English Tongue, with their Etymologies, Definitions, &c. Also, The
Terms of Divinity, Law, Physick, Mathematics, History, Agriculture,
Logick, Metaphysicks, Grammar, Poetry, Musick, Heraldry, Architecture,
Painting, War, and all other Arts and Sciences are herein explain'd,
from the best Modern Authors, as, Sir Isaac Newton, Dr. Harris, Dr.
Gregory, Mr. Lock, Mr. Evelyn, Mr. Dryden, Mr. Blunt, &c.''</ref>
London, 1707]]
```

(a) Text   (b) Image

Fig. 11. An example of articles from the low ANNR clusters. The text and image from the article are not directly related - the text discusses topic on 'image processing' and is not related to 'cat', which contributes to the low ANNR score in the cluster.

5   Conclusion

In this paper, being motivated by the challenges in current image retrieval exercise, we propose to explore two problems. The first one is whether text similarity implies image similarity. If the answer is no, the second one is when the implication holds or fails. Based on the experimental study, there are two findings. Firstly, the implication does not always hold. Secondly, the implication highly depends on the properties of data. If images have multiple meanings, or the concept of text is weakly correlated with that of image, the implication is violated. The conclusions drawn here is based on experimental study of Wikipedia dataset. For dataset from other domains, such as technical reports, the observation might be different. However, ANNR proposed in this experiment can serve as an index for the strength of implication in general. Researchers can use ANNR as a supplementary tool for observing the dataset property and determining appropriate Image Retrieval methods. For Future work, we will explore different dataset to further validate the assumptions here and consider from the other direction which is how image similarity implies text similarity.

Acknowledgement. This work was supported by the Economic Development Board Industrial Postgraduate Programme Fund under Grant R-263-000-A70-592.

References

1. Benavent, X., Garcia-Serrano, A., Granados, R., Benavent, J., de Ves, E.: Multimedia Information Retrieval Based on Late Semantic Fusion Approaches: Experiments on a Wikipedia Image Collection. IEEE Transactions on Multimedia 15(8), 2009–2021 (2013)
2. Gass, T., Weyand, T., Deselaers, T., Ney, H.: FIRE in imageCLEF 2007: Support vector machines and logistic models to fuse image descriptors for photo retrieval. In: Peters, C., Jijkoun, V., Mandl, T., Müller, H., Oard, D.W., Peñas, A., Petras, V., Santos, D. (eds.) CLEF 2007. LNCS, vol. 5152, pp. 492–499. Springer, Heidelberg (2008)
3. Benavent, J., Benavent, X., de Ves, E., Granados, R., Serrano, G.: Experiences at ImageCLEF 2010 using CBIR and TBIR Mixing Information Approaches. In: Working Notes of the ImageCLEF 2010 Lab. Padua, Italy (2010)
4. Csurka, G., Clinchant, S., Popescu, A.: XRCE and CEA LISTs Participation at Wikipedia Retrieval of ImageCLEF 2011. In: CLEF 2011 Working Notes (2011)
5. Depeursinge, A., Müller, H.: Fusion techniques for combining textual and visual information retrieval. In: ImageCLEF, pp. 95–114. Springer, Heidelberg (2010)
6. Clinchant, S., Ah-Pine, J., Csurka, G.: Semantic combination of textual and visual information in multimedia retrieval. In: Proceedings of the 1st ACM International Conference on Multimedia Retrieval, p. 44. ACM (2011)
7. Zhou, X., Depeursinge, A., Muller, H.: Information fusion for combining visual and textual image retrieval. In: 2010 20th International Conference on Pattern Recognition (ICPR), pp. 1590–1593. IEEE Press (2010)
8. Granados, R., Benavent, J., Benavent, X., de Ves, E., Garca-Serrano, A.: Multimodal Information Approaches for the Wikipedia Collection at ImageCLEF 2011. In: Working Notes of the CLEF 2011 (2011)
9. Aizawa, A.: An information-theoretic perspective of tf–idf measures. Information Processing & Management 39(1), 45–65 (2003)
10. Landauer, T.K., Foltz, P.W., Laham, D.: An introduction to latent semantic analysis. Discourse Processes 25(2–3), 259–284 (1998)
11. Sermanet, P., Eigen, D., Zhang, X., Mathieu, M., Fergus, R., LeCun, Y.: Overfeat: Integrated recognition, localization and detection using convolutional networks. In: ICLR (2014)
12. Bay, H., Tuytelaars, T., Van Gool, L.: SURF: Speeded up robust features. In: Leonardis, A., Bischof, H., Pinz, A. (eds.) ECCV 2006, Part I. LNCS, vol. 3951, pp. 404–417. Springer, Heidelberg (2006)
13. Popescu, A., Tsikrika, T., Kludas, J.: Overview of the Wikipedia Retrieval Task at ImageCLEF 2010. In: CLEF (2010)
14. Tsikrika, T., Popescu, A., Kludas, J.: Overview of the Wikipedia Image Retrieval Task at ImageCLEF 2011. In: CLEF (2011)
15. Porter, M.: Snowball: A language for stemming algorithms (2001), http://snowball.tartarus.org/texts/introduction
16. ImageCLEF Wikipedia Image Retrieval Datasets, http://www.imageclef.org/wikidata
17. Scikit-Learn, http://scikit-learn.org/stable/
18. Salton, G., Wong, A., Yang, C.S.: A vector space model for automatic indexing. Communications of the ACM 18(11), 613–620 (1975)

Author Index